Current Topics in Behavioral Neurosciences

Volume 32

About this Series

Current Topics in Behavioral Neurosciences provides critical and comprehensive discussions of the most significant areas of behavioral neuroscience research, written by leading international authorities. Each volume offers an informative and contemporary account of its subject, making it an unrivalled reference source. Titles in this series are available in both print and electronic formats.

With the development of new methodologies for brain imaging, genetic and genomic analyses, molecular engineering of mutant animals, novel routes for drug delivery, and sophisticated cross-species behavioral assessments, it is now possible to study behavior relevant to psychiatric and neurological diseases and disorders on the physiological level. The *Behavioral Neurosciences* series focuses on "translational medicine" and cutting-edge technologies. Preclinical and clinical trials for the development of new diagnostics and therapeutics as well as prevention efforts are covered whenever possible.

More information about this series at http://www.springer.com/series/7854

Michael H. Baumann • Richard A. Glennon •
Jenny L. Wiley
Editors

Neuropharmacology of New Psychoactive Substances (NPS)

The Science Behind the Headlines

Springer

Editors
Michael H. Baumann
Designer Drug Research Unit (DDRU)
National Institute on Drug Abuse
Baltimore, Maryland
USA

Richard A. Glennon
Department of Medicinal Chemistry
Virginia Commonwealth University
Richmond, Virginia
USA

Jenny L. Wiley
RTI International
Research Triangle Park
North Carolina
USA

ISSN 1866-3370 ISSN 1866-3389 (electronic)
Current Topics in Behavioral Neurosciences
ISBN 978-3-319-52442-9 ISBN 978-3-319-52444-3 (eBook)
DOI 10.1007/978-3-319-52444-3

Library of Congress Control Number: 2017937529

Printed on acid-free paper

This Springer imprint is published by Springer Nature
The registered company is Springer International Publishing AG
The registered company address is: Gewerbestrasse 11, 6330 Cham, Switzerland

Preface

Drug abuse and addiction are persistent problems in modern society, and an alarming new trend is the nonmedical use of so-called "designer drugs" or "legal highs," more formally known as "new psychoactive substances" (NPS). By definition, NPS are drugs of abuse that are not controlled by the 1961 Convention on Narcotic Drugs or the 1971 Convention on Psychotropic Substances, but which might pose a public health threat [1]. The chemical structures of many NPS are based on compounds extracted from the biomedical or patent literature, whereas others are analogs of illicit drugs or prescribed medications. In all cases, the substances are engineered to evade existing drug control laws. At the present time, there are NPS designed to mimic most major types of abused drugs – stimulants (e.g., bath salts), cannabinoids (e.g., spice), and hallucinogens (e.g., NBOMes). NPS produce subjective effects resembling those of their progenitors, but life-threatening adverse effects are well established and include tachycardia, hyperthermia, agitation, psychosis, violent behavior, coma, and even death. Most NPS are synthesized by Asian companies and are marketed for worldwide distribution via the Internet. The United Nations Office of Drugs and Crime (UNODC) reported that between 2008 and 2015, more than 600 NPS were identified by 102 countries and territories, and this number is expected to rise [2]. NPS represent a serious global public health threat, since there is no quality control in their manufacturing or packaging, and their biological effects are unknown when they first emerge into the recreational drug marketplace.

The purpose of this book is to provide the most up-to-date knowledge about the neuropharmacology, structure-activity relationships, and toxicology of NPS. The initial idea for the volume was based on a symposium entitled, "Bath salts, spice and related designer drugs: the science behind the headlines," held at the 2014 annual meeting of the Society for Neuroscience, in Washington DC [3]. The number of contributors for the book grew from the original symposium participants to include an international panel of experts in the field of NPS. Eighteen peer-reviewed chapters provide a rich source of information about the neurobiological effects of synthetic cathinones, cannabinoids, and hallucinogens. The topics presented range from molecular mechanisms of action to behavioral effects and include preclinical and clinical findings. The collective data demonstrate that NPS can produce effects that are similar to the drugs they intend to mimic. However, higher potency, enhanced efficacy, and idiosyncratic metabolism can render certain NPS much more dangerous than traditional drugs of abuse. The editors are indebted to each of the principal authors, and their coauthors, who committed time and expertise to craft seminal chapters for the book; we are also grateful to Springer publishing for guidance and support throughout the publication process. Our understanding of NPS is only just beginning, yet we hope this volume provides useful information to scientists, clinicians, law enforcement agencies, and policymakers who are engaged in responding to the growing phenomenon of NPS. We believe that disseminating unbiased scientific information about NPS is a key first step for increasing public awareness about the risks associated with these substances, thereby decreasing demand and avoiding potential harms.

Baltimore, MD, USA — Michael H. Baumann
Richmond, VA, USA — Richard A. Glennon
Research Triangle Park, NC, USA — Jenny L. Wiley

References

1. Brandt SD, King LA, Evans-Brown M (2014) The new drug phenomenon. Drug Test Analysis 6:587–597
2. United Nations Office on Drugs and Crime, World Drug Report 2016 (United Nations publication, Sales No. E.16.XI.7)
3. Baumann MH, Solis E Jr, Watterson LR, Marusich JA, Fantegrossi WE, Wiley JL (2014) Bath salts, spice, and related designer drugs: the science behind the headlines. J Neurosci 34:15150–15158

Contents

The Growing Problem of New Psychoactive Substances (NPS)

Bertha K. Madras

Abstract The term "new psychoactive substances" (NPS) can be defined as individual drugs in pure form or in complex preparations that are not scheduled under the Single Convention on Narcotic Drugs (1961) or the Convention on Psychotropic Substances (1971). NPS may be categorized by chemical structure, by psychoactive properties, by biological targets, or by source (plant, synthetic, or combined). The emergence of hundreds of NPS in the past decade is challenging for public health and drug policies globally. The novelty of NPS, their ambiguous legal status, ability to evade toxicological tests, swift adaptation to legal restrictions, global Internet marketing, and scant public knowledge of their adverse effects are among the key drivers of this twenty-first century phenomenon. Multi-disciplinary research in areas of biology, epidemiology, prevention, and web analytics are needed to develop effective responses in a domain capable of overwhelming current international conventions and national drug control policies. Ultimately, research-guided prevention education will fortify societies against this tidal wave.

Keywords Cathinones • New psychoactive substances • Synthetic cannabinoids

Contents

B.K. Madras (✉)
Harvard Medical School, Department of Psychiatry, Boston, MA, USA

Division of Alcohol and Drug Abuse, McLean Hospital, Oaks Building, Room 342, 115 Mill Street, Belmont, MA 02478, USA
e-mail: bmadras@partners.org; bertha_madras@hms.harvard.edu

Curr Topics Behav Neurosci (2017) 32: 1–18
DOI 10.1007/7854_2016_34
Published Online: 27 August 2016

1 Introduction

Foraging for food over millennia, humans serendipitously discovered that certain plants and fungi could produce diverse sensations distinct from satiety. A few were pleasantly arousing (tobacco, tea leaves, and coffee beans), and the liquid of fermented plants relaxed, dulled stress or melancholy, elevated mood, and intoxicated. One plant extract reduced pain, promoted euphoria, and induced sleep (opium) while others engendered euphoria and energy (coca and ephedra), or intoxicated, relaxed, heightened sensory perception and impaired thinking (marijuana). Some generated hallucinations and delusions (peyote and mushrooms). With the dawn of modern chemistry in the late 1700s, it became feasible to purify and identify the chemical structures of the psychoactive components in plants and fungi. Inspired by scientific curiosity or the drive to optimize medicinal properties of these compounds, chemists then synthesized variations of these and many other naturally occurring compounds. The unintended consequences of this inquiry and medical progress were not predictable: electronic sources of articles in medicinal chemistry, pharmacology, and biology journals, of patents, and failed candidate therapeutics became a treasure trove for entrepreneurs to craft psychoactive substances destined for furtive markets. This glut of new psychoactive substances has overwhelmed public health services, and created paroxysms in global public policy and legal systems. The spread of new psychoactive substances conceivably poses a public health challenge greater than that of substances listed in current drug conventions.

The term "new psychoactive substances" (NPS) can be defined as individual drugs in pure form or in complex preparations that are not scheduled under the Single Convention on Narcotic Drugs (1961) or the Convention on Psychotropic Substances (1971). NPS may be categorized by chemical structure, by psychoactive properties, by biological targets, or by source (plant, synthetic, or combined). The designation "new" is not necessarily limited to newly designed compounds with no historical precedent, but may also include compounds modified from progenitors or substances previously conceived of, some many decades ago. The majority are chemical analogs of drugs in restricted categories (e.g., THC or tetrahydrocannabinol, cocaine, cathinone, amphetamine, or methamphetamine, ketamine, LSD or lysergic acid diethylamide, and methaqualone), and may elicit psychoactive effects similar to the parent drug, or a more amplified response. Others may evoke unique or complex sensations because of their hybrid structures, or because several compounds with differing pharmacological profiles are amalgamated and sold as a unit. This diverse array includes phenethylamine derivatives such as synthetic cathinones and their pyrovalerone analogs, synthetic cannabinoids, piperazines, ketamine analogs, tryptamines, benzofurans, and opioids [1, 2].

At present, synthetic cathinone analogs and synthetic cannabinoids occupy a major share of this market.

The rapid expansion of products containing NPS in the past decade is fueled by a convergence of the information revolution, vague legal status, uncertain detectability, and financial incentives combined with guileful marketing.

2 What Drives Expanding Use of NPS?

2.1 Information Revolution

The Internet is a "global neural network" that can be exploited to disseminate promotion and distribution of these drugs instantly. The venues are chat rooms, blogs, instant messaging sites, social networking, or multimedia sites. At minimal cost, descriptions of new drugs, their positive psychoactive effects, doses, synthetic routes, and purchasing sites are accessible worldwide on computers, or mobile devices such as smart phones or smart watches. A blunt snapshot of the global reach of this market can be gleaned from the European Union (EU) funded Psychonaut Web Mapping Project, tasked with real-time identification of emerging NPS (sometimes known as "legal highs") through regular monitoring of the Internet. The project detected over 200 discussion forums, social media sites, online shops, websites, and other Internet resources on YouTube, eBay, Google, and Google Insight [3]. Many of the marketing sites are impervious to legal sanctions, as it takes time to deliberate the evidence and move newly emerging drugs into a legally restrictive zone, especially internationally. Imperfect international agreements and a gradual dissolution of international resolve to attenuate drug use confound solutions to this unique problem.

2.2 Vague Legal Status and Elusive Detection

More often than not, substances that imitate controlled drugs are unscheduled, unregulated, and not under the auspices of international law. Their nebulous legal status is an incentive for entrepreneurs to introduce new drugs quickly into the global market. The chemical structures of NPS differ from their progenitors (hallucinogens, stimulants, depressants, and euphoriants) that reside in restrictive drug schedules of the Controlled Substances Act (CSA) in the United States (USA), or in analogous schedules of other nations, and in international conventions. Reviving abandoned drugs by mining old sources (e.g., from chemical journals or patents) or creating new entities with slight or major structural variations can transform the restricted progenitor drug into an uncertain category of legal status, a "legal gray zone." The allure of NPS is magnified by current limitations in detecting them. Identifying these drugs for forensic, workplace, legal, and policy purposes is constrained by a lack of reference materials and the need for sophisticated detection methods which are not routinely available (e.g., mass spectroscopy). NPS tempt drug users who seek "legal highs" to circumvent the legal consequences of using standard drugs [4], desire drugs to be undetectable in drug screens, and attract poly-substance users seeking novelty in drug experiences. Despite the worldwide glut of marijuana, synthetic cannabinoid users report their reasons for using as curiosity or experimentation (91%), a desire to feel good or get high (89%), to relax (71%), and to get high without risking a positive drug test (71%) [5].

The chemical structures of NPS are designed to keep one step ahead of federal and international laws that restrict distribution and sale of specific chemicals. Law enforcement is in a perpetual race to outflank producers of NPS, a contest as old as the 1920s. During that era, chemists circumvented international drug laws by developing analogs of banned opioids. By the 1960s, a wave of new psychoactive drugs flooded American culture, some being absorbed into the culture to persist to this day. Other drugs lost popularity, because of safety concerns and undesirable psychoactive profiles. The incentives for producers are the same as they were 90 years ago, to evade legal sanctions and to profit before safety concerns precipitate scheduling. Nations respond differently to this challenge [4]. Some countries have introduced generic controls, controls on analogs, or imposed temporary restrictions on specific drugs until more data accumulates. Increasing surveillance of NPS has led to legislative actions taken by the Drug Enforcement Administration (DEA) of the USA, the World Health Organization (WHO), and other agencies of the United Nations. The WHO Expert Committee on Drug Dependence (ECDD) continues to review and render decisions on the scheduling of new substances [6, 7] in 2014 and 2015.

In the 1960s, the drug pandemonium in the USA catalyzed the formation of the DEA in 1973, a unified federal agency charged with regulating drugs with high abuse potential. Drugs were placed into five categories known as schedules. The most restrictive category, Schedule I, requires validation by a preponderance of evidence showing high abuse potential, no currently accepted medical use in treatment in the USA, and a lack of accepted safety for use. Schedule I controlled substances are regulated by administrative, civil, and criminal sanctions imposed on persons who handle (manufacture, distribute, import, export, engage in research, conduct instructional activities, and possess). Schedule II–V drugs have medicinal uses and their placement in each of the four categories is governed by relative abuse potential and safety profile. The DEA has emergency powers to temporarily schedule a drug for 36 months, a time frame to accumulate evidence for/against long-term drug scheduling. When poison control centers, emergency departments, or morgues become flooded with patients suffering from adverse effects of NPS, the legal "gray zone" can rapidly morph into a definitive Schedule I status. Automatic scheduling of novel drugs can be problematic without strong evidence for potential public harm, even if they are similar chemically and bind to the same receptors as do analogous scheduled drugs. These parameters frequently, but not uniformly, predict abuse liability. Examples in this regard include cannabidiol, a non-psychoactive analog of THC of marijuana, or non-amine nitrogen derivatives of the psychostimulants cocaine or CFT (WIN 35,428), which bind with high affinity to the dopamine transporter but do not penetrate the CNS [8]. In an effort to constrain the explosion of NPS, a Synthetic Drug Control Act of 2015 was introduced in the US Congress, to add more than 200 synthetic substances to Schedule I. Internationally, the WHO separately and in conjunction with other United Nations agencies conducts similar surveillance and recommends updates on scheduling.

Yet some have questioned the cost-benefit of drug scheduling and whether it effectively curtails NPS use. With curiosity and experimentation as primary motivators for NPS users of synthetic cannabinoids, despite a glut of marijuana, this contention is questionable [5]. It has been argued that an unintended consequence of drug scheduling may be the distribution of more dangerous drugs to replace the scheduled drug. An example is α-PVP (α-pyrrolidinovalerophenone, or "flakka") a demethylated derivative of pyrovalerone and analog of cathinone. α-PVP was gleaned from an early patent or perhaps from a more recent medicinal chemistry manuscript focused on medications for cocaine addiction [9]. More than 130 deaths have been associated with α-PVP, and hospitalizations were required for non-fatal acute intoxications. In cases where α-PVP use was established unambiguously by forensic verification, neurological and cardiovascular effects consistent with an extensive psychostimulant toxidrome have been observed and included cardiotoxicity, violent behavior, and display of psychotic behavior [10]. Emergency scheduling to ban methylone (3,4-methylenedioxy-*N*-methylcathinone) and MDPV (3,4-methylenedioxypyrovalerone) saw increases in methylone encounters with law enforcement, although whether prevalence of use increased in tandem is not clear [11].

On the other hand, mephedrone (4-methyl-*N*-methylcathinone) and related cathinones were controlled in the United Kingdom (UK) in 2010. Emergency department presentations of patients with acute toxicity related to mephedrone peaked prior to, and then fell significantly following, the control of mephedrone. The control of mephedrone in the UK may have been effective in reducing the acute harm associated with the drug [12].

2.3 Guileful Marketing

Wily packaging and labeling often blurs the authentic identity of NPS, reduces stigma, and attempts to evade legal sanctions with disclaimers. Packaging resembles standard quality products: "bath salts," "soap," and misleading labeling insinuates innocuous use: "air fresheners," "legal/herbal highs," "plant food," "insect repellent," "fireplace kindling," "bidet refreshers," and "humidity adsorbents." Disclaimers ("not for human consumption," "research purposes only," and "research chemicals") attract less legal attention and provide a veil of legitimacy on promotional materials. To entice consumption by young users, some synthetic cannabinoids, cathinones, and phenethylamines are sold in packages embellished with bright colors and cartoons and marketed with tasty varieties (blueberry, strawberry, mango, and bubblegum).

NSP are distributed in the USA in convenience stores, "head shops," stores catering to adult products, smoke shops, gas stations, and via the Internet. They may be displayed openly, or hidden from view to be sold only to trusted customers. Although the more common NPS are restricted, a small change in structure can transform a regulated into an unregulated chemical and nullify regulatory oversight. Legal constraints are less manageable if NPS are sold via the Internet, especially since their sources are mainly in Asia or unidentified, and may be beyond the reach of law enforcement. Financial incentives for producer and consumer are another driver of this market. The synthetic routes for producing most NPS are not challenging for competent chemists. The enterprise is lucrative, as the cost of starting materials is inconsequential compared with high markups in retail sales. Based on the cost of a dose unit, the user can purchase certain synthetic drugs at far lower cost than conventional drugs sold on street markets [1].

3 Scope of the NPS Problem

3.1 Prevalence and Use

Synthetic cathinones (mephedrone and MDPV) were among the first NPS to emerge and are frequently used interchangeably with other stimulants such as amphetamine and MDMA. Cathinones are primarily synthesized in Asia, exported, and then packaged. In Europe, more than 70 new cathinones have been recently identified. The EU Early Warning System (EWS) recorded the appearance of 418 NPS during the period of May 2005–December 2014 [13–15], with more than 450 of them currently monitored by the European Monitoring Center for Drugs and Drug Addiction [16]. In 2014, 101 new substances were detected for the first time and reported to the EWS, including 31 designer cathinones, 30 synthetic cannabinoids, and 9 phenethylamines. Sixteen public health alerts were issued in 2014. In the same year, the United Nations Office of Drugs and Crime documented the emergence of 540 different NPS in a worldwide survey of 80 countries [17]. It is estimated that 2.9 million people 15–24 years in the EU have tried NPS.

In the USA, NPS were first encountered in 2009 and since then, more than 250 new synthetic compounds have been identified. Synthetic cannabinoid use remains the most prevalent [18]. Synthetic cannabinoids are the fourth most popular drug class among 8th graders (after marijuana, inhalants, and amphetamines), the third most popular among 10th graders (after marijuana and amphetamines), and the fourth most popular among 12th graders (after marijuana, amphetamines, and Adderall®). Current Monitoring the Future survey data shows that there were no significant increases or decreases in use of "bath salts" in 2015. Use rates of MDMA (3,4-methylenedioxymethamphetamine), or ecstasy or Molly declined among 8th, 10th, and 12th graders since 2010, and continued to show significant declines in 2015 among 10th and 12th graders [19]. Despite these promising trends, indicators of use gleaned from the American Association of Poison Control Centers (AAPCC) show that in 2014, there were 3,677 calls to poison centers regarding *synthetic marijuana exposures*, a 37.8% increase from 2,668 in 2013. This represents the first increase since the number of calls peaked in 2011 at 6,968, with 2012 and 2013 showing a decline in the number of calls.

In contrast, AAPCC statistics show a declining number of calls to poison centers for cathinone exposure. For the year 2014, there were 580 calls, a 41.7% drop from the 995 calls in 2013. In the previous reporting period from 2012 to 2013, the number of calls dropped from 2,691 to 995, a 63% decrease. Although the data suggests that synthetic cathinone abuse is declining, the rebranding of these drugs as MDMA, "molly," or "flakka," to confuse or conceal their content as a synthetic cathinone, may compromise accuracy of self-reported survey data. Users may report MDMA use, when in fact the substance is a cathinone such as methylone or ethylone, or a pyrovalerone analog. Sophisticated analytical methods are the only procedures able to clarify trends in use of psychostimulant substances.

3.2 Medical Consequences

Most chemical classes of NPS can produce adverse psychiatric and medical consequences ([20]; see Schifano et al. this volume). Patients intoxicated with NPS present a significant burden to healthcare professionals, especially those involved with emergency medical care. The long-term neuropsychiatric consequences of NPS exposure are not known, but acute effects (e.g., agitation, hallucinations, psychosis, violent behaviors, and coma) are associated with their use. In the USA, an alarming spike in toxic exposures and fatalities associated with abuse of synthetic cannabinoids has occurred.

3.3 Purity and Quality

Quality control in manufacturing and a standard of purity do not exist for NPS. Cautious buyers may seek sellers who offer safety data or documented purity, but no regulatory bodies guarantee these claims. Each substance may harbor contaminants, or incorrectly identified compounds, to confer a potential health risk. The potent dopamine neurotoxin MPTP (1-methyl-4-phenyl-1,2,3,6-tetrahydropyridine) was a contaminant generated during clandestine synthesis of a meperidine analog in the 1980s. Had the target chemical been synthesized and purified according to procedures described in the source medicinal chemistry journal, the byproduct MPTP would not have produced severe parkinsonism in seven young heroin addicts [21]. Another repugnant example of indifference to purity, quality control, or safety in clandestine production is manganese contamination of ephedrone used in its synthesis [22]. Users can develop an "ephedrone parkinsonism" (EP) characterized by a complex, rapidly progressive, irreversible, and levodopa non-responsive parkinsonian and dystonic syndrome due to manganese toxicity.

NSP packets often include multiple substances. The chemical compositions of packets sold as "bath salts" (cathinones) vary widely, as do purity and safety. A convenience sample of 35 individual packets of "bath salt," purchased in six California cities and over the Internet, identified and quantified all substances in these products [23]. The majority of products (91%) contained either one ($n = 15$) or multiple cathinones ($n = 17$). Of the 14 different compounds identified, MDPV was the most common. Other cathinones detected were buphedrone, ethcathinone, ethylone, MDPBP (an MDPV analog), α-PBP (an α-PVP analog), other designer amines (ethylamphetamine and fluoramphetamine), and 5-IAI (5-iodo-2-aminoindane). Also detected was the antihistamine doxylamine, which had not been previously identified in the US "bath salt" products. In some cases, dramatic differences were found in either total cathinones or synthetic stimulants between products, even with the same declared weight and even between identically named and outwardly appearing products. These findings reveal not only inconsistencies in overall composition of "bath salts" from batch to batch, but significant qualitative and quantitative differences of cathinones and other drugs.

The cannabinoids in "Spice" or "K2" are also heterogeneous and contain a number of unregulated compounds [24]. In a 3-year study involving over 3,000 products described as vegetable material, powders, capsules, tablets, blotter paper, or drug paraphernalia, forensic testing confirmed the presence of 26 synthetic cannabinoids, 12 designer stimulants, and 5 hallucinogenic-like drugs. Overall, synthetic cannabinoids were significantly more prevalent than all the other designer drugs detected, but precise compositions were unpredictable and often formulated with multiple agents. The synthetic cannabinoids JWH-018, AM2201, JWH-122, JWH-210, and XLR11 were most commonly detected in green vegetable material and powder products. But tablets, capsules, and powders also contained designer stimulants such as MDPV, methylone, and pentedrone (-α-methylaminovalerophenone). Hallucinogenic drugs were rarely detected, but generally found on blotter paper products. Without quality assurance and with deceptive labeling, compounds vary from product to product, from batch to batch and even contain "hot spots" within each packet. This array of untested poly-pharmaceuticals places users at risk of adverse health consequences, and baffles emergency department physicians and staff who are powerless to identify the most significant threat to patient health and select effective antidotes.

4 Role of the Internet

4.1 Drug-Related Content Exists Across Social Media Sites

Drug policy, public health, and substance use research are being challenged by the emergence of the Internet to promote and market NPS anonymously. NPS conventionally were sold in buildings hosting "specialty shops," gas stations, or on the street, venues that limit sales, customer base, and expose the distributors to law enforcement. The Internet has recently evolved into a primary base of operations for NPS, changing the dynamics of marketing, reducing risk to suppliers and buyers, and expanding markets globally without personal contacts. It enables sellers and buyers to directly purchase precursors or products from source countries online.

Social networking sites, drug-themed apps, video- and picture-sharing services, and drug forums are venues for discussions, advertisements, and sales. Open websites distribute non-controlled substances or NPS with nebulous legal or international controls. "Dark net markets" which exist covertly on the Internet and are inaccessible through standard web browsers provide anonymity in buying and selling NPS. In 2013, EMCDDA identified 651 websites selling "legal highs" to Europeans [13–15]. These overt or covert sites may use untraceable currencies such as bitcoin and litecoin. Online, virtual drug markets, international sources, and cryptic websites challenge drug control policies and enforcement [14]. The evidence is insufficient on the role of social media in supply and use of NPS to formulate policies addressing these sites.

4.2 Harnessing Social Media

The Internet may be the driver of NPS, but it can also be used to counter its impact. Social media has been exploited to clarify patterns of drug use, reasons for using, and to improve prevention and treatment outcomes [25]. Regular multilingual qualitative assessments of websites, fora for drugs, and other online resources have been conducted using the Google search engine in eight languages from collaborating countries [26]. An online survey of the UK youth on a website found 31.4% of the respondents reported use of mephedrone (41.4%), *Salvia divinorum* (20.0%), "Spice drugs" (10.7%), methylone (1.4%), naphyrone (NRG) (2.1%), benzylpiperazine (BZP) (2.1%), with 15.7% not knowing what they were consuming. The majority (78.9%) considered these substances to be legal, while 50.8% were aware that illegal substances were included in the product.

A Recreational Drugs European Network (RDEN) project established itself as the first Europe-wide prevention program designed for NPS using novel communication technology-based forms of intervention. Prevention messages have been developed, tested, and disseminated via technological tools such as interactive websites, SMS alert, social networking (Facebook and Twitter), multimedia (YouTube), smartphone applications (iPhone), and virtual learning environments (Second Life). More than 650 NPS products and combinations were identified and relevant information disseminated to target populations. Advice given to the EU, international agencies, and national policy makers concluded that web-monitoring activities are needed to map the spread of NPS and match these data with targeted prevention programs. International partnerships were deemed fundamental for shaping a response to this international challenge.

5 Various Classes of NPS

5.1 Most Common Classes of NPS

There are a variety of NPS which include psychostimulant cathinones and their pyrovalerone derivatives, cannabinoids, hallucinogens, dissociative anesthetics, and opioids. The two most commonly used classes of drugs in the USA are synthetic cannabinoids and cathinones. Synthetic cannabinoids (commonly known as "Spice" and "K2") are synthesized in laboratories and simulate, but are not pharmacologically identical with THC, the main psychoactive ingredient in marijuana.

5.2 *Stimulant "Bath Salts": Cathinones and Pyrovalerone Analogs*

Cathinones, also commonly known as "bath salts," can produce pharmacological effects substantially similar to cathinone, methcathinone, MDMA, amphetamine, methamphetamine, and cocaine. The trace amine phenethylamine, found in the brain, is the backbone for most stimulant-type NPS. Analogs of cathinone and pyrovalerone (a pyrrolidine derivative of cathinone) are relatively easy to prepare and can be chemically fashioned in a myriad of ways to produce stimulants, stimulant-hallucinogens, or "entactogens" or "empathogens." Variants currently available represent a small fraction of conceivable structures. Among the more common ones detected recently in the USA are ethylone > MDMA > methylone > α-PVP > MDPV [11]. Mephedrone, methylone, ethylone, and pyrovalerone analogs, including MDPV, NRG, and α-PVP ("flakka"), are among the chemicals packaged as "bath salts," with substituted cathinones (synthetic derivatives of the stimulant cathinone in the plant khat) the most commonly found. These packets are sold as plant foods, insect repellent, bath salts, stain removers, under brand names such as Bliss, Blue Silk, Cloud Nine, Ivory Wave, and others. The products have been widely available in the UK for several years, but emerged in the USA more recently. They are typically manufactured in Asia and then imported into the USA through mail services, packaged and resold in stores or via the Internet.

Synthetic cathinones are usually insufflated or swallowed in their powder or crystal forms but can also be administered by injection, smoking, gingival delivery, or injection via intramuscular or other routes. Nationwide, typical male and female abusers of these substances range from teenagers to those in their 40s. Users often have an extensive history of drug abuse. Some abusers describe the effects as similar to methamphetamine, ecstasy, and cocaine, and have referred to the substances as "complete crank" while others use the term "fake cocaine" or "fake MDMA." Synthetic cathinones produce amphetamine-, MDMA-, or cocaine-like subjective effects by activating monoamine signaling in the brain and periphery via monoamine transporters (see Glennon and Dukat, this volume). These pharmacological effects are consistent with alterations in dopamine, serotonin, and norepinephrine biology [27]. The subjective effects of synthetic cathinones have previously been reviewed [28, 29], with the current book updating the literature. Clinical symptoms reported by healthcare providers involve the majority of organ systems: psychiatric, neurological, gastrointestinal, cardiac, pulmonary, renal, eyes, ear, nose, and throat. The spectrum of psychoactive effects includes aggression, dizziness, memory loss, seizures, blurred vision, anxiety, hallucinations, depression, dysphoria, euphoria, fatigue, increased energy and decreased concentration, panic, and paranoia. Other reported effects involve palpitations, shortness of breath, chest pain, dry mouth, abdominal pain, anorexia, vomiting, erectile dysfunction, discoloration of the skin, and muscular tension. Negative effects of synthetic cathinone use can include heart attacks, kidney and liver failure, paranoia, panic attacks, and rhabdomyolysis (breakdown of muscle tissue). They can also produce extreme agitation, which accounts for the steep rise in emergency department mentions. Not all cathinones are the same, with each eliciting a somewhat unique set of health risks and psychoactivities. Use continues, especially among youth, regardless of mounting evidence that they engender risks and adverse consequences, including emergency department mentions, slow clearance of adverse effects, addiction, psychiatric and cardiovascular effects, and even death. A paucity of information exists on the biological, physiological, and toxicological effects of many of these drugs, especially regarding their long-term effects after heavy and prolonged use.

5.3 *Synthetic Cannabinoids*

Synthetic cannabinoids were initially reported in the USA in December of 2008. The popularity and abuse of these substances and associated products has spread rapidly since then. Synthetic cannabinoids originally were limited to a few compounds (e.g., JWH-018 or 1-pentyl-3-(1-naphthoyl)indole), but others emerged rapidly, in parallel with the explosion of unique designer cathinones. JWH-018, JWH-073, JWH-200, CP-47,497, and cannabicyclohexanol were packaged and sold individually, or dusted on plant material and marketed with misleading designations. More recently identified cannabinoids include XLR11, AB-FUBINACA, and AB-PINACA [30]. Prior to being temporarily placed in Schedule I on March 1, 2011, "K2" and "Spice" were marketed under the guise of "herbal smoking mixtures," "incense," "herbal blends," "air freshener" and designated "not for human consumption." Promoted as legal alternatives to marijuana, they became widely available over the Internet, and sold in gas stations, convenience stores, tobacco and head shops to various populations.

Synthetic cannabinoids are distinctly different from the progenitor phytocannabinoids in *Cannabis sativa* or *Cannabis indica*. They are a conglomerate of a number of compounds designed to mimic the effects of THC in marijuana, and do so by targeting the cannabinoid receptors in brain. However, "Spice" or "K2" synthetic cannabinoids differ from marijuana because of their high potency and full efficacy at CB1 receptors, active metabolites, more robust and persistent effects, and the possibility of activating other non-cannabinoid brain receptors (see Wiley et al., this volume). Each year different cannabinoids emerge in the market, the chemical composition of "Spice" changes, and physiological and toxicological effects remain unknown in this shifting marketplace.

"Spice" has been implicated in numerous medical emergencies and reports of toxicity [30–33]. Symptoms may resolve spontaneously, but range from mild to moderate intoxication, nausea, emesis, weakness, tachycardia, hypertension to psychosis. Several reports have described users in "excited delirium," agitated, and sweating profusely. Severe symptoms include cardiac arrhythmias, myocardial infarction, hyperthermia, psychosis, respiratory depression, flaccid paralysis, rhabdomyolysis, seizures, coma, and even death. Protocols for emergency responses are "ad hoc" for each individual, with antidotes based not on a large body of pharmacological evidence, but on what is effective for the individual [34, 35]. Synthetic cannabinoids conceivably are addictive but the full spectrum of long-term consequences remains unknown.

5.4 *Other New Psychoactive Drugs*

The full spectrum of NPS is beyond the scope of this Introduction. Hundreds of other NPS exist, beyond the categories of synthetic cathinones and cannabinoids. These compounds can be classified by structure (e.g., piperazines, benzofurans, 2C-phenethylamines, tryptamines, NBOMe, methoxetamines, diphenidines, and synthetic opioids), or on the basis of their likely psychoactive effects (e.g., psychostimulants, hallucinogenic/psychedelics, cannabimimetics, dissociative anesthetics, and opioid-like) [36]. Each generation of NPS is not designed to improve safety but to increase markets. As these compounds change, as their doses remain unknown, and as the majority have not undergone systematic evaluation in laboratory animals or humans, their use amounts to a global human experiment without informed consent, safety standards, or safeguards [36].

6 Solutions

6.1 *Research Informed by Data-Sharing*

As witnessed by the opioid epidemic in the early twentieth century, the surge in NPS may overwhelm agencies and healthcare provisions globally before international and comprehensive strategies mature, or if social customs divert attention to different drugs. Synthetic drug producers rapidly adapt to shifting drug trends and legal status by modifying chemical structures to develop legions of new "legal" NPS. The advent of novel compounds is announced instantaneously on social media and other Internet sites, leading to quick adoption and significant profits before the legal gray zone evaporates. Some infrastructure exists to subdue this global challenge to public health; the US DEA, the WHO's Expert Committee on Drug Dependence, the EMCDDA, and the United Nations Office on Drugs and Crime (UNODC) monitor NPS sites. In the USA, a newly established National Drug Early Warning System (NDEWS, http://ndews.org/) uses state-of-the-art methodologies to track emerging drug trends and disseminate information. Yet exploitation of the Internet and other forms of social networks [37–40] for an effective NPS public education/prevention campaign has not materialized on an ambitious grand scale. Nor is there a research infrastructure developed to shape effective prevention messages that counter the appeal of NPS, and that targets appropriately user demographics, advertising methods, that account for the influence of interpersonal ties, and how to shape and deliver effective messages to educate potential or actual users on NPS.

The core of a prevention campaign is scientific evidence to document the potential consequences to users. Accumulation of such research data has been thwarted by the sheer number of current NPS, the complexity of marketed packets crammed with multiple drugs, and the complex pipelines for broadcasting and marketing NPS to evade legal restraints [41]. Research costs become prohibitive, considering the labor-intensive, time-consuming systematic evaluation of a single drug, multiplied by hundreds of unique substances, the swift emergence of others, and the complexity of exploring multiple drug combinations. These limitations clearly necessitate the use of large-scale biological screening methods and concentration on the most problematic substances. Integrated *real-time* Internet monitoring of trends can streamline the process.

6.2 Monitoring of Social Media

Research on NPS has been slow to adapt to social media as a form of communication. Improved methods of monitoring online social media content, possibly through real-time, well-constructed web analytics, can rapidly identify new trends. Research needs to progress from static identifiers of drug-related social media content to assessing how it affects drug use and how to exploit web analytics to shape prevention. Some examples of media monitoring include an NIDA-sponsored NDEWS which collects data from social media and web platforms to identify illicit drug trends and a program to interrogate the role of social media in drug use, addiction, prevention, and treatment. The EMCDDA also uses sophisticated techniques for monitoring web-based drug trends. Notwithstanding these important achievements, integration at an international level may be necessary as the trends in NPS apparently spread from different focal points in different nations.

6.3 Integrating Sources of NPS Information

Clinical cases, emergency department mentions, poison control centers, forensic lab reports (pathology and toxicology), medical reportage, and drug seizures provide critical information for emergency drug scheduling by international agencies and for public health responses. Is it possible to streamline this laborious, assimilative process in real-time and develop rapid responses in a timely manner? Efficient monitoring and responses would require real-time data entry, web analytics, integration of international databases to assist in developing guidelines for prioritizing prevention, in addressing medical emergencies, in forensics, and alerting national laboratories of the need for new chemical standards.

7 Gauging Biological Effects

7.1 Screening for and Testing NPS

The majority of NPS have not been subjected to extensive testing in controlled laboratory conditions. New compounds or analogs of known drugs can affect brain function unpredictably. Yet the responses they elicit in humans are gleaned largely from single case reports. An algorithm of key screening strategies in vitro and in vivo can inform the field and provide leads for emergency department antidotes. One effective method for predicting drug mechanisms is by broad automated screening at key elements of brain communication systems, the neuro-receptorome, which includes transporters, receptors, and ion channels [42]. Current neuroscience research has identified the biological substrates of "classical" drugs of abuse, which generally affect these three target categories [27]. With new or hybrid structures, it is important to be receptive to unpredictable targets. For example, the plant-based hallucinogen salvinorin A was presumed to function at the classic hallucinogenic receptor, the serotonin 5-HT$_{2A}$ subtype, until broad receptor screening identified its agonist actions at the kappa opioid receptor [43]. Deciphering the subtleties of target actions require further excavation of receptor agonist/antagonist, transporter substrate/inhibitor, or channel facilitator/blocker properties. Broad screening may also identify molecular targets contributing to side effects [44]. Preclinical behavioral, pharmacological, and physiological screening can offer limited but valuable information on the abuse liability of new compounds and potentially hazardous neurotoxic, cardiovascular, pulmonary, or temperature dysregulating effects, as well as pharmacokinetic properties, rates of metabolism, and pharmacology of metabolites. Psychiatric symptoms, which cannot be modeled adequately in animals, require clinical case reportage.

7.2 The Unknowns

The long-term consequences of continued use of NPS (brain and organ damage, cognitive impairment, addiction, psychosis, and psychiatric symptoms) remain essentially unknown for most drugs and require intense scrutiny, with defined tests that efficiently address this void. Other unknowns include the unpredictable responses elicited by a mixture of three or five compounds sold in the same packet, or in "hot spots" generated by spraying plant material, whether the pharmacological effects of a drug mixture will be additive, synergistic, antagonistic, or whether NPS synergistically or antagonistically interact with other drugs (e.g., alcohol or medications).

8 Public Education

8.1 Public Awareness and Research

Public awareness of the risks posed by NPS is scant and coordinated; international efforts to exploit social media are embryonic in nature. Public unawareness of specific hazards posed by NPS, of how drugs are approved as prescription medications, and of NPS misinformation proliferated via the Internet is not balanced by compelling counter-evidence. Factual online prevention videos inspire few views in comparison with videos and chat rooms that portray NPS in a positive light. Research on how to develop effective messages and increase traffic to Internet prevention sites is essential to drive scientifically based information towards Internet users at risk. Targeted messages may also offer NPS users opportunities to engage in bidirectional communication, that can tailor, if necessary, information on treatment and recovery support services.

9 Conclusions

The emergence of NPS is challenging for public health and drug policies globally. The novelty of NPS, their ambiguous legal status, ability to evade toxicological tests, swift adaptation to legal restrictions, global Internet marketing, and lack of public awareness are among the key drivers of this twenty-first century phenomenon. Multi-disciplinary research in areas of biology, epidemiology, prevention, and web analytics are needed to develop effective responses in a domain capable of overwhelming current international conventions and national drug control policies. Ultimately, research-guided prevention education will fortify societies against this tidal wave.

References

1. Nichols DE, Fantegrossi WE (2013) Emerging designer drugs. In: Madras B, Kuhar M (eds) The effects of drug abuse on the human nervous system. Academic Press, Cambridge, MA, pp 575–596
2. Papaseit E, Farré M, Schifano F, Torrens M (2014) Emerging drugs in Europe. Curr Opin Psychiatry 27:243–250

3. Deluca P, Davey Z, Corazza O et al (2012) Identifying emerging trends in recreational drug use; outcomes from the Psychonaut Web Mapping Project. Prog Neuropsychopharmacol Biol Psychiatry 39:221–226
4. Brandt SD, King LA, Evans-Brown M (2014) The new drug phenomenon. Drug Test Anal 6:587–597
5. Bonar EE, Ashrafioun L, Ilgen MA (2014) Synthetic cannabinoid use among patients in residential substance use disorder treatment: prevalence, motives, and correlates. Drug Alcohol Depend 143:268–271
6. WHO (2014) http://www.who.int/medicines/areas/quality_safety/36thecddmeet/en/
7. WHO (2015) http://www.who.int/medicines/access/controlled-substances/ecdd/en/
8. Madras BK, Fahey MA, Goulet M et al (2006) Dopamine transporter (DAT) inhibitors alleviate specific parkinsonian deficits in monkeys: association with DAT occupancy in vivo. J Pharmacol Exp Ther 319:570–585
9. Meltzer PC, Butler D, Deschamps JR, Madras BK (2006) 1-(4-Methylphenyl)-2-pyrrolidin-1-yl-pentan-1-one (Pyrovalerone) analogues: a promising class of monoamine uptake inhibitors. J Med Chem 49:1420–1432
10. WHO 5.3 (2015) http://www.who.int/medicines/access/controlled-substances/5.3_Alpha-PVP_CRev.pdf?ua=1
11. DEA (2014) Special report: synthetic cannabinoids and cathinones reported in NFLIS, 2010–2013. http://www.deadiversion.usdoj.gov/nflis/spec_rpt_CathCan_2013.pdf
12. Wood DM, Greene SL, Dargan PI (2013) Emergency department presentations in determining the effectiveness of drug control in the United Kingdom: mephedrone (4-methylmethcathinone) control appears to be effective using this model. Emerg Med J 30:70–71
13. EMCDDA (2015) New psychoactive substances in Europe: an update from the EU Early Warning System March 2015. http://www.emcdda.europa.eu/html.cfm/index44847EN.html
14. EMCDDA (2015) European Drug Report 2015: trends and developments. http://www.emcdda.europa.eu/attachements.cfm/att_239505_EN_TDAT15001ENN.pdf
15. EMCDDA (2015b) Internet and Drug Markets http://www.emcdda.europa.eu/system/files/publications/2155/TDXD16001ENN_FINAL.pdf
16. EMCDDA EU Early Warning System (2015) New psychoactive substances in Europe: an update from the EU early warning system March 2015. http://www.emcdda.europa.eu/attachements.cfm/att_235958_EN_TD0415135ENN.pdf
17. UNODC: United Nations Office of Drugs and Crime (2015) World Drug Report 2015. http://www.unodc.org/documents/wdr2015/World_Drug_Report_2015.pdf
18. DEA (2015) http://www.dea.gov/docs/2015%20NDTA%20Report.pdf
19. Johnston LD, O'Malley PM, Miech RA et al (2016) Monitoring the future national results on adolescent drug use: overview of key findings, 2015. Institute for Social Research, the University of Michigan, Ann Arbor, MI
20. Trecki J, Gerona RR, Schwartz MD (2015) Synthetic cannabinoid-related illnesses and deaths. N Engl J Med 373:103–107
21. Langston JW, Ballard P, Tetrud JW, Irwin I (1983) Chronic Parkinsonism in humans due to a product of meperidine-analog synthesis. Science 219:979–980
22. Bonnet C, Rusz J, Megrelishvili M et al (2014) Eye movements in ephedrone-induced parkinsonism. PLoS One 9, e104784
23. Schneir A, Ly BT, Casagrande K et al (2014) Comprehensive analysis of "bath salts" purchased from California stores and the internet. Clin Toxicol (Phila) 52:651–658
24. Seely KA, Patton AL, Moran CL et al (2013) Forensic investigation of K2, Spice, and "bath salt" commercial preparations: a three-year study of new designer drug products containing synthetic cannabinoid, stimulant, and hallucinogenic compounds. Forensic Sci Int 233:416–422
25. Carhart-Harris RL, King LA, Nutt DJ (2011) A web-based survey on mephedrone. Drug Alcohol Depend 118:19–22

26. Corazza O, Assi S, Simonato P et al (2013) Promoting innovation and excellence to face the rapid diffusion of novel psychoactive substances in the EU: the outcomes of the ReDNet project. Hum Psychopharmacol 28:317–323
27. Baumann MH, Solis E Jr, Watterson LR et al (2014) Baths salts, spice, and related designer drugs: the science behind the headlines. J Neurosci 34:15150–15158
28. Fass JA, Fass AD, Garcia AS (2012) Synthetic cathinones (bath salts): legal status and patterns of abuse. Ann Pharmacother 46:436–441
29. Prosser JM, Nelson LS (2012) The toxicology of bath salts: a review of synthetic cathinones. J Med Toxicol 8:33–42
30. NFLIS-DEA (2014) https://www.nflis.deadiversion.usdoj.gov/DesktopModules/ReportDownloads/Reports/NFLIS2014AR.pdf
31. Castaneto MS, Gorelick DA, Desrosiers NA et al (2014) Synthetic cannabinoids: epidemiology, pharmacodynamics, and clinical implications. Drug Alcohol Depend 144:12–41
32. Gunderson EW, Haughey HM, Ait-Daoud N et al (2012) "Spice" and "K2" herbal highs: a case series and systematic review of the clinical effects and biopsychosocial implications of synthetic cannabinoid use in humans. Am J Addict 21:320–326
33. Seely KA, Lapoint J, Moran JH, Fattore L (2012) Spice drugs are more than harmless herbal blends: a review of the pharmacology and toxicology of synthetic cannabinoids. Prog Neuropsychopharmacol Biol Psychiatry 39:234–243
34. Hermanns-Clausen M, Kithinji J, Spehl M et al (2016) Adverse effects after the use of JWH-210 - a case series from the EU Spice II plus project. Drug Test Anal 2016 [Epub ahead of print]
35. Tait RJ, Caldicott D, Mountain D et al (2016) A systematic review of adverse events arising from the use of synthetic cannabinoids and their associated treatment. Clin Toxicol (Phila) 54:1–13
36. Zawilska JB, Andrzejczak D (2015) Next generation of novel psychoactive substances on the horizon - A complex problem to face. Drug Alcohol Depend 157:1–17
37. Coviello L, Sohn Y, Kramer AD et al (2014) Detecting emotional contagion in massive social networks. PLoS One 12;9(3):e90315
38. Kim DA, Hwong AR, Stafford D, Hughes et al (2015) Social network targeting to maximise population behaviour change: a cluster randomised controlled trial. Lancet 386:145–153
39. Shakya HB, Christakis NA, Fowler JH (2012) Parental influence on substance use in adolescent social networks. Arch Pediatr Adolesc Med 166:1132–1139
40. YouTube https://www.youtube.com/watch?v = -WWNwW0aDh4
41. King LA, Ujváry I, Brandt SD (2014) Drug laws and the 'derivative' problem. Drug Test Anal 7–8:879–883
42. Jensen NH, Roth BL (2008) Massively parallel screening of the receptorome. Comb Chem High Throughput Screen 11:420–426
43. Sheffler DJ, Roth BL (2003) Salvinorin A: the "magic mint" hallucinogen finds a molecular target in the kappa opioid receptor. Trends Pharmacol Sci 24:107–109
44. Setola V, Roth BL (2005) Screening the receptorome reveals molecular targets responsible for drug-induced side effects: focus on 'fen-phen'. Expert Opin Drug Metab Toxicol 1:377–387

Structure-Activity Relationships of Synthetic Cathinones

Richard A. Glennon and Małgorzata Dukat

Abstract Until recently, there was rather little interest in the structure-activity relationships (SARs) of cathinone analogs because so few agents were available and because they represented a relatively minor drug abuse problem. Most of the early SAR was formulated on the basis of behavioral (e.g., locomotor and drug discrimination) studies using rodents. With the emergence on the clandestine market in the last few years of a large number of new cathinone analogs, termed "synthetic cathinones", and the realization that they likely act at dopamine, norepinephrine, and/or serotonin transporters as releasing agents (i.e., as substrates) or reuptake inhibitors (i.e., as transport blockers), it has now become possible to better examine their SAR and even their quantitative SAR (QSAR), in a more effective and systematic manner. An SAR picture is beginning to emerge, and key structural features, such as the nature of the terminal amine, the size of the α-substituent, stereochemistry, and the presence and position of aromatic substituents, are being found to impact action (i.e., as releasing agents or reuptake inhibitors) and transporter selectivity.

Keywords DAT • Methcathinone • Monoamine transporters • NET • QSAR • SAR • SERT

Contents

R.A. Glennon (✉) and M. Dukat
Department of Medicinal Chemistry, School of Pharmacy, Virginia Commonwealth University, Richmond, VA 23298, USA
e-mail: glennon@vcu.edu

Curr Topics Behav Neurosci (2017) 32: 19–48
DOI 10.1007/7854_2016_41
Published Online: 20 November 2016

1 Introduction

Structure-activity relationship (SAR) studies aim to define the qualitative influence of chemical structure on a given biological action and are focused on identifying *what*, and to what extent, substituents and – where applicable – the stereochemistry of substituents alters activity (i.e., action, potency). Recognized, but not widely acknowledged, is that more than a single SAR might be formulated for a given series of agents [1]. Consider that the behavioral actions of a series of agents might be related to their ability to activate a specific neurotransmitter receptor in the brain. An SAR might be formulated for these agents to bind at the target receptor using an in vitro radioligand binding assay (often referred to as a *structure-affinity relationship* or *SAFIR* study), whereas a different SAR might be formulated for their ability to act as agonists in an in vitro functional assay (e.g., some of the agents that display affinity for the receptor might function as weak partial agonists or antagonists rather than as agonists at the receptor of interest). Additionally, compounds that fail to bind at the receptor might act in an *allosteric* manner. Furthermore, the SARs for these actions might differ from SAR derived from their behavioral actions because some of the agents might be rapidly metabolized in vivo, or might be unable to penetrate the blood-brain barrier to reach their intended target. SAR is essentially linked to the assay from which the biological data were obtained, and the formulated SAR is not always conveniently extrapolated to different pharmacological actions/assays.

Quantitative structure-activity relationship (QSAR) studies attempt to explain *why/how* certain structural features influence the actions of a given series of agents. Once SAR studies have been conducted, a QSAR study can be performed using correlational analysis (often referred to as a *Hansch analysis*) to identify whether action/potency within a series of agents might be related to a specific physicochemical property of the substituent being altered. Measures include, but are not limited to, electronic character (as measured by the Taft steric parameter E_s or Hammett σ value), steric size (e.g., volume), overall or specific shape (e.g., Verloop parameters), and lipophilicity (π values). Other parameters consider the molecule as a whole.

Typically, SAR and QSAR studies are not an end unto themselves; rather, they are a means to an end. For example, the results of such studies can be employed (1) in drug design, to enhance the potency or selectivity of an agent; to reduce side (or off-target) effects; to reduce toxicity, or to alter metabolism; and (2) to investigate mechanisms of drug action.

As a general caveat, SAR and QSAR studies should focus on data derived from a common assay – ideally, data generated from the same laboratory and obtained

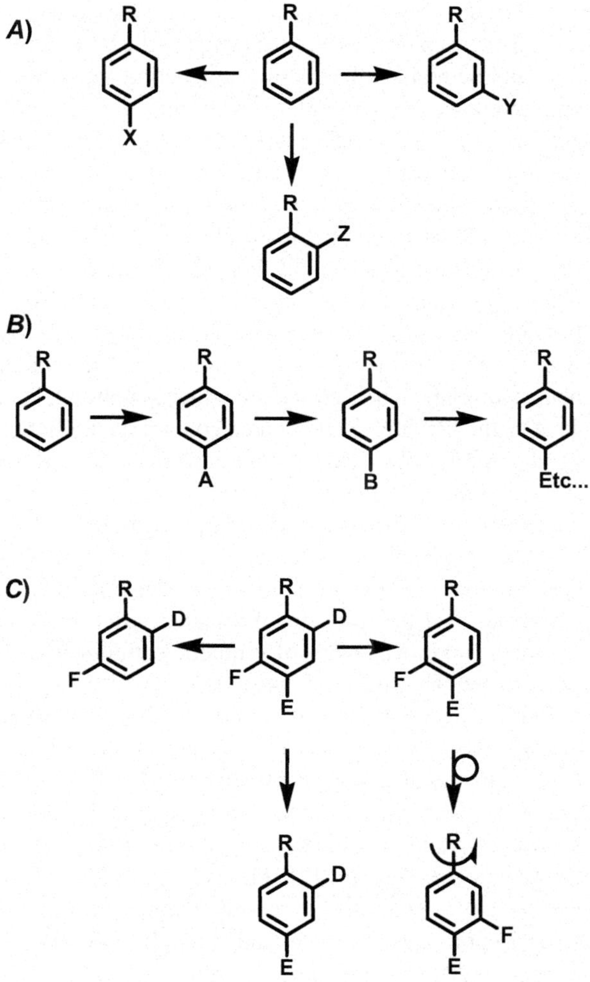

Fig. 1 Examples of three types of SAR studies that can be pursued. Panel (**A**) depicts a nonlinear SAR study where the structure of a molecule is modified one substituent at a time; results can be related back to a common molecule in a systematic manner. Panel (**B**) exemplifies how structures can be related to one another by a single and, usually, position-consistent, structural alteration. Panel (**C**) is a combination of these two approaches and depicts the concept of "deconstruction"

under similar conditions. Employing biological data for one compound from one study with data from (an)other compound(s) from different laboratories or from unrelated studies does not provide reliable SAR results other than from, perhaps, a simple qualitative perspective. The latter approach is not uncommon when a new agent has been identified and attempts are being made to learn something of interest. Certainly, such data are not amenable to QSAR studies. Better yet are data generated where the structure of a molecule is modified one substituent at a time whereby results can be related back to a common molecule in a systematic manner (i.e., nonlinear SAR; see Fig. 1A), or where structures can be related to one another by a single structural alteration (linear SAR; see Fig. 1B). Another approach to SAR is to "deconstruct" a molecule by removing one substituent at a time to identify its influence on a particular action (see Fig. 1C); the latter is actually a combination of the above two approaches.

All of the above approaches are commonly used to formulate SAR, and, indeed, most have been employed to investigate the SAR of synthetic cathinones. However, it should be realized that results emanating from these SAR studies are not always unambiguous. For example, considering the "deconstruction" approach, substituents E and F (Fig. 1C) might influence one another. That is, the presence of substituent E (or F) might be impacted by the adjacent substituent (i.e., one of these substituents might alter the orientation or role of the other – e.g., E might hydrogen bond with F, but in the absence of F it cannot – or E and F might sterically repel one another, changing their steric orientation, which would not be the case when one of the two substituents has been removed). The possibility of rotomeric binding also exists. For example, when the E and F substituents are present along with the D substituent, binding might occur in a manner dictated by the D substituent. However, when the D substituent is absent, rotomeric binding might occur. That is, the E/F (only)-substituted compound might bind as shown in the lower right hand corner of Fig. 1C. As a consequence, SAR studies, certainly "deconstruction" studies, must take these possibilities to the heart, and SAR should be formulated conservatively; alternate SAR interpretations could be possible.

This chapter attempts to capture the SAR and, where possible, to examine the QSAR, of the synthetic cathinones. Unfortunately, due to the variety of assays employed over the last several decades, strict SAR comparisons are often difficult to make, and most studies of synthetic cathinones have been typically of an *agent-only* nature. That is, most investigators, and rightfully so, have tended to focus on individual agents appearing on the clandestine market in order to characterize and classify them. Sometimes, several agents might have been examined in the same study, but, because the agents possessed multiple structural alterations, it can be difficult, if not impossible, to formulate reliable SAR. However, the number of new synthetic cathinones is increasing at an alarming rate. In a United Nations Office of Drugs and Crime report [2], 44 synthetic cathinones had already been identified as having appeared on the clandestine market prior to 2013. According to a European Drug Report, there were greater than 10,000 seizures of synthetic cathinones in Europe in 2013 alone, and 70 were identified as being "new" analogs that had not been encountered hitherto [3]. In 2014, of 101 new psychoactive substances not previously identified, synthetic cathinones represented the largest single category (31%). And new synthetic cathinones continue to appear. Hence, although an agent-by-agent investigation would be required to fully characterize the pharmacological actions of these 150 or so entities, SAR and QSAR studies on a more limited number of agents might assist in providing tentative classifications and guidance as to what might be expected in terms of action(s) and potency. One intent of SAR/QSAR studies with synthetic cathinones is to forecast the actions of agents that have yet to appear (or that have recently appeared) on the clandestine market. It is not the intent of this chapter to review the overall pharmacology of synthetic cathinones; rather, it is to examine SAR. For the most part, only those studies addressing SAR, or studies where some SAR can be formulated in a retrospective manner, will be cited. The information will be provided in, more or less, chronological order so that the reader can appreciate some of the problems that were encountered along the way. The remainder of this chapter is divided into three sections. The first section deals with early SAR studies involving cathinone culminating in the identification of methcathinone; the second section focusses on initial SAR investigations involving methcathinone, and the final section describes the most recent SAR and some QSAR findings.

2 Early SAR of Cathinone

These studies represent those conducted between the time of discovery of cathinone (**1**) and progress into the mid- to late 1990s. Cathinone, specifically *S*(−)cathinone or ***S*(−)1** (see Fig. 2 for chemical structures) was identified in 1975 as the active stimulant component of the shrub *Catha edulis*. Upon its discovery, *S*(−)cathinone (***S*(−)1**) was simply referred to as "cathinone"; *note*: early studies used the term "cathinone" to refer only to the *S*(−)-isomer, whereas more recent studies use the term "cathinone" to refer to racemic or (±)cathinone unless stereochemistry is specifically defined. Prior to 1975, it was thought that (+)cathine (**2**, Fig. 2) represented the major stimulant constituent of the plant. Not unexpectedly, then, some of the first investigations focused on a pharmacological comparison of these two agents, cathinone and (+)cathine, and on the stereochemical aspects of cathinone. The United Nations Narcotics Laboratory synthesized cathinone and its optical isomers in 1978 and made samples available shortly thereafter. It was soon shown that cathinone (**1**) was more potent than (+)cathine (**2**) as a locomotor stimulant in rodents and in other behavioral studies (reviewed: [4]). For example, in tests of stimulus generalization using rats trained to discriminate *S*(+)amphetamine (AMPH, ***S*(+)3**, Fig. 2), *S*(−)cathinone (***S*(−)1**) was several times more potent than (+)cathine (**2**), and nearly as potent, typically more potent, than AMPH (***S*(+)3**). Likewise, *S*(−)cathinone was more potent than (±)cathinone and *R*(+)cathinone both as a locomotor stimulant and in drug discrimination studies with AMPH-

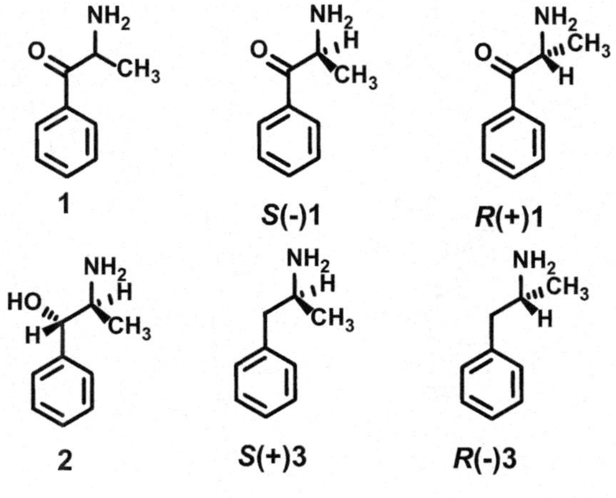

Fig. 2 Structures of (±) cathinone (**1**), S(−) cathinone or **S(−)1**, *R*(+) cathinone or ***R*(+)1**, (+) cathine (**2**), *S*(+) amphetamine or ***S*(+)3**, and *R*(−)amphetamine or ***R*(−)3**

trained and cathinone-trained rats. Hence, two of the earliest relevant SAR findings were that oxidation of the hydroxyl group of (+)cathine (**2**) to its corresponding keto analog (i.e., cathinone, **1**) resulted in retention of stimulant action and an increase in behavioral potency and that *S*(−)cathinone was more potent as a stimulant or amphetamine-like agent than *R*(+)cathinone.

The reduced in vivo potency of (+)cathine (**2**) relative to cathinone (**1**) might be attributed to its lower lipophilicity (*c*Log*P* = 0.81 and 1.16, respectively) and the consequent decreased ability of **2** to penetrate the blood-brain barrier, and/or because the hydroxyl group of **2** is simply not pharmacologically tolerated by its target protein.

Although the (+)-isomer of amphetamine is the more potent of its two optical isomers (i.e., it represents the *eutomer*), it is the (−)-isomer that represents the cathinone eutomer. Regardless, it is the absolute configuration of these isomers, the *S*-isomer in both cases (Fig. 2), not their optical rotation as designated by + or −, that most accurately describes their structural relationship in three-dimensional space.

Some of the next SAR questions addressed were (1) does the α-methyl group of cathinone contribute to its stimulant/stimulus actions, (2) what is the effect of aryl substitution, and (3) will *N*-alkylation alter the potency of cathinone? Initially, because cathinone was a central stimulant, these questions were addressed by examining their locomotor actions in rodents and discriminative stimulus effects in rats (such studies are still being employed). As time went on, it was demonstrated that cathinone, like amphetamine, was a dopamine (DA) releasing agent, and later studies turned in that direction to investigate SAR.

α-Demethylation: As a locomotor stimulant in mice, α-desmethylcathinone (**4**, Fig. 3), where the *α*-methyl group of cathinone has been eliminated, was inactive at

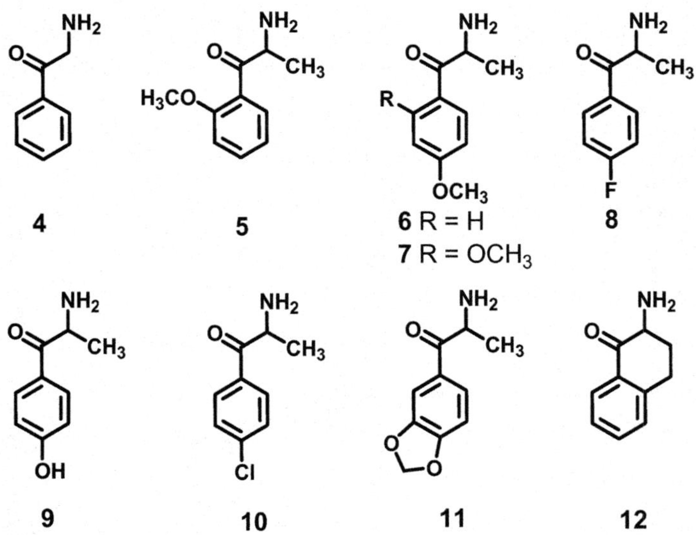

Fig. 3 Structures of some cathinone analogs involved in early SAR investigations

several times the effective dose of cathinone [5]. In drug discrimination studies, **4** failed to substitute (i.e., produced vehicle-appropriate responding) at doses of up to 10 times the ED_{50} dose of *S*(−)cathinone (***S*(−)1**) both in rats trained to discriminate AMPH [6] or (±)cathinone from saline vehicle [7, 8]. As a releasing agent of tritiated dopamine ([^{3}H]DA) from rat caudate nucleus, α-desmethylcathinone (**4**) was about one-fourth as potent as *S*(−)cathinone [9]. Much more recently, it was shown that **4** is about one-third as potent as cathinone (EC_{50} = 208 nM and 83 nM, respectively) as a DA releasing agent in a rat brain synaptosome preparation [10]. If the behavioral effects of cathinone are related to the release of dopamine via the dopamine transporter (DAT), as are those of amphetamine, some effect might have been expected for the doses examined; it would appear, then, that α-desmethylcathinone (**4**) either does not readily enter the brain and/or it is rapidly metabolized in vivo. It might be noted that β-phenylethylamine (PEA), the α-desmethyl analog of amphetamine, also failed to substitute in AMPH-trained rats [7] even though it is only about one-third as potent as AMPH as a depolarizing agent (i.e., the signature of a releasing agent) at the human dopamine transporter [11]. Reith et al. [10] demonstrated that PEA is about one-fourth as potent as AMPH as a dopamine releasing agent. Chain extension of **4** also resulted only in weakly active compounds [9].

Aryl-substitution: Relatively few ring-substituted cathinone analogs have been examined. 2-Methoxycathinone, 4-methoxycathinone, 2,4-dimethoxycathinone, and 4-fluorocathinone (**5–8**, respectively) failed to produce locomotor stimulation in mice at several times as an active dose of *S*(−)cathinone [5]. In rats trained to discriminate (±)cathinone from vehicle, 4-methoxycathinone (**6**) and 4-hydroxycathinone (**9**) produced saline-like effects, and 4-chlorocathinone (**10**) produced only partial generalization [7].

In AMPH-trained rats, 3,4-methylenedioxycathinone (MDC, **11**) elicited partial (50%) substitution, followed by disruption of the animals' behavior at slightly higher doses [12]. This was an indication that MDC (**11**) likely produces central effects other than, or in addition to, its AMPH-like action. Indeed, MDC (**11**) fully substituted in rats trained to discriminate the empathogen MDMA (i.e., 1-(3,4-methylenedioxyphenyl)-2-aminopropane, Ecstasy) from vehicle [12]. MDC (**11**) is the β-keto or β-carbonyl counterpart of MDA, and drug discrimination studies hinted that MDC might possess MDMA-like behavioral qualities.

Conformational constraint: 2-Amino-1-tetralone (**12**, Fig. 3) is an example of a conformationally constrained analog of cathinone. In rats trained to discriminate AMPH from saline vehicle, **12** produced saline-like responding at 10 times the ED_{50} dose of *S*(−)cathinone; however, as a releasing agent of [^{3}H]DA from rat caudate nucleus, it was only about five times less potent than *S*(−)cathinone [9].

N-Alkylation: Similar to amphetamine, cathinone is a central stimulant and a DA releasing agent. Because *N*-monomethylation of amphetamine, to afford methamphetamine, enhances its stimulant potency, *N*-(mono)methylcathinone was prepared and termed "methcathinone" (MCAT, **13**, Fig. 4) analogous to amphetamine/methamphetamine terminology [13]. Although this entity was first synthesized in the early twentieth century, and its locomotor stimulant actions in

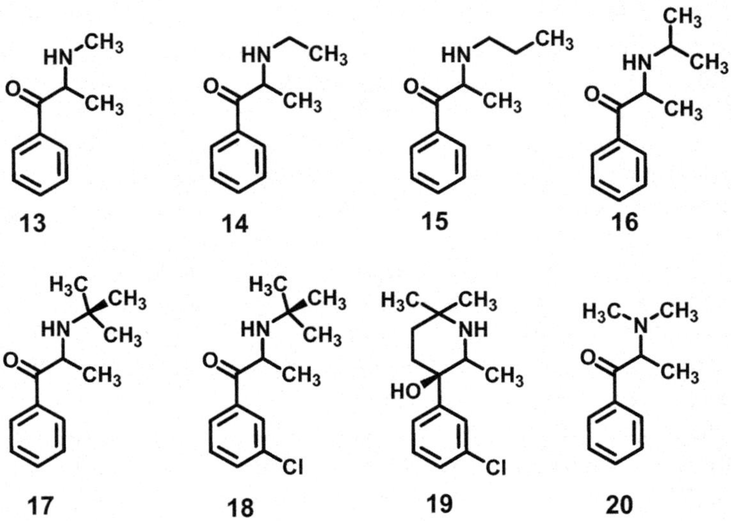

Fig. 4 Some amine-modified methcathinone (**13**) analogs

rodents were noted (reviewed: [4]), it was not until the term "methcathinone" was coined that **13** became a serious target of investigation. Shortly thereafter, it was found that methcathinone (**13**) constituted a serious drug abuse problem in the former Soviet Union where it was known as ephedrone; however, reports of its use had yet to be disseminated in the scientific literature (reviewed: [4]).

Methcathinone (**13**) was identified as a potent DA releasing agent, a locomotor stimulant in rodents, and as an agent more potent than methamphetamine in tests of stimulus generalization in drug discrimination studies (see Table 1) using AMPH-trained rats [12]. *S*(−)Methcathinone also substituted in *S*(+)methamphetamine and (−)ephedrine-trained rats [14, 15]. Behaviorally, *S*(−)methcathinone was found more potent than its *R*(+)-enantiomer in drug discrimination (Table 1) and mouse locomotor studies when enantiomeric comparisons were made.

Homologation of the *N*-methyl group of methcathinone (**13**) to an ethyl (i.e., ethcathinone) or *n*-propyl group (**14** and **15**, respectively, Fig. 4) resulted in small declines in potency in drug discrimination studies with rats trained to discriminate AMPH from vehicle (Table 1).

More recently, ethcathinone (**14**) was shown to behave both as a weak DA reuptake inhibitor (IC_{50} = 1014 nM) and a weak DA releasing agent (EC_{50} = 2118 nM), whereas its propyl homolog **15** was found inactive as a DA reuptake inhibitor [10].

N-Isopropylcathinone (**16**) and *N*-*tert*-butylcathinone (**17**) (Fig. 4) produced stimulant-characteristic hyperlocomotion in rats at doses of 7.5 mg/kg and 10 mg/kg, respectively, that were approximately half that seen following administration of methcathinone at 5 mg/kg [16]. Compound **17** is the des-chloro analog of the antidepressant bupropion (**18**, Fig. 4), and bupropion was also found to be a locomotor stimulant in the same study mentioned above at doses of 10 and

Table 1 Comparison of stimulus generalization potencies of several amphetamine and cathinone analogs to substitute for training drug using rats trained to discriminate *S*(+) amphetamine from saline vehicle [12]

Agent	ED_{50} (mg/kg)
(±)Amphetamine (**3**)	0.71
(±)Methamphetamine	0.49
(±)Methcathinone (**13**)	0.37
S(−)**13**	0.25
R(+)**13**	0.66
(±)Ethcathinone (**14**)	0.77
(±)*N*-*n*-Propylcathinone (**15**)	2.02
(±)*N,N*-Dimethylcathinone (**20**)	0.61
S(−)**20**	0.44

15 mg/kg. In rats trained to discriminate AMPH from saline vehicle, bupropion (**18**) ($ED_{50} = 5.4$ mg/kg) fully generalized to (+)amphetamine, but was 18 times lower in potency; of six bupropion metabolites, only one substituted for the AMPH stimulus: *S,S*-hydroxybupropion (**19**, Fig. 4) ($ED_{50} = 4.4$ mg/kg) [17].

A number of years later, Carroll et al. [18] conducted a very thorough SAR investigation of three dozen bupropion analogs by examining their binding at the DAT, norepinephrine transporter (NET), and the serotonin (5-HT) transporter (SERT), their actions as reuptake inhibitors at all three major transporters, their locomotor stimulant actions in mice, and their stimulus generalization properties in rats trained to discriminate cocaine from vehicle. The majority of the analogs possessed the *tert*-butyl amine substituent common to bupropion (**18**). The des-chloro analog of bupropion (i.e., **17**) displayed about one-sixth the affinity of bupropion (**18**) at DAT, was equipotent at NET and inactive at SERT, was about half as potent as **18** as a locomotor stimulant, and substituted in cocaine-trained rats. Another interesting observation was that extension of the α-methyl side chain had a pronounced influence on affinity at DAT with the following rank order of potency: -methyl < -ethyl < -*n*-propyl > *n*-butyl >*n*-pentyl > *n*-hexyl, with the *n*-hexyl analog still binding with an affinity comparable to bupropion (**18**). A conformationally constrained bupropion analog (i.e., the *N*-*tert*-butyl 5-chloro counterpart of **12**) displayed little affinity for DAT (or NET or SERT) but was twice as potent as **18** as a locomotor stimulant.

Together, these findings suggested that fairly bulky substituents are accommodated on the terminal amine of cathinone, but that they tended to decrease the potency of the resultant agents as amphetamine-like stimulants. At the time, little was known about the mechanism(s) of action of these agents.

N,N-Dimethylation of cathinone (i.e., *N,N*-dimethylcathinone or *N*-methylmethcathinone, **20**, Fig. 4) resulted in retention of AMPH-like stimulus action, but with about half the potency of methcathinone (**13**); the *S*(−)-isomer of **20** was more potent than racemic **20** (Table 1). It might be mentioned that **20**, also termed dimethylpropion, metamfepramone, and DMCN, was once examined as an anorectic agent; investigation of its metabolism in human subjects revealed that nearly half of an orally administered dose of **20** was metabolically demethylated to what is now termed methcathinone (**13**) [19]. Hence, some of the behavioral actions ascribed to **20** might be the result of its metabolism to **13**.

Summary: Learned from early SAR studies was that (1) oxidation of the β-hydroxyl group of (+)cathine to cathinone enhances its stimulant character; (2) cathinone is an amphetamine-like central stimulant with potency nearly comparable to, or greater than, that of amphetamine; (3) the *S*(−)-isomer of cathinone is more potent than its *R*(+)enantiomer in various behavioral assays; (4) α-demethylation abolishes cathinone-like behavioral actions up to the doses evaluated, but results in an agent that retains DA releasing action; (5) aryl substitution, at least for those analogs examined, reduces or abolishes amphetamine-like stimulant/stimulus actions; (6) conformational constraint of the side chain of cathinone, as in 2-amino-1-tetralone (**12**), diminishes stimulant character; (7) *N*-monomethylation (viz. methcathinone, **13**) enhances the potency of cathinone in behavioral assays; (8) *S*(−)methcathinone is more potent than *R*(+)methcathinone in behavioral assays; (9) homologation and/or increasing the bulk of the *N*-methyl group of methcathinone to an ethyl, *n*-propyl, isopropyl, or *tert*-butyl group results in retention of stimulant character but in a reduction in potency; and (10) *N*,*N*-dimethylation of cathinone results in retention of amphetamine-like character, but with a slight decrease in potency. Early studies also provided evidence that cathinone (**1**) and methcathinone (**13**) behaved, at least in part, as DA releasing agents.

3 The Methcathinone Years

Until the mid-1990s, studies focused primarily on structural modification of cathinone; however, once methcathinone was identified as a potent central stimulant, there was a shift in attention to analogs of the latter. There was also a better understanding that methcathinone and some of its analogs might be acting as DA releasing agents. SAR derived from behavioral studies began to decline somewhat in favor of studies that focused more on the SAR for releasing action at the DAT. Eventually, methcathinone became a US Schedule I substance.

Due to the substantial potency of methcathinone (**13**) as a central stimulant, investigations were conducted on the stimulus generalization of methcathinone and its optical isomers ***S*(−)13** and ***R*(+)13** (Fig. 5) in cocaine-trained rats, the use of methcathinone as a potential training drug in drug discrimination studies, and on rodent locomotor studies on a few aryl-substituted methcathinone analogs.

In a locomotor assay in rats, 3-bromomethcathinone (**21**, Fig. 5) produced hyperlocomotion at doses of 7.5 and 10 mg/kg comparable to that produced by methcathinone at 5 mg/kg, whereas 4-bromomethcathinone (**22**), now termed brephedrone or 4-BMC, was inactive at these doses [16]. Subsequently, in a related assay in rats, 4-trifluoromethylmethcathinone (4-CF_3 MCAT, **23**) was inactive as a locomotor stimulant [20].

In cocaine-trained rats, stimulus generalization occurred with the following order of potency (ED_{50} values given in parenthesis): *S*(−)MCAT (0.18 mg/kg) > (±)MCAT (0.39 mg/kg) > *R*(+)MCAT (0.51 mg/kg) > cocaine (2.6 mg/

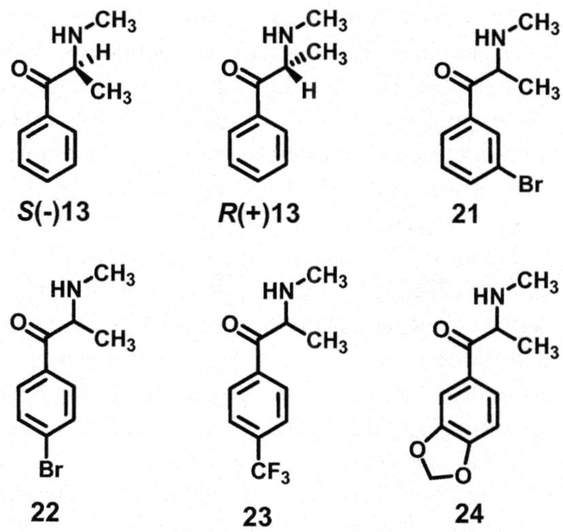

Fig. 5 Some early methcathinone analogs

Table 2 Biogenic amine releasing potency of several agents of interest [24]

Agent	EC_{50} (nM)		
	DAT	NET	SERT
S(+)Amphetamine ***S*(+)3**	24.8	7.1	1765
S(+)Methamphetamine	24.5	12.3	736
S(−)Cathinone ***S*(−)1**	18.5	12.4	2386
S(−)Methcathinone ***S*(−)13**	14.8	13.1	1772
(+)Cathine **2**	88.3	15.0	Inactive
(−)Cathine **25**	1371.0	137.0	Inactive

kg) [21]. Later, Kohut et al. [22] showed that methcathinone also substituted for cocaine in monkeys.

Unlike MDC (**11**), methylenedioxymethcathinone (MDMC, methylone, **24**), the *N*-methyl analog of MDC, substituted in AMPH-trained rats; however, it was six times less potent than methcathinone [12]. MDMC (**24**) also substituted in MDMA-trained rats [12].

S(−)Methcathinone was demonstrated to serve as a training drug in rats, and stimulus generalization occurred upon administration of other central stimulants with the following order of potency (ED_{50} values given in parentheses): *S*(−)MCAT (0.11 mg/kg; 0.5 μM/kg) > *S*(+)methamphetamine (0.17 mg/kg, 0.9 μM/kg) > (±)MCAT (0.25 mg/kg, 1.2 μM/kg) > *R*(+)MCAT (0.43 mg/kg, 2.1 μM/kg) ~ (±)cathinone (0.41 mg/kg, 2.2 μM/kg) [23]. As an aside, it might be noted that the *S*(−)methcathinone stimulus also generalized to cocaine ($ED_{50} = 1.47$ mg/kg, 4.3 μM/kg) [23].

Because methcathinone (**13**) had been shown to act as a releasing agent at the DAT, a SAR study was conducted, and several agents were compared for their ability to release DA, norepinephrine, and 5-HT from rat brain synaptosomes [24]. Some of the data are shown in Table 2.

The behaviorally more potent *S*-isomers of cathinone (**1**), methcathinone (**13**), amphetamine (**3**), and methamphetamine displayed comparable potencies, and similar potencies as releasing agents, at DAT and NET (Table 2). All displayed lower potencies as 5-HT releasing agents. The potency of (+)cathine (**2**) as a norepinephrine releasing agent was comparable to that of the above agents; however, **2** was severalfold less potent than the others at DAT and inactive at SERT (Table 2). Data for (−)cathine (**25**) is shown in Table 2 for comparison. Interesting is that an AMPH stimulus generalized to all of the agents in Table 2 except for (−) cathine (**25**) [25] supporting a possible role for DA and/or norepinephrine in their actions. A decade later, Cozzi et al. [20] compared the actions of 4-CF_3 MCAT (**23**) with those of its 2- and 3-substituted positional isomers; introduction of the –CF_3 group had a deleterious effect on DAT and NET release and resulted in enhanced selectivity for SERT, and **23** was found to lack activity as a locomotor stimulant in rats (the other two positional isomers were not examined).

From these investigations, it was (1) confirmed that *S*(−)methcathinone is more potent than *R*(+)methcathinone as a central stimulant/stimulus, (2) shown that *S*(−) methcathinone is a potent DA and norepinephrine releasing agent, (3) demonstrated that introduction of aryl substituents decreases the stimulant and DAT releasing effects of methcathinone, and (4) shown that *S*(−)methcathinone can be used as a training drug to examine the stimulus effects of other cathinone and non-cathinone central stimulants.

4 Current SAR Studies

In 2010, Iversen [26] submitted a report to the British Home Office on the alarming emergence of synthetic cathinone analogs on the European clandestine market. This ushered in a new era in cathinone research. A recent PubMed search (accessed December 20, 2015) for "synthetic cathinones" yielded 189 papers published during the 5-year period between January 2011 and December 2015.

One early drug combination popular around 2010 was referred to as *bath salts*; it was also known by several other names. *Bath salts* contained either methylone (MDMC, **24**), mephedrone (**26**), methylenedioxypyrovalerone (MDPV, **27**) (Fig. 6), or a combination of one, two, or more of these and/or other agents

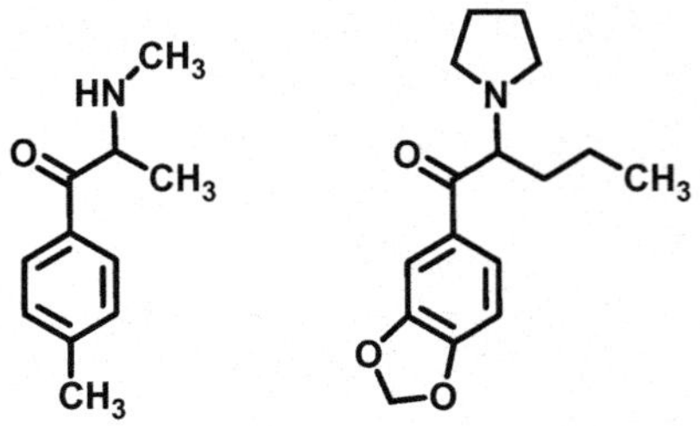

26 27

Fig. 6 Two early "bath salts" constituents: mephedrone (**26**) and MDPV (**27**)

(reviewed: [4]). Some information was already available about methylone (**24**) (vide supra), but little was known about the other two agents. Based on SAR already formulated for cathinone and amphetamine analogs, it was suspected that mephedrone (**26**) would be a DA releasing agent with reduced potency and selectivity relative to methcathinone (**13**). MDPV (**27**) represented something "new," although Meltzer et al. [27] had already examined several related pyrovalerones (but not MDPV) as DA reuptake inhibitors for their therapeutic potential.

Various investigators demonstrated that these three synthetic cathinones (methylone, mephedrone, MDPV), among others (see below), behaved as locomotor stimulants in rodents (e.g., [28–34]), were self-administered by rats [29, 35, 36], and produced discriminative stimulus effects similar to other central stimulants using methamphetamine-trained and cocaine-trained rats [32, 37] and MDPV-trained mice [31]. There is substantial difficulty in formulating a reliable SAR for these agents (not the intent of the original investigations and very difficult in retrospect) because the studies were not focused on SAR; this is related to the paucity and structural diversity of agents examined in the individual studies, the different animal species employed, the different routes (e.g., i.v. versus i.p. or i.m.) of drug administration, and the various temporal parameters employed. Gatch et al. [37] made the observation that although locomotor activity was a good predictor for dose ranges to be examined in their subsequent drug discrimination studies, the magnitude of locomotor stimulation might not be a good predictor of abuse liability for the agents they examined. They also speculated that reaching a threshold level of neurotransmitter release or reuptake might be sufficient to produce behavioral effects, but that subtle differences might not be important.

Attention turned to the ability of synthetic cathinones to act at biogenic amine neurotransmitter transporters (i.e., DAT, NET, SERT) that might underlie their behavioral effects. Mephedrone (**26**) was shown to behave as a DA releasing agent. Within a year after the Iversen [26] report, MDPV (**27**) was shown to display characteristics of a DA reuptake inhibitor. Using a frog oocyte preparation transfected with hDAT, mephedrone (**26**) produced dopamine-like depolarization, whereas MDPV (**27**) produced cocaine-like hyperpolarization [38–40]. These are the signatures of a releasing agent and a reuptake inhibitor, respectively. Simmler et al. [41] later examined several synthetic cathinones and found that mephedrone (**26**) was nearly equipotent as an inhibitor and releaser of DA and 5-HT, but substantially more potent as an inhibitor of NET. In contrast, MDPV (**27**) was a potent reuptake inhibitor at DAT and NET, a very weak inhibitor of SERT, but did not release DA or 5-HT [41]. Eshleman et al. [42] reported similar results. These overall findings were similar to what was reported by Baumann and co-workers [30] (Table 3), except that methylone (MDMC, **24**) was more potent as a DA releasing agent than a reuptake inhibitor. Different investigators employed different assays (procedures, radioligands, etc.,) – hence, differences exist with respect to the results shown here. So again, there are problems to formulate reliable SAR between studies.

Table 3 Effect of several synthetic cathinones on synaptosomal release and reuptake inhibition at biogenic transporters; data for amphetamine and cocaine included for comparison [30]

	Release EC_{50} (nM)			Reuptake inhibition IC_{50} (nM)		
	DAT	NET	SERT	DAT	NET	SERT
Methylone (**24**)	117	140	234	1232	1031	1017
Mephedrone (**26**)	51	58	122	762	487	422
MDPV (**27**)	n.a.	n.a.	n.a.	4.1	26	3349
S(+)Amphetamine	5.8	6.6	698	93	67	3418
Cocaine	n.a.	n.a.	n.a.	211	292	313

n.a. not active, demonstrating release <35%

Expansion of the methylenedioxy ring of methylone (**24**) to an ethylenedioxy ring (i.e., ethylenedioxymethcathinone, EDMC, **30**, Fig. 7) decreased its potency as a releasing agent at all three transporters by two- to threefold [43]. Homolgation of the α-methyl group of methylone (**24**) to an α-ethyl group (i.e., butylone, **31**) and replacement of the methylenedioxy group of MDPV (**27**) with a fused phenyl ring (i.e., naphyrone, **32**) resulted in reuptake inhibitors at the three transporters [41, 42]. Naphyrone (**32**) was five- to tenfold more potent than butylone (**31**) at DAT, NET, and SERT and five- to tenfold less potent than MDPV (**27**) at DAT and NET, but 10 times more potent than MDPV (**27**) at SERT. It was apparent from these (and other concurrent or subsequent) investigations that substituents on the terminal amine and/or at the α-position of cathinone analogs have a significant impact both on the actions of these agents as releasing agents (i.e., as substrates) versus reuptake inhibitors, and on their selectivity for the major biogenic amine transporters.

Synthetic cathinones as reuptake inhibitors: MDPV (**27**) was unique among the synthetic cathinones appearing on the clandestine market because it was the first of them to be identified as a DA reuptake inhibitor. Figure 8 shows a systematic SAR deconstruction of MDPV (**27**) to determine which of, and to what extent, its various structural features contribute to its actions as a DA reuptake inhibitor [44]. All of the compounds shown in Fig. 8 behaved as reuptake inhibitors but varied appreciably with respect to potency. Removal of the carbonyl group, converting MDPV to its amphetamine analog **33**, reduced its potency by about eightfold, whereas removal of the methylenedioxy substituent (i.e., α-PVP, **34**) had a negligible effect. The length of the α-side chain would appear to be critical; shortening the side chain (i.e., MDPPP, **35**) resulted in a > 25-fold decrease in potency. With an intact side chain, the next most important feature was the amine. Conversion of the amine from the simplest tertiary amine (i.e., **36**, dimethylone) to a secondary or primary amine (i.e., **37** or pentylone and **38**, respectively) ultimately resulted in a 200-fold decrease in potency).

Because the methylenedioxy group played a minimal role in the ability of MDPV (**27**) to act as a DAT inhibitor, a series of analogs lacking this functionality was examined with a focus on SAR [47]. These analogs might be viewed as being derived from α-PVP (*flakka*, **34**) (Fig. 9 and Table 4) – currently, a very popular

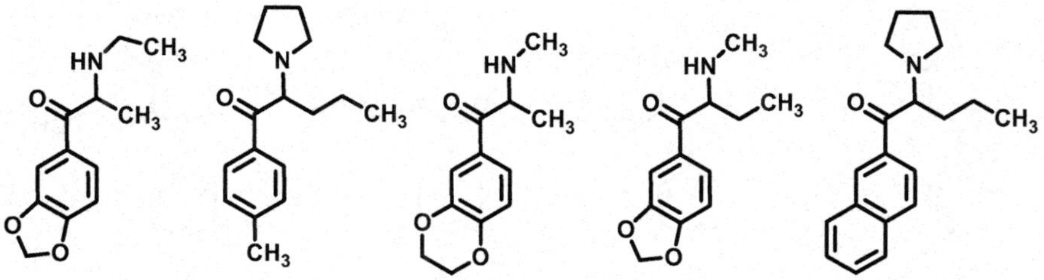

28 29 30 31 32

Fig. 7 Structures of ethylone (**28**), pyrovalerone (**29**), EDMC (**30**), butylone (**31**), and naphyrone (**32**)

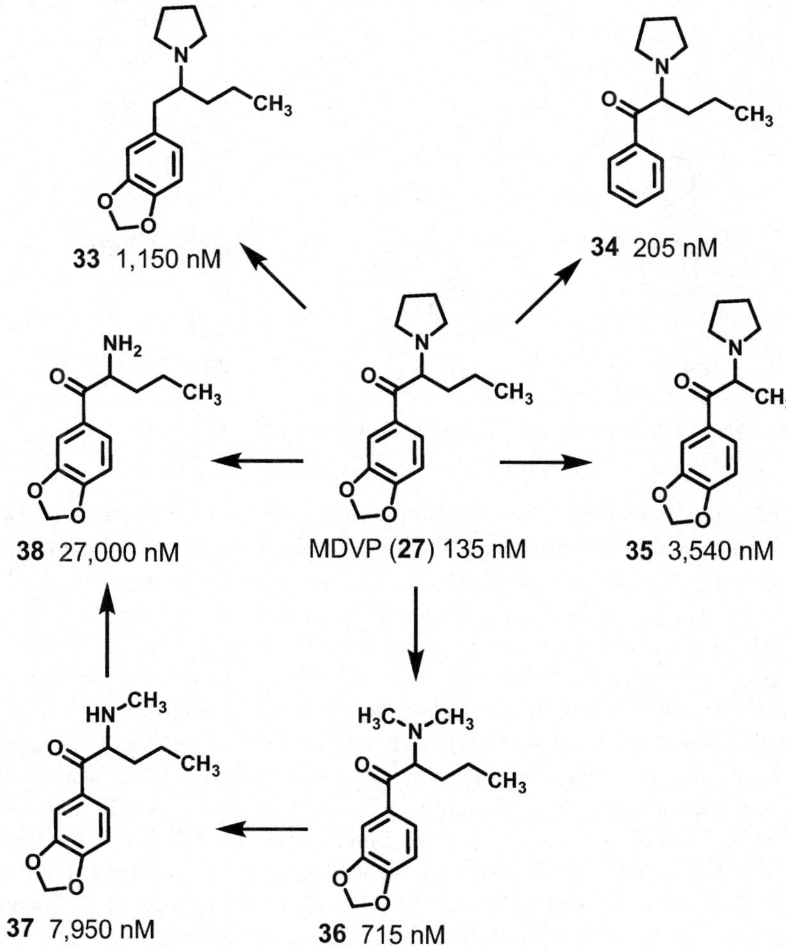

Fig. 8 Deconstruction of MDPV (**27**). Values represent the potency of the analogs to block the reuptake of DA [44]

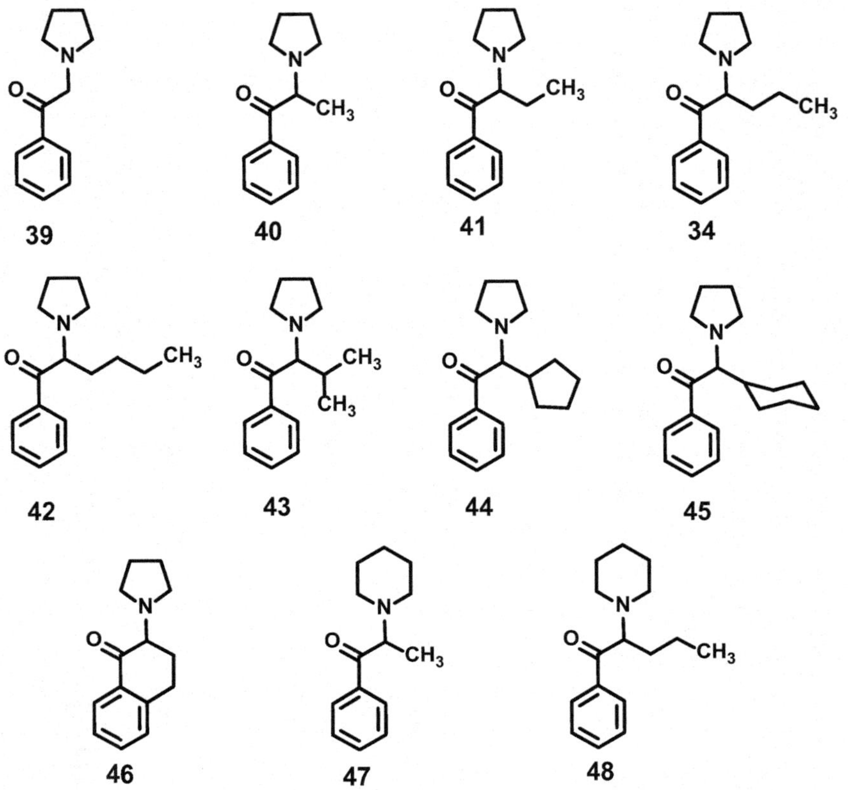

Fig. 9 Deconstructed (**39–41**) and elaborated (**42–48**) analogs of α-PVP (**34**)

drug of abuse. Using the deconstruction process, the α-*n*-propyl group of **34** was shortened in a stepwise manner [47]. Truncation of the α-propyl substituent to an α-ethyl group (i.e., α-PBP, **41**, Table 4) reduced potency by about threefold, and further contraction to an α-methyl group (i.e., α-PPP, **40**) resulted in an overall >tenfold decrease in potency. Elimination of the α-*n*-propyl group altogether – that is, replacement by –H (i.e., **39**, Table 4) – resulted in a nearly 200-fold decrease in potency. Nevertheless, all of the analogs behaved as DA reuptake inhibitors. The findings support those shown in Fig. 8 in that the α-substituent of synthetic cathinones plays a major role in their actions as reuptake inhibitors at DAT when the amine substituent is held constant as a pyrrolidine moiety. In addition, none of the analogs was effective as a reuptake inhibitor at SERT ($IC_{50} > 10{,}000$ nM). From these data, it can be surmised that the length of the α-substituent is influential for DAT action, but that action at SERT might not readily accommodate a pyrrolidine moiety regardless of the length of the α-substituent.

Because the length/bulk of the α-substituent seemed important for these agents to act as reuptake inhibitors at DAT, additional compounds were examined (i.e., an

Table 4 Potency of deconstructed and elaborated α-PVP (**34**) analogs to inhibit synaptosomal reuptake at DAT [47]

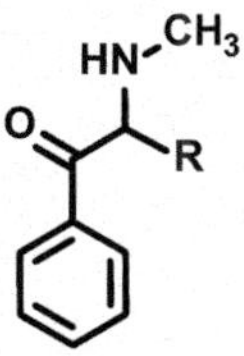

Agent	R	IC_{50} (nM)
39	–H	3250
40	$-CH_3$	196.7
41	$-CH_2CH_3$	63.3
34	$-CH_2CH_2CH_3$	17.5
42	$-CH_2CH_2CH_2CH_3$	11.6
43	$-CH(CH_3)_2$	92.3
44	$-C_5H_9$	17.1
45	$-C_6H_{11}$	8.3

"elaboration" investigation – see Glennon and Young [1] for conceptual details). Increasing the length of the α-substituent from *n*-propyl (i.e., **34**) to *n*-butyl (i.e., α-PHP, **42**) resulted in a slight increase in potency (Table 4). Branching of the α-ethyl side chain of **41** to its isopropyl counterpart **43** reduced potency by about 50% (Table 4). But, as might now be expected based on the information provided in Table 4, increasing the bulk of this branched analog should result in increased potency. Elaborated analogs **44** and **45** (IC_{50} = 17.1 and 8.3 nM, Table 4) were at least as potent as α-PVP (**34**) [47].

Compound **46** is a conformationally constrained analog of **41**; its potency (IC_{50} = 12,900 nM) is >200-fold less than that of **41**. Ring expansion of the pyrrolidine ring of α-PVP (**34**) to a piperidine ring (i.e., **48** IC_{50} = 128 nM) resulted in a sevenfold decrease in potency; likewise, the piperidine analog of **40** (i.e., **47** IC_{50} = 2490 nM) also displayed reduced potency. The overall results of these studies are not inconsistent with results published by Meltzer et al. [27] on a related series of pyrovalerones although, in their studies, most of the compounds possessed aryl substituents such that direct SAR comparisons are difficult to make.

Pyrovalerone (**29**) possesses a chiral center and two optical isomers are possible; the *S*-isomer was 100 times more potent than its *R*-enantiomer as a reuptake inhibitor at DAT [27]. Both isomers were less potent at NET and SERT. MDPV (**27**) also exists as a pair of optical isomers (Fig. 10), and both were prepared and examined [45] with respect to their neurochemical actions on neurotransmitter reuptake and behavioral effects in an assay of intracranial self-stimulation (ICSS) in rats – a behavioral procedure used to evaluate abuse potential. In assays of DAT uptake inhibition, *S*(+)MDPV (IC_{50} = 2.13 nM) was twice as potent as (±)MDPV (IC_{50} = 4.85 nM) and 180-fold more potent than *R*(−)MDPV (IC_{50} = 382.80 nM); as such, the *S*-isomer was 100-fold more potent than cocaine (IC_{50} = 198.80 nM).

Table 5 Potency of selected 4-substituted methcathinone analogs as releasing agents at DAT and SERT, and in a rat ICSS assay, used in a QSAR study [46]

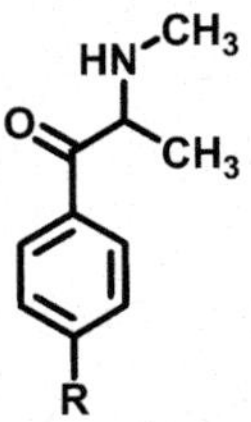

Agent	R	DAT EC_{50} (nM)[a]	SERT EC_{50} (nM)[a]	DAT selectivity[b]	ICSS maximum % baseline facilitation[c]
Methcathinone (**13**)	–H	12.5	3860	309	191.9
Flephedrone (**50**)	–F	83.4	1290	15.4	156.3
Methedrone (**51**)	$–OCH_3$	506	120	0.24	110.9
4-Chloromethcathinone (**52**)	–Cl	42.2	144	3.40	114.9
4-Bromomethcathinone (**22**)	–Br	59.4	60	1.01	118.0
Mephedrone (**26**)	$–CH_3$	49.1	118	2.41	102.5
4-Trifluoromethylmethcathinone (**23**)	$–CF_3$	2700	190	0.07	90.9

[a]EC_{50} values and ICSS data are from Bonano et al. [49]
[b]DAT selectivity calculated as $(\text{DAT EC}_{50})^{-1} \div (\text{SERT EC}_{50})^{-1}$; higher values indicate greater DAT selectivity

The three were less potent at NET uptake inhibition, but with the same rank order of potency, $IC_{50} = 9.86$ nM, 16.84 nM, and 726 nM, respectively, relative to cocaine $IC_{50} = 395.9$ nM. Neither (±)MDPV nor either of its optical isomers inhibited the reuptake at SERT. *S*(+)MDPV produced an abuse-related and dose-dependent facilitation of ICSS in rats, and the potency of *S*(+)MDPV (significant facilitation at doses ≥0.1 mg/kg) was greater than that of (±)MDPV, whereas *R*(−)-MDPV failed to alter ICSS at doses up to 100 times greater than the lowest effective dose of *S*(+)MDPV [45].

Another preliminary finding, although additional studies are required, was that the *N*-methyl quaternary amine counterpart of MDPV (i.e., Q-MDPV, **49**, Fig. 10) produced hyperpolarization in frog oocytes transfected with hDAT ([48], Sakloth, Solis, DeFelice, Glennon, *unpublished data*). These results suggest that **49** can act as a DAT reuptake inhibitor.

One of the first QSAR studies published on synthetic cathinones to inhibit reuptake at DAT indicated that potency is related to the "size" of the α-side chain [46, 48]. That is, using the data shown for the eight agents in Table 4, potency was significantly correlated both with the volume ($r = 0.909$) and the lipophilicity (π value, $r = 0.917$) of their α-substituents. However, for the substituents in this set,

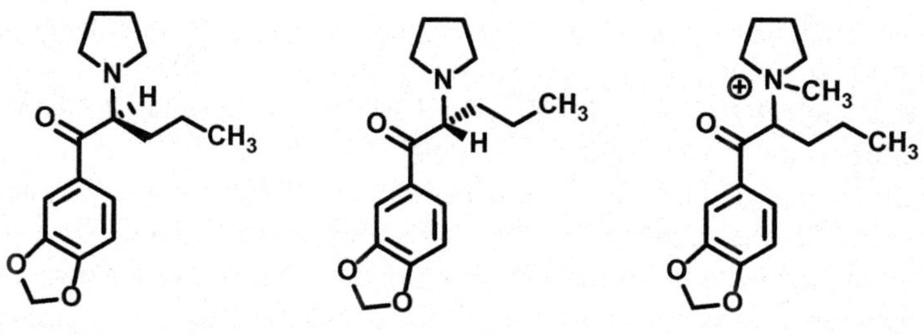

Fig. 10 Newer MDPV (**27**) analogs: *S*(+)MDPV, *R*(−)MDPV, and Q-MDPV (**49**)

volume and π were highly intercorrelated ($r = 0.997$). Additional studies will be required to determine which of these two parameters is more important for activity and, obviously, to determine the optimal volume and/or lipophilicity for this action.

Synthetic cathinones as releasing agents: As already mentioned, mephedrone (**26**), although not as potent or selective as MCAT (**13**), was found to act as a releasing agent at DAT. The same is true of a number of related methcathinone analogs. The literature has described a number of studies on such agents (e.g., see [4] for a review and more recent references on individual synthetic cathinones). SAR and QSAR endpoints were not the focus of most of these investigations, and only those that specifically addressed the topic will be mentioned here.

Seven methcathinone analogs that differed only with respect to their 4-position substituents were examined for their ability to modulate in vivo ICSS in rats and to act as substrates (i.e., releasers) at DAT and SERT. The potencies and selectivities of these agents varied over a broad range (Table 5). The most potent analog in the ICSS assay was MCAT (**13**), whereas the least potent was 4-trifluoromethymethcathinone (**23**); the potencies of the other agents fell somewhere in between. In vitro DAT versus SERT selectivity correlated with in vivo efficacy to produce ICSS facilitation ($r = 0.92$). Furthermore, the Taft steric parameter (i.e., E_S value) of the 4-position substituents correlated both with DAT versus SERT selectivity ($r = 0.78$) and magnitude of ICSS facilitation ($r = 0.81$) [49]. There was no relationship either in ICSS facilitation, DAT potency, or SERT potency and either the electronic (σ) or lipophilic (π) character of the substituents. In a follow-up study, more specific steric parameters were examined including substituent volume and Verloop size (i.e., substituent length, L, and minimum or maximum substituent width, B1 and B5, respectively). Maximal ICSS facilitation was negatively correlated with the four steric parameters: volume (Å^3) ($r = -0.915$), length (L) ($r = -0.773$), minimum width (B1) ($r = -0.778$), and maximum width (B5) ($r = -0.814$). Internal correlations were found between certain parameters: volume and both substituent length ($r = 0.814$) and maximal width ($r = 0.935$), as well as between length and maximal width ($r = 0.798$) [46]. The potency of the agents to promote in vitro monoamine release via DAT was negatively correlated with increasing volume ($r = -0.803$) and maximal substituent width ($r = -0.807$), whereas potency at SERT was positively correlated with increasing volume ($r = 0.825$) and length ($r = 0.903$) of the 4-position substituent. Selectivity for DAT vs. SERT was correlated with volume ($r = -0.972$) and maximal width ($r = -0.917$).

It might be expected that multiple structural features such as the terminal amine and the aromatic ring contribute to the interactions of MCAT analogs at DAT and SERT; but these structural features are common to all the analogs shown in Table 5. That is, these analogs varied only by the substituent at the 4-position; in this respect, they represent a matched set with only a single variable substituent among them. Found was that as the size of the 4-position substituent increased, potency at DAT decreased, whereas potency at SERT increased, and this reciprocal relationship was also seen with DAT versus SERT selectivity. It would seem, then, that the steric properties of the 4-position substituent play a major role in the actions of the investigated MCAT analogs. However, because of internal correlations between some of the steric parameters, it was not possible to identify a single steric parameter as being the most relevant. Nevertheless, it was speculated that because of the consistent identification and the higher correlation coefficients associated with steric volume (i.e., $Å^3$), the total volume of the 4-position substituent is likely the most important feature for these interactions [46].

Homology modeling studies provided new models for hDAT and hSERT using the dDAT crystal structure as a template. Docking studies with the agents in Table 5 showed that large substituents at the MCAT (**13**) 4-position were better accommodated by hSERT than hDAT [46]. The results were consistent with the results of the QSAR studies described above. Furthermore, a hydropathic interaction (HINT) analysis that considered potential interactions of the 4-position substituents with specific nearby transporter amino acid residues suggested that hydrophobic interactions provided by the 4-position substituent are necessary for potency at hSERT, whereas unfavorable polar interactions at the 4-position might play a role in determining potency at hDAT. The overall conclusion was that in the MCAT binding pocket associated with the 4-position substituent in hDAT, bulky substituents are not readily accommodated, whereas the larger and less polar binding pocket in hSERT more readily accommodated them. The overall conclusions of these QSAR and modeling/docking investigations are that (1) MCAT analogs with small 4-position substituents favor binding at DAT versus SERT, (2) larger substituents favor binding at SERT, and (3) the hydrophobic nature of these substituents modulates potency at SERT.

In a follow-up study, the 4-tert-butyl analog of methcathinone was prepared and examined. Although the potency of this agent as a DA releasing agent ($EC_{50} = 942$ nM) was greater than that of its corresponding 4-trifluoromethyl counterpart **23**, it was only a partial (ca 50%) releasing agent; furthermore, this agent now acted as a weak reuptake inhibitor at DAT ($IC_{50} = 2207$ nM) and lacked action as either a releasing agent or reuptake inhibitor at SERT up to concentrations of 1000 and 10,000 nM) ([48], Sakloth, Partilla, Bauman, Glennon, *unpublished data*). The results support the finding that large 4-position substituents are not tolerated at DAT and that there is a limit to the size of this substituent that can be accommodated by SERT.

Table 6 Potencies of stereoisomers of several simple cathinone analogs for the synaptosomal release of neurotransmitter from DAT, NET, and SERT [50, 51]

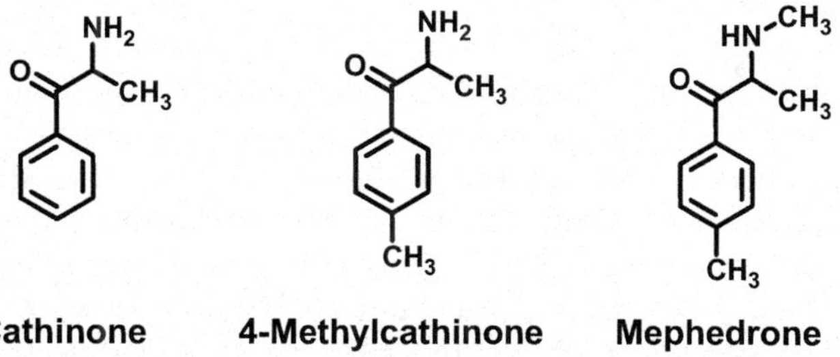

	EC_{50} (nM)[a]			DAT vs. SERT selectivity
	DAT	NET	SERT	
S(−)Cathinone ***S*(−)1**	25	14	9267	370
R(+)Cathinone ***R*(+)1**	184	72	>10,000	>50
S(−)4-Methylcathinone ***S*(−)53**	150	89	179	1.2
R(+)4-Methylcathinone ***R*(+)53**	391	115	1592	4.1
S(−)Mephedrone ***S*(−)26**	74	–	61	0.8
R(+)Mephedrone ***R*(+)26**	31	–	1470	47

[a]Some values have been rounded off to the nearest whole number

In a QSAR study, using in vivo microdialysis to examine the relationship between the volume of the 4-position substituent and the in vivo neurochemical selectivity of cathinone analogs to alter nucleus accumbens (NAc) DA and 5-HT levels, rats were implanted with bilateral guide cannulae targeting the NAc and were administered MCAT (**13**) and five of the 4-substituted MCAT analogs shown in Table 5 (i.e., **22**, **26**, **50–52**). All six compounds produced dose-dependent increases in NAc DA and/or 5-HT levels. In vivo selectivity (determined as the dose required to increase peak 5-HT levels by 250% divided by the dose required to increase peak DA levels by 250%) was correlated with in vitro selectivity to promote monoamine release via DAT and SERT ($r = 0.95$) and the molecular volume (i.e., $Å^3$) of the 4-position substituent ($r = -0.85$). The results further support a relationship between these molecular, neurochemical, and behavioral measures described above [52].

Racemates, by definition, consist of an *equal* mixture of two optical isomers or antipodes of a given agent. Often, one isomer is the "active" isomer (i.e., the eutomer), whereas the other is less active or "inactive" (i.e., the distomer). When the distomer is inactive, the action is termed *stereospecific* (or, *enantiospecific*), and the eutomer is twice as potent as the racemate. The "inactive" isomer simply dilutes the potency of the "active" isomer by 50%, and the maximal theoretical potency of the eutomer is twice that of the racemate. In other cases, both isomers – the eutomer and the distomer – are "active," but one is more potent than the other; the optical isomers are then said to produce a *stereoselective* (or *enantioselective*) effect.

As described above, methcathinone (**13**) is stereoselective with respect to its behavioral actions and potency as a DA releasing agent; *S*(−)methcathinone represents the eutomer. Mephedrone (**26**) is a less-selective releasing agent (Table 5). Mephedrone (**26**), which is 4-methylmethcathinone, can be deconstructed to 4-methylcathinone (**53**) and, by removal of the *N*-methyl group, to cathinone (**1**). An examination of the optical isomers of these agents can provide insight as to the stereoselective versus stereospecific nature of their actions. *S*(−) Cathinone (Table 6) is stereoselective, but not selective with respect to release at DAT versus NET. That is, both isomers are "active," but the *S*(−)-isomer is more potent than its *R*(+)enantiomer by severalfold (Table 6). However, both cathinone isomers are essentially inactive at SERT. Introduction of the 4-methyl group (i.e., 4-methylcathinone, **53**) resulted both in decreased (<threefold) stereoselectivity for DAT/NET release and decreased selectivity for DAT/NET release versus release at SERT (Table 6). Introduction of an *N*-methyl group (i.e., mephedrone, **26**) resulted in further decreased stereoselectivity at DAT with the two optical isomers being nearly equipotent (i.e., the difference in potency for *S*(−)- and *R*(+)mephedrone is not much more than twofold). In addition, *S*(−)mephedrone is no longer selective for DAT versus SERT.

The abuse-related potential of biogenic amine releasing agents appears to be related to their ability to promote greater release via DAT than via SERT such that DAT-selective agents possess higher abuse liability (see [51, 53] for extended discussion). Consistent with this concept is that *R*(+)mephedrone with 50-fold selectivity for DAT over SERT produced in rats greater locomotor stimulation and rewarding properties as measured by conditioned place preference and facilitation of ICSS than its *S*(−)enantiomer which produced weak locomotor stimulation and lacked rewarding properties [50]. For 4-methylcathinone (**53**), ***R*(+)53** was less potent than its *S*-enantiomer to promote release at DAT, but displayed slightly greater DAT versus SERT selectivity and produced abuse-related effects in ICSS [51]. These studies (i.e., those reported by [50, 51]) were the first to show the subtle stereochemical relationship between the ability of optical isomers of cathinone analogs to behave as substrates at DAT and SERT and their stimulant or rewarding actions. It would seem that future studies should focus greater attention on the optical isomers of related agents.

As already alluded to, the nature of the terminal amine, α-substituents, aryl substituents, and stereochemistry can alter the actions of synthetic cathinones. That is, methcathinone (**13**) is primarily a releasing agent at DAT, whereas introduction of a 4-methyl group (i.e., mephedrone, **26**) enhances its potency as a releasing agent at SERT such that methcathinone displays >300-fold selectivity for DAT, whereas mephedrone displays only slightly more than twofold selectivity (Table 5). Increasing the bulk on the terminal amine of cathinone analogs tends to enhance action as a DAT reuptake inhibitor (see Fig. 8). Also, the *N*-ethyl homolog of methcathinone, ethcathinone (**14**), is both a weak DA releasing agent and reuptake inhibitor [10]. Hence, by "mixing and matching" of appropriate substituents, certain cathinone analogs might possess "mixed" or "hybrid" actions. Saha et al. [54] examined this by evaluating three agents with gradually increasing steric bulk on the terminal amine: mephedrone (**26**), its *N*-ethyl homolog 4-methyl-*N*-ethylcathinone (**54**, 4-MEC), and its pyrrolidine counterpart 4′-methyl-α-pyrrolidinopropiophenone (**55**, 4-MePPP) (Fig. 11).

Fig. 11 A structural comparison of mephedrone (**26**), 4-MEC (**54**), and 4-MePPP (**55**)

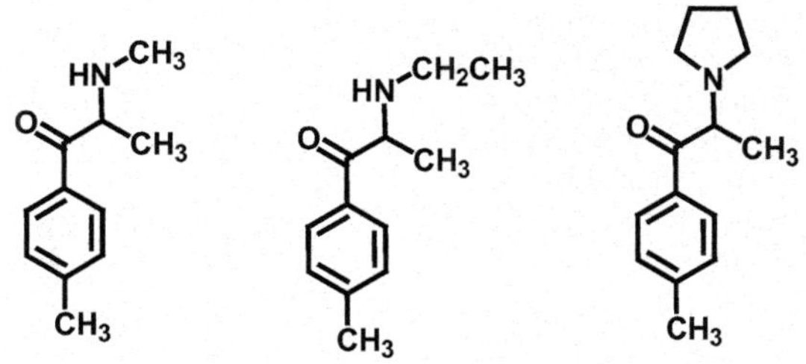

26 54 55

Mephedrone (**26**) and 4-MEC (**54**) were nearly equipotent at inhibiting reuptake at DAT (IC_{50} ca. 800 nM) and SERT (IC_{50} ca. 500 nM), whereas 4-MePPP (**55**) was more potent as an inhibitor at the former ($IC_{50} = 215$ nM) as opposed to the latter ($IC_{50} > 10{,}000$ nM). However, in a synaptosomal release assay, mephedrone (**26**) and 4-MEC (**54**) were similar in potency to evoke release from SERT (EC_{50} ca. 100 nM), whereas 4-MePPP (**55**) was inactive. In contrast, mephedrone was an effective releaser at DAT ($EC_{50} = 39$ nM), whereas 4-MEC (**54**) and 4-MePPP (**55**) were inactive. The overall conclusion was that changing the nature of the *N*-alkyl substituent of cathinone analogs has a profound influence on their actions; 4-MEC (**54**) is a SERT releasing agent/DAT blocker, whereas 4-MePPP (**55**) is a selective DAT blocker [54].

5 Overall Conclusions

From a practical perspective, the results of SAR and QSAR studies are typically used to investigate mechanisms of drug action and to forecast the action(s)/potency of novel agents. With respect to synthetic cathinones, SAR studies, and certainly QSAR studies, are still in their infancy. Only in the last year or two have QSAR methods (e.g., Hansch analyses, homology modeling and docking studies, HINT analyses) been applied to these agents.

SAR studies with synthetic cathinones began in the late 1970s, following the discovery of cathinone (**1**) as the active constituent of *khat*, with the simple findings that cathinone (**1**) was more potent than (+)cathine (**2**) as a central stimulant, that *S*(−)cathinone was more potent than its *R*(+)-enantiomer, and of the identification (i.e., the "rediscovery") of what is now termed methcathinone (**13**). Early SAR studies examined the behavioral effects (i.e., locomotor stimulation, discriminative stimulus properties) of cathinone analogs because their mechanism of action was unknown. Once it was shown that these agents might be producing their effects by acting as DAT releasing agents, attention slowly shifted focus. Some agents also displayed releasing action at NET and/or SERT. Subsequently, MDPV (**27**) was identified as a drug of abuse that acted primarily as a reuptake inhibitor at DAT.

Substituents on the phenylpropanonamine scaffold of cathinone can influence these actions. Figure 12 summarizes some of structural aspects of cathinone analogs that have been investigated in SAR studies, with an emphasis on DAT.

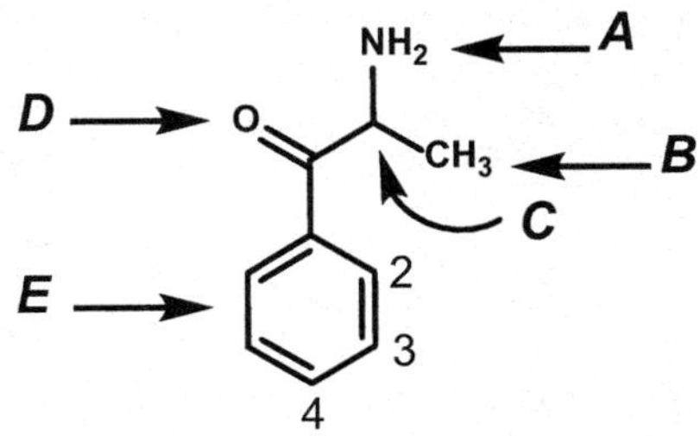

Fig. 12 The structure of cathinone and the various structural alterations (at positions *A*–*E*) that have been examined in SAR studies to be summarized below

Synopsis of some SAR findings on synthetic cathinones:

(1) As reuptake inhibitors at DAT, optimal potency and selectivity are associated with a tertiary amine (*A*, Fig. 12) and an extended α-side chain (*B*); with an extended side chain, the amine can be tertiary, secondary, or even primary – although potency generally decreases in this same rank order. To date, the most potent DA reuptake inhibitors are specifically associated with a pyrrolidine moiety as *A* and a fairly large α (i.e., *B*)-substituent (e.g., *n*-butyl, cyclohexyl). Stereochemistry at *C* has not been extensively investigated, but for MDPV (**27**) and pyrovalerone (**29**), the *S*-isomers are substantially (i.e., >100-fold) more potent than their *R*-enantiomers. The carbonyl group at *D* generally has minimal influence on stimulant or DAT action, and its elimination converts the cathinone analog (a phenylpropanonamine) to its amphetamine analog (a phenylisopropylamine, e.g., compare **27** and **33** in Fig. 8). However, reduction of the carbonyl group of pyrovalerone (**29**) to a hydroxyl group results in diastereomers that lack affinity for DAT, NET, and SERT [27]. Substitution on the aryl ring (i.e., *E*) has not been extensively examined; however, there are preliminary indications that certain substituents might influence potency. This requires further investigation.

Typically, substituents optimal for action as a DAT reuptake inhibitor decrease or abolish actions at SERT.

(2) As releasing agents at DAT, a primary amine seems optimal with a simple *N*-methyl secondary amine being nearly as potent, and sometimes slightly more potent, than the primary amine. The increased behavioral potency of *N*-methyl analogs of cathinones over their primary amine counterparts might be related to the slightly higher affinity of the latter for DAT (although few comparisons are available), their greater resistance to metabolism, and/or their enhanced ability to penetrate the blood-brain barrier due to their increased lipophilicity. As the size/bulk of the amine (i.e., *A*) substituent increases, potency as a releasing agent (and perhaps selectivity) decreases. Increasing bulk at the terminal amine shifts the action of a DAT releasing agent to a DAT reuptake inhibitor. An α-methyl group at *B* would seem optimal; increasing the length of the substituent can also reverse action from a releasing agent to a reuptake inhibitor at DAT. When the α-substituent is a methyl group, *S*-isomers (i.e., *C*) are typically more potent (or equipotent) at DAT and SERT than their *R*-enantiomers; however, DAT/SERT

selectivity as governed by their *S*/*R* ratio can influence behavioral actions, and aryl substituents at *E* also play a role here. The carbonyl group at *D* plays a minimal role; however, reduction of the carbonyl group to a hydroxyl group (i.e., a phenylpropanolamine) reduces both its potency as a DAT substrate and as a centrally acting stimulant. Aryl substituents (i.e., *E*) at the ring 4-position generally decrease potency at DAT and can shift selectivity toward SERT; the larger the substituent, the greater the likelihood that it will favor SERT versus DAT. Substituents at the ring 2- and 3-positions have not been thoroughly investigated from an SAR perspective.

(3) Most SAR studies on synthetic cathinones have focused on DAT and SERT action; however, a role for the NET should not be overlooked. However, too few studies have been reported to allow for general SAR to be formulated. Nevertheless, for a series of seven synthetic cathinones (i.e., those shown in Table 5), it was found that their potency to act as substrates at DAT and NET was significantly ($r = 0.906$) correlated (Fig. 13) [48]. Although SAR might not be identical at DAT and NET, there appear to be some similarities; additional agents will need to be examined.

The time has finally arrived where multiple agents are being investigated in the same study, under similar conditions, to allow reasonable SAR/QSAR conclusions to be formulated. It should be appreciated that the emerging SAR/QSAR results described here are based on a limited number of agents and investigations. At almost any time, novel agents might appear that will question or challenge these relationships. Nevertheless, the advent of novel agents will only strengthen and refine the current SAR.

Within the past several months, several new synthetic cathinones have been confiscated, or purchased from Internet sources, for purpose of physicochemical and spectral characterization (e.g., [55, 56]). These agents include α-EAPP (**56**), 4-MeEAPP (**57**), α-PHP (**58**), α-POP (**59**), 4-fluoro-PV-9 (**60**), 3,4-dimethoxy-α-PVP (**61**), 4-fluoro-α-PVP (**62**), and MPHP (**63**) (Fig. 14).

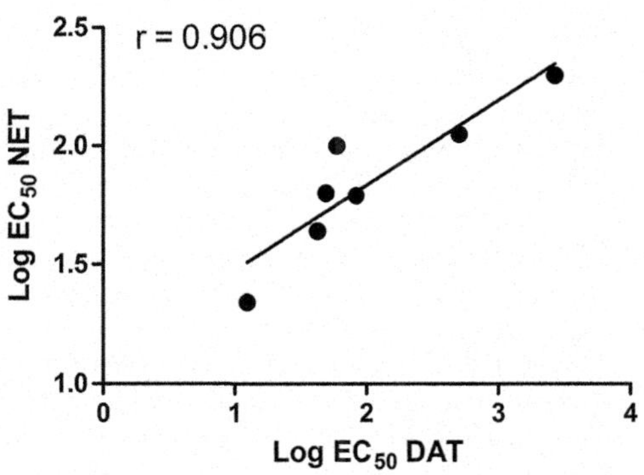

Fig. 13 Relationship between potencies of seven 4-substituted cathinone analogs (i.e., those in Table 6) to act as releasing agents at DAT and NET (from [48])

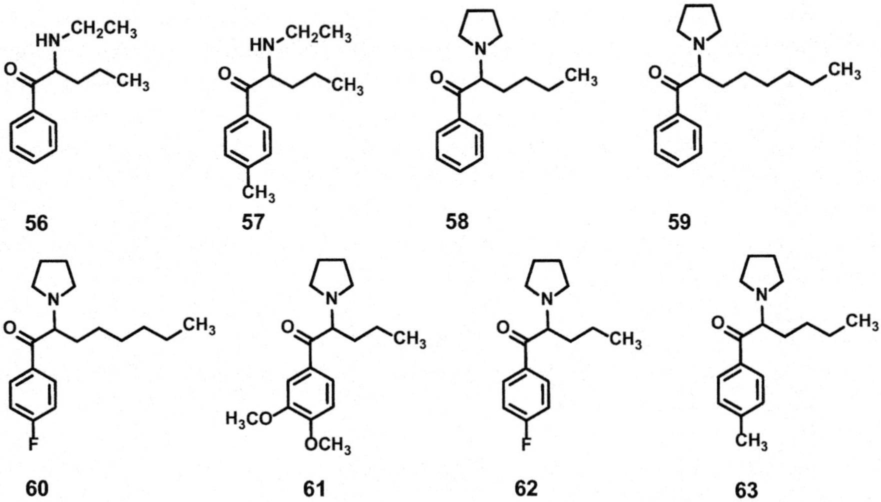

Fig. 14 Structures of some novel synthetic cathinones that have not been thoroughly investigated

These agents (Fig. 14) have yet to be pharmacologically evaluated. However, it can be seen that they represent variations of common structural themes discussed herein. On the basis of the SAR reviewed above, it should now be possible to make some educated guesses as to the actions and approximate potencies of these novel substances.

Acknowledgments Work from the authors' laboratory was supported by US Public Health Services grants DA-01642 and DA-033930.

References

1. Glennon RA, Young R (2011) Drug discrimination: applications to medicinal chemistry and drug studies. Wiley, Hoboken
2. UNODC (2013) The challenge of new psychoactive substances. United Nations Office of Drugs and Crime, Vienna
3. EMCDDA (2015) European Drug Report. Trends and Developments. European Monitoring Centre for Drugs and Drug Addiction. Luxembourg
4. Glennon RA (2014) Bath salts, mephedrone, and methylenedioxypyrovalerone as emerging illicit drugs that will need targeted therapeutic intervention. Adv Pharmacol 69:581–620
5. Glennon RA, Showalter D (1981) The effect of cathinone and several related derivatives on locomotor activity. Res Commun Subst Abus 2:186–192
6. Glennon RA, Schechter MD, Rosecrans JA (1984) Discriminative stimulus properties of S(−)- and R(+)-cathinone, (+)-cathine and several structural modifications. Pharmacol Biochem Behav 21:1–3
7. Glennon RA, Young R, Hauck AE, et al. (1984) Structure-activity studies on amphetamine analogs using drug discrimination methodology. Pharmacol Biochem Behav 21:895–901

8. Goudie AJ, Atkinson J, West CR (1986) Discriminative properties of the psychostimulant dl-cathinone in a two lever operant task. Lack of evidence for dopaminergic mediation. Neuropharmacology 25:85–94
9. Kalix P, Glennon RA (1986) Further evidence for an amphetamine-like mechanism of action of the alkaloid cathinone. Biochem Pharmacol 35:3015–3019
10. Reith MEA, Blough BE, Hong WC, et al. (2015) Behavioral, biological, and chemical perspectives on a typical agents targeting the dopamine transporter. Drug Alcohol Depend 147:1–19
11. Tang QY, Kolanos R, De Felice LJ, et al. (2015) Structural analysis of dopamine- and amphetamine-induced depolarization currents in the human dopamine transporter. ACS Chem Neurosci 6:551–558
12. Dal Cason TA, Young R, Glennon RA (1997) Cathinone: an investigation of several N-alkyl and methylenedioxy-substituted analogs. Pharmacol Biochem Behav 58:1109–1116
13. Glennon RA, Yousif M, Naiman N, et al. (1987) Methcathinone: a new and potent amphetamine-like agent. Pharmacol Biochem Behav 26:547–551
14. Bondareva TS, Young R, Glennon RA (2002) Central stimulants as discriminative stimuli: asymmetric generalization between (−)ephedrine and S(+)methamphetamine. Pharmacol Biochem Behav 74:157–162
15. Young R, Glennon RA (1998) Discriminative stimulus properties of (−)ephedrine. Pharmacol Biochem Behav 60:771–775
16. Foley KF, Cozzi NV (2003) Novel aminopropiophenones as potential antidepressants. Drug Dev Res 60:252–260
17. Bondarev ML, Bondareva TS, Young R, et al. (2003) Behavioral and biochemical investigations of bupropion metabolites. Eur J Pharmacol 474:85–93
18. Carroll FI, Blough BE, Abraham P, et al. (2009) Synthesis and biological evaluation of bupropion analogues as potential pharmacotherapies for cocaine addiction. J Med Chem 52:6768–6781
19. Markantonis SL, Kyroudis A, Beckett AH (1986) The stereoselective metabolism of dimethylpropion and monomethylpropion. Biochem Pharmacol 35:529–532
20. Cozzi NV, Brandt SD, Daley PF, et al. (2013) Pharmacological examination of trifluoromethyl ring-substituted methcathinone analogs. Eur J Pharmacol 699:180–187
21. Young R, Glennon RA (1993) Cocaine-stimulus generalization to two new designer drugs: methcathinone and 4-methylaminorex. Pharmacol Biochem Behav 45:229–231
22. Kohut SJ, Fivel PA, Blough BE, et al. (2013) Effects of methcathinone and 3-Cl-methcathinone (PAL-434) in cocaine discrimination or self-administration in rhesus monkeys. Int J Neuropsychopharmacol 16:1985–1998
23. Young R, Glennon RA (1998) Discriminative stimulus effects of S(−)-methcathinone (CAT): a potent stimulant drug of abuse. Psychopharmacology (Berl) 140:250–256
24. Rothman RB, Vu N, Partilla JS, et al. (2003) In vitro characterization of ephedrine-related stereoisomers at biogenic amine transporters and the receptorome reveals selective action as norepinephrine transporter substrates. J Pharmacol Exp Ther 307:138–145
25. Young R, Glennon RA (2000) Stimulus effects of phenylpropanolamine optical isomers in (+) amphetamine-trained rats. Pharmacol Biochem Behav 66:489–494
26. Iversen LE (2010) Consideration of the cathinones. Advisory Council on the Misuse of Drugs. A report submitted to the Home Secretary of the UK (March 31, 2010)
27. Meltzer PC, Butler D, Deschamps JR, et al. (2006) 1-(4-Methylphenyl)-2-pyrrolidin-1-yl-pentan-1-one (pyrovalerone) analogues: a promising class of monoamine uptake inhibitors. J Med Chem 49:1420–1432
28. Aarde SM, Creehan KM, Vandewater SA, et al. (2015) In vivo potency and efficacy of the novel cathinone α-pyrrolidinopentiophenone and 3,4-methylenedioxypyrovalerone: self-administration and locomotor stimulation in male rats. Psychopharmacology (Berl) 232:3045–3055

29. Aarde SM, Huang PK, Creehan KM, et al. (2013) The novel recreational drug 3,4-methylenedioxypyrovalerone (MDPV) is a potent psychomotor stimulant: self-administration and locomotor activity in rats. Neuropharmacology 71:130–140
30. Baumann MH, Partilla JS, Lehner KR, et al. (2013) Powerful cocaine-like actions of 3,4-methylenedioxypyrovalerone (MDPV), a principal constituent of psychoactive 'bath salts' products. Neuropsychopharmacology 38:552–562
31. Fantegrossi WE, Gannon BM, Zimmerman SM, et al. (2013) In vivo effects of abused 'bath salt' constituent 3,4-methylenedioxypyrovalerone (MDPV) in mice: drug discrimination, thermoregulation, and locomotor activity. Neuropsychopharmacology 38:563–573
32. Gatch MB, Dolan SB, Forster MJ (2015) Comparative behavioral pharmacology of three pyrrolidine-containing synthetic cathinone derivatives. J Pharmacol Exp Ther 354:103–110
33. Marusich JA, Antonazzo KR, Wiley JL, et al. (2014) Pharmacology of novel synthetic stimulants structurally related to the "bath salts" constituent 3,4-methylenedioxypyrovalerone (MDPV). Neuropharmacology 87:206–213
34. Marusich JA, Grant KR, Blough BE, et al. (2012) Effects of synthetic cathinones contained in "bath salts" on motor behavior and a functional observational battery in mice. Neurotoxicology 33:1305–1313
35. Watterson LR, Olive MF (2014) Synthetic cathinones and their rewarding and reinforcing effects in rodents. Adv Neurosci 2014:209875
36. Watterson LR, Kufahl PR, Nemirovsky NE, et al. (2014) Potent rewarding and reinforcing effects of the synthetic cathinone 3,4-methylenedioxypyrovalerone (MDPV). Addict Biol 19:165–174
37. Gatch MB, Taylor CM, Forster MJ (2013) Locomotor stimulant and discriminative stimulus effects of 'bath salt' cathinones. Behav Pharmacol 24:437–447
38. Cameron K, Kolanos R, Solis E, et al. (2013) Bath salts components mephedrone and methylenedioxypyrovalerone (MDPV) act synergistically at the human dopamine transporter. Br J Pharmacol 168:1750–1757
39. Cameron K, Kolanos R, Vekariya R, et al. (2013) Mephedrone and methylenedioxypyrovalerone (MDPV), major constituents of "bath salts," produce opposite effects at the human dopamine transporter. Psychopharmacology (Berl) 227:493–499
40. Kolanos R, Cameron KN, Vekariya RH et al (2011) "Bath salts": an imitation of methamphetamine plus cocaine? Abstract #249. Southeast Regional American Chemical Society Meeting. Richmond, VA
41. Simmler LD, Buser TA, Donzelli M, et al. (2013) Pharmacological characterization of designer cathinones in vitro. Br J Pharmacol 168:458–470
42. Eshleman AJ, Wolfrum KM, Hatfield MG, et al. (2013) Substituted methcathinones differ in transporter and receptor interactions. Biochem Pharmacol 85:1803–1815
43. Del Bello F, Sakloth F, Partilla JS, et al. (2015) Ethylenedioxy homologs of N-methyl-(3,4-methylenedioxyphenyl)-2-aminopropane (MDMA) and its corresponding cathinone analog methylenedioxymethcathinone: interactions with transporters for serotonin, dopamine, and norepinephrine. Bioorg Med Chem 23:5574–5579
44. Kolanos R, Solis Jr E, Sakloth F, et al. (2013) "deconstruction" of the abused synthetic cathinone methylenedioxypyrovalerone (MDPV) and an examination of effects at the human dopamine transporter. ACS Chem Neurosci 4:1524–1529
45. Kolanos R, Partilla JS, Baumann MH, et al. (2015) Stereoselective actions of methylenedioxypyrovalerone (MDPV) to inhibit dopamine and norepinephrine transporters and facilitate intracranial self-stimulation in rats. ACS Chem Neurosci 20:771–777
46. Sakloth F, Kolanos R, Mosier PD, et al. (2015) Steric parameters, molecular modeling and hydropathic interaction analysis of the pharmacology of para-substituted methcathinone analogues. Br J Pharmacol 172:2210–2218
47. Kolanos R, Sakloth F, Jain AD, et al. (2015) Structural modification of the designer stimulant α-pyrrolidinovalerophenone (α-PVP) influences potency at dopamine transporters. ACS Chem Neurosci 6:1726–1731

48. Sakloth F (2015) Psychoactive synthetic cathinones (or 'bath salts'): investigation of mechanisms of action. PhD dissertation, Virginia Commonwealth University, Richmond, VA
49. Bonano JS, Banks ML, Kolanos R, et al. (2015) Quantitative structure-activity relationship analysis of the pharmacology of para-substituted methcathinone analogues. Br J Pharmacol 172:2433–2444
50. Gregg RA, Baumann MH, Partilla JS, et al. (2015) Stereochemistry of mephedrone neuropharmacology: enantiomer-specific behavioural and neurochemical effects in rats. Br J Pharmacol 172:883–894
51. Hutsell BA, Baumann MH, Partilla JS, et al. (2016) Abuse-related neurochemical and behavioral effects of cathinone and 4-methylcathinone stereoisomers in rats. Eur Neuropsychopharmacol 26:288–297
52. Suyama JA, Sakloth F, Kolanos R, et al. (2016) Abuse-related neurochemical effects of para-substituted methcathinone analogs in rats: microdialysis studies of nucleus accumbens dopamine and serotonin. J Pharmacol Exp Ther 356:182–190
53. Negus SS, Mello NK, Blough BE, et al. (2007) Monoamine releasers with varying selectivity for dopamine/norepinephrine versus serotonin release as candidate "agonist" medications for cocaine dependence: studies in assays of cocaine discrimination and cocaine self-administration in rhesus monkeys. J Pharmacol Exp Ther 320:627–636
54. Saha K, Partilla JS, Lehner KR, et al. (2015) 'Second-generation' mephedrone analogs, 4-MEC and 4-MePPP, differentially affect monoamine transporter function. Neuropsychopharmacology 40:1321–1331
55. Majchrzak M, Rojkiewicz M, Celiński R, et al. (2016) Identification and characterization of new designer drug 4-fluoro-PV9 and α-PHP in the seized materials. Forensic Toxicol 34:115–124
56. Uchiyama N, Matsuda S, Kawamura M, et al. (2014) Characterization of four new designer drugs, 5-chloro-NNEI, NNEI indazole analog, α-PHPP and α-POP, with 11 newly distributed designer drugs in illegal products. Forensic Sci Int 243:1–13

Interactions of Cathinone NPS with Human Transporters and Receptors in Transfected Cells

Linda D. Simmler and Matthias E. Liechti

Abstract Pharmacological assays carried out in transfected cells have been very useful for describing the mechanism of action of cathinone new psychoactive substances (NPS). These in vitro characterizations provide fast and reliable information on psychoactive substances soon after they emerge for recreational use. Well-investigated comparator compounds, such as methamphetamine, 3,4-methylenedioxymethamphetamine, cocaine, and lysergic acid diethylamide, should always be included in the characterization to enhance the translation of the in vitro data into clinically useful information. We classified cathinone NPS according to their pharmacology at monoamine transporters and receptors. Cathinone NPS are monoamine uptake inhibitors and most induce transporter-mediated monoamine efflux with weak to no activity at pre- or postsynaptic receptors. Cathinones with a nitrogen-containing pyrrolidine ring emerged as NPS that are extremely potent transporter inhibitors but not monoamine releasers. Cathinones exhibit clinically relevant differences in relative potencies at serotonin vs. dopamine transporters. Additionally, cathinone NPS have more dopaminergic vs. serotonergic properties compared with their non-β-keto amphetamine analogs, suggesting more stimulant and reinforcing properties. In conclusion, in vitro pharmacological assays in heterologous expression systems help to predict the psychoactive and toxicological effects of NPS.

Keywords Cathinones • Efflux • Heterologous expression systems • In vitro • New psychoactive substances • Pharmacology • Transporters • Uptake

---

L.D. Simmler and M.E. Liechti (✉)
Division of Clinical Pharmacology and Toxicology, Department of Biomedicine and Department of Clinical Research, University Hospital Basel, Basel, Switzerland
e-mail: matthias.liechti@usb.ch

Curr Topics Behav Neurosci (2017) 32: 49–72
DOI 10.1007/7854_2016_20
Published Online: 1 June 2016

Contents

1 Introduction

In 2014, the European Union Early Warning System [1] reported the emergence of 101 new psychoactive substances (NPS). The variety of largely unknown NPS is still increasing compared with recent years. With this high number of new substances, rapid testing systems are needed to obtain an immediate understanding of the mechanism of action of these NPS. Animal studies that utilize behavioral paradigms (e.g., to test abuse liability) or neurochemical assessments (e.g., microdialysis and voltammetry) to investigate the pharmacology and toxicology of new compounds in vivo are relatively expensive and require weeks or months to conduct. Moreover, typically only a small number of substances can be tested. In contrast, rapid first characterizations of new compounds can be performed within days in a laboratory with a set of well-established in vitro assays and using reference data from well-known substances. Typically, relatively simple in vitro pharmacological assays with transfected cell lines have limited significance in neuroscientific research because more complex behavioral and circuit-wide conclusions are required for a comprehensive understanding of the mechanism of action of psychoactive substances in the brain. Transfected cell lines in heterologous expression systems only reveal the mechanism of action of drugs on specific targets that are expressed by the host cell. Therefore, any complex whole-brain interactions are lacking. However, to elucidate the pharmacology of a larger set of unknown compounds, in vitro assays are highly valuable as the first screening tools. Through decades of intensive animal and clinical experimental studies on various psychoactive substances (e.g., cocaine, methamphetamine, 3,4-methylenedioxymethamphetamine [MDMA], and lysergic acid diethylamide [LSD]), their mechanism of action in vitro and pharmacological effects in vivo are relatively well known, thus allowing translational interpretations of in vitro data on NPS [2]. Thus, the clinical pharmacology of NPS can be predicted based on similarities between the in vitro mechanisms of action of NPS and well-known and also clinically characterized comparator compounds.

Our in vitro characterization of cathinone NPS has allowed the rapid characterization of these newly emerging substances at known human targets of psychoactive compounds [3–5]. In the context of in vitro and in vivo studies in other laboratories [6–8] and clinical reports, we found that in vitro characterizations are consistent with in vivo data but allow for the faster initial characterization of larger numbers of newly emerging compounds. Cathinone NPS have striking differences in pharmacological potencies to inhibit monoamine transporters, which are relevant to appraisals of the type of psychoactivity, abuse liability, and to some extent clinical toxicity. For example, in vitro testing has shown that 3,4-methylenedioxypyrovalerone (MDPV) inhibits the dopamine transporter (DAT) and norepinephrine transporter (NET) far more potently when compared with classic psychostimulants, such as cocaine and methamphetamine [4, 7] (see Baumann et al. 2016, this volume), suggesting that small doses may exert large clinical effects and enhance the risk of overdose. This information is essential for users of these compounds and clinicians who treat overdose cases. However, pharmacological properties, such as bioavailability and blood–brain barrier permeability, are also important for determining the potency of a substance in vivo. Additional pharmacological studies are thus needed for a more comprehensive characterization. Overall, in vitro profiling is particularly helpful for systematic comparative characterizations of a large number of substances, in which basic and rapid information on the compounds' pharmacological characteristics is essential, such as with the current NPS problem.

In this article, we discuss the principles of in vitro pharmacological assays that are used to characterize the primary mechanisms of action of cathinone NPS. We discuss the advantages and limitations of such assays with regard to the rapid emergence of NPS in recent years. We also highlight methodological issues and discuss the main characteristics of cathinone NPS in these assays.

2 Methods for Studying Transporter and Receptor Pharmacology in Transfected Cells

Stably transfected cells represent a heterologous expression system in which the protein of interest is expressed in a host cell that does not endogenously express the respective protein. For the pharmacological profiling of cathinone NPS, the respective monoamine transporter or pre- and postsynaptic receptor genes are introduced into neutral cell lines [9–11]. Human embryonic kidney (HEK) 293 cells are very commonly used for stable transfections and subsequent pharmacological assays. For stable transfections, a plasmid with the cDNA sequence of the target protein from any species is introduced into the cells [12]. The co-introduction of a geneticin-resistance gene ensures that only transfected cells are maintained in culture [13]. The stable expression of a target protein is not necessarily required for in vitro pharmacological assays [14], but stable cell lines simplify the workflow because the step of transiently transfecting cells before each assay can be omitted. Transfected cell cultures are a standard procedure for molecular biology laboratories. With recent technological improvements (e.g., CRISPR/Cas9 technology), transfections are becoming even easier [15]. Once stably transfected, the cells express the protein in high abundance both in the membrane, which is essential for functional assays, and in the cytoplasm [16, 17]. For assays that are used for investigations of cathinone NPS, only one gene of interest is introduced per cell line, thus ensuring selectivity in the pharmacological assessment. Non-transfected cells can serve as a control for nonspecific drug action (i.e., nonspecific binding to the cell membrane; [9]).

To comprehensively characterize psychoactive compounds at their typical neuronal target sites in vitro, the effects of these compounds on the different monoaminergic neurotransmitter uptake transporters and various neurotransmitter G-protein-coupled receptors need to be determined in a battery of assays. Therefore, individual cell lines that overexpress the respective target protein after transfection are used to determine binding affinity, uptake transport inhibition, and transporter-mediated efflux in separate assay setups. For transporters, uptake inhibition (e.g., in the case of cocaine) and the transport-mediated efflux of transmitter (e.g., in the case of most amphetamines) are determined in different assays. For the relevant receptors, functional assays are performed to determine agonistic or antagonistic properties, including information about full or partial agonist effects. Binding affinities at both transporters and receptors are also frequently determined, but functional tests are considered more conclusive than binding affinities. The assay principles are described in more detail later in this chapter. Briefly, transport assays require a radiolabeled substrate of the transporters, usually endogenous neurotransmitters [9]. Through quantification of the transported radiolabeled substrates, the inhibition potencies or efflux characteristics of a specific substance can be determined. To determine binding affinities, a radioligand displacement principle is applied, in which the substance's ability to compete with the radioligand for the binding site is quantified [18]. For receptor coupling activity, cyclic adenosine monophosphate (cAMP) levels can be quantified [19]. This downstream factor indicates signaling that is induced by G-protein-coupled receptors, in which cAMP levels increase upon activation of the receptors or decrease upon inhibition of the receptors [20]. For all of the assays, classic enzyme kinetics are the basis for calculating pharmacological determinants (i.e., IC_{50}, EC_{50}, and K_i values; [21, 22]).

Heterologous expression systems for monoaminergic neurotransmitter transporters have been relevant in neuropsychopharmacology research since these transporters were first cloned. Transporter-expressing cell lines allow the characterization of psychoactive compounds [11] and are also a useful tool for discovering psychoactive therapeutic drugs [23]. Furthermore, in vitro experiments with transfected cells formed the basis for many genetic mutations that were later engineered in mice, which now serve for in vivo investigations of psychoactive drugs or as preclinical models of mental disorders [14, 24, 25]. For example, in vitro experiments allowed the construction of a transgenic mouse model with a 5-hydroxytryptamine (5-HT [serotonin]) transporter (SERT) mutation for the in vivo assessment of SERT-mediated effects of antidepressants or cocaine [26, 27] or to shed light on functional abnormalities of the DAT variant Val559, which is being investigated as a potential mouse model of attention-deficit hyperactivity disorder [28].

Today, heterologous expression systems are a relatively simple tool for use in any laboratory with basic cell culture and molecular biology setups. Furthermore, once cell lines stably express a specific receptor, these lines can be maintained by freezing stocks, and such stocks can then be used over decades. One of the greatest strengths of in vitro screening assays that use transfected cells is the high selectivity for the pharmacological targets of interest. For example, for DAT uptake inhibition, cells that overexpress DAT are used, while for SERT inhibition a different cell line overexpressing SERT is used. Due to separation of the targets in different runs no unspecific action at the second target can affect the result. Furthermore, human proteins can be overexpressed to assess pharmacological profiles directly with targets of the human species [11]. Species differences could be a concern in ex vivo or in vivo experiments because target proteins may exhibit distinct substance recognition between rodents/nonhuman primates and humans or show differential expression patterns. For example, the antidepressant imipramine is more potent at the human SERT than at the rat SERT, whereas cocaine inhibits both rat and human SERT with equal potencies [29]. The most common variant of the respective target is usually expressed in NPS screening, but it is also feasible to generate cell lines with different variants of human transporters or receptors to specifically assess the pharmacological and toxicological effects of psychoactive substances on less common gene variants. While many advantages are evident for the use of heterologous expression systems to screen NPS pharmacological profiles, there are also limitations and disadvantages compared to similar experimental approaches. Synaptosomes or brain slices are frequently used ex vivo preparations to assess the pharmacology of psychoactive substances. In brain slices substantial cellular characteristics are still intact, and synaptosomes contain the full complement of synaptic proteins and synaptic vesicles [30]. Synaptosomes resemble the natural environment of the site of psychostimulant action more than transfected cell lines. Interpretations from experiments in transfected cells are limited since they lack elements of the protein machinery of intact neuronal membranes that could be critical for certain protein/substance interactions and consequences. However, for target-selective assays typically used for the determination of pharmacological constants unintended targets have to be pharmacologically blocked in synaptosomes [31, 32]. In this regard, both transfected cell lines and ex vivo preparations (e.g., synaptosomes) have their advantages and limitations for the screening of NPS pharmacology and should always be kept in mind when interpreting results. Nevertheless, pharmacological profiles of NPS assessed in transfected cells have largely been in accordance with data obtained from synaptosomes.

It is self-evident that there are limitations to in vitro screenings with transfected cells or ex vivo preparations and various consequences of NPS use can only be assessed by in vivo testing, particularly behavior or long-term toxicity. With regards to pharmacological profiles, however, we would like to point out that the possibility of active metabolites should be considered. Heterologous cell lines for in vitro screenings of NPS pharmacology are largely unable to detect the possible contribution of active metabolites that could, however, be relevant in vivo. For example, 3,4-methylenedioxyamphetamine (MDA) is an active metabolite of

MDMA and likely contributes to the subjective drug experience and toxicity associated with MDMA [33]. Cathinone NPS may also have active metabolites that should be taken into account in more comprehensive pharmacological substance characterizations. For example, β-keto MDA is a metabolite of methylone [34] and interacts with monoamine transporters similarly to MDA in in vitro tests [35]. In vitro testing for active metabolites requires knowledge of the metabolic pathway and synthesis of possibly active metabolites or the use of cell systems that contain metabolic enzymes. To elaborate the metabolites for every single NPS would be a very labor-intensive process. In vivo neurochemical studies that utilize microdialysis can be performed more easily and may include possible contributing effects of active metabolites on neurotransmission.

The specific assay setups for uptake and efflux transport assays vary considerably between laboratories. In the most widely used experimental setup for in vitro pharmacology, transfected cell lines are grown to adherence in well plates or small culture dishes. Adherence of the cells allows for the removal of uptake buffer and washing with ice-cold buffer to stop substrate transport. However, if timing is an essential factor in uptake experiments (which is usually more essential for substrate kinetics than for inhibition potencies [IC_{50} values]), then the possibility of rapid and timely termination of the uptake process is crucial. With suspended synaptosome preparations, the use of a Brandel tissue harvester allows for the timely termination of 24–96 vials at once. It becomes more difficult when the assay is conducted on adherent cell cultures. Even with an automated wash station for cell culture plates, achieving satisfactory accuracy to terminate the uptake process can be either challenging or impossible. When we established the assay that is currently used in our laboratory, we chose to use a silicone-oil-centrifugation method. We perform the uptake assay in cell suspensions that are prepared from adherent cells. Centrifuging the cells through a silicone oil layer allows for rapid and precise termination of the uptake process and the cleaning of cells from the buffer [36]. Silicone oil is used as a middle layer in a tube. In the centrifugation step, the cells but not radioactive uptake buffer transfer to the lower layer (consisting of 3 M KOH, which lyses the cells). We have found that this method is very reliable and precise, but handling can be more elaborate and more difficult than working with adherent cells or synaptosomes. No conclusive recommendation has been made for the ideal assay setup. In fact, every laboratory needs to establish and validate its own assay setup for transport assays. If the assay follows the rules of enzyme kinetics and if reproducibility within the laboratory can be demonstrated, then the specific details of the assay are of less concern.

Between uptake assays for different pharmacological targets (e.g., SERT vs. DAT uptake inhibition), direct comparisons even within a laboratory and setup cannot be guaranteed if only IC_{50} and not K_i values are determined. However, the inclusion of a set of comparator compounds (e.g., methamphetamine, MDMA, and cocaine) with widely reported pharmacological characteristics should serve to set the standard for comparisons of IC_{50} values between targets. For example, calculating the DAT/SERT ratio for well-known compounds like MDMA can be the reference for unknown compounds [6, 35]. This again shows the importance of including well-known reference compounds in screening and that the value of a study increases according to the number of substances that are included.

Reproducibility within a laboratory is essential for the extensive characterization of multiple compounds. In general, for comparable IC_{50} values in large screenings within one laboratory requires strict adherence to the established protocol since IC_{50} values depend on substrate concentration, in addition to temperature and incubation times. We regularly test the reproducibility of IC_{50} values for our standard compounds and find that the values are very consistent across both time and experimenters. This regular validation ensures that the data for all substances that are reported from our laboratory can be directly compared with our previously reported data. Another issue to consider is possible fluctuations in target protein expression in heterologous expression systems that could account for inconsistent IC_{50} values within one laboratory [37]. However, if in vitro assays are set up with a targeted protein concentration within a linear range in a protein concentration vs. substrate transport relationship, moderate changes in cell number used for an individual assay or in target protein expression are usually tolerated and do not affect the reproducibility of IC_{50} values within laboratory, always given a linear relationship of target protein vs. substrate transport. As a side note, this is in contrast to transport kinetics (i.e., Michaelis-Menten kinetics), in which the maximal velocity is highly dependent on the expression levels of the transporter. With these considerations in mind, comparison of IC_{50} values within one laboratory is usually not a problem. For direct comparison of pharmacological constants between different laboratories K_i values should be assessed, since IC_{50} but not K_i values depend significantly on assay conditions [21]. The determination of K_i values is more complex because it requires knowledge or assessment of the mode of inhibition (e.g., competitive, noncompetitive, or mixed; [21]). Although K_i values would be the best constants to determine, the rapid and extensive characterization of the effects of a large set of cathinone NPS on multiple targets usually does not allow the labor-intensive determination of K_i values. Given these limitations, in vitro screenings assessing IC_{50} values are most useful when a large number of substances is assessed within one laboratory, or if well-known comparator drugs are included as reference compounds that allow for an interpretation of pharmacological profiles relative to the reference compounds.

Different setups for monoamine efflux assays have been described, all resulting in similar qualitative characterizations of compounds. Although different setups are valid, establishing an efflux assay can be difficult. Efflux can be measured using electrophysiological methods [38, 39], which allow the very reliable determination of transporter-mediated monoamine release and its associated currents that are induced by compounds (see Solis 2016, this volume). However, because patch-clamp electrophysiology requires specialized recording equipment, we only discuss radiolabeled substrate transport assays herein. Rothman et al. [31] reported the use of efflux assays with rat synaptosomes, in which synaptosomes were first preloaded to steady-state with the radioactive substrate via transporter-mediated uptake. Release was then induced without removing the radioactive uptake buffer. Using this method, a high signal-to-noise ratio was reported, but efflux potency values could be determined. Verrico et al. adapted this protocol for transfected HEK293 cells in suspension [40]. We initially followed this protocol [41] but later adapted it according to the principles reported by Scholze et al. [42], who used a superfusion system. The superfusion system is preferentially used for rodent tissue slices that are preloaded with radioactive transporter substrates [28], but it can also be adapted for transfected cells [42, 43]. Transfected cells are grown on coverslips and loaded with radioactive substrates. They are then moved to superfusion chambers where the cells are constantly superfused with non-radioactive buffer [42]. The advantage of this method is that the radioactive substrates that are released are transported away from the cells or tissue [44] so that the reuptake of released substrate should not occur. We adapted this principle to our laboratory but used well plates instead of a superfusion system. To achieve a similar effect as superfusion with regard to the immediate removal of released substrate, we took advantage of the dilution effect. Using a high buffer-to-cell ratio, the monoamine substrate that is released by the cells is distributed in a large volume of buffer, resulting in negligible extracellular substrate concentrations. To achieve a high buffer-to-cell ratio, we used special 24-well plates (XF24, Seahorse Biosciences, North Billerica, MA, USA), which fit 1 mL of buffer per well, but the area for cell growth is as small as the one from a regular 96-well plate. Therefore, the buffer-to-cell ratio is much higher than the one in a standard cell culture 96-well plate or 24-well plate, thus providing an optimal assay setup for testing substance-induced monoamine efflux. Release is quantified by assessing the monoamine radioactivity that remains in the cells after incubation with the test substance and compared with a vehicle control. Additionally, radioactivity that is associated with the released monoamine can be measured in the supernatant. In transfected cells, an apparent release of approximately 20% for pure uptake inhibitors is observed even with the superfusion method, most likely because of the high expression levels of transporters that transport nonspecifically released monoamines back into the cells [42]. Thus, uptake inhibitors need to be included as a negative control condition to account for apparent release. Apparent release can be lowered if ^{3}H-MPP+ is used for DAT and NET instead of the endogenous substrates DA and NE, but one caveat is the difference in transport kinetics between MPP+ and the endogenous substrates [45]. In our hands, apparent release was less with our well-plate method than with cells in suspension. Nevertheless, we chose to focus on determining qualitative release instead of release potencies, which are more difficult to determine. The precise determination of apparent release-corrected efflux potencies would require knowledge of the respective apparent release percentage for each concentration in the concentration/release curve. This would require a perfect match of uptake potencies of the control substance to measure apparent efflux and the actually releasing substance, which is practically unfeasible. Therefore, we determined release qualitatively by inducing it with high concentrations of a drug to determine whether the drug is a releaser and thus a transporter substrate or not.

Binding affinity can be determined for any ligand/protein interaction. For binding affinity, the ability of a substance to displace a radiolabeled ligand at the receptor or transporter is assessed, which requires competition between two compounds at the binding site. To assess the mode of action of NPS, binding can be determined for receptors and transporters [4, 11]. However, for both receptors and transporters, the functional assays are considered to have higher predictive validity with regard to in vivo effects. For the transporter, functional information is derived from the uptake and efflux assays. Specifically for substances that are releasers and thus substrates of the transporters, the binding properties or even the binding sites can differ from the radioligand that is to be displaced. Additionally, the substrates are transported and thus removed from competition with the radioligand. Binding affinity values do not necessarily reflect the functional uptake inhibition potency [4]. This is a common phenomenon for binding studies that use ligands that are also transporter substrates because transport of the substrate can alter the apparent binding affinity [46–48]. Thus, if a substance is a substrate-type releaser, then its binding affinity, when assessed by the described displacement assay, is not representative. This discrepancy between binding affinities and uptake inhibition potencies can even be used to characterize a substance as substrate-type releaser or pure uptake blocker [49, 50].

The determination of binding affinity is more common for receptors than for transporters. However, it is also important for receptor pharmacology to distinguish between functional activity and binding affinity [19]. The concepts for assessing activity and affinity in heterologous expression systems are different. To determine binding affinity, only the target protein from the expression is required. Therefore, isolated membrane preparations that can be stored in a frozen state are usually made from transfected cells. In radioligand displacement assays, the binding affinities of compounds at the binding site of the radioligand are determined. Functional information with regard to activation or inactivation of a G-protein-coupled receptor can be gained from cAMP measurements in living transfected cells using convenient, commercially available kits that do not require radioactivity. The activation of G-protein-coupled receptors results in a concentration-dependent increase in cAMP levels, the activation potency of which can be determined (EC_{50} value). Similarly, the activation of G-protein-coupled receptors can be assessed by measuring intracellular calcium changes [51]. With the inclusion of a known full agonist (typically an endogenous ligand) in the assay, the maximal efficacy can be determined. Full agonists induce maximal efficacy, whereas partial agonists induce only partial efficacy compared to endogenous ligands.

With regard to the translational relevance of in vitro screenings, setting the data in an informative clinical context is essential. Comparisons with well-known psychoactive substances inform about the similarity of NPS to these substances with known subjective effects, toxicity, and abuse liability. Furthermore, data on the link between pharmacological targets and subjective/physiological effects are needed. Several rodent and human studies have contributed to our understanding of the roles of DAT, SERT, and NET inhibition in the mode of action of psychoactive drugs. In rodents, particularly mice, genetic modification allows the elimination of a specific target and assessment of the behavioral and molecular impacts of the knockout. Constitutive knockout mouse models generally have the limitation of compensatory alterations that can occur, thus resulting in distinct phenotypes that are not ideal for finding explicit target-mediated effects [52–54]. Nevertheless, several knockout studies have implicated the DAT and SERT in the actions of psychostimulants. For example, SERT knockout mice exhibit greater rewarding effects of cocaine in the conditioned place preference paradigm compared with wildtype mice [55]. More sophisticated genetic models with a triple amino acid mutation in the DAT gene showed that DAT inhibition is necessary for cocaine-induced conditioned place preference [56] and cocaine-evoked synaptic plasticity [57]. Clinical studies that assess pharmacological interactions between a psychostimulant and receptor-selective antagonists or well-characterized transporter ligands shed light on specific molecular target mediating subjective effects and acute toxicity in humans. For example, our laboratory investigated the mode of action of MDMA in humans by blocking the NET, SERT, or DAT or combinations thereof [41, 58–62]. These studies showed that NET and α_1-adrenergic stimulation are crucially involved in MDMA-induced sympathomimetic activation, including elevations of blood pressure and body temperature [58, 62–64] and that the SERT-mediated release of 5-HT is involved in the subjective entactogenic/empathogenic effects of MDMA [41, 60, 65]. Interactions with the DAT and activation of the DA system are generally considered responsible for the reinforcing and addictive properties of a substance [66]. Accordingly, NPS that mostly interact with the SERT can be expected to produce more empathogenic MDMA-like effects, in contrast to NPS that mostly interact with the NET and DAT and are thus expected to produce more stimulant-type effects and addiction similar to methamphetamine [2, 4]. Additionally, we noted that substances, such as MDMA, that primarily release endogenous monoamines via the transporter may have a shorter duration of action despite having a long plasma half-life [41] than substances that only inhibit a transporter (e.g., pyrovalerone cathinones; [67]) or interact with postsynaptic receptors (e.g., hallucinogens; [51, 68]).

In vivo studies in rodents and humans increase our knowledge of the effects and toxicity that are related to individual targets that mediate the complex actions of psychostimulants and help predict the toxicity of NPS. Dissecting the clinical roles of different neurotransmitter systems and attributing specific effects to specific targets or pharmacological profiles (e.g., DAT/SERT ratio; [2, 4]) support the meaningful translation of in vitro NPS pharmacology to expected subjective effects and toxicity in humans. Newer techniques, such as optogenetic approaches, for dissecting brain circuitry or sophisticated transgenic animal models without compensatory alterations that can isolate target-mediated effects in vivo will continue to shape our understanding of psychoactive drug actions with regard to specific targets, which will also impact interpretations of the in vitro pharmacology of NPS.

3 Effects on Cathinone Analogs on Transporter-Mediated Uptake

All cathinone NPS inhibit transporter-mediated monoaminergic uptake but with different selectivity and relative potencies. The precise profile of relative DAT, SERT, and NET inhibition potencies likely determines the different experiences that are described by drug users. In the screening from our laboratory, most cathinone NPS are potent NET inhibitors, with uptake inhibition potencies in the submicromolar range (Table 1). *N,N*-dimethylcathinone, ethylone, methedrone, and 4-methylethcathinone are the exceptions with NET inhibition IC_{50} in the low micromolar range. High potency for NET inhibition relative to DAT and SERT were also reported from other laboratories [8, 69, 70], but with less prominent fold-shifts compared to DAT inhibition. This likely arises from different assay conditions that determine the IC_{50} values. However, the general high inhibition potency of NET for most cathinones NPS are consistent across laboratories. Drug-induced increases in NE markedly contribute to the psychostimulation of a drug and sympathomimetic toxicity [41, 58]. We compared the common recreational doses that are taken in a single drug session and uptake inhibition potencies at the NET, SERT, or DAT and found that the recreational doses correlated mainly with NET inhibition potencies [4]. This is in agreement with Rothman et al. [31] who found a linear correlation between release-induction potency in synaptosomes and oral doses producing Therefore, the in vitro inhibition potency at NET best predicts clinical potency and the doses that are likely to be used recreationally.

Significant differences in DAT and SERT inhibition potencies among cathinone NPS are evident [4, 5, 8]. Many cathinone NPS are potent DAT inhibitors that are comparable to methamphetamine or cocaine, and some cathinone NPS are weak DAT inhibitors that are more comparable to MDMA. In our assays, methamphetamine and cocaine, which are well-known psychostimulants that act on the DAT, exhibit DAT inhibition potencies (IC_{50} values) around 1 μM. Many pyrovalerone cathinones are extremely potent DAT inhibitors. The most popular pyrovalerone cathinone, MDPV, is 30-times more potent in inhibiting the DAT in heterologous expression systems than cocaine [4, 69]. Similarly in synaptosomes, 40–50-fold differences in DAT inhibition potency between MDPV and cocaine were reported [7]. MDPV is also called "super coke," and small doses may have strong and long-lasting effects because of its high potency and pure uptake inhibition [71]. Severe toxicity and even deaths have resulted from the recreational use of this substance [72, 73]. To avoid such cases, warnings could be issued for extremely potent substances like MDPV as soon as they emerge as recreationally used substances. Therefore, testing newly emerged NPS in in vitro pharmacological screenings as fast as possible is crucial for detecting substances with high potencies at monoaminergic targets that are relevant to stimulant or other psychotropic actions.

Inhibition of the SERT is generally less represented among the cathinone derivatives but is characteristic for such substances as benzofurans [35], aminoindanes, benzylpiperazines [74], and para ring-substituted amphetamines

Table 1 Uptake inhibition potencies of cathinone NPS and the respective non-β-keto analogs

	Pharmacology cathinone analogs				Pharmacology amphetamine analogs			
Cathinone analogs	NET IC_{50} (μM) (95% CI)	DAT IC_{50} (μM) (95% CI)	SERT IC_{50} (μM) (95% CI)	Amphetamine analogs	NET IC_{50} (μM) (95% CI)	DAT IC_{50} (μM) (95% CI)	SERT IC_{50} (μM) (95% CI)	Values published in
4-Bromomethcathinone	0.41 (0.30–0.57)	5.6 (2.7–12)	2.2 (1.7–2.8)					(3)
Buphedrone	0.65 (0.51–0.81)	4.24 (3.3–5.5)	70 (2–2700)					(2)
Buthylone	2.02 (1.5–2.7)	2.90 (2.5–3.4)	6.22 (4.3–9.0)	MBDB	2.80 (1.9–4.1)	22 (20–26)	2.04 (1.4–3.0)	(1)
Cathinone	0.199 (0.15–0.26)	14.0 (10–20)	>100	Amphetamine	0.094 (0.06–0.14)	1.30 (0.83–2.0)	>10	(1)
N,N-Dimethylcathinone	7.71 (5–12)	27 (21–36)	>500					(2)
Ethcathinone	0.44 (0.34–0.56)	5.00 (3.7–6.8)	48 (4–529)	*N*-Ethylamphetamine	0.20 (0.15–0.27)	5.86 (4.8–7.1)	8.77 (6–13)	(2)
4-Ethylmethcathinone	2.5 (1.7–3.7)	31 (13–72)	4.3 (3.2–5.9)					(3)
Ethylone	2.54 (2.0–3.2)	5.68 (4.9–6.5)	4.46 (3.8–5.2)	MDEA	1.02 (0.78–1.3)	9.3 (8.0–11)	1.27 (0.93–1.7)	(1)
Flephedrone	0.246 (0.16–0.37)	6.35 (4.2–9.5)	>10	4-Fluoromethamphetamine	0.22 (0.14–0.35)	7.7 (2.5–24)	8.7 (3.8–20)	(1),(3)
3-Fluoromethcathinone	0.19 (0.13–0.29)	1.7 (1.0–3.0)	56 (7–472)					(2)
β-keto MDA	1.6 (1.1–2.3)	14 (10–18)	21 (15–28)	MDA	0.42 (0.3–0.6)	20.5 (20.3–20.6)	4.9 (3.5–6.8)	(4)
MDPBP	0.16 (0.11–0.24)	0.11 (0.07–0.16)	15 (5.4–39)					(3)
MDPPP	0.97 (0.62–1.5)	0.53 (0.27–1.1)	75 (49–114)					(3)

MDPV	0.044 (0.03–0.07)	0.031 (0.03–0.04)	9.30 (6.8–12.8)					(1)
Mephedrone	0.254 (0.22–0.30)	3.31 (2.6–4.2)	4.64 (3.7–5.9)					(1)
Methcathinone	0.085 (0.06–0.17)	1.12 (0.83–1.5)	>10	Methamphetamine	0.064 (0.04–0.09)	1.05 (0.74–1.5)	>10	(1)
Methedrone	2.24 (1.4–3.5)	35 (15–79)	4.73 (3.2–6.9)	PMMA	1.20 (0.75–1.8)	49 (18–135)	1.77 (1.1–2.9)	(2)
4-Methylethcathinone	2.23 (1.6–3.2)	4.28 (3.4–5.4)	7.93 (3.5–18)					(2)
Methylone	0.542 (0.39–0.75)	4.82 (3.8–6.1)	15.5 (10–26)	MDMA	0.447 (0.33–0.60)	17 (12–24)	1.36 (1.0–2.0)	(1)
Naphyrone	0.25 (0.20–0.32)	0.47 (0.40–0.55)	0.96 (0.85–1.09)					(1)
Pentedrone	0.61 (0.52–0.72)	2.50 (2.0–3.2)	135 (5–3700)					(2)
Pentylone	0.99 (0.72–1.4)	1.34 (1.0–1.7)	8.37 (5.4–13)					(2)
Pyrovalerone	0.043 (0.03–0.06)	0.035 (0.03–0.04)	13.0 (10.8–15.8)					(1)
α-PVP	0.02 (0.01–0.03)	0.04 (0.01–0.1)	>100					(3)

(1) Simmler et al. [4], Br J Pharmacol; (2) Simmler et al. [5], Neuropharmacology; (3) Rickli et al. [3], Eur Neuropsychopharmacol; Rickli et al. [35], Br J Pharmacol

[3], which have MDMA-like psychoactive properties. Compared with the serotonergic drug MDMA, only naphyrone among the cathinone NPS is equally potent in inhibiting the SERT [4, 8]. However, methedrone has a similar DAT/SERT inhibition ratio to MDMA, thus predicting a similar effect profile to MDMA, in addition to predicting high risk of hyperthermia because of its similarity to para-methoxyamphetamine [2, 5]. Other cathinone NPS inhibit the SERT with lower potencies, resulting in relatively more dopaminergic properties, or their SERT inhibition is negligible.

Ideally, the SERT inhibition potency of substances is set relative to their DAT inhibition. Relative activity at the DAT vs. SERT can serve as an indicator of the abuse liability of a psychoactive substance because potent SERT activity relative to DAT activity can be protective against the abuse of a drug [75–77] (see Negus and Banks 2016, this volume). Substances with potent SERT inhibition are less reinforcing than substances with low SERT vs. DAT activity [75, 77, 78]. Using uptake inhibition potencies, we calculated DAT/SERT ratios ($IC_{50,SERT}/IC_{50,DAT}$). Note that the calculation with the reciprocal formula $IC_{50,SERT}/IC_{50,DAT}$ results in high DAT/SERT ratios for substances that inhibit DAT more potently (lower IC_{50} value) than SERT (higher IC_{50} value) and vice versa. In our hands, where cocaine has a DAT/SERT ratio of ~1, substances with a DAT/SERT ratio >1 can be considered to have high abuse liability. Substances with a DAT/SERT ratio close to that of MDMA (0.1) likely have lower abuse liability. For example, we predicted particularly high abuse potential for MDPV based on its high DAT/SERT inhibition ratio [4]. Animal studies and clinical observations confirmed the potent reinforcing and rewarding properties of MDPV, confirming in vitro study-based predictions of abuse potential [79, 80] (see Watterson and Olive 2016, this volume).

For some cathinone NPS in our screening studies, we determined the profile of respective structural amphetamine analogs that lack the β-keto group [4, 5, 35]. Adding a β-keto group to MDMA to form methylone resulted in a higher DAT/SERT ratio and thus higher predicted abuse liability. The shift in the DAT/SERT inhibition ratio that results from the addition of a β-keto group was less pronounced for amphetamines with an already high DAT/SERT inhibition ratio, such as methamphetamine. Notably, a small change in the molecular structure of some amphetamines can result in a significantly different pharmacological profile.

4 Effects of Cathinones on Transporter-Mediated Efflux

Substances that inhibit monoamine transporters are either pure uptake inhibitors or releasers [31]. If they are monoamine releasers, then they induce transporter-mediated efflux, which should not be confused with exocytotic calcium-dependent vesicular monoamine release. Transporter-mediated efflux occurs when drugs act as substrates of the transporters [81]. As substrates, the substances are transported into the cell. Because amphetamine analogs, such as MDMA and methamphetamine, are releasers [31, 82], it is of interest to characterize cathinone NPS as releasers or pure uptake inhibitors. All releasers or substrates, including the endogenous substrates (i.e., DA, NE, and 5-HT), present uptake inhibition properties because of competition for transport [31]. Therefore, uptake assays cannot determine whether a substance is an inhibitor or a substrate releaser, but separate efflux assays can determine whether a drug is a releaser or pure uptake inhibitor. Interestingly, pyrovalerone cathinones are pure uptake inhibitors (Table 2), although they are amphetamine-type substances. Most other cathinone NPS are releasers like their amphetamine analogs (Table 2).

We distinguish monoamine-releasing substances from pure monoamine uptake inhibitors, but the impact of release vs. pure uptake inhibition on psychoactive effects is unclear and likely less relevant than the DAT/SERT inhibition ratio [2]. This distinction is less relevant for subjective and stimulant effects than for cellular toxicity. Because release-inducing substances enter nerve terminals via transporters, they are more likely to exert intracellular effects and toxicity compared with pure uptake inhibitors [81]. Typically, releasers act on vesicular monoamine transporters and deplete vesicles, which can have short- or long-term toxic consequences [83].

With the large numbers of NPS reported in the recent years, there is need for a classification of NPS. NPS can be classified by their chemical structures. For example, Hill and Thomas [84] classified MDMA as ring-substituted methylenedioxyphenethylamine, mephedrone as β-ketonated amphetamine, and MDPV as β-ketonated substituted methylenedioxyphenethylamine. A structural classification is very useful for an audience with an interest in the chemical structure of NPS. An audience with a clinical focus might mainly be interested in anticipated subjective effects and toxicology. A classification according to pharmacological profiles are likely more meaningful for clinicians than chemical structures, particularly also since structural similarities not necessarily result in comparable pharmacological profiles. In our NPS screenings, we classify cathinone derivatives according to the similarity of their in vitro profile to methamphetamine, cocaine, and MDMA [2, 4]. DAT/NET-selective pyrovalerone cathinones represent a separate group since they are extremely potent inhibitors. Importantly, small structural changes can markedly alter the pharmacological profile of substances, sometimes in an unpredicted manner, resulting in different psychoactive and toxicological effects. For example methylone, the β-keto analog of MDMA, presents a prominent increase in DAT/SERT ratio, suggesting a higher abuse potential of methylone compared to MDMA [4, 6]. Classification according to pharmacology may thus be more conclusive as a reference for clinical applications than structural analogies.

Table 2 Qualitative characterization of cathinone NPS and the respective non-β-keto analogs as releasers at NET, DAT, and SERT

	Cathinone analogs				Amphetamine analogs			
Cathinone analogs	NE efflux	DA efflux	5-HT efflux	Amphetamine analogs	NE efflux	DA efflux	5-HT efflux	Values published in
4-Bromomethcathinone	Yes	Yes	No					(3)
Buphedrone	Yes	No	No					(2)
Buthylone	*NA*	No	Yes	MBDB	NA	No	Yes	(1)
Cathinone	*NA*	Yes	No	Amphetamine	Yes	Yes	Yes	(1), (3)
N,N-Dimethylcathinone	No	No	No					(2)
Ethcathinone	Yes	No	Yes	*N*-Ethylamphetamine	Yes	Yes	Yes	(2)
4-Ethylmethcathinone	Yes	Yes	Yes					(3)
Ethylone	*NA*	No	Yes	MDEA	NA	No	Yes	(1)
Flephedrone	Yes	Yes	Yes*	4-Fluoromethamphetamine	Yes	Yes	Yes	(1), (3)
3-Fluoromethcathinone	Yes	Yes	Yes					(2)
β-keto MDA	Yes	No	Yes	MDA	Yes	Yes	Yes	(4)
MDPBP	No	No	No					(3)
MDPPP	No	No	No					(3)
MDPV	No	No	No					(1), (3)
Mephedrone	Yes	Yes	Yes					(1), (3)
Methcathinone	Yes	Yes	Yes*	Methamphetamine	Yes	Yes	Yes	(1), (3)
Methedrone	Yes	No	Yes	PMMA	Yes	Yes	Yes	(2)
4-Methylethcathinone	No	No	Yes					(2)
Methylone	*NA*	No	Yes	MDMA	Yes	Yes	Yes	(1), (3)
Naphyrone	No	No	No					(1), (3)
Pentedrone	No	No	No					(2)

Pentylone	No	No	Yes					(2)
Pyrovalerone	No	No	No					(1), (3)
α-PVP	No	No	No					(3)

(1) Simmler et al. [4], Br J Pharmacol; (2) Simmler et al. [5], Neuropharmacology; (3) Rickli et al. [3], Eur Neuropsychopharmacol; (4) Rickli et al. [35], Br J Pharmacol

NA not assessed

*Not significant in Rickli et al. [3], Eur Neuropsychopharmacol

5 Drug Interactions with G-Protein-Coupled Receptors

In addition to transporter pharmacology, assessing receptor interactions is necessary for a comprehensive pharmacological characterization of psychoactive substances. The major implications would be for the assessment of any hallucinogenic properties of NPS. LSD has high affinity for the 5-HT_{2A} receptor [51, 85], which is associated with its hallucinogenic properties. Other drugs with potent 5-HT_{2A} activity have been shown to substitute for LSD in drug-discrimination studies [86]. In vitro activity at the 5-HT_{2A} receptor is a good predictor of possible hallucinogenic effects and is likely the most relevant receptor/NPS interaction that is assessed in in vitro screening, particularly for potentially hallucinogenic compounds [51]. The activation of DA D_1 receptors but not D_2 receptors might be sufficient for a substance to be rewarding [87]. Noradrenergic receptors are involved in sympathomimetic toxicity, leading to vasoconstriction, hyperthermia, increased blood pressure, and increased heart rate [63, 64].

The main targets of amphetamine analogs are typically monoamine transporters, but some substances have weak affinity for monoamine receptors. However, it is questionable if direct receptor affinity contributes markedly to the overall drug effect of substances that foremost are transporter inhibitors. The rise in extracellular monoamine concentrations that is evoked by drug actions at the transporters results in neurotransmitter binding to postsynaptic receptors, which suggests that direct agonism has only negligible contribution to overall drug effect. Direct antagonistic receptor activation might, to some extent, counteract neurotransmitter binding at postsynaptic receptors. We and others did not find any cathinones or amphetamines with relevant affinity at D_1, D_2, or D_3 receptors [3–5, 8]. However, some cathinone analogs exhibit weak affinity for 5-HT_{2A} or 5-HT_{2C} receptors and are low-potency 5-HT_{2A} antagonists [69]. Compared with hallucinogens that exert their psychoactive effects mainly via 5-HT receptors (e.g., the NPS benzodifuran 2C-B-Fly or novel *N*-2-methoxybenzyl-derivatives), with receptor binding values in the submicromolar range [35, 51], the weak binding affinities of cathinones at these targets are likely irrelevant.

In our pharmacological characterization of NPS, we also include the trace amine-associated receptor 1 (TAAR1; [3–5, 88]). Methamphetamine and other amphetamine-type drugs have been shown to activate the TAAR1, and the TAAR1 could be a target for the pharmacological treatment of addiction [89]. -Substance-mediated agonist effects at the TAAR1 may reduce the stimulant properties of MDMA and methamphetamine [90, 91]. In contrast, cathinone NPS do not display affinity for the TAAR1, and may thus have more stimulant-like effects and be more addictive than their amphetamine analogs because of the lack of this TAAR1-mediated "auto-inhibition," in addition to their greater dopaminergic properties. TAAR-1 activation could be relevant for experiments conducted in rodents. In humans, however, direct affinity of psychoactive substances at TAAR1 is probably not important since most psychostimulants have weak activity at the human TAAR1 as determined by in vitro screenings [88]. Nevertheless,

TAAR1 presents a promising target that could be relevant for psychostimulant treatment.

6 Summary

NPS continue to emerge and are used recreationally without much knowledge about their pharmacology or toxicology. In vitro characterizations of psychoactive compounds that utilize transfected cell lines are useful for gaining fast and translationally important information on cathinone NPS. The in vitro pharmacological profiles of cathinone NPS have predicted considerable abuse liability of these drugs and identified pyrovalerone cathinones with extremely high potencies for DAT inhibition. Small structural changes, such as the β-keto group in the amphetamine-basic structure, can substantially change the pharmacological profile of substances with regard to their potency and relative activity at different monoaminergic targets.

References

1. EMCDDA (2015). European Drug Report 2015. Luxembourg: European Monitoring Centre for Drugs and Drug Addiction (EMCDDA)
2. Liechti M (2015) Novel psychoactive substances (designer drugs): overview and pharmacology of modulators of monoamine signaling. Swiss Med Wkly 145:w14043
3. Rickli A, Hoener MC, Liechti ME (2015) Monoamine transporter and receptor interaction profiles of novel psychoactive substances: para-halogenated amphetamines and pyrovalerone cathinones. Eur Neuropsychopharmacol 25:365–376
4. Simmler L, Buser T, Donzelli M, Schramm Y, Dieu LH, Huwyler J et al (2013) Pharmacological characterization of designer cathinones in vitro. Br J Pharmacol 168:458–470
5. Simmler LD, Rickli A, Hoener MC, Liechti ME (2014) Monoamine transporter and receptor interaction profiles of a new series of designer cathinones. Neuropharmacology 79:152–160
6. Baumann MH, Ayestas MA Jr, Partilla JS, Sink JR, Shulgin AT, Daley PF et al (2012) The designer methcathinone analogs, mephedrone and methylone, are substrates for monoamine transporters in brain tissue. Neuropsychopharmacology 37:1192–1203
7. Baumann MH, Partilla JS, Lehner KR, Thorndike EB, Hoffman AF, Holy M et al (2013) Powerful cocaine-like actions of 3,4-methylenedioxypyrovalerone (MDPV), a principal constituent of psychoactive 'bath salts' products. Neuropsychopharmacology 38:552–562
8. Iversen L, Gibbons S, Treble R, Setola V, Huang XP, Roth BL (2013) Neurochemical profiles of some novel psychoactive substances. Eur J Pharmacol 700:147–151
9. Ramamoorthy S, Bauman AL, Moore KR, Han H, Yang-Feng T, Chang AS et al (1993) Antidepressant- and cocaine-sensitive human serotonin transporter: molecular cloning, expression, and chromosomal localization. Proc Natl Acad Sci U S A 90:2542–2546
10. Revel FG, Moreau JL, Gainetdinov RR, Bradaia A, Sotnikova TD, Mory R et al (2011) TAAR1 activation modulates monoaminergic neurotransmission, preventing hyperdopaminergic and hypoglutamatergic activity. Proc Natl Acad Sci U S A 108:8485–8490

11. Tatsumi M, Groshan K, Blakely RD, Richelson E (1997) Pharmacological profile of antidepressants and related compounds at human monoamine transporters. Eur J Pharmacol 340:249–258
12. Groskreutz D, Schenborn ET (1997) Reporter systems. Methods Mol Biol 63:11–30
13. Chaudhary S, Pak JE, Gruswitz F, Sharma V, Stroud RM (2012) Overexpressing human membrane proteins in stably transfected and clonal human embryonic kidney 293S cells. Nat Protoc 7:453–466
14. Henry LK, Field JR, Adkins EM, Parnas ML, Vaughan RA, Zou MF et al (2006) Tyr-95 and Ile-172 in transmembrane segments 1 and 3 of human serotonin transporters interact to establish high affinity recognition of antidepressants. J Biol Chem 281:2012–2023
15. Ran FA, Hsu PD, Wright J, Agarwala V, Scott DA, Zhang F (2013) Genome engineering using the CRISPR-Cas9 system. Nat Protoc 8:2281–2308
16. Chamba A, Holder MJ, Barnes NM, Gordon J (2008) Characterisation of the endogenous human peripheral serotonin transporter SLC6A4 reveals surface expression without N-glycosylation. J Neuroimmunol 204:75–84
17. Marazziti D, Mandillo S, Di Pietro C, Golini E, Matteoni R, Tocchini-Valentini GP (2007) GPR37 associates with the dopamine transporter to modulate dopamine uptake and behavioral responses to dopaminergic drugs. Proc Natl Acad Sci U S A 104:9846–9851
18. Maguire JJ, Kuc RE, Davenport AP (2012) Radioligand binding assays and their analysis. Methods Mol Biol 897:31–77
19. Zhang R, Xie X (2012) Tools for GPCR drug discovery. Acta Pharmacol Sin 33:372–384
20. Tate CG (2012) A crystal clear solution for determining G-protein-coupled receptor structures. Trends Biochem Sci 37:343–352
21. Burlingham BT, Widlanski TS (2003) An intuitive look at the relationship of K-i and IC50: A more general use for the Dixon plot. J Chem Educ 80:214–218
22. Cheng Y, Prusoff WH (1973) Relationship between the inhibition constant (K1) and the concentration of inhibitor which causes 50 per cent inhibition (I50) of an enzymatic reaction. Biochem Pharmacol 22:3099–3108
23. Bang-Andersen B, Ruhland T, Jorgensen M, Smith G, Frederiksen K, Jensen KG et al (2011) Discovery of 1-[2-(2,4-dimethylphenylsulfanyl)phenyl]piperazine (Lu AA21004): a novel multimodal compound for the treatment of major depressive disorder. J Med Chem 54:3206–3221
24. Mazei-Robison MS, Bowton E, Holy M, Schmudermaier M, Freissmuth M, Sitte HH et al (2008) Anomalous dopamine release associated with a human dopamine transporter coding variant. J Neurosci 28:7040–7046
25. Prasad HC, Zhu CB, McCauley JL, Samuvel DJ, Ramamoorthy S, Shelton RC et al (2005) Human serotonin transporter variants display altered sensitivity to protein kinase G and p38 mitogen-activated protein kinase. Proc Natl Acad Sci U S A 102:11545–11550
26. Prosser RA, Stowie A, Amicarelli M, Nackenoff AG, Blakely RD, Glass JD (2014) Cocaine modulates mammalian circadian clock timing by decreasing serotonin transport in the SCN. Neuroscience 275:184–193
27. Thompson BJ, Jessen T, Henry LK, Field JR, Gamble KL, Gresch PJ et al (2011) Transgenic elimination of high-affinity antidepressant and cocaine sensitivity in the presynaptic serotonin transporter. Proc Natl Acad Sci U S A 108:3785–3790
28. Mergy MA, Gowrishankar R, Gresch PJ, Gantz SC, Williams J, Davis GL et al (2014) The rare DAT coding variant Val559 perturbs DA neuron function, changes behavior, and alters in vivo responses to psychostimulants. Proc Natl Acad Sci U S A 111:E4779–E4788
29. Barker EL, Blakely RD (1996) Identification of a single amino acid, phenylalanine 586, that is responsible for high affinity interactions of tricyclic antidepressants with the human serotonin transporter. Mol Pharmacol 50:957–965
30. Wilhelm BG, Mandad S, Truckenbrodt S, Krohnert K, Schafer C, Rammner B et al (2014) Composition of isolated synaptic boutons reveals the amounts of vesicle trafficking proteins. Science 344:1023–1028

31. Rothman RB, Baumann MH, Dersch CM, Romero DV, Rice KC, Carroll FI et al (2001) Amphetamine-type central nervous system stimulants release norepinephrine more potently than they release dopamine and serotonin. Synapse 39:32–41
32. Rothman RB, Lewis B, Dersch C, Xu H, Radesca L, de Costa BR et al (1993) Identification of a GBR12935 homolog, LR1111, which is over 4,000-fold selective for the dopamine transporter, relative to serotonin and norepinephrine transporters. Synapse 14:34–39
33. de la Torre R, Farre M, Ortuno J, Mas M, Brenneisen R, Roset PN et al (2000) Non-linear pharmacokinetics of MDMA ('ecstasy') in humans. Br J Clin Pharmacol 49:104–109
34. Mueller DM, Rentsch KM (2012) Generation of metabolites by an automated online metabolism method using human liver microsomes with subsequent identification by LC-MS(n), and metabolism of 11 cathinones. Anal Bioanal Chem 402:2141–2151
35. Rickli A, Kopf S, Hoener MC, Liechti ME (2015) Pharmacological profile of novel psychoactive benzofurans. Br J Pharmacol 172:3412–3425
36. Torok M, Huwyler J, Drewe J, Gutmann H, Fricker G (1998) Transport of the beta-lactam antibiotic benzylpenicillin and the dipeptide glycylsarcosine by brain capillary endothelial cells in vitro. Drug Metab Dispos 26:1144–1148
37. Ukairo OT, Ramanujapuram S, Surratt CK (2007) Fluctuation of the dopamine uptake inhibition potency of cocaine, but not amphetamine, at mammalian cells expressing the dopamine transporter. Brain Res 1131:68–76
38. Hilber B, Scholze P, Dorostkar MM, Sandtner W, Holy M, Boehm S et al (2005) Serotonin-transporter mediated efflux: a pharmacological analysis of amphetamines and non-amphetamines. Neuropharmacology 49:811–819
39. Khoshbouei H, Wang H, Lechleiter JD, Javitch JA, Galli A (2003) Amphetamine-induced dopamine efflux. A voltage-sensitive and intracellular Na+-dependent mechanism. J Biol Chem 278:12070–12077
40. Verrico CD, Miller GM, Madras BK (2007) MDMA (Ecstasy) and human dopamine, norepinephrine, and serotonin transporters: implications for MDMA-induced neurotoxicity and treatment. Psychopharmacology (Berl) 189:489–503
41. Hysek CM, Simmler LD, Nicola VG, Vischer N, Donzelli M, Krahenbuhl S et al (2012) Duloxetine inhibits effects of MDMA ("Ecstasy") in vitro and in humans in a randomized placebo-controlled laboratory study. PLoS One 7, e36476
42. Scholze P, Zwach J, Kattinger A, Pifl C, Singer EA, Sitte HH (2000) Transporter-mediated release: a superfusion study on human embryonic kidney cells stably expressing the human serotonin transporter. J Pharmacol Exp Ther 293:870–878
43. Pifl C, Drobny H, Reither H, Hornykiewicz O, Singer EA (1995) Mechanism of the dopamine-releasing actions of amphetamine and cocaine: plasmalemmal dopamine transporter versus vesicular monoamine transporter. Mol Pharmacol 47:368–373
44. Raiteri M, Angelini F, Levi G (1974) A simple apparatus for studying the release of neurotransmitters from synaptosomes. Eur J Pharmacol 25:411–414
45. Johnson RA, Eshleman AJ, Meyers T, Neve KA, Janowsky A (1998) [3H]substrate- and cell-specific effects of uptake inhibitors on human dopamine and serotonin transporter-mediated efflux. Synapse 30:97–106
46. Marcusson JO, Backstrom IT, Ross SB (1986) Single-site model of the neuronal 5-hydroxytryptamine uptake and imipramine-binding site. Mol Pharmacol 30:121–128
47. Nelson PJ, Rudnick G (1979) Coupling between platelet 5-hydroxytryptamine and potassium transport. J Biol Chem 254:10084–10089
48. Talvenheimo J, Nelson PJ, Rudnick G (1979) Mechanism of imipramine inhibition of platelet 5-hydroxytryptamine transport. J Biol Chem 254:4631–4635
49. Eshleman AJ, Carmolli M, Cumbay M, Martens CR, Neve KA, Janowsky A (1999) Characteristics of drug interactions with recombinant biogenic amine transporters expressed in the same cell type. J Pharmacol Exp Ther 289:877–885

50. Rothman RB, Ayestas MA, Dersch CM, Baumann MH (1999) Aminorex, fenfluramine, and chlorphentermine are serotonin transporter substrates. Implications for primary pulmonary hypertension. Circulation 100:869–875
51. Rickli A, Luethi D, Reinisch J, Buchy D, Hoener MC, Liechti ME (2015) Receptor interaction profiles of novel N-2-methoxybenzyl (NBOMe) derivatives of 2,5-dimethoxy-substituted phenethylamines (2C drugs). Neuropharmacology 99:546–553
52. Kalueff AV, Olivier JD, Nonkes LJ, Homberg JR (2010) Conserved role for the serotonin transporter gene in rat and mouse neurobehavioral endophenotypes. Neurosci Biobehav Rev 34:373–386
53. Viggiano D, Ruocco LA, Sadile AG (2003) Dopamine phenotype and behaviour in animal models: in relation to attention deficit hyperactivity disorder. Neurosci Biobehav Rev 27:623–637
54. Xu F, Gainetdinov RR, Wetsel WC, Jones SR, Bohn LM, Miller GW et al (2000) Mice lacking the norepinephrine transporter are supersensitive to psychostimulants. Nat Neurosci 3:465–471
55. Sora I, Hall FS, Andrews AM, Itokawa M, Li XF, Wei HB et al (2001) Molecular mechanisms of cocaine reward: combined dopamine and serotonin transporter knockouts eliminate cocaine place preference. Proc Natl Acad Sci U S A 98:5300–5305
56. O'Neill B, Tilley MR, Han DD, Thirtamara-Rajamani K, Hill ER, Bishop GA et al (2014) Behavior of knock-in mice with a cocaine-insensitive dopamine transporter after virogenetic restoration of cocaine sensitivity in the striatum. Neuropharmacology 79:626–633
57. Brown MT, Bellone C, Mameli M, Labouebe G, Bocklisch C, Balland B et al (2010) Drug-driven AMPA receptor redistribution mimicked by selective dopamine neuron stimulation. PLoS One 5:e15870
58. Hysek CM, Simmler LD, Ineichen M, Grouzmann E, Hoener MC, Brenneisen R et al (2011) The norepinephrine transporter inhibitor reboxetine reduces stimulant effects of MDMA ("ecstasy") in humans. Clin Pharmacol Ther 90:246–255
59. Hysek CM, Simmler LD, Schillinger N, Meyer N, Schmid Y, Donzelli M et al (2014) Pharmacokinetic and pharmacodynamic effects of methylphenidate and MDMA administered alone or in combination. Int J Neuropsychopharmacol 17:371–381
60. Liechti ME, Baumann C, Gamma A, Vollenweider FX (2000) Acute psychological effects of 3,4-methylenedioxymethamphetamine (MDMA, "Ecstasy") are attenuated by the serotonin uptake inhibitor citalopram. Neuropsychopharmacology 22:513–521
61. Schmid Y, Hysek CM, Simmler LD, Crockett MJ, Quednow BB, Liechti ME (2014) Differential effects of MDMA and methylphenidate on social cognition. J Psychopharmacol 28:847–856
62. Schmid Y, Rickli A, Schaffner A, Duthaler U, Grouzmann E, Hysek CM et al (2015) Interactions between bupropion and 3,4-methylenedioxymethamphetamine in healthy subjects. J Pharmacol Exp Ther 353:102–111
63. Hysek C, Schmid Y, Rickli A, Simmler L, Donzelli M, Grouzmann E et al (2012) Carvedilol inhibits the cardiostimulant and thermogenic effects of MDMA in humans. Br J Pharmacol 166:2277–2288
64. Hysek CM, Fink AE, Simmler LD, Donzelli M, Grouzmann E, Liechti ME (2013) alpha(1)-Adrenergic receptors contribute to the acute effects of 3,4-methylenedioxymethamphetamine in humans. J Clin Psychopharmacol 33:658–666
65. Hysek CM, Domes G, Liechti ME (2012) MDMA enhances "mind reading" of positive emotions and impairs "mind reading" of negative emotions. Psychopharmacology (Berl) 222:293–302
66. Howell LL, Wilcox KM (2002) Functional imaging and neurochemical correlates of stimulant self-administration in primates. Psychopharmacology (Berl) 163:352–361
67. Derungs A, Schietzel S, Meyer MR, Maurer HH, Krahenbuhl S, Liechti ME (2011) Sympathomimetic toxicity in a case of analytically confirmed recreational use of naphyrone (naphthylpyrovalerone). Clin Toxicol (Phila) 49:691–693

68. Dolder PC, Schmid Y, Haschke M, Rentsch KM, Liechti ME (2015) Pharmacokinetics and concentration-effect relationship of oral LSD in humans. Int J Neuropsychopharmacol 19
69. Eshleman AJ, Wolfrum KM, Hatfield MG, Johnson RA, Murphy KV, Janowsky A (2013) Substituted methcathinones differ in transporter and receptor interactions. Biochem Pharmacol 85:1803–1815
70. Rosenauer R, Luf A, Holy M, Freissmuth M, Schmid R, Sitte HH (2013) A combined approach using transporter-flux assays and mass spectrometry to examine psychostimulant street drugs of unknown content. ACS Chem Neurosci 4:182–190
71. Ross EA, Reisfield GM, Watson MC, Chronister CW, Goldberger BA (2012) Psychoactive "bath salts" intoxication with methylenedioxypyrovalerone. Am J Med 125:854–858
72. Borek HA, Holstege CP (2012) Hyperthermia and multiorgan failure after abuse of "bath salts" containing 3,4-methylenedioxypyrovalerone. Ann Emerg Med 60:103–105
73. Murray BL, Murphy CM, Beuhler MC (2012) Death following recreational use of designer drug "bath salts" containing 3,4-Methylenedioxypyrovalerone (MDPV). J Med Toxicol 8:69–75
74. Simmler LD, Rickli A, Schramm Y, Hoener MC, Liechti ME (2014) Pharmacological profiles of aminoindanes, piperazines, and pipradrol derivatives. Biochem Pharmacol 88:237–244
75. Bauer CT, Banks ML, Blough BE, Negus SS (2013) Use of intracranial self-stimulation to evaluate abuse-related and abuse-limiting effects of monoamine releasers in rats. Br J Pharmacol 168:850–862
76. Schindler CW, Thorndike EB, Goldberg SR, Lehner KR, Cozzi NV, Brandt SD et al (2015) Reinforcing and neurochemical effects of the "bath salts" constituents 3,4-methylenedioxypyrovalerone (MDPV) and 3,4-methylenedioxy-N-methylcathinone (methylone) in male rats. Psychopharmacology (Berl)
77. Wee S, Anderson KG, Baumann MH, Rothman RB, Blough BE, Woolverton WL (2005) Relationship between the serotonergic activity and reinforcing effects of a series of amphetamine analogs. J Pharmacol Exp Ther 313:848–854
78. Rothman RB, Baumann MH (2006) Balance between dopamine and serotonin release modulates behavioral effects of amphetamine-type drugs. Ann N Y Acad Sci 1074:245–260
79. Watterson LR, Kufahl PR, Nemirovsky NE, Sewalia K, Grabenauer M, Thomas BF et al (2014) Potent rewarding and reinforcing effects of the synthetic cathinone 3,4-methylenedioxypyrovalerone (MDPV). Addict Biol 19:165–174
80. Watterson LR, Olive MF (2014) Synthetic cathinones and their rewarding and reinforcing effects in rodents. Adv Neurosci 2014:209875
81. Sulzer D, Sonders MS, Poulsen NW, Galli A (2005) Mechanisms of neurotransmitter release by amphetamines: a review. Prog Neurobiol 75:406–433
82. Rudnick G, Wall SC (1992) The molecular mechanism of "ecstasy" [3,4-methylenedioxymethamphetamine (MDMA)]: serotonin transporters are targets for MDMA-induced serotonin release. Proc Natl Acad Sci U S A 89:1817–1821
83. Steinkellner T, Freissmuth M, Sitte HH, Montgomery T (2011) The ugly side of amphetamines: short- and long-term toxicity of 3,4-methylenedioxymethamphetamine (MDMA, 'Ecstasy'), methamphetamine and D-amphetamine. Biol Chem 392:103–115
84. Hill SL, Thomas SH (2011) Clinical toxicology of newer recreational drugs. Clin Toxicol (Phila) 49:705–719
85. Nichols DE (2004) Hallucinogens. Pharmacol Ther 101:131–181
86. Eshleman AJ, Forster MJ, Wolfrum KM, Johnson RA, Janowsky A, Gatch MB (2014) Behavioral and neurochemical pharmacology of six psychoactive substituted phenethylamines: mouse locomotion, rat drug discrimination and in vitro receptor and transporter binding and function. Psychopharmacology (Berl) 231:875–888
87. Caine SB, Thomsen M, Gabriel KI, Berkowitz JS, Gold LH, Koob GF et al (2007) Lack of self-administration of cocaine in dopamine D1 receptor knock-out mice. J Neurosci 27:13140–13150

88. Simmler LD, Buchy D, Chaboz S, Hoener MC, Liechti ME (2016) In vitro characterization of psychoactive substances at rat, mouse, and human trace amine-associated receptor 1. J Pharmacol Exp Ther. doi:10.1124/jpet.115.229765
89. Jing L, Li JX (2015) Trace amine-associated receptor 1: A promising target for the treatment of psychostimulant addiction. Eur J Pharmacol 761:345–352
90. Achat-Mendes C, Lynch LJ, Sullivan KA, Vallender EJ, Miller GM (2012) Augmentation of methamphetamine-induced behaviors in transgenic mice lacking the trace amine-associated receptor 1. Pharmacol Biochem Behav 101:201–207
91. Di Cara B, Maggio R, Aloisi G, Rivet JM, Lundius EG, Yoshitake T et al (2011) Genetic deletion of trace amine 1 receptors reveals their role in auto-inhibiting the actions of ecstasy (MDMA). J Neurosci 31:16928–16940

Electrophysiological Actions of Synthetic Cathinones on Monoamine Transporters

Ernesto Solis, Jr.

Abstract Products containing psychoactive synthetic cathinones, such as mephedrone and 3,4-methylenedioxypyrovalerone (MDPV) are prevalent in our society. Synthetic cathinones are structurally similar to methamphetamine, and numerous synthetics have biological activity at dopamine, serotonin, and norepinephrine transporters. Importantly, monoamine transporters co-transport sodium ions along with their substrate, and movement of substrates and ions through the transporter can generate measurable ionic currents. Here we review how electrophysiological information has enabled us to determine how synthetic cathinones affect transporter-mediated currents in cells that express these transporters. Specifically, drugs that act as transporter substrates induce inward depolarizing currents when cells are held near their resting membrane potential, whereas drugs that act as transporter blockers induce apparent outward currents by blocking an inherent inward leak current. We have employed the two-electrode voltage-clamp technique in *Xenopus laevis* oocytes overexpressing monoamine transporters to determine whether synthetic cathinones found in the so-called bath salts products behave as blockers or substrates. We also examined the structure–activity relationships for synthetic cathinone analogs related to the widely abused compound MDPV, a common constituent in "bath salts" possessing potent actions at the dopamine transporter.

Keywords Bath salts • Dopamine transporter • Serotonin transporter • Two-electrode voltage-clamp

---

E. Solis, Jr. (✉)
In Vivo Electrophysiology Unit, Behavioral Neuroscience Research Branch, National Institute on Drug Abuse – Intramural Research Program, National Institutes of Health, Triad Technology Center, 333 Cassell Drive, Suite 2200, Baltimore, MD 21224, USA
e-mail: ernesto.solisjr@nih.gov

Curr Topics Behav Neurosci (2017) 32: 73–92
DOI 10.1007/7854_2016_39
Published Online: 27 September 2016

Contents

1 Synthetic Cathinones Hit the Streets

Not long ago, legal amphetamine-related drugs suddenly emerged in Western Europe and the United States (USA). Many human drug users were sent to emergency rooms with severe neurological, psychiatric, and cardiovascular effects after being exposed to products marketed as "legal high," "bath salts," "insect repellant," "plant food," etc. that were purchased legally in convenience stores or on-line. To avoid regulatory control, these drugs were labeled "not for human consumption." American poison centers assessed drug content in blood and urine samples from patients who had ingested *bath salts* and discovered that the common ingredients were β-keto amphetamine (cathinone) derivatives (i.e., synthetic cathinones) [1]. In 2011, three synthetic cathinones commonly found in bath salts were identified and classified by the DEA as Schedule I controlled substances. Cathinone consumption can be traced back for at least hundreds of years to people ranging from South Africa to the Arabian Peninsula who have chewed on the plant *Catha edulis* (or *khat*) for its mild central stimulant effects [2, 3]. The main stimulant effects in *khat* derive from cathinone, a β-keto amphetamine (AMPH) analog (see Fig. 1 for common synthetic cathinone structures). The β-keto methamphetamine (METH) analog (called methcathinone or MCAT) was highly abused in the former Soviet Union and Eastern Europe dating back to the early 1980s. Clandestine chemists realized that by making simple chemical modifications to cathinone and MCAT, a vast number of completely legal "designer" synthetic cathinones could be synthesized that would elicit a range of behavioral effects in people. In 2009, synthetic cathinones surfaced as an abuse problem in the United Kingdom [4] where primarily the *para*-methyl analog of MCAT, 4-methylmethcathinone (4MMC or mephedrone) appeared. Shortly thereafter (as mentioned above), synthetic cathinones found their way to the USA disguised as home products with ulterior

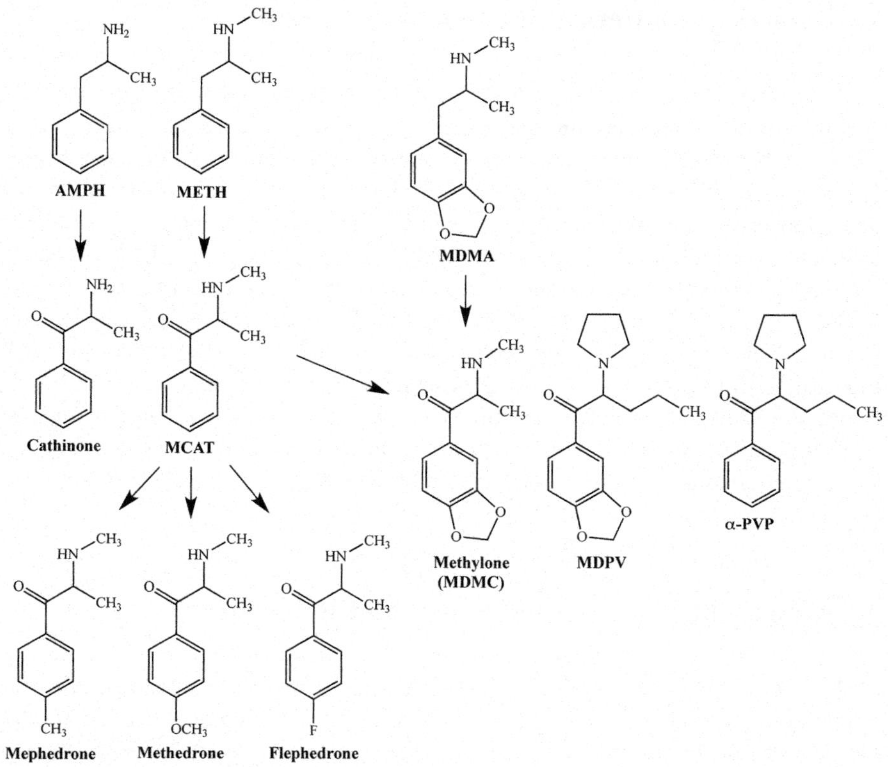

Fig. 1 Structural relationship between AMPH (and related drugs) and synthetic cathinones. Cathinone is the beta-keto analog of AMPH and similarly MCAT is the beta-keto analog of METH. Substitutions to the para position of MCAT yields additional synthetic analogs, including mephedrone, methedrone, and flephedrone. The beta-keto analog of MDMA is methylone (or MDMC). Further modifications to methylone results in the potent substituted cathinones MDPV and α-PVP. Abbreviations: AMPH, amphetamine; METH, methamphetamine; MCAT, methcathinone; MDMA, methylenedioxymethamphetamine; MDMC, methylenedioxymethcathinone; MDPV, methylenedioxypyrovalerone; α-pyrrolidinovalerophenone, α-PVP

uses. The most notorious product containing synthetic cathinones was *bath salts*, and the prevalent synthetic cathinones identified in *bath salts* were mephedrone, methylone (the β-keto analog of 3,4-methylenedioxymethamphetamine or MDMA), and a more complex synthetic cathinone called 3,4-methylenedioxypyrovalerone (MDPV), which possesses a pyrrolidine ring and was ubiquitous in *bath salts* concoctions taken by people who presented to the ER. Following drug scheduling of these synthetic cathinones by the DEA, an MDPV analog recently emerged called α-pyrrolidinovalerophenone (α-PVP or *flakka*), which has become an abuse problem in Florida and other states [5].

2 Neurotransmitters at the Synapse

In the central nervous system (CNS), the neurotransmitters norepinephrine (NE), dopamine (DA), and serotonin (5HT) are ordinarily released into the synaptic cleft via vesicular fusion in response to presynaptic depolarization. After release, neurotransmitters diffuse and activate postsynaptic and presynaptic neurons, then neurotransmission is terminated by reuptake of the transmitter into the presynaptic terminal via transporters, or in some cases hyperpolarization of the presynaptic terminal via neurotransmitter auto-receptors. The respective reuptake transporters for NE, DA, and 5HT (i.e., NET, DAT, and SERT) are located at perisynaptic sites [6, 7], whence monoamines are re-packaged into synaptic vesicles via vesicular monoamine transporters (VMATs). In particular, the vesicular monoamine transporter 2 (VMAT2), which is primarily found in the CNS, is responsible for neurotransmitter reuptake into synaptic vesicles that are poised for docking and release in dopaminergic, serotonergic, and noradrenergic neurons [8].

3 Monoamine Neurotransmitters and Behavior

5HT plays a role in regulating many behaviors, such as mood, sleep, appetite, temperature, sexual behavior, and aggression [9, 10]. Disturbances in the serotonergic system are implicated in mental diseases, including depression, bipolar disorder, autism, and a spectrum of psychiatric disorders, such as anorexia nervosa, bulimia, and obsessive-compulsive disorder (OCD) [11–14]. Behaviors that are regulated by DA include cognition, attention, working memory, motivation, and voluntary movement. Disturbances in the dopaminergic system have been implicated in Huntington's chorea, Parkinson's disease, schizophrenia, attention-deficit/hyperactivity-disorder (ADHD), depression, and addiction [15–18]. NE plays a role in attention, emotion, learning, and memory, and dysregulation of the noradrenergic system can lead to severe physiological effects [19]. In addition, dysfunction of the adrenergic system is linked to medical conditions, such as depression, post-traumatic stress disorder, and hypertension [20]. Behaviors modulated by monoamine neurotransmitter systems have been linked to synthetic cathinone use; for instance the components in *bath salts* can cause intense euphoria, alertness, increased concentration, heightened libido, as well as anorexia, anxiety, increased heart rate, and memory problems [21–23]. Whereas cardiovascular symptoms and stimulant effects caused by MDPV can be linked to the dopaminergic and adrenergic systems [23, 24], the emphathogenic symptoms induced by mephedrone and methylone are associated with the serotonergic system [22].

4 Therapeutic and Abused Drugs Target Monoamine Transporters

To treat medical conditions associated with disturbances in the serotoninergic system, several classes of drugs targeting monoamine transporters have been developed, such as tricyclic antidepressants (TCAs) and selective serotonin reuptake inhibitors (SSRIs) that inhibit reuptake of 5HT into presynaptic terminals and prolong neurotransmitter action at the synapse [25]. In addition to increasing the extracellular levels of 5HT in the brain, a TCA can exert effects as a NE reuptake inhibitor, an anticholinergic-antimuscarinic agent, an alpha1-adrenergic antagonist, an antihistamine, and a Na^+ channel inhibitor, which can potentially cause lethal cardiac arrhythmias and seizures [10]. The adverse side effects of TCAs have led to the development of SSRIs targeting SERT. Fluoxetine (FLX, Prozac) was the first drug of this kind approved as a therapeutic agent by regulatory authorities in the USA. Other SSRIs have been synthesized to lessen the adverse side effects of FLX; these include citalopram, escitalopram, fluvoxamine, and sertraline that lack adverse side effects, such as insomnia, anxiety, and tremors and display fewer gastrointestinal side effects, such as nausea, diarrhea, anorexia, and vomiting [26]. Presently, SSRIs are the most widely prescribed drugs for the treatment of depression, OCD, and bipolar disorder, as well as anxiety, anorexia, and panic disorders [27]. In addition, the recreational drug MDMA (i.e., ecstasy) has been effectively used to treat anxiety disorders, including post-traumatic stress disorder (PTSD) [28, 29]. MDMA targets all monoamine transporters but possesses greater potency for SERT. It is thought that MDMA increases monoamine transmitter levels in the brain by reverse transport via SERT, DAT, and NET of the corresponding endogenous transmitters [9].

The human dopamine transporter (hDAT) is a major molecular target for therapeutic agents, such as methylphenidate hydrochloride (MPH, Ritalin) and AMPH (Adderall). Both compounds, which are often prescribed to treat attention-deficit/hyperactivity disorder (ADHD), act directly on hDAT to increase extracellular DA levels, but they do so via different actions on hDAT. Whereas MPH is a DAT reuptake inhibitor, AMPH is a DAT substrate thought to stimulate DA release through non-vesicular reverse DA transport through DAT [30, 31]. Certain drugs of abuse, such as cocaine (COC) and METH, increase DA levels via the same mechanisms. Like MPH, COC inhibits DA transport and thus increases extracellular DA. On the other hand, METH behaves like AMPH and is transported by DAT to release DA by reversing DAT transport. Although abnormal increases in DA may underlie psychiatric disorders, drug abuse liability [32], and might cause adverse reactions such as psychosis, some of these compounds that elevate transmitter levels are effective treatments for certain mental illnesses [33]. In particular, since the psychostimulants AMPH and METH lead to the release of catecholamines (DA and NE) in the frontal lobe and limbic system (by transmitter reuptake inhibition at DAT and NET and transmitter efflux by DAT and NET), they have been used clinically to treat medical conditions such as ADHD and narcolepsy

[33, 34]. The improved potency to release DA through hDAT by the dextrorotary AMPH isomer (S(+)AMPH) over the levorotary isomer (R(−)AMPH) [35–37] underlies the therapeutic efficacy of agents composed primarily of S(+)AMPH. For example, Adderall is composed of 3:1 S(+):R(−)AMPH [38], and Vyvanse (lisdexamphetamine) is a pro-drug composed of S(+)AMPH conjugated to L-lysine, which is metabolized to S(+)AMPH [39, 40]. The dextrorotary isomer of METH is marketed as Desoxyn for the treatment of ADHD and narcolepsy [41]. Clinical manifestations associated with the abuse of AMPH, its precursors, or derivatives, such as phenethylamine or METH, are well documented [42–44]. In an attempt to bypass the reward system, the selective NE reuptake inhibitor (NRI) atomoxetine (Strattera) is used to treat ADHD [39]. Lastly, bupropion (Wellbutrin) is used to treat depression [31, 45] through its action as a dual DA and NE reuptake inhibitor [46].

5 Functional Mechanisms of Monoamine Transporters

5.1 Human Serotonin, Dopamine, and Norepinephrine Transporters

Biogenic amine transporters – including hDAT, human SERT (hSERT), and human NET (hNET) – are classified as Na^+/Cl^--coupled co-transporters since they require both Na^+ and Cl^- to transport substrates; however, the role of Cl^- as a transported ion is less established [47]. hSERT, hNET, and related proteins belonging to the SLC6 gene family that includes GABA, glycine, and taurine transporters are also termed neurotransmitter sodium symporters [48], reflecting the limitation of knowledge about the ionic contribution for substrate transport [49]. Co-transporters use existing ion gradients to concentrate their substrate against their own concentration gradient, e.g., Na^+ levels are ten times higher outside than inside cells [50–53]. Historically, alternating access models describe transport in which ions (Na^+ and Cl^-) and substrate (5HT or NE) bind to the transporter in its outward-facing conformation, catalyze an inward-facing conformational change, and transport the neurotransmitter from outside to inside the cell. In some cases a counter-ion, either a proton (H^+) or a K^+, is transported from inside to outside of the plasma membrane returning the transporter to the outward-facing conformation. This model is supported by biochemical and radiolabeled neurotransmitter uptake data [54, 55] and is consistent with recent structural data for co-transporters [56–61]. In particular, the hDAT is described by the alternating access model in which DA transport is coupled with fixed stoichiometry to the downhill movement of two Na^+ and one Cl^- coupling to each DA in the outward-facing hDAT conformation, and either a K^+ or H^+ binds to the inward-facing conformation to return hDAT to the outward-facing conformation [52, 53, 62–64].

5.2 Reverse Transport (Efflux) Mechanism of Monoamine Transporters

The reward and addiction properties of AMPH, METH, and MDMA rely on their ability to increase extracellular DA, NE, and 5HT levels by mechanisms as yet only partially understood. These agents competitively inhibit monoamine transporters, leading to diminished uptake and increased neurotransmitters levels at the synaptic cleft. Additionally, AMPH and related compounds may increase neurotransmitter levels via "reverse transport" or efflux [65]. DAT is the predominant transporter studied presumably due to its implications for addiction. The principal proposed mechanisms for AMPH-induced DAT-mediated DA efflux are: (1) facilitated exchange diffusion [34], (2) channel-in-transporter DA efflux model [66], (3) -oligomer-based counter-transport [67], and (4) vesicular depletion (or weak-base model), in which interaction of the releasing substrate (AMPH) with the vesicular monoamine transporter disrupts vesicular storage leading to an increase in free cytoplasmic transmitter levels [68, 69]. Regulation of DAT-mediated DA efflux includes protein kinase C (PKC)-activated DA efflux [70] and Ca^{++}/calmodulin-dependent protein kinase II (CaMKII) facilitated phosphorylation [71].

5.3 Transport-Associated Currents of Monoamine Transporters

Early biochemical and radiolabeled flux data led to the alternating access model for SERT, NET, and DAT transport; however, subsequent studies uncovered uncoupled currents and channel-like activity in transporters [47, 72–74]. Since the early 1990s, currents associated with substrate transport were found to be larger than alternating access models predicted [51, 64, 75–84]. Uncoupled currents in transporters are largely unexplained structurally and their function is speculative; one possibility is that these currents may depolarize or hyperpolarize neurons to a sufficient extent to produce changes in neuronal excitability [74, 85–89]. Most evidence for channels in transporters comes from heterologous expression systems; however, large 5HT-induced currents are generated in SERT at native serotonergic synapses [90, 91]. In two studies, Cl^- is reported to contribute to the ionic composition of DAT substrate-induced currents [85, 86]; however, Na^+ seems to be a major contributor to these DAT currents [92].

5.4 The Leak Current

Mager and colleagues established the existence of endogenous leak currents at monoamine transporters as revealed with use of transporter inhibitors [82]. For

SERT, studies that employed the two-electrode voltage-clamp (TEVC) technique in SERT-expressing *Xenopus laevis* oocytes demonstrated that a variety of inhibitors could uncover the SERT leak current. This response is seen as an outward current relative to baseline but is actually the inhibition of a constitutive inward current that is thought to be mediated primarily by Na^+. Compounds that helped uncover the SERT leak current include FLX [93], citalopram (both R and S isomers) [94], and the tricyclic antidepressants desipramine [95] and imipramine [96]. The majority of SERT inhibitors elicit long-lasting electrophysiological effects after their removal, a distinct action on SERT compared to natural substrates that are easily washed out. Wang and colleagues showed that exposure to FLX leads to a greatly diminished 5HT-induced hSERT-mediated current response, measured by the peak current time constant: "the time constant for 5HT-induced current became much greater than that for the first 5HT perfusion." In other words, after exposure to FLX, there is a much weaker inward current produced by 5HT (or other substrates) at hSERT as compared to an initial 5HT response at hSERT. Lastly, under physiologically relevant experimental concentrations, FLX-induced outward currents supersede inward 5HT-induced currents when 5HT and FLX are simultaneously applied – even at high 5HT concentrations. This result suggests FLX inhibits the endogenous leak current *and* disables substrate-induced currents at SERT. These results are consistent with other SERT inhibitors, including desipramine [95], imipramine [96], and paroxetine (unpublished data). Storustovu and colleagues demonstrate that applying either citalopram enantiomer (especially the S-isoform) during the 5HT-induced SERT current results in an outward current [94]. Cocaine and cocaine analogs also reveal leak currents in DAT-expressing, voltage-clamped *Xenopus laevis* oocytes [64]. However, the DAT-mediated outward current elicited by COC washes out more slowly than the substrate (DA)-induced inward current, which is attributed to its action as an inhibitor – rather than a substrate – at DAT. Inhibitors with much higher affinity, such as the cocaine analog (1R)-2beta-Carbomethoxy-3beta-(4-iodophenyl)tropane (β-CIT), are also more difficult to wash out, similar to FLX on SERT. The leak current has been observed in NET overexpressed in HEK cells by desipramine [80], but the technical limitation to overexpress NET in oocytes has precluded extensive NET research [20].

5.5 DAT and SERT Display an Induced Persistent Current

Recent studies have uncovered a novel mechanism of the action of AMPH on DAT based on electrophysiological data. In this model, AMPH is transported by DAT and concentrated inside the cell where the drug persists and is available to bind to the transporter at an internal site. The binding of AMPH at this internal site may maintain the transporter in a conductive state even when the external substrate is removed, leading to a persistent leak or "shelf" inward current; furthermore, it is proposed that external DA and other substrates can hold DAT in a constitutively active state once internal AMPH is present [92]. The induced persistent current can be elicited with additional select releasing substrates and in different monoamine transporters; in particular S(+)METH can produce a persistent leak current in hDAT, and para-chloroamphetamine can produce the same response in hSERT [97]. This mechanism could have consequences on synaptic transmission, as the persistent current would depolarize neurons long after exposure to the drug. More work needs to be done to address the importance and implications of this novel monoamine transporter mechanism [98].

6 Mechanism of Action of Synthetic Cathinones

6.1 Synthetic Cathinones Target Monoamine Transporters

Using different techniques, several groups have sought to understand the pharmacology and action of synthetic cathinones; particularly, ones found in *bath salts*. Since amphetamine and related drugs act on the monoamine transporters by inhibiting reuptake of endogenous neurotransmitters and by releasing endogenous transmitter, studies of structurally related synthetic cathinones are being carried out to determine their precise mechanisms of action. Experiments using rat brain synaptosomes pre-loaded with radiolabeled transmitter show that mephedrone exhibits similar releasing properties as MDMA at NET and DAT but weaker at SERT. Furthermore, methylone exhibits similar albeit weaker behavior [99]. In rat brain synaptosomes, MDPV was much stronger as an uptake inhibitor than AMPH, COC, or MEPH through DAT and NET, whereas MEPH and COC were more potent at inhibiting SERT uptake than MDPV or AMPH [100]. AMPH and MEPH, but not COC or MDPV, behave as releasers at DAT, NET, and SERT [100]. Further uptake inhibition assays in rat brain synaptosomes employing MDPV analogs confirmed that the α-alkyl chain is essential for hDAT affinity, whereas the methylenedioxy group does not affect affinity [101]. Substituting a trifluoromethyl on the 3 or 4 position of MCAT's ring increases the compound's selectivity towards SERT over NET and DAT [102]. Second generation MEPH analogs also elicit release through SERT and DAT [5]. Studies employing HEK293 cells expressing monoamine transporters confirmed MDPV is an uptake blocker without release properties at hDAT, hSERT, and hNET, whereas MEPH (along with methylone and 4-fluoromethcathinone) behave as uptake inhibitors and METH-like releasers at all three transporters, but with highest potency at hNET [103]. Another study employing a number of synthetic cathinones classified the compounds based on their actions on monoamine transporters, including compounds that: (1) exhibit relatively non-selective actions as uptake inhibitors and display "MDMA-like" 5HT release through SERT, (2) show preferential catecholamine transporter actions as DAT and NET uptake inhibitors and induce DA release (like METH), and (3) are potent and selective catecholamine transporter uptake inhibitors but do not induce release (MDPV) [104]. A similar study assessing synthetic cathinones on monoamine transporters showed that: (1) most of the compounds were more potent inhibitors of NET uptake, (2) addition of a β-keto group tends to enhance the DAT uptake inhibition over SERT uptake inhibition, (3) ring substitutions enhance serotonergic uptake inhibition, (4) some synthetic cathinones behave as pure uptake inhibitors, whereas others act as substrate releasers, and (5) there is weak binding of synthetic cathinones to a panel of receptors [105].

6.2 Electrophysiological Actions of Synthetic Cathinones on hDAT

The electrophysiological effects of a drug can provide a molecular signature of the specific interaction between a drug and the transporter. In a cell voltage-clamped to −60 mV, compounds that produce transporter-mediated inward currents are considered transported substrates (or releasers), and compounds that produce outward currents (interpreted as a block of an endogenous leak current) are considered as non-transported inhibitors. To illustrate this, currents from hDAT overexpressed in *Xenopus laevis* oocytes were recorded in response to DA, METH, the synthetic cathinones MEPH and MDPV, and COC (Fig. 2). DA, METH, and MEPH induced hDAT-mediated inward currents, in agreement with previous studies that determined that METH and MEPH act as releasers [99, 100]. On the other hand, MDPV

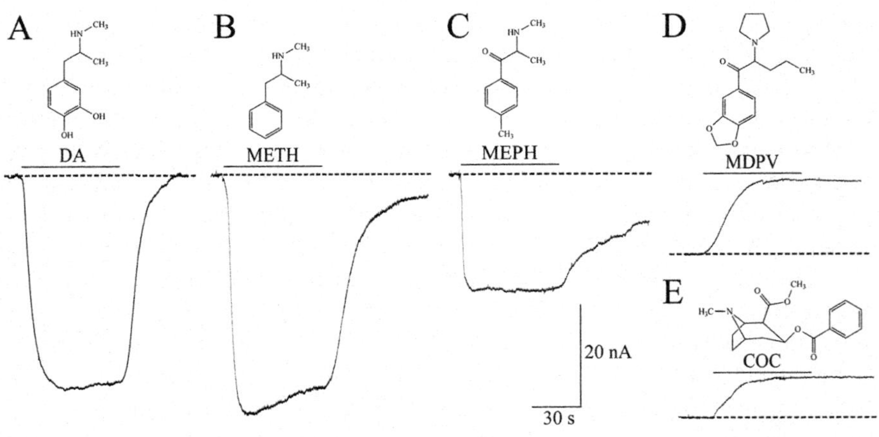

Fig. 2 Electrophysiological signature of DA, METH, MEPH, MDPV, and cocaine at hDAT. (**a–e**) By employing the two-electrode voltage-clamp (TEVC) technique, currents through hDAT elicited by external drug application (60 s duration, 10 μM) are measured in hDAT-expressing *Xenopus laevis* oocytes voltage clamped to −60 mV. (**a**) DA induces a large inward peak current that returns to baseline when DA is removed. (**b**) METH and (**c**) MEPH elicit inward peak currents and induced persistent currents (enhanced current in the absence of drug). (**d**) MDPV exposure produces an outward current similar to the current induced by (**e**) COC, which is a known reuptake inhibitor. The responses induced by MDPV and COC reveal the presence of an endogenous inward leak current typically uncovered with inhibitors. Figure adapted from [106]

and COC possess the signature of a non-transported blocker at hDAT, which qualitatively is seen as an upward deflection (Fig. 2) [106]. The mixture of MEPH and MDPV commonly found in *bath salts*, in combination with these findings, indicate that *bath salts* may contain a DA releasing agent and a DA reuptake inhibitor. The two drugs have different kinetics and rather than cancel each other they would exacerbate the effect of either drug taken alone. Further recordings showed that MDPV produces a long-lasting effect at hDAT, that is, washout fails to return the outward current to baseline in contrast to COC, which is more easily washed out [106]. In congruence, MDPV proved to be 10–35 times more potent than COC as an uptake inhibitor for DAT [100, 106].

6.3 Electrophysiological Actions of Synthetic Cathinones on hSERT

There are limited studies on the electrophysiological actions of synthetic cathinones on hSERT, which require more elaboration. Mephedrone analogs (second-generation cathinones) are stimulants that induce euphoria and elicit inward currents [5]. The METH analog, para-methoxymethamphetamine (PMMA), which induces behavioral effects similar to MDMA without stimulant effects, and the β-keto analog methedrone, found in *bath salts*, display potent effects at SERT and NET [105]. In particular, methedrone is the synthetic cathinone with the highest selectivity for SERT, and it behaves as a substrate that induces monoamine efflux [105]. In agreement, employing the TEVC technique in voltage-clamped (−60 mV) oocytes overexpressing hSERT show that the S(+) isomer of PMMA or methedrone elicit inward currents through hSERT comparable to the 5HT-induced inward current (Fig. 3). This is the signature of substrates (or releasers) and, interestingly, methedrone produces a large persistent inward current after washout, which confirms its potent effect at hSERT.

6.4 Structural Determinants for Potency of MDPV on hDAT

Deconstruction of MDPV into analogs allowed the determination of the moieties in MDPV responsible for its potency. In hDAT-expressing *Xenopus laevis* oocytes clamped to −60 mV, MDPV and its analogs induced comparable outward currents (block of an endogenous inward leak) that, after drug washout, did not return to the baseline before drug application. Furthermore, DA-induced currents obtained following application of either MDPV or its analogs displayed amplitude recovery profiles relative to the initial DA-induced currents (see Fig. 4) that highly correlated with the compounds' potency to inhibit DA uptake via hDAT. In this study, the combination of uptake inhibition assays and an electrophysiological protocol

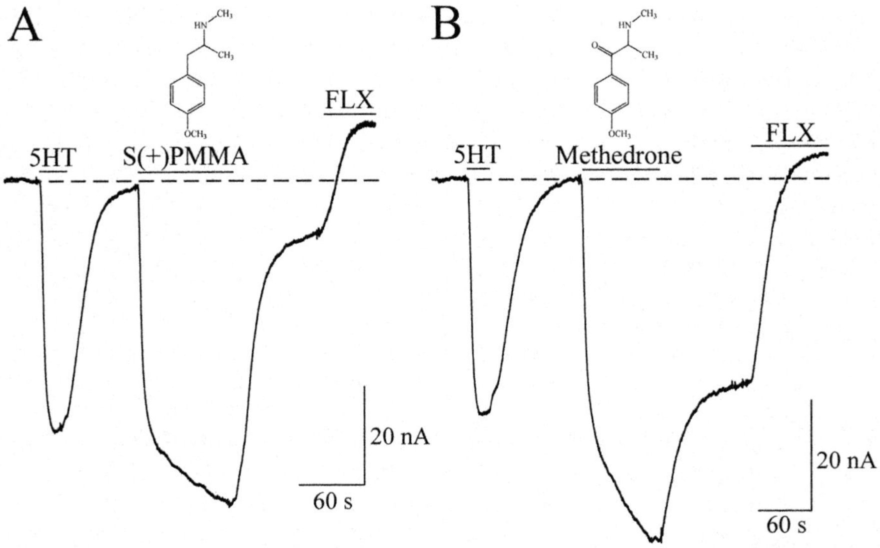

Fig. 3 Electrophysiological signature of S(+)PMMA and the synthetic cathinone methedrone at hSERT. By utilizing the TEVC technique currents are measured in *Xenopus laevis* oocytes overexpressing hSERT in response to 5 μM 5HT followed by 10 μM of either the S(+) isomer of para-methoxy-N-methylamphetamine (S(+)PMMA) (**a**) or methedrone (**b**). While the 5HT-induced hSERT response returns to baseline after washout, the washout following exposure to S(+)PMMA or methedrone for 100 s results in a persistent inward current. The hSERT inhibitor fluoxetine (FLX) reveals the endogenous leak current

revealed the major contributor for MDPV's potency at hDAT to be the extended α-alkyl group, followed by the carbonyl group and a tertiary amine, whereas the methylenedioxy group made a minimal contribution [107].

6.5 A Distinct Site for Action of MDPV Analogs on hDAT

All compounds used in the Kolanos et al. study of the electrophysiological effects of MDPV analogs at hDAT shifted the baseline to more positive values, which is attributed to the block of the leak current (see Fig. 4). Interestingly, for the MDPV analogs 1-(benzo[d][1,3]dioxol-5-yl)-2-(dimethylamino)propan-1-one (bk-MDDMA) (Fig. 4g) and 2-amino-1-(benzo[d]-1,3-dioxol-5-yl)pentan-1-one (ABDP) (Fig. 4h), the second DA response recovered to 100 % of the first DA response relative to the new baseline. The baseline shift following 10 μM of either bk-MDDMA or ABDP application cannot be washed out and is unaffected by a second exposure to DA. In subsequent recordings from hDAT oocytes, bk-MDDMA was perfused and removed while extracellular DA was present (Fig. 5a); however, a shift in baseline still occurred. Similarly, ABDP produces a shift in baseline even in the presence of constant extracellular DA (Fig. 5c). The shift in baseline produced by these analogs was not

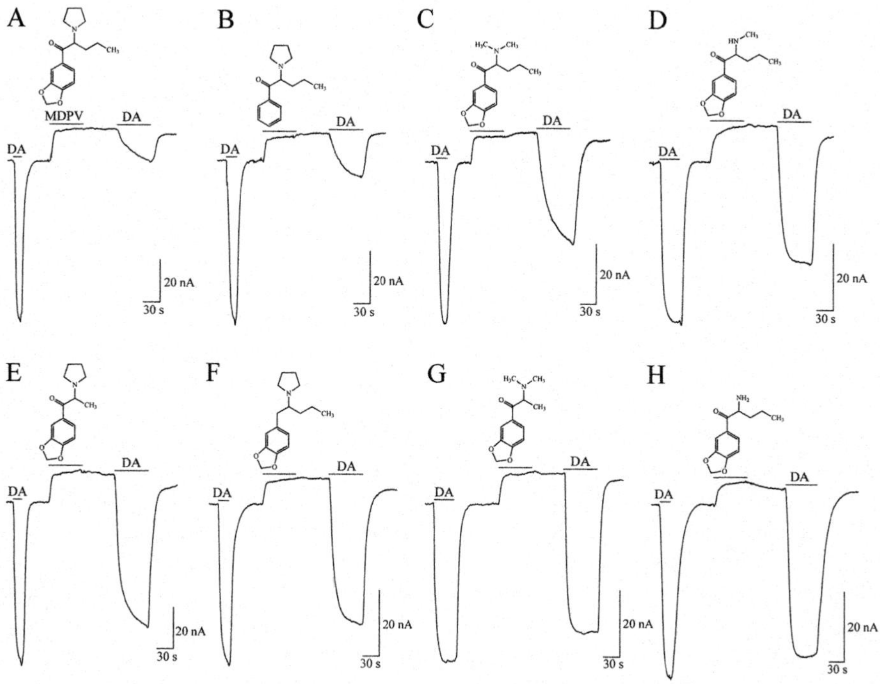

Fig. 4 Currents induced by MDPV and its analogs in voltage-clamped (−60 mV) *Xenopus laevis* oocytes expressing hDAT. (**a–h**) Initial exposure to DA (5 μM) yields an hDAT-mediated inward current. Subsequent exposure to MDPV (**a**) or any of its analogs (**b–h**) (10 μM, 1 min) produces the typical response associated with the block of the endogenous current at hDAT. The upward deflection does not return to baseline when any of the compounds are washed out for 1 min. (**a**) Following exposure to MDPV a 5 μM DA application induces a diminished hDAT-mediated inward current (as compared to the current produced in response to the initial DA exposure). The protocol is repeated for the MDPV analogs (**b–H**). Exposure to the different MDPV analogs elicits variable DA current recovery (compare second DA exposure to first DA exposure). For example, in contrast to the diminished DA-induced hDAT-mediated inward current following MDPV exposure (**a**), after exposing hDAT to the last two compounds (**g–h**) application of DA results in large inward currents that fully recover to the level of the current elicited by the first DA exposure. Figure adapted from [107]

impeded by the presence of a high concentration of dopamine (data not shown). At hSERT, bk-MDDMA elicits a weak, reversible block of the inward leak current (Fig. 5b), and ABDP produces an inward, substrate-like current and a persistent inward current (Fig. 5d). These results suggest two distinct sites of action for MDPV analogs targeting hDAT, and the distinct effect of the compounds at hSERT differentiates these compounds at the two transporters. There are only a few reports of a secondary site of action on monoamine transporters [108], but none employing electrophysiology. This secondary site of action of drugs targeting hDAT could have important implications for drug development to treat addiction and other disorders.

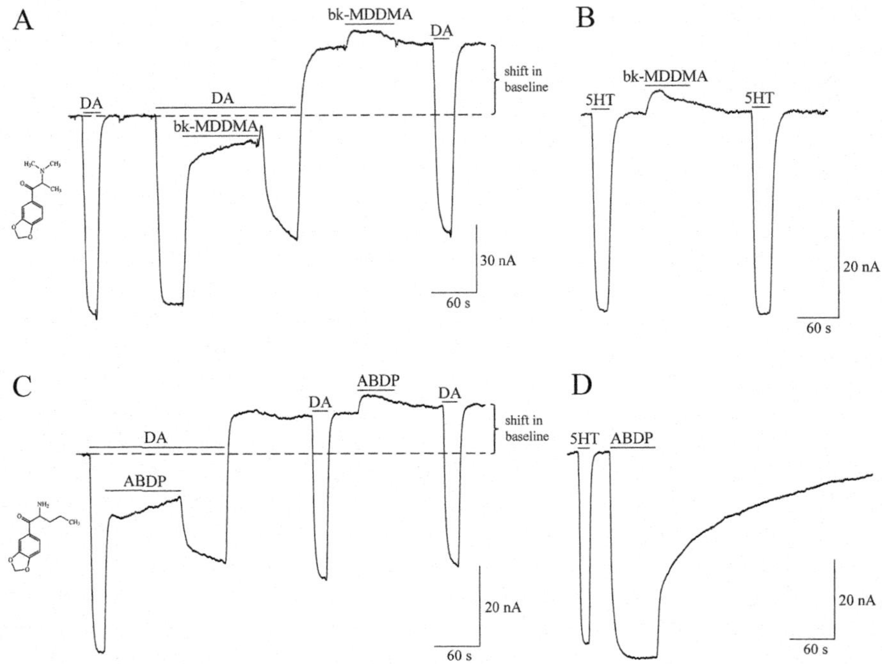

Fig. 5 Currents induced by MDPV analogs at hDAT and hSERT in voltage-clamped (−60 mV) *Xenopus laevis* oocytes. (**a**) Application of bk-MDDMA (20 μM) during constant perfusion of DA counteracts the inward DA-induced current. After removal of bk-MDDMA, a DA-induced inward current is observed; however, after DA is washed out, the current goes above baseline (above *dashed line*). A second exposure to bk-MDDMA induces a small outward current that returns to the level the baseline was shifted to. A subsequent DA-induced hDAT current is similar to the initial DA response, and after the last DA application is washed out, the baseline remains shifted. (**b**) At hSERT, bk-MDDMA (20 μM) induces a small block of the endogenous leak current, but in contrast to what happens at hDAT, after washing out bk-MDDMA the holding current at hSERT returns to its original level. (**c**) Application of ABDP shifts the baseline even in the presence of DA. ABDP application (20 μM) is applied to hDAT in the presence of DA, which elicits a counteracting hDAT-mediated current. After removal of ABDP application, the DA present induces an hDAT-mediated inward current; however, after DA is washed out, the current goes above baseline. A second exposure to ABDP application induces a small outward current. A subsequent DA-induced hDAT current is similar to the initial DA response, and after the last DA application is washed out, the baseline remains shifted. (**d**) ABDP (20 μM) induces an hSERT-mediated inward current that does not return to baseline. Note: All DA and 5HT challenges are 5 μM

7 Conclusion

Electrophysiological methods can characterize the actions of drugs on the monoamine transporters, determine whether they act as releasers or inhibitors, can quantify potency and efficacy, and evaluate structure–activity relationships of new compounds. Correlative studies with neurotransmitter uptake/release and fluorescent microscopy will enhance our understanding of drug action. Lastly, structure–function analysis of monoamine transporter protein structures as they become available can be combined with functional information to uncover the molecular mechanisms underlying drug–transporter interactions.

Acknowledgments I would like to acknowledge Louis J. De Felice and Michael H. Baumann for valuable input in the writing of the chapter. The work described in this chapter was supported by NIH/NIDA R01DA033930 and R01DA033930-S2.

References

1. Spiller HA, Ryan ML, Weston RG, Jansen J (2011) Clinical experience with and analytical confirmation of "bath salts" and "legal highs" (synthetic cathinones) in the United States. Clin Toxicol (Phila) 49:499–505
2. De Felice LJ, Glennon RA, Negus SS (2014) Synthetic cathinones: chemical phylogeny, physiology, and neuropharmacology. Life Sci 97:20–26
3. Schechter MD, Glennon RA (1985) Cathinone, cocaine and methamphetamine: similarity of behavioral effects. Pharmacol Biochem Behav 22:913–916
4. Iversen L, White M, Treble R (2014) Designer psychostimulants: pharmacology and differences. Neuropharmacology 87:59–65
5. Saha K, Partilla JS Lehner KR, Seddik A, Stockner T, Holy M, Sandtner W, Ecker GF, Sitte HH, Baumann MH (2015) 'Second-generation' mephedrone analogs, 4-MEC and 4-MePPP, differentially affect monoamine transporter function. Neuropsychopharmacology 40:1321–1331
6. Hoffman BJ, Hansson SR, Mezey E, Palkovits M (1998) Localization and dynamic regulation of biogenic amine transporters in the mammalian central nervous system. Front Neuroendocrinol 19:187–231
7. Tao-Cheng JH, Zhou FC (1999) Differential polarization of serotonin transporters in axons versus soma-dendrites: an immunogold electron microscopy study. Neuroscience 94:821–830
8. Wimalasena K (2011) Vesicular monoamine transporters: structure-function, pharmacology, and medicinal chemistry. Med Res Rev 31:483–519
9. Schloss P, Williams DC (1998) The serotonin transporter: a primary target for antidepressant drugs. J Psychopharmacol 12:115–121
10. Stahl SM (1998) Basic psychopharmacology of antidepressants, part 1: antidepressants have seven distinct mechanisms of action. J Clin Psychiatry 59(Suppl 4):5–14
11. Coppen A, Shaw DM, Herzberg B, Maggs R (1967) Tryptophan in the treatment of depression. Lancet 2:1178–1180
12. Feighner JP (1994) Clinical effects of serotonin reuptake inhibitors--a review. Fortschr Neurol Psychiatr 62(Suppl 1):9–15
13. Vaswani M, Kalra H (2004) Selective serotonin re-uptake inhibitors in anorexia nervosa. Expert Opin Investig Drugs 13:349–357
14. Vaswani M, Linda FK, Ramesh S (2003) Role of selective serotonin reuptake inhibitors in psychiatric disorders: a comprehensive review. Prog Neuropsychopharmacol Biol Psychiatry 27:85–102
15. Barbeau A (1970) Dopamine and disease. Can Med Assoc J 103:824–832
16. Gainetdinov RR (2008) Dopamine transporter mutant mice in experimental neuropharmacology. Naunyn Schmiedebergs Arch Pharmacol 377:301–313

17. Gainetdinov RR, Caron MG (2003) Monoamine transporters: from genes to behavior. Annu Rev Pharmacol Toxicol 43:261–284
18. Javitch JA, Snyder SH (1984) Uptake of MPP(+) by dopamine neurons explains selectivity of parkinsonism-inducing neurotoxin, MPTP. Eur J Pharmacol 106:455–456
19. Foote SL, Aston-Jones G, Bloom FE (1980) Impulse activity of locus coeruleus neurons in awake rats and monkeys is a function of sensory stimulation and arousal. Proc Natl Acad Sci U S A 77:3033–3037
20. Schwartz JW, Piston D, DeFelice LJ (2006) Molecular microfluorometry: converting arbitrary fluorescence units into absolute molecular concentrations to study binding kinetics and stoichiometry in transporters. Handb Exp Pharmacol (175):23–57
21. Banks ML, Worst TJ, Rusyniak DE, Sprague JE (2014) Synthetic cathinones ("bath salts"). J Emerg Med 46:632–642
22. German CL, Fleckenstein AE, Hanson GR (2014) Bath salts and synthetic cathinones: an emerging designer drug phenomenon. Life Sci 97:2–8
23. Miotto K, Striebel J, Cho AK, Wang C (2013) Clinical and pharmacological aspects of bath salt use: a review of the literature and case reports. Drug Alcohol Depend 132:1–12
24. Valente MJ, Guedes de Pinho P, de Lourdes Bastos M, Carvalho F, Carvalho M (2014) Khat and synthetic cathinones: a review. Arch Toxicol 88:15–45
25. White KJ, Walline CC, Barker EL (2005) Serotonin transporters: implications for antidepressant drug development. AAPS J 7:E421–E433
26. Sghendo L, Mifsud J (2012) Understanding the molecular pharmacology of the serotonergic system: using fluoxetine as a model. J Pharm Pharmacol 64:317–325
27. Stahl SM (1998) Mechanism of action of serotonin selective reuptake inhibitors. Serotonin receptors and pathways mediate therapeutic effects and side effects. J Affect Disord 51:215–235
28. Johansen PO, Krebs TS (2009) How could MDMA (ecstasy) help anxiety disorders? A neurobiological rationale. J Psychopharmacol 23:389–391
29. Mithoefer MC, Wagner MT, Mithoefer AT, Jerome L, Doblin R (2011) The safety and efficacy of {+/-}3,4-methylenedioxymethamphetamine-assisted psychotherapy in subjects with chronic, treatment-resistant posttraumatic stress disorder: the first randomized controlled pilot study. J Psychopharmacol 25:439–452
30. Wall SC, Gu H, Rudnick G (1995) Biogenic amine flux mediated by cloned transporters stably expressed in cultured cell lines: amphetamine specificity for inhibition and efflux. Mol Pharmacol 47:544–550
31. Wu X, Gu HH (1999) Molecular cloning of the mouse dopamine transporter and pharmacological comparison with the human homologue. Gene 233:163–170
32. Henry LK, Blakely RD (2008) Distinctions between dopamine transporter antagonists could be just around the bend. Mol Pharmacol 73:616–618
33. Fleckenstein AE, Volz TJ, Riddle EL, Gibb JW, Hanson GR (2007) New insights into the mechanism of action of amphetamines. Annu Rev Pharmacol Toxicol 47:681–698
34. Burnette WB, Bailey MD, Kukoyi S, Blakely RD, Trowbridge CG, Justice JB Jr (1996) Human norepinephrine transporter kinetics using rotating disk electrode voltammetry. Anal Chem 68:2932–2938
35. Holmes JC, Rutledge CO (1976) Effects of the d- and l-isomers of amphetamine on uptake, release and catabolism of norepinephrine, dopamine and 5-hydroxytryptamine in several regions of rat brain. Biochem Pharmacol 25:447–451
36. Kuczenski R, Segal DS, Cho AK, Melega W (1995) Hippocampus norepinephrine, caudate dopamine and serotonin, and behavioral responses to the stereoisomers of amphetamine and methamphetamine. J Neurosci 15:1308–1317
37. Phillips AG, Brooke SM, Fibiger HC (1975) Effects of amphetamine isomers and neuroleptics on self-stimulation from the nucleus accumbens and dorsal noradrenergic bundle. Brain Res 85:13–22

38. Cody JT, Valtier S, Nelson SL (2003) Amphetamine enantiomer excretion profile following administration of Adderall. J Anal Toxicol 27:485–492
39. Heal DJ, Cheetham SC, Smith SL (2009) The neuropharmacology of ADHD drugs in vivo: insights on efficacy and safety. Neuropharmacology 57:608–618
40. Najib J (2009) The efficacy and safety profile of lisdexamfetamine dimesylate, a prodrug of d-amphetamine, for the treatment of attention-deficit/hyperactivity disorder in children and adults. Clin Ther 31:142–176
41. Mendelson J, Uemura N, Harris D, Nath RP, Fernandez E, Jacob P 3rd, Everhart ET, Jones RT (2006) Human pharmacology of the methamphetamine stereoisomers. Clin Pharmacol Ther 80:403–420
42. Potkin SG, Karoum F, Chuang LW, Cannon-Spoor HE, Phillips I, Wyatt RJ (1979) Phenylethylamine in paranoid chronic schizophrenia. Science 206:470–471
43. Romanelli F, Smith KM (2006) Clinical effects and management of methamphetamine abuse. Pharmacotherapy 26:1148–1156
44. Winslow BT, Voorhees KI, Pehl KA (2007) Methamphetamine abuse. Am Fam Physician 76:1169–1174
45. Mazei-Robinson MS, Blakely RD (2006) ADHD and the dopamine transporter: are there reasons to pay attention? Handb Exp Pharmacol (175):373–415
46. Arias HR (2009) Is the inhibition of nicotinic acetylcholine receptors by bupropion involved in its clinical actions? Int J Biochem Cell Biol 41:2098–2108
47. De Felice LJ (2016) Chloride requirement for monoamine transporters. Pflugers Arch 468:503–511
48. Singh SK (2008) LeuT: A prokaryotic stepping stone on the way to a eukaryotic neurotransmitter transporter structure. Channels (Austin) 2
49. Ramamoorthy S, Bauman AL, Moore KR, Han H, Yang-Feng T, Chang AS, Ganapathy V, Blakely RD (1993) Antidepressant- and cocaine-sensitive human serotonin transporter: molecular cloning, expression, and chromosomal localization. Proc Natl Acad Sci U S A 90:2542–2546
50. DeFelice LJ, Blakely RD (1996) Pore models for transporters? Biophys J 70:579–580
51. Galli A, Blakely RD, DeFelice LJ (1996) Norepinephrine transporters have channel modes of conduction. Proc Natl Acad Sci U S A 93:8671–8676
52. Rudnick G (1998) Bioenergetics of neurotransmitter transport. J Bioenerg Biomembr 30:173–185
53. Rudnick G (1998) Ion-coupled neurotransmitter transport: thermodynamic vs. kinetic determinations of stoichiometry. Methods Enzymol 296:233–247
54. Naftalin RJ (1984) The thermostatics and thermodynamics of cotransport. Biochim Biophys Acta 778:155–175
55. Stein WD, Lieb WR (1986) Transport and diffusion across cell membranes. Academic, Orlando
56. Abramson J, Smirnova I, Kasho V, Verner G, Iwata S, Kaback HR (2003) The lactose permease of Escherichia coli: overall structure, the sugar-binding site and the alternating access model for transport. FEBS Lett 555:96–101
57. Abramson J, Smirnova I, Kasho V, Verner G, Kaback HR, Iwata S (2003) Structure and mechanism of the lactose permease of Escherichia coli. Science 301:610–615
58. Penmatsa A, Wang KH, Gouaux E (2015) X-ray structures of Drosophila dopamine transporter in complex with nisoxetine and reboxetine. Nat Struct Mol Biol 22:506–508
59. Wang KH, Penmatsa A, Gouaux E (2015) Neurotransmitter and psychostimulant recognition by the dopamine transporter. Nature 521:322–327
60. Yamashita A, Singh SK, Kawate T, Jin Y, Gouaux E (2005) Crystal structure of a bacterial homologue of Na+/Cl--dependent neurotransmitter transporters. Nature 437:215–223
61. Yernool D, Boudker O, Jin Y, Gouaux E (2004) Structure of a glutamate transporter homologue from Pyrococcus horikoshii. Nature 431:811–818

62. Gu H, Wall SC, Rudnick G (1994) Stable expression of biogenic amine transporters reveals differences in inhibitor sensitivity, kinetics, and ion dependence. J Biol Chem 269:7124–7130
63. McElvain JS, Schenk JO (1992) A multisubstrate mechanism of striatal dopamine uptake and its inhibition by cocaine. Biochem Pharmacol 43:2189–2199
64. Sonders MS, Zhu SJ, Zahniser NR, Kavanaugh MP, Amara SG (1997) Multiple ionic conductances of the human dopamine transporter: the actions of dopamine and psychostimulants. J Neurosci 17:960–974
65. Khoshbouei H, Wang H, Lechleiter JD, Javitch JA, Galli A (2003) Amphetamine-induced dopamine efflux. A voltage-sensitive and intracellular Na+-dependent mechanism. J Biol Chem 278:12070–12077
66. Kahlig KM, Binda F, Khoshbouei H, Blakely RD, McMahon DG, Javitch JA, Galli A (2005) Amphetamine induces dopamine efflux through a dopamine transporter channel. Proc Natl Acad Sci U S A 102:3495–3500
67. Seidel S, Singer EA, Just H, Farhan H, Scholze P, Kudlacek O, Holy M, Koppatz K, Krivanek P, Freissmuth M, Sitte HH (2005) Amphetamines take two to tango: an oligomer-based counter-transport model of neurotransmitter transport explores the amphetamine action. Mol Pharmacol 67:140–151
68. Sulzer D, Chen TK, Lau YY, Kristensen H, Rayport S, Ewing A (1995) Amphetamine redistributes dopamine from synaptic vesicles to the cytosol and promotes reverse transport. J Neurosci 15:4102–4108
69. Sulzer D, Maidment NT, Rayport S (1993) Amphetamine and other weak bases act to promote reverse transport of dopamine in ventral midbrain neurons. J Neurochem 60:527–535
70. Khoshbouei H, Sen N, Guptaroy B, Johnson L, Lund D, Gnegy ME, Galli A, Javitch JA (2004) N-terminal phosphorylation of the dopamine transporter is required for amphetamine-induced efflux. PLoS Biol 2, E78
71. Fog JU, Khoshbouei H, Holy M, Owens WA, Vaegter CB, Sen N, Nikandrova Y, Bowton E, McMahon DG, Colbran RJ, Daws LC, Sitte HH, Javitch JA, Galli A, Gether U (2006) Calmodulin kinase II interacts with the dopamine transporter C terminus to regulate amphetamine-induced reverse transport. Neuron 51:417–429
72. DeFelice LJ (2004) Going against the flow. Nature 432:279
73. DeFelice LJ, Goswami T (2007) Transporters as channels. Annu Rev Physiol 69:87–112
74. Quick MW (2003) Regulating the conducting states of a mammalian serotonin transporter. Neuron 40:537–549
75. Adams SV, DeFelice LJ (2002) Flux coupling in the human serotonin transporter. Biophys J 83:3268–3282
76. Adams SV, DeFelice LJ (2003) Ionic currents in the human serotonin transporter reveal inconsistencies in the alternating access hypothesis. Biophys J 85:1548–1559
77. DeFelice LJ, Galli A (1998) Electrophysiological analysis of transporter function. Adv Pharmacol 42:186–190
78. DeFelice LJ, Galli A (1998) Fluctuation analysis of norepinephrine and serotonin transporter currents. Methods Enzymol 296:578–593
79. Galli A, Blakely RD, DeFelice LJ (1998) Patch-clamp and amperometric recordings from norepinephrine transporters: channel activity and voltage-dependent uptake. Proc Natl Acad Sci U S A 95:13260–13265
80. Galli A, DeFelice LJ, Duke BJ, Moore KR, Blakely RD (1995) Sodium-dependent norepinephrine-induced currents in norepinephrine-transporter-transfected HEK-293 cells blocked by cocaine and antidepressants. J Exp Biol 198:2197–2212
81. Li C, Zhong H, Wang Y, Wang H, Yang Z, Zheng Y, Liu K, Liu Y (2006) Voltage and ionic regulation of human serotonin transporter in Xenopus oocytes. Clin Exp Pharmacol Physiol 33:1088–1092

82. Mager S, Min C, Henry DJ, Chavkin C, Hoffman BJ, Davidson N, Lester HA (1994) Conducting states of a mammalian serotonin transporter. Neuron 12:845–859
83. Petersen CI, DeFelice LJ (1999) Ionic interactions in the Drosophila serotonin transporter identify it as a serotonin channel. Nat Neurosci 2:605–610
84. Ramsey IS, DeFelice LJ (2002) Serotonin transporter function and pharmacology are sensitive to expression level: evidence for an endogenous regulatory factor. J Biol Chem 277:14475–14482
85. Carvelli L, McDonald PW, Blakely RD, Defelice LJ (2004) Dopamine transporters depolarize neurons by a channel mechanism. Proc Natl Acad Sci U S A 101:16046–16051
86. Ingram SL, Prasad BM, Amara SG (2002) Dopamine transporter-mediated conductances increase excitability of midbrain dopamine neurons. Nat Neurosci 5:971–978
87. Quick MW (2002) Role of syntaxin 1A on serotonin transporter expression in developing thalamocortical neurons. Int J Dev Neurosci 20:219–224
88. Ryan RM, Mindell JA (2007) The uncoupled chloride conductance of a bacterial glutamate transporter homolog. Nat Struct Mol Biol 14:365–371
89. Sonders MS, Amara SG (1996) Channels in transporters. Curr Opin Neurobiol 6:294–302
90. Bruns D (1998) Serotonin transport in cultured leech neurons. Methods Enzymol 296:593–607
91. Bruns D, Engert F, Lux HD (1993) A fast activating presynaptic reuptake current during serotonergic transmission in identified neurons of Hirudo. Neuron 10:559–572
92. Rodriguez-Menchaca AA, Solis E Jr, Cameron K, De Felice LJ (2012) S(+)amphetamine induces a persistent leak in the human dopamine transporter: molecular stent hypothesis. Br J Pharmacol 165:2749–2757
93. Wang HW, Li CZ, Yang ZF, Zheng YQ, Zhang Y, Liu YM (2006) Electrophysiological effect of fluoxetine on Xenopus oocytes heterologously expressing human serotonin transporter. Acta Pharmacol Sin 27:289–293
94. Storustovu S, Sanchez C, Porzgen P, Brennum LT, Larsen AK, Pulis M, Ebert B (2004) R-citalopram functionally antagonises escitalopram in vivo and in vitro: evidence for kinetic interaction at the serotonin transporter. Br J Pharmacol 142:172–180
95. Lin F, Lester HA, Mager S (1996) Single-channel currents produced by the serotonin transporter and analysis of a mutation affecting ion permeation. Biophys J 71:3126–3135
96. Barker EL, Moore KR, Rakhshan F, Blakely RD (1999) Transmembrane domain I contributes to the permeation pathway for serotonin and ions in the serotonin transporter. J Neurosci 19:4705–4717
97. Sandtner W, Schmid D, Schicker K, Gerstbrein K, Koenig X, Mayer FP, Boehm S, Freissmuth M, Sitte HH (2014) A quantitative model of amphetamine action on the 5-HT transporter. Br J Pharmacol 171:1007–1018
98. De Felice LJ, Cameron KN (2015) Comments on 'A quantitative model of amphetamine action on the serotonin transporter', by Sandtner et al., Br J Pharmacol 171: 1007–1018. Br J Pharmacol 172:4772–4774
99. Baumann MH, Ayestas MA Jr, Partilla JS, Sink JR, Shulgin AT, Daley PF, Brandt SD, Rothman RB, Ruoho AE, Cozzi NV (2012) The designer methcathinone analogs, mephedrone and methylone, are substrates for monoamine transporters in brain tissue. Neuropsychopharmacology 37:1192–1203
100. Baumann MH, Partilla JS, Lehner KR, Thorndike EB, Hoffman AF, Holy M, Rothman RB, Goldberg SR, Lupica CR, Sitte HH, Brandt SD, Tella SR, Cozzi NV, Schindler CW (2013) Powerful cocaine-like actions of 3,4-methylenedioxypyrovalerone (MDPV), a principal constituent of psychoactive 'bath salts' products. Neuropsychopharmacology 38:552–562
101. Marusich JA, Antonazzo KR, Wiley JL, Blough BE, Partilla JS, Baumann MH (2014) Pharmacology of novel synthetic stimulants structurally related to the "bath salts" constituent 3,4-methylenedioxypyrovalerone (MDPV). Neuropharmacology 87:206–213

102. Cozzi NV, Brandt SD, Daley PF, Partilla JS, Rothman RB, Tulzer A, Sitte HH, Baumann MH (2013) Pharmacological examination of trifluoromethyl ring-substituted methcathinone analogs. Eur J Pharmacol 699:180–187
103. Eshleman AJ, Wolfrum KM, Hatfield MG, Johnson RA, Murphy KV, Janowsky A (2013) Substituted methcathinones differ in transporter and receptor interactions. Biochem Pharmacol 85:1803–1815
104. Simmler LD, Buser TA, Donzelli M, Schramm Y, Dieu LH, Huwyler J, Chaboz S, Hoener MC, Liechti ME (2013) Pharmacological characterization of designer cathinones in vitro. Br J Pharmacol 168:458–470
105. Simmler LD, Rickli A, Hoener MC, Liechti ME (2014) Monoamine transporter and receptor interaction profiles of a new series of designer cathinones. Neuropharmacology 79:152–160
106. Cameron KN, Kolanos R, Solis E Jr, Glennon RA, De Felice LJ (2013) Bath salts components mephedrone and methylenedioxypyrovalerone (MDPV) act synergistically at the human dopamine transporter. Br J Pharmacol 168:1750–1757
107. Kolanos R, Solis E Jr, Sakloth F, De Felice LJ, Glennon RA (2013) "Deconstruction" of the abused synthetic cathinone methylenedioxypyrovalerone (MDPV) and an examination of effects at the human dopamine transporter. ACS Chem Neurosci 4:1524–1529
108. Bulling S, Schicker K, Zhang YW, Steinkellner T, Stockner T, Gruber CW, Boehm S, Freissmuth M, Rudnick G, Sitte HH, Sandtner W (2012) The mechanistic basis for noncompetitive ibogaine inhibition of serotonin and dopamine transporters. J Biol Chem 287:18524–18534

Neuropharmacology of 3,4-Methylenedioxypyrovalerone (MDPV), Its Metabolites, and Related Analogs

Michael H. Baumann, Mohammad O. Bukhari, Kurt R. Lehner, Sebastien Anizan, Kenner C. Rice, Marta Concheiro, and Marilyn A. Huestis

Abstract 3,4-Methylenedioxypyrovalerone (MDPV) is a psychoactive component of so-called bath salts products that has caused serious medical consequences in humans. In this chapter, we review the neuropharmacology of MDPV and related analogs, and supplement the discussion with new results from our preclinical experiments. MDPV acts as a potent uptake inhibitor at plasma membrane transporters for dopamine (DAT) and norepinephrine (NET) in nervous tissue. The MDPV formulation in bath salts is a racemic mixture, and the *S* isomer is much more potent than the *R* isomer at blocking DAT and producing abuse-related effects. Elevations in brain extracellular dopamine produced by MDPV are likely to underlie its locomotor stimulant and addictive properties. MDPV displays rapid pharmacokinetics when injected into rats (0.5–2.0 mg/kg), with peak plasma concentrations achieved by 10–20 min and declining quickly thereafter. MDPV is metabolized to 3,4-dihydroxypyrovalerone (3,4-catechol-PV) and 4-hydroxy-3-

M.H. Baumann (✉), M.O. Bukhari, and K.R. Lehner
Designer Drug Research Unit of the Intramural Research Program, National Institute on Drug Abuse (NIDA), National Institutes of Health, Baltimore, MD, USA
e-mail: mbaumann@mail.nih.gov

S. Anizan and M.A. Huestis
Chemistry and Drug Metabolism Section of the Intramural Research Program, National Institute on Drug Abuse (NIDA), National Institutes of Health, Baltimore, MD, USA

K.C. Rice
Drug Design and Synthesis Section of the Intramural Research Program, National Institute on Drug Abuse (NIDA), National Institutes of Health, Baltimore, MD, USA

M. Concheiro
Chemistry and Drug Metabolism Section of the Intramural Research Program, National Institute on Drug Abuse (NIDA), National Institutes of Health, Baltimore, MD, USA

Department of Sciences, John Jay College of Criminal Justice, City University of New York, New York, NY, USA

Curr Topics Behav Neurosci (2017) 32: 93–118
DOI 10.1007/7854_2016_53
Published Online: 8 November 2016

methoxypyrovalerone (4-OH-3-MeO-PV) in vivo, but motor activation produced by the drug is positively correlated with plasma concentrations of parent drug and not its metabolites. 3,4-Catechol-PV is a potent uptake blocker at DAT in vitro but has little activity after administration in vivo. 4-OH-3-MeO-PV is the main MDPV metabolite but is weak at DAT and NET. MDPV analogs, such as α-pyrrolidinovalerophenone (α-PVP), display similar ability to inhibit DAT and increase extracellular dopamine concentrations. Taken together, these findings demonstrate that MDPV and its analogs represent a unique class of transporter inhibitors with a high propensity for abuse and addiction.

Keywords Addiction • Dopamine • MDPV • Pyrrolidinophenones • Synthetic cathinones • Transporter • Uptake • α-PVP

Contents

1 Introduction

Drug abuse and addiction are persistent public health concerns, and an alarming new trend is the increased non-medical use of so-called designer drugs, legal highs, or research chemicals [1] [2, this volume], [3]. These drugs, collectively known as "new psychoactive substances" (NPS), are synthetic alternatives to more traditional illegal drugs of abuse. At the present time, there are popular NPS which mimic the effects of most types of abused drugs, including stimulants (e.g., "bath salts"), cannabinoids (e.g., "spice"), and hallucinogens (e.g., "NBOMes"). Most are manufactured by Asian laboratories and sold to consumers via the Internet or shipped to locations in Europe, the United States America (US), and elsewhere to be packaged for retail sale [4, 5]. NPS are marketed as non-drug products, given innocuous names, and labeled "not for human consumption" as a means to avoid legal scrutiny. Compared to traditional drugs of abuse, NPS are cheap, easy to obtain, and often not detectable by standard toxicology screens. As governments pass laws to ban specific NPS, clandestine chemists respond by quickly creating novel "replacement" analogs to stay one step ahead of law enforcement [6, 7]. The abuse of NPS is a global phenomenon fueled by information freely available on the

Internet. Recent data from the United Nations indicate that 540 different NPS have been identified worldwide as of 2014, and this number is expected to rise [8].

The first stimulant-like NPS to appear in the US were found in the so-called bath salts products which flooded the recreational drug marketplace beginning in late 2010 [9]. By early 2011, there was a dramatic spike in reports of bath salts intoxications to poison control centers, and an influx of patients admitted to emergency departments with toxic exposures [10–12]. Bath salts consist of powders or crystals that are administered intra-nasally or orally to produce their psychoactive effects. Low doses of bath salts induce typical psychomotor stimulant effects such as increased energy and mood elevation, but high doses or binge use can cause severe symptoms including hallucinations, psychosis, increased heart rate, high blood pressure and hyperthermia, often accompanied by combative or violent behaviors [9, 13]. The most serious syndrome induced by bath salts is known as "excited delirium," a constellation of symptoms including elevated body temperature, delirium, agitation, breakdown of muscle tissue, and kidney failure, sometimes culminating in death [14, 15]. Forensic analysis of bath salts products in 2010 and 2011 identified three main synthetic compounds: 4-methyl-*N*-methylcathinone (mephedrone), 3,4-methylenedioxy-*N*-methylcathinone (methylone), and 3,4-methylenedioxypyrovalerone (MDPV) (Spiller et al. 2011; [7, 12]). These compounds are chemically similar to the naturally occurring substance cathinone, an amphetamine-like stimulant found in the khat plant, *Catha edulis*. Legislation passed in 2013 placed mephedrone, methylone, and MDPV into permanent Schedule I control, making the drugs illegal in the US [16]. Figure 1 depicts the chemical structures of bath salts cathinones compared to the related compounds amphetamine and cathinone.

Although a number of different cathinones are found in bath salts products (e.g., [7, 12, 17]), MDPV appears more apt to cause life-threatening medical consequences (see [18]). For example, in the first study of patients reported to US poison control centers for "bath salts" overdose, the majority of subjects with blood and urine toxicology data were positive for MDPV but not mephedrone or methylone [12]. A more recent interrogation of a US clinical toxicology database found that all patients with confirmed synthetic cathinone exposure tested positive for MDPV [19]. Perhaps more importantly, MDPV was found in blood and urine from many fatal cases of drug overdose in the US and Europe [12, 14, 20–22]. Collectively, the clinical case data point to MDPV as the chief culprit in causing serious medical consequences. Given the widespread popularity of MDPV and the risks associated with its use, the purpose of the present chapter is to describe the neuropharmacology of MDPV, its metabolites, and related analogs. We review the literature on this topic and supplement the discussion with new data from preclinical experiments carried out at the Intramural Research Program (IRP) of the National Institute on Drug Abuse (NIDA).

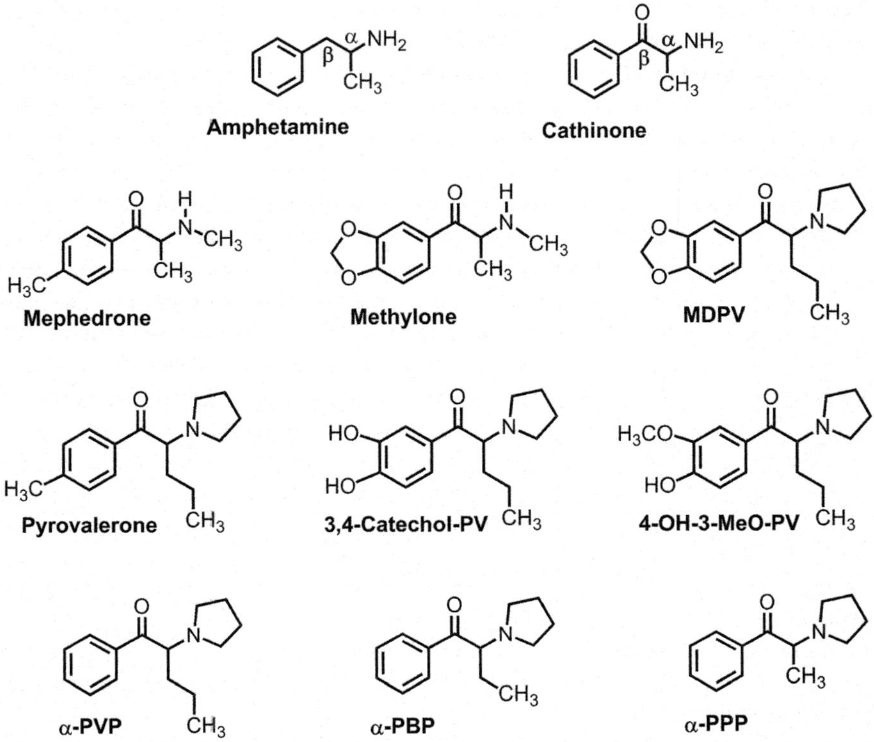

Fig. 1 Chemical structures of MDPV and related pyrrolidinophenones, and their relationship to amphetamine and cathinone

2 Pharmacology of MDPV and Its Stereoisomers

2.1 Stimulant Drugs and Monoamine Transporters

As noted above, the psychoactive constituents of bath salts are chemically related to the parent compound cathinone, the β-keto analog of amphetamine (see Fig. 1 for structures). Mephedrone and methylone have functional groups attached to the phenyl ring and are considered ring-substituted cathinones, whereas MDPV is structurally more complex with a bulky nitrogen-containing pyrrolidine ring and a flexible alkyl chain extending from the α-carbon. MDPV and related compounds containing a pyrrolidine ring are collectively known as pyrrolidinophenones. Like other stimulant drugs, bath salts cathinones exert their effects by binding to transporter proteins on the surface of nerve cells that synthesize the monoamine neurotransmitters dopamine, norepinephrine, and serotonin (5-HT) [23] [24, this volume]. In order to understand the precise mechanism of action for cathinone analogs at the molecular level, it is essential to first consider the physiological role of monoamine transporters and the types of drugs targeting these proteins.

Under normal circumstances, the solute carrier 6 (SLC6) transporters for dopamine (DAT), norepinephrine (NET), and serotonin (SERT) are responsible for translocating previously released neurotransmitter molecules from the extracellular medium back into the neuronal cytoplasm, a process known as neurotransmitter "uptake" [25]. Transporter-mediated uptake is the principal mechanism for terminating the action of monoamine neurotransmitters, so drugs targeting these transporter proteins can have profound effects on cell-to-cell monoamine signaling. Accordingly, monoamine transporters are the principal sites of action for medications used to treat a range of psychiatric diseases such as depression, anxiety, and attention-deficit hyperactivity disorder [26, 27]. Drugs which preferentially interact at SERT are widely prescribed as efficacious treatments for major depression and anxiety disorders. By contrast, drugs which preferentially act at DAT and NET, like amphetamine and methamphetamine, have powerful psychomotor stimulant and addictive properties [28, 29].

Drugs that bind to monoamine transporters can be divided into two types based on their precise molecular mechanisms of action: (1) cocaine-like "inhibitors" – which bind to the neurotransmitter binding site on the transporter (i.e., orthosteric site), thereby blocking uptake of neurotransmitters from the extracellular medium, and (2) amphetamine-like "substrates" – which also bind to the orthosteric site, but are subsequently translocated through the transporter channel into the neuronal cytoplasm and trigger efflux of intracellular neurotransmitter molecules (i.e., transporter-mediated release) [30, 31]. Drugs that act as transporter substrates are often called "releasers" because they induce non-exocytotic transporter-mediated neurotransmitter release from neurons. Irrespective of molecular mechanism, all drugs which bind to transporters can dramatically increase extracellular concentrations of monoamines in vivo, amplifying cell-to-cell chemical signaling in various brain circuits. It is important to distinguish between transporter inhibitors versus substrates because substrates display a number of unique properties: they are translocated into cells along with sodium ions, they induce inward depolarizing sodium currents, and they reverse the normal direction of transporter flux to trigger non-exocytotic release of neurotransmitters (i.e., reverse transport) [30, 31]. Finally, because transporter substrate-type drugs are accumulated into the neuronal cytoplasm, they can produce intracellular deficits in monoamine neurons such as inhibition of neurotransmitter synthesis and disruption of vesicular storage, leading to long-term neurotransmitter depletions [32, this volume] [33] [34].

In our laboratory, we developed in vitro functional assays to assess the ability of test drugs to act as inhibitors or substrates at DAT, NET, and SERT [35, 36]. We employ two types of assays: (1) uptake inhibition and (2) release. The assays are carried out in synaptosomes derived from rat brain tissue and are designed to rapidly assess potency and efficacy of drugs at all three transporters under similar conditions. Synaptosomes consist of sealed vesicle-filled nerve endings with their plasma membrane leaflets oriented in a manner akin to neurons in vivo. For the uptake inhibition assays, radiolabeled substrate (i.e., [^{3}H]neurotransmitter) and test drug are co-incubated with synaptosomes for a brief period, and the reaction is stopped by vacuum filtration. If test drugs are transporter inhibitors, the accumulation of [^{3}H]neurotransmitter into the synaptosomes (i.e., uptake) is blocked because the test drug and neurotransmitter compete for the same orthosteric binding site on the transporter protein. It is noteworthy that uptake inhibition assays cannot distinguish between inhibitors and substrates because both types of drugs will effectively reduce the accumulation of [^{3}H]neurotransmitter into synaptosomes.

In order to definitively identify substrate-type drugs, we use release assays. For the release assays, synaptosomes are first incubated with radiolabeled substrate molecules in order to fill or "preload" the interior of the synaptosomes. [^{3}H]1-Methyl-4-phenylpyridinium ([^{3}H]MPP$^+$) is used as the radiolabeled substrate for DAT and NET release assays, whereas [^{3}H]5-HT is used for SERT release assays. Once synaptosomes are preloaded, test drug is added for a brief incubation period, and the reaction is stopped by vacuum filtration. If test drugs are transporter substrates, efflux of [^{3}H]MPP$^+$ or [^{3}H]5-HT out of the synaptosomes is induced (i.e., release) by reversal of the normal direction of transporter flux. Drugs that act as pure transporter inhibitors will not evoke substantial release of [^{3}H]MPP$^+$ or [^{3}H]5-HT from preloaded synaptosomes. Therefore, by testing drugs in the combined uptake inhibition and release assay procedures, the precise molecular mechanism of drug action can be ascertained.

2.2 *Molecular Mechanisms of Action*

Prior to the bath salts phenomenon in 2010–2011, few scientific investigations had examined the pharmacology of ring-substituted cathinones or pyrrolidinophenones. Studies from the 1980s demonstrated that cathinone and methcathinone release dopamine from rat brain tissue by an amphetamine-like mechanism [37, 38], and subsequent reports revealed methcathinone acts as a substrate for DAT, NET, and SERT [36, 39]. Cozzi et al. [39] first demonstrated that methylone acts as an uptake inhibitor at monoamine transporters [39], while other investigations showed the drug is a transporter substrate capable of releasing dopamine, norepinephrine, and 5-HT from rat brain tissue [40]. Studies from the 1990s revealed that pyrovalerone, a structural analog of MDPV (see Fig. 1), is a potent dopamine uptake blocker which produces psychomotor stimulant effects when administered to rodents [41, 42]. A comprehensive study by Meltzer et al. [43] examined the monoamine transporter activities for several pyrovalerone analogs and showed these agents are potent inhibitors of DAT and NET with minimal activity at SERT [43]. Importantly, the study of Meltzer and colleagues did not address the possibility of whether pyrovalerone analogs might act as transporter substrates, and no assessment of MDPV activity was included.

Hadlock et al. [44] carried out the first detailed investigation of mephedrone pharmacology, and found the drug inhibits dopamine uptake and stimulates dopamine release from rat brain synaptosomes [44]. López-Arnau et al. [45] reported that mephedrone and methylone both inhibit uptake at DAT and SERT, but no transporter release data were reported in their study [45]. Our laboratory extended the findings of Lopez-Arnau and colleagues by showing that mephedrone and methylone act as non-selective transporter substrates that evoke release of [^{3}H]MPP$^+$ from DAT and NET, and release of [^{3}H]5-HT from SERT [46]. The non-selective substrate activity of mephedrone and methylone at monoamine transporters is similar to the molecular mechanism of action for the club drug 3,4-methylenedioxy-*N*-methylamphetamine (MDMA). In assay systems using human transporters expressed in human embryonic kidney (HEK) cells, mephedrone and methylone act as substrates for DAT, NET, and SERT [47, 48], consistent with the findings in synaptosomes. Taken together, results from studies using rat and human transporters agree that ring-substituted cathinones like mephedrone and methylone are non-selective transporter substrates capable of inducing transmitter release via DAT, NET, and SERT.

We examined the in vitro transporter activity of MDPV in rat brain synaptosomes and showed the drug displays potent uptake inhibition at DAT ($IC_{50} = 4.1$ nM) and NET ($IC_{50} = 26$ nM), with much weaker activity at SERT ($IC_{50} = 3{,}349$ nM) [49]. Table 1 summarizes the uptake inhibition potencies at DAT, NET, and SERT for MDPV and a number of other stimulant drugs discussed in this chapter. The in vitro results with MDPV agree with prior data of Meltzer et al. (2006) showing that pyrovalerone analogs are potent inhibitors of DAT and NET. When compared to the prototypical uptake inhibitor cocaine, MDPV is 50-fold more potent as an inhibitor at DAT, tenfold more potent at NET and tenfold

Table 1 Effects of MDPV and related analogs on the uptake of [^{3}H]neurotransmitters at DAT, NET, and SERT in rat brain synaptosomes

Test drug	DAT uptake inhibition IC_{50} (nM)	NET uptake inhibition IC_{50} (nM)	SERT uptake inhibition IC_{50} (nM)	DAT/SERT ratio
Cocaine	211 ± 19	292 ± 34	313 ± 17	1.48
Amphetamine[a]	93 ± 17	67 ± 16	3,418 ± 314	36.75
Mephedrone[a]	762 ± 79	487 ± 66	422 ± 26	0.55
Methylone[a]	1,323 ± 133	1,031 ± 162	1,017 ± 59	0.77
MDPV	4.1 ± 0.5	26 ± 8	3,349 ± 305	816.82
S-MDPV	2.1 ± 0.2	9.8 ± 1.0	n.d.	–
R-MDPV	382 ± 53	726 ± 150	n.d.	–
3,4-Catechol-PV	11 ± 1	11 ± 1	>10,000	>900
4-OH-3-MeO-PV	784 ± 87	407 ± 43	>10,000	>12
α-PVP	12 ± 1	14 ± 1	>10,000	>833
α-PBP	63 ± 6	92 ± 13	>10,000	>159
α-PPP	197 ± 10	445 ± 39	>10,000	>50

Values are mean ± SD for $N = 3$ experiments each repeated in triplicate. IC_{50} indicates drug concentration at which uptake is inhibited to 50 percent of control uptake

DAT/SERT ratio is $(\text{DAT } IC_{50})^{-1}/(\text{SERT } IC_{50})^{-1}$; higher values indicate greater DAT selectivity

Data are taken from [49–51]

[a]These test drugs act as substrates for monoamine transporters and are included for comparison

less potent at SERT. We found that MDPV does not act as a substrate for monoamine transporters, probably because the drug molecule is sterically too bulky to fit through the transporter channel. In an informative structure-activity study, Kolanos et al. [52] "deconstructed" the MDPV molecule piece-by-piece to determine which structural features govern activity at DAT. They found that the bulky pyrrolidine ring and the flexible α-carbon chain are critical attributes for potent uptake inhibition at DAT, whereas the 3,4-methylenedioxy ring moiety is of little consequence in this regard.

In mouse striatal slices, MDPV is a potent and efficacious inhibitor of DAT-mediated dopamine clearance (i.e., dopamine uptake) as measured by fast-scan cyclic voltammetry [49]. In assays using HEK cells expressing human transporters, Eshleman et al. [47] and Simmler et al. [48] confirmed that MDPV is a potent inhibitor at DAT and NET, but not SERT, and the drug does not evoke transporter-mediated release. These same investigators examined the potency of MDPV at various G protein-coupled receptor subtypes and found no significant affinity of the drug for non-transporter sites of action [47, 48]. Cameron et al. [53] provided definitive evidence that MDPV is not a substrate at DAT by comparing the electrophysiological effects of mephedrone and MDPV in *Xenopus* oocytes expressing human DAT [53]. They found that mephedrone induces a DAT-mediated inward depolarizing current, consistent with the action of a transportable substrate, whereas MDPV does not produce this effect. In fact, MDPV induces a DAT-mediated outward hyperpolarizing current due to the inhibition of an inward "leak" current. Overall, the in vitro findings from a variety of different assay methods in native tissues and transporter-expressing cells indicate that MDPV is a potent inhibitor at DAT and NET, which lacks significant activity at SERT and non-transporter sites of action.

The formulation of MDPV available in the recreational drug marketplace is a racemic mixture of *S* and *R* isomers, which poses a logical question about whether these isomers have stereoselective biological effects. Meltzer et al. [43] showed that *S*-pyrovalerone is much more potent as an inhibitor at DAT and NET when compared to *R*-pyrovalerone, suggesting MDPV isomers might exhibit a similar degree of transporter selectivity. Kolanos et al. [50] reported the stereoselective synthesis of MDPV enantiomers using *S*- and *R*-norvaline as starting materials [50], whereas Suzuki et al. [54] resolved MDPV enantiomers from the racemic mixture [54]. In the study of Kolanos et al. [50], *S*-MDPV was 100-times more potent at inhibiting DAT when compared to *R*-MDPV (see Table 1). Therefore, similar to the findings reported for pyrovalerone, the biological activity of racemic MDPV resides primarily with the *S* isomer. In agreement with the in vitro transporter results, *S*-MDPV is much more potent than *R*-MDPV in eliciting locomotor stimulant and reinforcing effects in both rats and mice [50, 55].

2.3 In Vivo Pharmacological Effects

Drugs which act as inhibitors or substrates at DAT, NET, and SERT increase the extracellular concentrations of dopamine, norepinephrine, and 5-HT in the brain to enhance monoamine signaling [28, 29]. In our laboratory, we developed in vivo methods to simultaneously examine the neurochemical and behavioral effects of transporter ligands in rats [56, 57]. Specifically, we use in vivo microdialysis perfusion to collect samples of extracellular fluid (i.e., dialysate samples) from the brains of conscious freely behaving rats. The microdialysis probes are placed into the nucleus accumbens, a brain region implicated in the locomotor stimulant and reinforcing effects of abused drugs [58, 59], and dialysate samples are analyzed for concentrations of dopamine and 5-HT using high-performance liquid chromatography coupled to electrochemical detection (HPLC-ECD). Rats undergoing microdialysis are housed in chambers equipped with photo-beam arrays sensitive to locomotor activity in the horizontal plane (i.e., ambulation) and repetitive back-and-forth movements of the head, trunk, and limbs (i.e., stereotypy). Our methods allow for the assessment of relationships between extracellular monoamines and behavior. For example, in previous studies, we found a significant positive correlation between the amount of dialysate dopamine in the nucleus accumbens and the extent of locomotor activation produced by stimulant drugs [56, 60]. Furthermore, data reveal that elevations in dialysate 5-HT alone are not sufficient to produce locomotor activation [61], but elevations in extracellular 5-HT can dampen the motor stimulant effects mediated by concurrent elevations in extracellular dopamine [56, 57].

Kehr et al. [62] first reported that subcutaneous (s.c.) administration of mephedrone to rats evokes concurrent elevations in extracellular dopamine and 5-HT in the nucleus accumbens [62], and other research groups confirmed these findings in rats receiving s.c. or intraperitoneal (i.p.) mephedrone injections [63, 64]. Intravenous (i.v.) administration of mephedrone or methylone produces dose-related increases in extracellular dopamine and 5-HT in rat nucleus accumbens, with mephedrone slightly more potent than methylone [46]. Interestingly, all microdialysis studies with mephedrone and methylone have found that the magnitude of increase in dialysate 5-HT exceeds the accompanying increase in dialysate dopamine. The profile of in vivo neurochemical effects produced by mephedrone and methylone is consistent with the substrate activity of these drugs at DAT and SERT, and mimics the known neurochemical effects of MDMA [46, 62, 65]. We reported that i.v. MDPV administration to rats produces dose-related increases in extracellular dopamine but not 5-HT, and MDPV is tenfold more potent than cocaine in its ability to increase dialysate dopamine [49, 66]. The selective rise in extracellular dopamine produced by MDPV is consistent with the potent inhibition of dopamine uptake produced by the drug in vitro. Figure 2 depicts unpublished data showing the effects of MDPV administration on extracellular dopamine and 5-HT, along with concurrent measures of ambulation. In these experiments, rats undergoing in vivo microdialysis in the nucleus accumbens were housed in

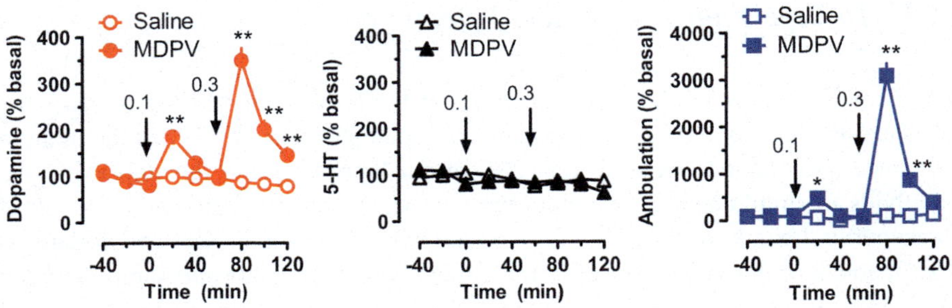

Fig. 2 Neurochemical and behavioral effects of MDPV in male Sprague–Dawley rats undergoing microdialysis in nucleus accumbens. Rats received i.v. injection of 0.1 mg/kg at time zero, followed by 0.3 mg/kg 60 min later. Extracellular concentrations of monoamine transmitters (dopamine, 5-HT) and forward locomotion (ambulation) are expressed as % basal, determined from three time points prior to injection. Data are mean ± SEM, for $N = 6$–7 rats/group. *$p < 0.05$, **$p < 0.01$ compared to saline control at a given time point

chambers equipped with photobeams to allow for measurement of locomotor behaviors. After three baseline dialysate samples were obtained, rats received i.v. injection of 0.1 mg/kg MDPV at time zero, followed by 0.3 mg/kg 60 min later. Dialysate samples were collected at 20 min intervals before, during, and after drug injections. Data were analyzed by two-way (drug × time) ANOVA followed by Bonferroni post-hoc tests. The results show that MDPV produces significant dose-related increases in extracellular dopamine ($F_{1,8} = 157.3$, $p < 0.0001$), but not 5-HT ($F_{1,8} = 1.6$, NS), along with a parallel increases in ambulation ($F_{1,8} = 198.7$, $p < 0.0001$).

The behavioral effects produced by MDPV have been recently reviewed [67], enabling brief consideration here, focusing on locomotor activity and drug self-administration studies. All of the synthetic cathinones examined thus far are known to stimulate locomotor activity when administered to rats [46, 68, 69] or mice [70–72]. In a representative study, Marusich et al. [72] showed that mephedrone, methylone, and MDPV produce dose-dependent increases in ambulation in mice, but MDPV is much more potent in this regard. We found that MDPV is about tenfold more potent than cocaine as a locomotor stimulant in rats, and MDPV is also more efficacious than cocaine, stimulating an overall greater magnitude of motor activation [49]. When MDPV is administered across a broad range of doses, the dose–response relationship for ambulation is an inverted U-shaped curve [68, 71]; the reduction in forward locomotion at higher MDPV doses is due to the emergence of focused stereotypies, such as in-place perseverative sniffing and head bobbing, as dose increases. In mice, the locomotor stimulation produced by MDPV is reduced by pretreatment with the dopamine D1 receptor antagonist SCH23390 [51]. Taken together with the microdialysis data, the available evidence indicates that MDPV elevates extracellular dopamine in critical brain circuits via DAT inhibition, and subsequent activation of D1 dopamine receptors by endogenous dopamine is responsible for locomotor stimulant effects of the drug.

The role of extracellular 5-HT in modulating the dopaminergic effects of synthetic cathinones is a topic of great interest. To this end, a recent investigation compared the neurochemical and locomotor effects of MDPV and methylone in rats [66]. It was found that i.v. doses of 0.3 mg/kg MDPV and 3.0 mg/kg methylone produce nearly identical threefold elevations in extracellular dopamine, whereas only methylone produces a dramatic tenfold elevation in extracellular 5-HT. At these same doses, MDPV elicits a much greater stimulation of ambulation and stereotypy when compared to methylone. One interpretation of these findings is that elevations in extracellular 5-HT tend to reduce locomotor stimulant effects mediated by extracellular dopamine. Indeed, substantial evidence indicates that high-affinity 5-HT_{2C} receptor sites in the brain provide a strong inhibitory influence over dopamine-mediated behavioral effects [73]. Thus, MDPV's powerful locomotor effects could be related to its potent DAT inhibition, coupled with its lack of activity at SERT and failure to increase extracellular 5-HT.

Drug self-administration is considered the "gold standard" behavioral test for determining the addictive potential of drugs, as most drugs self-administered by laboratory animals are abused by humans [74, this volume] [75]. In the rat drug self-administration paradigm, animals with surgically implanted i.v. catheters are trained to lever-press or nose-poke to obtain i.v. drug injections which are delivered via a computer-controlled infusion pump. A number of studies have shown that rats will self-administer mephedrone [44, 76, 77] and methylone [78–80], indicating these drugs have abuse liability. With regard to MDPV, Aarde et al. [68] reported that MDPV is readily self-administered by rats at i.v. training doses ranging from 0.01 to 0.5 mg/kg, and the drug is more potent and efficacious than methamphetamine, a known stimulant drug of abuse. Watterson et al. [82] found similar results with rats self-administering MDPV, and also showed the amount of drug administered displays robust escalation if rats are allowed prolonged access to the drug. Schindler et al. [66] directly compared the acquisition of i.v. self-administration behavior for MDPV (0.05 mg/kg) and methylone (0.5 mg/kg) in rats. It was found that MDPV self-administration is rapidly acquired within the first few days of training, whereas methylone self-administration takes much longer to develop. Additionally, the number of infusions per session is significantly greater for MDPV when compared to methylone. Based on the neurochemical effects of MDPV and methylone already mentioned, it is tempting to speculate that serotonergic effects of methylone function to counteract the positive reinforcing effects of this drug when compared to MDPV. In agreement with this idea, Bonano et al. [83] showed that MDPV is much more potent than methylone at facilitating intracranial self-stimulation (ICSS) in rats, an index of reinforcing effects of drugs. Furthermore, MDPV produces only abuse-related effects while methylone produces a mixture of abuse-related and abuse-limiting actions. Overall, the self-administration and ICSS data demonstrate that MDPV is a potent and efficacious reinforcer in rats, indicating the drug has a high potential for abuse and addiction in humans.

3 MDPV Pharmacokinetics and Metabolism

Pharmacokinetics (PK) describes the time course of drug concentrations in blood and tissues. Investigating the PK of synthetic cathinones and other NPS is important for the forensic detection of these substances and for evaluating their pharmacological/toxicological effects. When NPS first appear in the recreational drug marketplace, they must be identified and quantified in confiscated drug products and in biological specimens from subjects exposed to the drugs. As mentioned in the Introduction, most NPS are not detected by traditional toxicology screening methods, which rely on antibody-based technology (i.e., immunoassays) and recognize specific drugs and metabolites. Given the rapid increase in number and variety of NPS, the slow and cumbersome process of developing new immunoassays cannot keep pace with the appearance of new substances [84, 85]. Consequently, alternative analytical methods, particularly liquid chromatography (LC) coupled to mass spectrometry (MS) or high-resolution mass spectrometry (HRMS), are now being implemented to detect and quantify newly emerging drugs of abuse [86–88]. In vitro strategies using liver microsomes or hepatocytes are being exploited to quickly identify metabolites of NPS, since certain metabolites may be bioactive or have a much longer half-life than the parent compound, thereby serving as more persistent markers of drug exposure [89, 90]. Finally, because there are few controlled clinical studies examining the effects of NPS in humans, experiments in animal models must be employed to characterize in vivo PK and metabolism [85].

The chemical structure of MDPV displays a 3,4-methylenedioxy group on the phenyl ring, similar to the structure of methylone and MDMA. It is well established that the methylenedioxy moiety of MDMA is a primary target for metabolism by hepatic cytochrome P450 (CYP) enzymes, particularly CYP 2D6 [91–93]. Strano-Rossi et al. [94] reported the first description of MDPV metabolism in human liver microsomes in vitro. These investigators employed gas chromatography with MS for metabolite identification and LC-HRMS for definitive structural elucidation. It was found that MDPV is metabolized in a manner analogous to MDMA by *O*-demethylenation of the 3,4-methylenedioxy ring to form 3,4-dihydroxypyrovalerone (3,4-catechol-PV), followed by *O*-methylation to yield 4-hydroxy-3-methoxypyrovalerone (4-OH-3-MeO-PV) (see Fig. 1 for structures). Both of the phase I metabolites are conjugated to form phase II sulfates or glucuronides, which are subsequently excreted in urine. Meyer et al. [95] found that MDPV is metabolized in vitro by a number of mechanisms including demethylenation, aromatic and side-chain hydroxylation, and oxidation of the pyrrolidine ring, but 4-OH-3-MeO-PV is the major metabolite found in urine samples from rats and humans exposed to MDPV administration. Importantly, multiple hepatic enzymes including CYP 2C19, CYP 2D6, and CYP 1A2 were found to catalyze the primary *O*-demethylenation reaction forming 3,4-catechol-PV [95].

In our laboratory, we are interested in examining the in vivo PK and metabolism of MDPV in rats because data from controlled drug administration studies in humans are lacking. We have previously evaluated pharmacodynamic and PK parameters for MDMA in rats [96, 97], and used similar methods for examining the effects of MDPV [98]. As a first step, Anizan et al. [99] developed a fully validated analytical procedure to simultaneously detect and quantify MDPV, 3,4-catechol-PV and 4-OH-3-MeO-PV using LC-HRMS. The method involves specimen hydrolysis to cleave conjugated 3,4-catechol-PV and 4-OH-3-MeO-PV to their free forms, followed by protein precipitation prior to analysis. Limits of detection are 0.1 μg/L and the linear range is 0.25–1,000 μg/L. The high sensitivity for the assay is essential in order to quantify low analyte concentrations in the small volume of plasma obtained from catheterized rats. To examine PK of MDPV and its metabolites, Anizan et al. [98] administered s.c. doses of MDPV (0.5, 1, 2 mg/kg) to rats bearing surgically implanted i.v. catheters. Rats were placed into chambers equipped with photobeams to measure locomotor parameters, and connected to a tethering system which allowed free movement within the chamber. The i.v. catheters were attached to extension tubing that was threaded through the tether to facilitate stress-free blood withdrawal without any disturbance to the animal. Repeated blood samples (300 μL) were withdrawn via the catheter at various time points before and after injection. Blood samples were centrifuged and plasma specimens assayed for MDPV, 3,4-catechol-PV and 4-OH-3-MeO-PV using the LC-HRMS methods described above. Utilizing this strategy, we were able to simultaneously obtain pharmacodynamic measures (i.e., ambulation and stereotypy) and circulating concentrations of MDPV and its metabolites.

Results from the study of Anizan et al. [98] demonstrated that s.c. MDPV engenders rapid PK in rats, with maximal concentrations (C_{max}) in plasma occurring within 15–20 min of injection and decreasing quickly thereafter. Upon injection of 2 mg/kg s.c. MDPV, the plasma C_{max} for the drug is 271 μg/L (~1 μM) and the half-life ($t_{1/2}$) is about 80 min. Plasma concentrations of the metabolites 3,4-catechol-PV and 4-OH-3-MeO-PV increase at a much slower rate, reaching C_{max} between 3 and 4 h post-injection. Based on area-under-the-curve (AUC) values, 4-OH-3-MeO-PV is the major metabolite in rat plasma, in agreement with the findings of Meyer et al. [95], who found this to be the predominant metabolite in rat urine. As expected, s.c. MDPV produces dose-related stimulation of ambulation and stereotypy in catheterized rats, and plasma concentrations of MDPV are positively correlated with the extent of motor activation. Two additional findings from the study of Anizan et al. [98] are worth noting: (1) plasma MDPV concentrations display linear dose-proportional kinetics and (2) plasma MDPV metabolite concentrations are negatively correlated with locomotor activation produced by the drug. We found it surprising that MDPV displays linear PK in rats because other drugs exhibiting the 3,4-methylenedioxy moiety (e.g., MDMA) are known to cause sustained inhibition of CYP 2D6 in humans and CYP 2D1 in rats [100, 101], thereby leading to nonlinear accumulation of the parent drug in both species [97, 102, 103]. Indeed, recent evidence shows that MDPV inhibits CYP 2D6 in vitro with an IC_{50} of 1.3 μM [104]. The fact that MDPV metabolites are

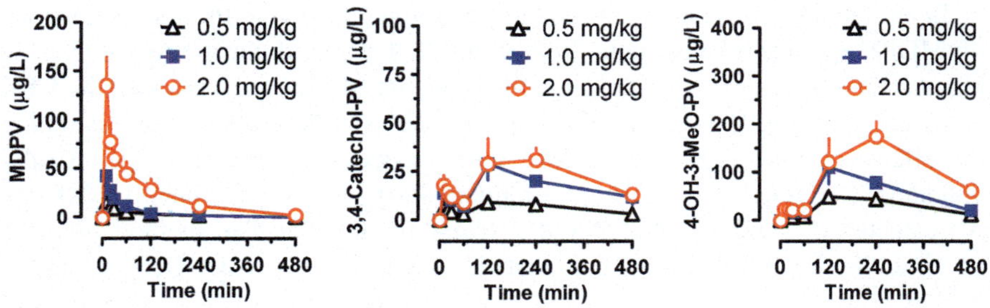

Fig. 3 Concentration-time profiles for MDPV and its metabolites, 3,4-catechol-PV and 4-OH-3-MeO-PV, after i.p. injection of MDPV in male Sprague–Dawley rats. Rats received i.p. injection of MDPV (0.5, 1 or 2 mg/kg) at time zero, and repeated blood samples (300 μL) were withdrawn immediately before and at 10, 20, 30, 60, 120, 240, and 480 min post-injection. Plasma specimens were assayed for concentrations of MDPV and its metabolites by LC-HRMS. Data are mean ± SEM for $N = 6$–7 rats/group

negatively correlated with locomotor stimulation suggests the compounds might be bioactive and counteract effects of the parent compound.

As a means to further explore the in vivo PK and metabolism of MDPV, we carried out a follow-up set of experiments to examine effects of i.p. MDPV administration in rats. The i.p. route of administration is expected to induce greater MDPV metabolism, leading to lower concentrations of the parent compound but higher concentrations of its metabolites. In these experiments, rats received i.p. MDPV (0.5, 1, 2 mg/kg), repeated blood samples were withdrawn at various time points, and all other aspects of the experiments were identical to those described above by Anizan and coworkers [98]. Figure 3 depicts new data showing the concentration-time profiles for MDPV and its metabolites after i.p. MDPV administration, while Table 2 summarizes the relevant PK parameters. Similar to the results with s.c. administration, i.p. MDPV engenders rapid PK, with C_{max} being achieved within 10 min of injection. After 2 mg/kg i.p. MDPV, the C_{max} for the drug is 135 μg/L (~0.5 μM) and plasma $t_{1/2}$ is about 90 min. Our data demonstrate that i.p. MDPV yields circulating drug concentrations in rats which are about half that observed after s.c. administration of equivalent doses. It is noteworthy that MDPV C_{max} values reported here for rats are in the same range as MDPV blood concentrations reported in human cases of non-fatal bath salts intoxication [12], but below those associated with fatal overdose [14, 22]. In contrast to the data with s.c. MDPV administration, i.p. administration appears to induce nonlinear PK. The results in Table 2 demonstrate that a fourfold increase in MDPV dose from 0.5 to 2.0 mg/kg is associated with an eightfold increase in MDPV AUC from 1,114 to 8,726 min μg/L, much greater than dose-proportional. The i.p. route of administration facilitates greater interaction of MDPV with hepatic enzymes when compared to the s.c. route. Thus, high i.p. doses of MDPV may produce nonlinear PK because in vivo drug concentrations in hepatic portal blood are close to the IC_{50} for inhibition of CYP 2D1. Future preclinical studies should explore PK parameters after the administration of higher doses of MDPV to rats.

Table 2 Pharmacokinetic parameters for plasma 3,4-methylenedioxypyrovalerone (MDPV), 3,4-dihydroxypyrovalerone (3,4-catechol-PV), and 4-hydroxy-3-methoxypyrovalerone (4-OH-3-MeO-PV) after intraperitoneal MDPV administration

Analyte	Dose (mg/kg)	C_{max} (μg/L)	T_{max} (min)	AUC (min μg/L)	$t_{1/2}$ (min)
MPDV	0.5	20 ± 5	10 ± 0	1,114 ± 330	92 ± 7
	1.0	54 ± 18	10 ± 0	2,858 ± 859	79 ± 13
	2.0	135 ± 29	10 ± 0	8,726 ± 1,877	99 ± 14
3,4-Catechol-PV	0.5	14 ± 2	160 ± 25	2,822 ± 415	n.d.
	1.0	36 ± 4	189 ± 59	8,317 ± 589	n.d.
	2.0	46 ± 9	206 ± 25	12,762 ± 1,625	n.d.
4-OH-3-MeO-PV	0.5	57 ± 7	168 ± 27	14,657 ± 2,577	n.d.
	1.0	135 ± 23	260 ± 24	28,168 ± 3,373	n.d.
	2.0	198 ± 31	240 ± 54	51,925 ± 7,946	n.d.

Data are expressed as mean ± SEM for $N = 6$ rats/group

C_{max} maximum concentration, T_{max} time of maximum concentration, *AUC* area-under-the-curve, $t_{1/2}$ plasma half-life

n.d. = not determined due to insufficient data from descending limb of the concentration-time profile

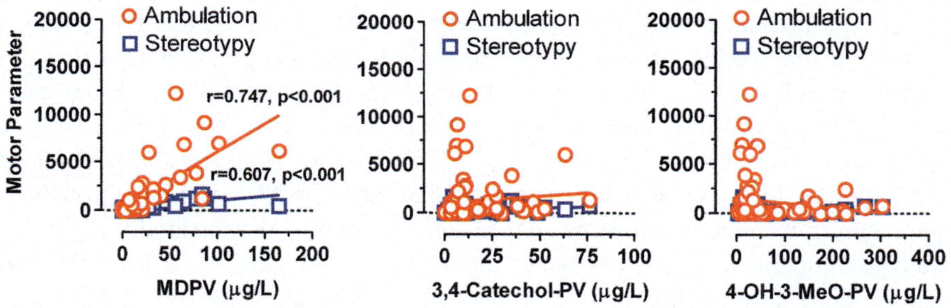

Fig. 4 Correlations between motor parameters and plasma concentrations of MDPV and its metabolites after i.p. MDPV administration. To construct correlation plots, ambulation (cm) and stereotypy (episodes) measures obtained at 20, 60, 120, and 240 min post-injection were plotted against plasma concentrations of MDPV, 3,4-catechol-PV or 4-OH-3-MeO-PV (μg/L) at the same time points. Data were subjected to Pearson correlation analysis. Ambulation ($r = 0.747$, $p < 0.001$) and stereotypy ($r = 0.607$, $p < 0.001$) were significantly correlated with plasma MDPV but not its metabolites

The data in Table 2 show that plasma concentrations of the metabolites 3,4-catechol-PV and 4-OH-3-MeO-PV display slow PK after i.p. MDPV administration, achieving C_{max} between 3 and 4 h post-injection. Based on AUC values shown in Table 2, 4-OH-3-MeO-PV is the major metabolite in rat plasma. Intraperitoneal MDPV produces dose-related stimulation of ambulation and stereotypy, and the data in Fig. 4 show that both locomotor parameters are significantly correlated with circulating MDPV concentrations but not its metabolites. To generate the correlation plots depicted in Fig. 4, the pharmacodynamic data from the 20, 60, 120, and 240 min time points were plotted against simultaneously measured plasma MDPV or metabolite concentrations. The data matrix was subjected to Pearson correlation analysis. It was found that circulating MDPV concentrations positively correlate with the magnitude of ambulation ($r = 0.747$, $p < 0.001$) and stereotypy ($r = 0.067$, $p < 0.001$), whereas metabolites show no significant relationships with motor endpoints. The findings with i.p. MDPV indicate that the parent compound is the major factor contributing to the locomotor stimulant effects of the drug. In agreement with this idea, Novellas et al. [105] recently reported that MDPV concentrations in rat striatum are positively correlated with the extent of locomotor activation produced after MDPV administration [105]. These authors further speculated that MDPV-induced elevations in extracellular dopamine in striatal regions underlie behavioral effects observed in rats.

The data considered thus far indicate that hydroxylated MDPV metabolites are probably not contributing to in vivo effects of systemically administered MDPV, especially since these metabolites exist as conjugated forms and are not "free" in the circulation. Nonetheless, we examined the possible biological activity of these metabolites because our previous work showed the 3,4-dihydroxy metabolite of MDMA is bioactive [106]. The effects of 3,4-catechol-PV and 4-OH-3-MeO-PV were first examined in uptake inhibition assays for DAT, NET, and SERT. Data in Table 1 demonstrate that 3,4-catechol-PV is a potent uptake blocker at DAT ($IC_{50} = 11$ nM) and NET ($IC_{50} = 11$ nM), whereas 4-OH-3-MeO-PV is much weaker in this regard. Neither of the metabolites displays measurable activity at inhibiting SERT, even at doses up to 10 μM. Data shown in Table 1 for 3,4-catechol-PV agree with previous findings of Meltzer et al. [43], who found that this compound is an uptake inhibitor at DAT and NET, with potency similar to pyrovalerone [43]. We next tested the metabolites of MDPV in the microdialysis paradigm to examine possible in vivo actions. Neither of the metabolites affected dialysate dopamine or behavior when administered at i.v. doses of 0.1 and 0.3 mg/kg; the same doses of MDPV elicit robust effects on both parameters (see Fig. 2). Given the in vitro potency of 3,4-catechol-PV at DAT, we examined the effects of higher doses of this metabolite in vivo. Figure 5 depicts new data that show

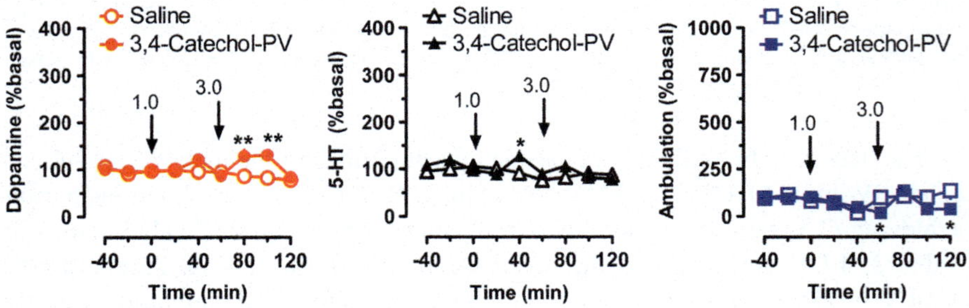

Fig. 5 Neurochemical and behavioral effects of 3,4-catechol-PV in male Sprague–Dawley rats undergoing microdialysis in nucleus accumbens. Rats received i.v. injection of 1.0 mg/kg at time zero, followed by 3.0 mg/kg 60 min later. Extracellular concentrations of monoamine transmitters (dopamine, 5-HT) and forward locomotion (ambulation) are expressed as % basal, determined from three time points prior to injection. Data are mean ± SEM, for $N = 6$–7 rats/group. $^{*}p < 0.05$, $^{**}p < 0.01$ compared to saline control at a given time point

i.v. administration of 3 mg/kg 3,4-catechol-PV produces small, albeit significant, elevations in extracellular dopamine but no change in ambulation. Taken together, the in vitro and in vivo findings with 3,4-catechol-PV indicate this compound may be too polar to readily penetrate the blood–brain barrier and achieve robust neurochemical effects. In support of this hypothesis, the total polar surface area of 3,4-catechol-PV is 60.77 compared to 38.78 for MDPV. The findings with 3,4-catechol-PV shown here serve as a cautionary reminder that inferring the mechanism of drug action should not rely on results from in vitro transporter/receptor profiling alone.

4 Pharmacology of "Replacement" Analogs of MDPV

As mentioned in the Introduction, legislation enacted in the US placed mephedrone, methylone, and MDPV into Schedule I control, rendering these drugs illegal. In response to this legislation, a number of "replacement" analogs appeared in the recreational drug marketplace, including several pyrrolidinophenone compounds. Perhaps the most notorious replacement analog of MDPV is α-pyrrolidinovalerophenone (α-PVP) (see Fig. 1). With regard to chemical structure, α-PVP is distinguished from MDPV by the absence of the 3,4-methylenedioxy moiety on the phenyl ring. α-PVP first appeared in the street drug marketplace in 2012 and quickly became a problematic drug of abuse in the US [6], especially in south Florida where the drug is known as "flakka" [107]. Many clinical cases of serious intoxication and death were attributed to overdose from α-PVP in the US and elsewhere [17, 40, 63, 108]. Meltzer et al. [43] first demonstrated that α-PVP is an inhibitor of dopamine and norepinephrine uptake, with potencies at DAT and NET in the same range as pyrovalerone. More recently, Marusich et al. [51] showed that α-PVP inhibits uptake at DAT and NET with IC_{50} values of 12 and 14 nM, respectively (Table 1). In studies carried out in HEK cells transfected with human transporters, α-PVP and a number of ring-substituted pyrrolidinophenones act as potent inhibitors of DAT and NET, but do not evoke release of preloaded [^{3}H] substrates [109]. Thus, data from synaptosomes and cell systems agree that cathinone-related compounds which possess a pyrrolidine ring act as transporter inhibitors and not substrates.

The data in Table 1 show that removing the 3,4-methylenedioxy moiety from the phenyl ring of MDPV has little effect on drug potency at catecholamine transporters, consistent with the earlier findings of Kolanos and coworkers [52]. However, decreasing alkyl chain length at the α-carbon of α-PVP from propyl to ethyl for α-pyrrolidinobutiophenone (α-PBP), or methyl for α-pyrrolidinopropiophenone (α-PPP), produces a corresponding decrease in potencies at DAT and NET, but no change in transporter selectivity [51]. In a study which examined the structure–activity relationships for a series of α-PVP analogs, Kolanos et al. [110] found that increasing alkyl chain length at the α-carbon to four carbons, or even adding a hexane ring to this position, results in potent DAT inhibitors. Overall, the volume and lipophilicity of the α-carbon substituent of pyrrolidinophenone analogs are positively correlated with potency at DAT, indicating structural modifications at this position have a profound impact on biological activity of the compounds.

In one of the first investigations to examine in vivo α-PVP actions, Kaizaki et al. [111] reported that oral administration of 25 mg/kg α-PVP to male mice produces elevations in striatal extracellular dopamine, along with stimulation of ambulation. It was also found that motor stimulant effects of α-PVP are significantly reduced by pretreatment with antagonists for D1 or D2 dopamine receptor subtypes, implicating dopaminergic mechanisms in mediating behavioral activation. Subsequent reports confirmed α-PVP produces dose-related stimulation of ambulation in mice and rats [51, 112, 113]. In our laboratory, we recently examined the neurochemical effects of α-PVP in male rats undergoing microdialysis in the nucleus accumbens. For our experiments, rats received i.v. injection of 0.1 mg/kg α-PVP at time zero, followed by i.v. injection of 0.3 mg/kg 60 min later. Control rats received i.v. injections of saline vehicle on the same schedule. Microdialysis samples were collected at 20 min intervals before, during, and after drug injections, and dialysate concentrations of dopamine and 5-HT were assayed by HPLC-ECD. Data were analyzed by two-way ANOVA (drug × time) followed by Bonferroni post-hoc tests. The new data depicted in Fig. 6 illustrate that α-PVP causes dose-related increases in extracellular dopamine ($F_{1,8} = 126.6, p < 0.0001$) and concurrent stimulation of ambulation ($F_{1,8} = 213.8, p < 0.0001$) in rats. Interestingly, α-PVP also produces small, albeit significant, *decreases* in extracellular 5-HT in the same subjects ($F_{1,8} = 3.5, p < 0.01$). The increases in extracellular dopamine and motor activity produced by α-PVP are similar to the effects of MDPV, and are fully consistent with potent DAT blockade. While the decreases in 5-HT produced by α-PVP are more difficult to interpret, the drug is clearly not increasing serotonergic tone. Marusich et al. [51] showed that α-PVP, α-PBP, and α-PPP produce dose-related stimulation of locomotor activity in mice, and the rank order of in vivo

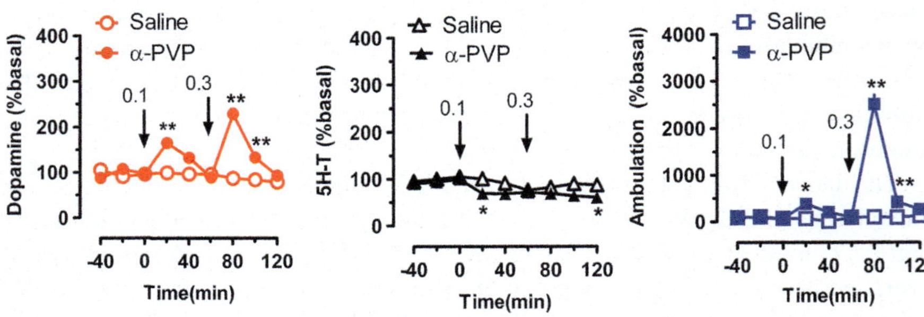

Fig. 6 Neurochemical and behavioral effects of α-PVP in male Sprague–Dawley rats undergoing microdialysis in nucleus accumbens. Rats received i.v. injection of 0.1 mg/kg at time zero, followed by 0.3 mg/kg 60 min later. Extracellular concentrations of monoamine transmitters (dopamine, 5-HT) and forward locomotion (ambulation) are expressed as % basal, determined from three time points prior to injection. Data are mean ± SEM, for $N = 6$–7 rats/group. *$p < 0.05$, **$p < 0.01$ compared to saline control at a given time point

potency (i.e., α-PVP > α-PBP > α-PPP) correlates with potency of the drugs at inhibiting DAT.

Recent studies examined the reinforcing effects of α-PVP using self-administration and ICSS assays in rats. Aarde et al. [112] directly compared effects of α-PVP and MDPV using i.v. self-administration in rats, and found a 0.05 mg/kg training dose produces similar patterns of acquisition for both drugs under a fixed-ratio schedule of reinforcement. In a progressive-ratio paradigm, it was shown that α-PVP and MDPV display nearly identical potency and efficacy, indicating similar abuse liability for the drugs. Watterson et al. [81] compared the effects of α-PVP and methamphetamine using ICSS, and noted both drugs produce dose-related reductions in self-stimulation thresholds, a measure of positive rewarding effects. Importantly, the potency of α-PVP in the ICSS model was identical to methamphetamine potency.

5 Summary

The findings reviewed in this chapter reveal that the pharmacology of MDPV differs substantially from the pharmacology of ring-substituted cathinones like mephedrone and methylone. MDPV is a potent inhibitor at DAT and NET, and the drug does not act as a transporter substrate like mephedrone and methylone. MDPV is highly selective for catecholamine transporters, whereas mephedrone and methylone are non-selective in this regard. The presence of a bulky pyrrolidine ring and a flexible α-carbon alkyl chain are the most critical structural elements governing potency of uptake inhibition at DAT and NET. *S*-MDPV is much more potent at inhibiting DAT and NET than *R*-MDPV, so the *S* isomer is responsible for pharmacological effects of the racemate. MDPV-induced increases in extracellular dopamine in mesolimbic reward circuits are likely responsible for the powerful stimulant and reinforcing actions of the drug. Upon systemic administration of MDPV, the circulating concentrations of the parent compound are positively correlated with the extent of locomotor activation, while concentrations of its metabolites are not. MDPV appears to induce nonlinear PK in rats after i.p. doses above 1 mg/kg, perhaps due to inhibition of CYP 2D1, and the phenomenon of nonlinear PK deserves further inquiry. Replacement analogs of MDPV like α-PVP, α-PBP, and α-PPP maintain potent and selective inhibition at DAT and NET, indicating these drugs have high abuse liability. Despite substantial knowledge about the pharmacology of MDPV and its analogs, a number of fundamental questions remain: What is the role of NET inhibition in the behavioral and cardiovascular effects of MDPV? Are there non-transporter targets of action for MDPV and its analogs? What are the molecular and cellular changes in the brain induced by chronic administration of MDPV, α-PVP, and related drugs? Finally, could certain pyrrolidinophenone analogs exhibit utility in treating dopamine deficit syndromes such as Parkinson's disease? These and other questions warrant further consideration.

Funding and Disclosures This research was generously supported by the Intramural Research Program (IRP) of the National Institute on Drug Abuse (NIDA) grant 1ZIADA000523-08. The authors have nothing to disclose.

References

1. Baumann MH, Solis E Jr, Watterson LR, Marusich JA, Fantegrossi WE, Wiley JL (2014) Baths salts, spice, and related designer drugs: the science behind the headlines. J Neurosci 34:15150–151158
2. Madras BK (2016) The growing problem of new psychoactive substances (NPS). Curr Top Behav Neurosci (in press)
3. Zawilska JB, Andrzejczak D (2015) Next generation of novel psychoactive substances on the horizon – a complex problem to face. Drug Alcohol Depend 157:1–17
4. Baumann MH, Volkow ND (2016) Abuse of new psychoactive substances: threats and solutions. Neuropsychopharmacology 41:663–665
5. Brandt SD, King LA, Evans-Brown M (2014) The new drug phenomenon. Drug Test Anal 6:587–597
6. Drug Enforcement Administration, Office of Diversion Control (2014) Special report: synthetic cannabinoids and cathinones reported in NFLIS, 2010–2013. http://www.deadiversion.usdoj.gov/nflis/spec_rpt_CathCan_2013.pdf
7. Shanks KG, Dahn T, Behonick G, Terrell A (2012) Analysis of first and second generation legal highs for synthetic cannabinoids and synthetic stimulants by ultra-performance liquid chromatography and time of flight mass spectrometry. J Anal Toxicol 36:360–371
8. United Nations Office of Drugs and Crime. World Drug Report (2015). http://www.unodc.org/documents/wdr2015/World_Drug_Report_2015.pdf
9. Prosser JM, Nelson LS (2012) The toxicology of bath salts: a review of synthetic cathinones. J Med Toxicol 8:33–34
10. Centers for Disease Control and Prevention (CDC) (2011) Emergency department visits after use of a drug sold as "bath salts"–Michigan, November 13, 2010-March 31, 2011. MMWR Morb Mortal Wkly Rep 60:624–627
11. Murphy CM, Dulaney AR, Beuhler MC, Kacinko S (2013) "Bath salts" and "plant food" products: the experience of one regional US poison center. J Med Toxicol 9:42–48
12. Spiller HA, Ryan ML, Weston RG, Jansen J (2011) Clinical experience with and analytical confirmation of "bath salts" and "legal highs" (synthetic cathinones) in United States. Clin Toxicol (Phila) 49:499–505
13. Ross EA, Reisfield GM, Watson MC, Chronister CW, Goldberger BA (2012) Psychoactive "bath salts" intoxication with methylenedioxypyrovalerone. Am J Med 125:854–858
14. Kesha K, Boggs CL, Ripple MG, Allan CH, Levine B, Jufer-Phipps R, Doyon S, Chi P, Fowler DR (2013) Methylenedioxypyrovalerone ("bath salts"), related death: case report and review of the literature. J Forensic Sci 58:1654–1659
15. Penders TM, Gestring RE, Vilensky DA (2012) Intoxication delirium following use of synthetic cathinone derivatives. Am J Drug Alcohol Abuse 38:616–617
16. Drug Enforcement Administration (2013) Establishment of drug codes for 26 substances. Final rule. Fed Regist 78:664–666
17. Marinetti LJ, Antonides HM (2013) Analysis of synthetic cathinones commonly found in bath salts in human performance and postmortem toxicology: method development, drug distribution and interpretation of results. J Anal Toxicol 37:135–146
18. Karch SB (2015) Cathinone neurotoxicity (The "3Ms"). Curr Neuropharmacol 13:21–25
19. Froberg BA, Levine M, Beuhler MC, Judge BS, Moore PW, Engebretsen KM, Mckeown NJ, Rosenbaum CD, Young AC, Rusyniak DE; ACMT Toxicology Investigators Consortium (ToxIC) (2015) Acute methylenedioxypyrovalerone toxicity. J Med Toxicol 11:185–194

20. European Monitoring Centre for Drugs and Drug Addiction (2014) EMCDDA–Europol Joint Report on a new psychoactive substance: MDPV (3,4-methylenedioxypyrovalerone). http://www.emcdda.europa.eu/system/files/publications/819/TDAS14001ENN_466653.pdf
21. Murray BL, Murphy CM, Beuhler MC (2012) Death following recreational use of designer drug "bath salts" containing 3,4-Methylenedioxypyrovalerone (MDPV). J Med Toxicol 8:69–75
22. Wyman JF, Lavins ES, Engelhart D, Armstrong EJ, Snell KD, Boggs PD, Taylor SM, Norris RN, Miller FP (2013) Postmortem tissue distribution of MDPV following lethal intoxication by "bath salts". J Anal Toxicol 37:182–185
23. Baumann MH (2014) Awash in a sea of 'bath salts': implications for biomedical research and public health. Addiction 109:1577–1579
24. Simmler LD, Liechti ME (2016) Interactions of cathinone NPS with human transporters and receptors in transfected cells. Curr Top Behav Neurosci (in press)
25. Kristensen AS, Andersen J, Jørgensen TN, Sørensen L, Eriksen J, Loland CJ, Strømgaard K, Gether U (2011) SLC6 neurotransmitter transporters: structure, function, and regulation. Pharmacol Rev 63:585–640
26. Gorman JM, Kent JM (1999) SSRIs and SMRIs: broad spectrum of efficacy beyond major depression. J Clin Psychiatry 60(Suppl 4):33–38
27. Iversen L (2006) Neurotransmitter transporters and their impact on the development of psychopharmacology. Br J Pharmacol 147(Suppl1):S82–S88
28. Howell LL, Kimmel HL (2008) Monoamine transporters and psychostimulant addiction. Biochem Pharmacol 75:196–217
29. Rothman RB, Baumann MH (2003) Monoamine transporters and psychostimulant drugs. Eur J Pharmacol 479:23–40
30. Reith ME, Blough BE, Hong WC, Jones KT, Schmitt KC, Baumann MH, Partilla JS, Rothman RB, Katz JL (2015) Behavioral, biological, and chemical perspectives on atypical agents targeting the dopamine transporter. Drug Alcohol Depend 147:1–19
31. Sitte HH, Freissmuth M (2015) Amphetamines, new psychoactive drugs and the monoamine transporter cycle. Trends Pharmacol Sci 36:41–50
32. Angoa-Perez M, Anneken JH, Kuhn DM (2016) Neurotoxicology of synthetic cathinone analogs. Curr Top Behav Neurosci (in press)
33. Baumann MH, Wang X, Rothman RB (2007) 3,4-Methylenedioxymethamphetamine (MDMA) neurotoxicity in rats: a reappraisal of past and present findings. Psychopharmacology (Berl) 189:407–424
34. Fleckenstein AE, Volz TJ, Riddle EL, Gibb JW, Hanson GR (2007) New insights into the mechanism of action of amphetamines. Annu Rev Pharmacol Toxicol 47:681–698
35. Rothman RB, Baumann MH, Dersch CM, Romero DV, Rice KC, Carroll FI, Partilla JS (2001) Amphetamine-type central nervous system stimulants release norepinephrine more potently than they release dopamine and serotonin. Synapse 39:32–41
36. Rothman RB, Vu N, Partilla JS, Roth BL, Hufeisen SJ, Compton-Toth BA, Birkes J, Young R, Glennon RA (2003) In vitro characterization of ephedrine-related stereoisomers at biogenic amine transporters and the receptorome reveals selective actions as norepinephrine transporter substrates. J Pharmacol Exp Ther 307:138–145
37. Glennon RA, Yousif M, Naiman N, Kalix P (1987) Methcathinone: a new and potent amphetamine-like agent. Pharmacol Biochem Behav 26:547–551
38. Kalix P, Glennon RA (1986) Further evidence for an amphetamine-like mechanism of action of the alkaloid cathinone. Biochem Pharmacol 35:3015–3019
39. Cozzi NV, Sievert MK, Shulgin AT, Jacob P 3rd, Ruoho AE (1999) Inhibition of plasma membrane monoamine transporters by beta-ketoamphetamines. Eur J Pharmacol 381:63–69
40. Nagai F, Nonaka R, Satoh Hisashi Kamimura K (2007) The effects of non-medically used psychoactive drugs on monoamine neurotransmission in rat brain. Eur J Pharmacol 559:132–137
41. Héron C, Costentin J, Bonnet JJ (1994) Evidence that pure uptake inhibitors including cocaine interact slowly with the dopamine neuronal carrier. Eur J Pharmacol 264:391–398

42. Vaugeois JM, Bonnet JJ, Duterte-Boucher D, Costentin J (1993) In vivo occupancy of the striatal dopamine uptake complex by various inhibitors does not predict their effects on locomotion. Eur J Pharmacol 230:195–201
43. Meltzer PC, Butler D, Deschamps JR, Madras BK (2006) 1(4-Methylphenyl)-2-pyrrolidin-1-yl-petan-1-one (pyrovalerone) analogues: a promising class of monoamine uptake inhibitors. J Med Chem 49:1420–1432
44. Hadlock GC, Webb KM, McFadden LM, Chu PW, Ellis JD, Allen SC, Andrenyak DM, Vieira-Brock PL, German CL, Conrad KM, Hoonakker AJ, Gibb JW, Wilkins DG, Hanson GR, Fleckenstein AE (2011) 4-Methylmethcathinone (mephedrone): neuropharmacological effects of a designer stimulant of abuse. J Pharmacol Exp Ther 339:530–536
45. López-Arnau R, Martínez-Clemente J, Pubill D, Escubedo E, Camarasa J (2012) Comparative neuropharmacology of three psychostimulant cathinone derivatives: butylone, mephedrone and methylone. Br J Pharmacol 167:407–420
46. Baumann MH, Ayestas MA Jr, Partilla JS, Sink JR, Shulgin AT, Daley PF, Brandt SD, Rothman RB, Ruoho AE, Cozzi NV (2012) The designer methcathinone analogs, mephedrone and methylone, are substrates for monoamine transporters in brain tissue. Neuropsychopharmacology 37:1192–1203
47. Eshleman AJ, Wolfrum KM, Hatfield MG, Johnson RA, Murphy KV, Janowsky A (2013) Substituted methcathinones differ in transporter and receptor interactions. Biochem Pharmacol 85:1803–1815
48. Simmler LD, Buser TA, Donzelli M, Schramm Y, Dieu LH, Huwyler J, Chaboz S, Hoener MC, Liechti ME (2013) Pharmacological characterization of designer cathinones in vitro. Br J Pharmacol 168:458–470
49. Baumann MH, Partilla JS, Lehner KR, Thorndike EB, Hoffman AF, Holy M, Rothman RB, Goldberg SR, Lupica CR, Sitte HH, Brandt SD, Tella SR, Cozzi NV, Schindler CW (2013) Powerful cocaine-like actions of 3,4-methylenedioxypyrovalerone (MDPV), a principal constituent of psychoactive 'bath salts' products. Neuropsychopharmacology 38:552–562
50. Kolanos R, Partilla JS, Baumann MH, Hutsell BA, Banks ML, Negus SS, Glennon RA (2015) Stereoselective actions of methylenedioxypyrovalerone (MDPV) to inhibit dopamine and norepinephrine transporters and facilitate intracranial self-stimulation in rats. ACS Chem Neurosci 6:771–777
51. Marusich JA, Antonazzo KR, Wiley JL, Blough BE, Partilla JS, Baumann MH (2014) Pharmacology of novel synthetic stimulants structurally related to the "bath salts" constituent 3,4-methylenedioxypyrovalerone (MDPV). Neuropharmacology 87:206–213
52. Kolanos R, Solis E Jr, Sakloth F, De Felice LJ, Glennon RA (2013) "Deconstruction" of the abused synthetic cathinone methylenedioxypyrovalerone (MDPV) and an examination of effects at the human dopamine transporter. ACS Chem Neurosci 4:1524–1529
53. Cameron KN, Kolanos R, Solis E Jr, Glennon RA, De Felice LJ (2013) Bath salts components mephedrone and methylenedioxypyrovalerone (MDPV) act synergistically at the human dopamine transporter. Br J Pharmacol 168:1750–1757
54. Suzuki M, Deschamps JR, Jacobson AE, Rice KC (2015) Chiral resolution and absolute configuration of the enantiomers of the psychoactive "designer drug" 3,4-methylenedioxypyrovalerone. Chirality 27:287–293
55. Gannon BM, Williamson A, Suzuki M, Rice KC, Fantegrossi WE (2016) Stereoselective effects of abused "bath salt" constituent 3,4-methylenedioxypyrovalerone in mice: drug discrimination, locomotor activity, and thermoregulation. J Pharmacol Exp Ther 356:615–623
56. Baumann MH, Clark RD, Woolverton WL, Wee S, Blough BE, Rothman RB (2011) In vivo effects of amphetamine analogs reveal evidence for serotonergic inhibition of mesolimbic dopamine transmission in the rat. J Pharmacol Exp Ther 337:218–2125
57. Rothman RB, Blough BE, Woolverton WL, Anderson KG, Negus SS, Mello NK, Roth BL, Baumann MH (2005) Development of a rationally designed, low abuse potential, biogenic amine releaser that suppresses cocaine self-administration. J Pharmacol Exp Ther 313:1361–1369
58. Ikemoto S (2007) Dopamine reward circuitry: two projection systems from the ventral midbrain to the nucleus accumbens-olfactory tubercle complex. Brain Res Rev 56:27–78

59. Willuhn I, Wanat MJ, Clark JJ, Phillips PE (2010) Dopamine signaling in the nucleus accumbens of animals self-administering drugs of abuse. Curr Top Behav Neurosci 3:29–71
60. Zolkowska D, Jain R, Rothman RB, Partilla JS, Roth BL, Setola V, Prisinzano TE, Baumann MH (2009) Evidence for the involvement of dopamine transporters in behavioral stimulant effects of modafinil. J Pharmacol Exp Ther 329:738–746
61. Cozzi NV, Brandt SD, Daley PF, Partilla JS, Rothman RB, Tulzer A, Sitte HH, Baumann MH (2013) Pharmacological examination of trifluoromethyl ring-substituted methcathinone analogs. Eur J Pharmacol 699:180–187
62. Kehr J, Ichinose F, Yoshitake S, Goiny M, Sievertsson T, Nyberg F, Yoshitake T (2011) Mephedrone, compared with MDMA (ecstasy) and amphetamine, rapidly increases both dopamine and 5-HT levels in nucleus accumbens of awake rats. Br J Pharmacol 164:1949–1958
63. Wright MJ Jr, Angrish D, Aarde SM, Barlow DJ, Buczynski MW, Creehan KM, Vandewater SA, Parsons LH, Houseknecht KL, Dickerson TJ, Taffe MA (2012) Effect of ambient temperature on the thermoregulatory and locomotor stimulant effects of 4-methylmethcathinone in Wistar and Sprague-Dawley rats. PLoS One 7, e44652
64. Suyama JA, Sakloth F, Kolanos R, Glennon RA, Lazenka MF, Negus SS, Banks ML (2016) Abuse-related neurochemical effects of para-substituted methcathinone analogs in rats: microdialysis studies of nucleus accumbens dopamine and serotine. J Pharmacol Exp Ther 356:182–190
65. Baumann MH, Clark RD, Franken FH, Rutter JJ, Rothman RB (2008) Tolerance to 3,4-methylenedioxymethamphetamine in rats exposed to single high-dose binges. Neuroscience 152:773–784
66. Schindler CW, Thorndike EB, Goldberg SR, Lehner KR, Cozzi NV, Brandt SD, Baumann MH (2016) Reinforcing and neurochemical effects of the "bath salts" constituents 3,4-methylenedioxypyrovalerone (MDPV) and 3,4-methylenedioxy-N-methylcathinone (methylone) in male rats. Psychopharmacology (Berl) 233:1981–1990
67. Glennon RA, Young R (2016) Neurobiology of 3,4-methylenedioxypyrovalerone (MDPV) and α-pyrrolidinovalerophenone (α-PVP). Brain Res Bull 126:111–126
68. Aarde SM, Huang PK, Creehan KM, Dickerson TJ, Taffe MA (2013) The novel recreational drug 3,4-methylenedioxypyrovalerone (MDPV) is a potent psychomotor stimulant: self-administration and locomotor activity in rats. Neuropharmacology 71:130–140
69. Huang PK, Aarde SM, Angrish D, Houseknecht KL, Dickerson TJ, Taffe MA (2012) Contrasting effects of d-methamphetamine, 3,4-methylenedioxymethamphetamine, 3,4-methylenedioxypyrovalerone, and 4-methylmethcathinone on wheel activity in rats. Drug Alcohol Depend 126:168–175
70. Fantegrossi WE, Gannon BM, Zimmerman SM, Rice KC (2013) In vivo effects of abused 'bath salt' constituent 3,4-methylenedioxypyrovalerone (MDPV) in mice: drug discrimination, thermoregulation, and locomotor activity. Neuropsychopharmacology 38:563–573
71. Gatch MB, Taylor CM, Forster MJ (2013) Locomotor stimulant and discriminative stimulus effects of 'bath salt' cathinones. Behav Pharmacol 24:437–447
72. Marusich JA, Grant KR, Blough BE, Wiley JL (2012) Effects of synthetic cathinones contained in "bath salts" on motor behavior and a functional observational battery in mice. Neurotoxicology 33:1305–1313
73. Howell LL, Cunningham KA (2015) Serotonin 5-HT2 receptor interactions with dopamine function: implications for therapeutics in cocaine use disorder. Pharmacol Rev 67:176–1797
74. Watterson LR, Olive MF (2016) Reinforcing effects of cathinone NPS in the i.v. drug self-administration paradigm. Curr Top Behav Neurosci (in press)
75. Watterson LR, Watterson E, Olive MF (2013) Abuse liability of novel 'legal high' designer stimulants: evidence from animal models. Behav Pharmacol 24:341–355
76. Aarde SM, Angrish D, Barlow DJ, Wright MJ Jr, Vandewater SA, Creehan KM, Houseknecht KL, Dickerson TJ, Taffe MA (2013) Mephedrone (4-methylmethcathinone) supports i.v. self-administration in Sprague-Dawley and Wistar rats. Addict Biol 18:786–799
77. Motbey CP, Clemens KJ, Apetz N, Winstock AR, Ramsey J, Li KM, Wyatt N, Callaghan PD, Bowen MT, Cornish JL, McGregor IS (2013) High levels of i.v. mephedrone (4-methylmethcathinone) self-administration in rats: neural consequences and comparison with methamphetamine. J Psychopharmacol 27:823–836

78. Creehan KM, Vandewater SA, Taffe MA (2015) I.v. self-administration of mephedrone, methylone and MDMA in female rats. Neuropharmacology 92:90–97
79. Vandewater SA, Creehan KM, Taffe MA (2015) I.v. self-administration of entactogen-class stimulants in male rats. Neuropharmacology 99:538–545
80. Watterson LR, Hood L, Sewalia K, Tomek SE, Yahn S, Johnson CT, Wegner S, Blough BE, Marusich JA, Olive MF (2012) The reinforcing and rewarding effects of methylone, a synthetic cathinone commonly found in "bath salts". J Addict Res Ther Suppl 9:002
81. Watterson LR, Burrows BT, Hernandez RD, Moore KN, Grabenauer M, Marusich JA, Olive MF (2014) Effects of α-pyrrolidinopentiophenone and 4-methyl-N-ethylcathinone, two synthetic cathinones commonly found in second-generation "bath salts," on intracranial self-stimulation thresholds in rats. Int J Neuropsychopharmacol 18 pii:pyu014
82. Watterson LR, Kufahl PR, Nemirovsky NE, Sewalia K, Grabenauer M, Thomas BF, Marusich JA, Wegner S, Olive MF (2014) Potent rewarding and reinforcing effects of the synthetic cathinone 3,4-methylenedioxypyrovalerone (MDPV). Addict Biol 19:165–174
83. Bonano JS, Glennon RA, De Felice LJ, Banks ML, Negus SS (2014) Abuse-related and abuse-limiting effects of methcathinone and the synthetic "bath salts" cathinone analogs methylenedioxypyrovalerone (MDPV), methylone and mephedrone on intracranial self-stimulation in rats. Psychopharmacology (Berl) 231:199–207
84. Tettey J, Crean C (2015) New psychoactive substances: catalysing a shift in forensic science practice? Philos Trans R Soc Lond B Biol Sci 370:20140265
85. Ellefsen KN, Concherio M, Huestis MA (2016) Synthetic cathinone pharmacokinetics, analytical methods, and toxicological findings from human performance and postmortem cases. Drug Metab Rev 48:237–265
86. Concheiro M, Castaneto M, Kronstrand R, Huestis MA (2015) Simultaneous determination of 40 novel psychoactive stimulants in urine by liquid chromatography-high resolution mass spectrometry and library matching. J Chromatogr A 1397:32–42
87. Favretto D, Pascali JP, Tagliaro F (2013) New challenges and innovation in forensic toxicology: focus on the "New Psychoactive Substances". J Chromatogr A 1287:84–95
88. Smith JP, Sutcliffe OB, Banks CE (2015) An overview of recent developments in the analytical detection of new psychoactive substances (NPSs). Analyst 140:4932–4948
89. Castaneto MS, Wohlfarth A, Desrosiers NA, Hartman RL, Gorelick DA, Huestis MA (2015) Synthetic cannabinoids pharmacokinetics and detection methods in biological matrices. Drug Metab Rev 47:124–174
90. Meyer MR, Maurer HH (2012) Current applications of high-resolution mass spectrometry in drug metabolism studies. Anal Bioanal Chem 403:1221–1231
91. Kreth K, Kovar K, Schwab M, Zanger UM (2000) Identification of the human cytochromes P450 involved in the oxidative metabolism of "Ecstasy"-related designer drugs. Biochem Pharmacol 59:1563–1571
92. Meyer MR, Peters FT, Maurer HH (2008) The role of human hepatic cytochrome P450 isozymes in the metabolism of racemic 3,4-methylenedioxy-methamphetamine and its enantiomers. Drug Metab Dispos 36:2345–2354
93. Tucker GT, Lennard MS, Ellis SW, Woods HF, Cho AK, Lin LY, Hiratsuka A, Schmitz DA, Chu TY (1994) The demethylenation of methylenedioxymethamphetamine ("ecstasy") by debrisoquine hydroxylase (CYP2D6). Biochem Pharmacol 47:1151–1156
94. Strano-Rossi S, Cadwallader AB, de la Torre X, Botrè F (2010) Toxicological determination and in vitro metabolism of the designer drug methylenedioxypyrovalerone (MDPV) by gas chromatography/mass spectrometry and liquid chromatography/quadrupole time-of-flight mass spectrometry. Rapid Commun Mass Spectrom 24:2706–2714
95. Meyer MR, Du P, Schuster F, Maurer HH (2010) Studies on the metabolism of the α-pyrrolidinophenone designer drug methylenedioxy-pyrovalerone (MDPV) in rat and human urine and human liver microsomes using GC-MS and LC-high-resolution MS and its detectability in urine by GC-MS. J Mass Spectrom 45:1426–1442

96. Baumann MH, Zolkowska D, Kim I, Scheidweiler KB, Rothman RB, Huestis MA (2009) Effects of dose and route of administration on pharmacokinetics of 3,4-methylenedioxymethamphetamine in the rat. Drug Metab Dispos 37:2163–2170
97. Concheiro M, Baumann MH, Scheidweiler KB, Rothman RB, Marrone GF, Huestis MA (2014) Nonlinear pharmacokinetics of (+/-)3,4-methylenedioxymethamphetamine (MDMA) and its pharmacodynamic consequences in the rat. Drug Metab Dispos 42:119–125
98. Anizan S, Concheiro M, Lehner KR, Bukhari MO, Suzuki M, Rice KC, Baumann MH, Huestis MA (2016) Linear pharmacokinetics of 3,4-methylenedioxypyrovalerone (MDPV) and its metabolites in the rat: relationship to pharmacodynamic effects. Addict Biol 21:339–347
99. Anizan S, Ellefsen K, Concheiro M, Suzuki M, Rice KC, Baumann MH, Huestis MA (2014) 3,4-Methylenedioxypyrovalerone (MDPV) and metabolites quantification in human and rat plasma by liquid chromatography-high resolution mass spectrometry. Anal Chim Acta 827:54–63
100. Delaforge M, Jaouen M, Bouille G (1999) Inhibitory metabolite complex formation of methylenedioxymethamphetamine with rat and human cytochrome P450. Particular involvement of CYP 2D. Environ Toxicol Pharmacol 7:153–158
101. Heydari A, Yeo KR, Lennard MS, Ellis SW, Tucker GT, Rostami-Hodjegan A (2004) Mechanism-based inactivation of CYP2D6 by methylenedioxymethamphetamine. Drug Metab Dispos 32:1213–1217
102. Chu T, Kumagai Y, DiStefano EW, Cho AK (1996) Disposition of methylenedioxymethamphetamine and three metabolites in the brains of different rat strains and their possible roles in acute serotonin depletion. Biochem Pharmacol 51:789–796
103. de la Torre R, Farré M, Ortuño J, Mas M, Brenneisen R, Roset PN, Segura J, Camí J (2000) Non-linear pharmacokinetics of MDMA ('ecstasy') in humans. Br J Clin Pharmacol 49:104–109
104. Dinger J, Meyer MR, Maurer HH (2016) In vitro cytochrome P450 inhibition potential of methylenedioxy-derived designer drugs studied with a two-cocktail approach. Arch Toxicol 90:305–318
105. Novellas J, López-Arnau R, Carbó ML, Pubill D, Camarasa J, Escubedo E (2015) Concentrations of MDPV in rat striatum correlate with the psychostimulant effect. J Psychopharmacol 29:1209–1218
106. Schindler CW, Thorndike EB, Blough BE, Tella SR, Goldberg SR, Baumann MH (2014) Effects of 3,4-methylenedioxymethamphetamine (MDMA) and its main metabolites on cardiovascular function in conscious rats. Br J Pharmacol 171:83–91
107. Crespi C (2016) Flakka-induced prolonged psychosis. Case Rep Psychiatry 2016:3460849
108. Sykutera M, Cychowska M, Bloch-Boguslawska E (2015) A fatal case of pentedrone and α-pyrrolidinovalerophenone poisoning. J Anal Toxicol 39:324–329
109. Rickli A, Hoener MC, Liechti ME (2015) Monoamine transporter and receptor interaction profiles of novel psychoactive substances: para-halogenated amphetamines and pyrovalerone cathinones. Eur Neuropsychopharmacol 25:365–376
110. Kolanos R, Sakloth F, Jain AD, Partilla JS, Baumann MH, Glennon RA (2015) Structural modification of the designer stimulant α-pyrrolidinovalerophenone (α-PVP) influences potency at dopamine transporters. ACS Chem Neurosci 6:1726–1731
111. Kaizaki A, Tanaka S, Numazawa S (2014) New recreational drug 1-phenyl-2-(1-pyrrolidinyl)-1-pentanone (alpha-PVP) activates central nervous system via dopaminergic neuron. J Toxicol Sci 39:1–6
112. Aarde SM, Creehan KM, Vandewater SA, Dickerson TJ, Taffe MA (2015) In vivo potency and efficacy of the cathinone α-pyrrolidinopentiophenone and 3,4-methylenedioxypyrovalerone: self-administration and locomotor stimulation in male rats. Psychopharmacology (Berl) 232:3045–3055
113. Gatch MB, Dolan SB, Forster MJ (2015) Comparative behavioral pharmacology of three pyrrolidine-containing synthetic cathinone derivatives. J Pharmacol Exp Ther 354:103–110

Decoding the Structure of Abuse Potential for New Psychoactive Substances: Structure–Activity Relationships for Abuse-Related Effects of 4-Substituted Methcathinone Analogs

S. Stevens Negus and Matthew L. Banks

Abstract Many cathinone analogs act as substrates or inhibitors at dopamine, norepinephrine, and serotonin transporters (DAT, NET, SERT, respectively). Drug selectivity at DAT vs. SERT is a key determinant of abuse potential for monoamine transporter substrates and inhibitors, such that potency at DAT > SERT is associated with high abuse potential, whereas potency at DAT < SERT is associated with low abuse potential. Quantitative structure–activity relationship (QSAR) studies with a series of 4-substituted methcathinone analogs identified volume of the 4-position substituent on the methcathinone phenyl ring as one structural determinant of both DAT vs. SERT selectivity and abuse-related behavioral effects in an intracranial self-stimulation procedure in rats. Subsequent modeling studies implicated specific amino acids in DAT and SERT that might interact with 4-substituent volume to determine effects produced by this series of cathinone analogs. These studies illustrate use of QSAR analysis to investigate pharmacology of cathinones and function of monoamine transporters.

Keywords Dopamine transporter • Flephedrone • Intracranial self-stimulation • Mephedrone • Methcathinone • Methedrone • Microdialysis • Serotonin transporter • Structure–activity relationship

Contents

S.S. Negus (✉) and M.L. Banks
Department of Pharmacology and Toxicology, Virginia Commonwealth University, Richmond, VA, USA
e-mail: sidney.negus@vcuhealth.org

Curr Topics Behav Neurosci (2017) 32: 119–132
DOI 10.1007/7854_2016_18
Published Online: 11 July 2016

1 Introduction

Synthetic cathinone analogs are new members of an old family of drugs with high abuse potential [1, 2]. Most drugs in this family share a common effectiveness to either traverse or block dopamine transporters (DAT) and ultimately to increase extracellular dopamine (DA) levels in key brain reward structures such as the nucleus accumbens. In addition to their effects on the DAT, many drugs in this family also act on two related transporter proteins, the norepinephrine transporter (NET) and serotonin transporter (SERT), to modulate extracellular levels of their respective monoamine neurotransmitters norepinephrine (NE) and serotonin (5HT). A growing body of evidence supports the general hypothesis that abuse potential of drugs in this family is determined by their relative selectivity to act at DAT vs. SERT. As a prelude to discussing the relationship between structure and abuse potential of novel methcathinone analogs, this chapter will begin by reviewing evidence that implicates DAT/SERT selectivity as a determinant of abuse potential. This evidence provides a framework for interpreting effects of new psychoactive substances.

2 Amphetamine, MDMA, and Fenfluramine as Prototype Monoamine Releasers

2.1 Neurochemical Effects

The drugs amphetamine, 3,4-methylenedioxymethamphetamine (MDMA), and fenfluramine illustrate the range of effects that can be produced by drugs with different profiles of selectivity for DAT vs. SERT. All three drugs can traverse monoamine transporters and trigger a series of intracellular events that promote monoamine neurotransmitter release [3–6]. As a group, these drugs are sometimes called "transporter substrates," because like the endogenous neurotransmitters, they can pass from the extracellular space through the transporter channel to the intracellular space. They are also often called "monoamine releasers," because one consequence of their transport is the release of monoamine neurotransmitter stored in synaptic terminals. Although these drugs share a similar general mechanism of action as transporter substrates and monoamine releasers, they differ in their

Table 1 EC_{50} values (nM ± SD) for (+)amphetamine, (+)MDMA, and (±)fenfluramine to promote monoamine release from rat brain synaptosomes

Drug	EC_{50} values		DAT vs. SERT Selectivity[a]
	DA release	5HT release	
(+)Amphetamine[b]	25 ± 4	1765 ± 94	71
(+)MDMA[c]	142 ± 4	74 ± 3	0.52
(±)Fenfluramine[b]	>10,000	79 ± 12	<0.01

[a]Selectivity calculated as SERT EC_{50}/DAT EC_{50}
[b]Rothman et al. [7]
[c]Setola et al. [8]

relative potencies at DAT and SERT. For example, Table 1 shows the relative in vitro potency of each drug to promote monoamine release via DAT or SERT from rat brain synaptosomes loaded with radiolabeled monoamine [7, 8]. By this metric, (+)amphetamine is DAT selective, (±)fenfluramine is SERT selective, and (+)MDMA displays similar potencies to act at both transporters. (Note: The potency of each compound is slightly greater to act at NET than DAT, but effects at NET are not addressed further here because other evidence suggests a minimal role for NE in abuse potential.) These in vitro neurochemical effects mirror effects of these drugs on brain neurochemistry in vivo. For example, Fig. 1 shows the effects of behaviorally active doses of (+)amphetamine and (±)fenfluramine on extracellular DA and 5HT levels measured in nucleus accumbens of rats using in vivo microdialysis [9]. (+)Amphetamine selectively increases DA levels, whereas (±)fenfluramine selectively increases 5HT levels. By contrast, MDMA increases both DA and 5HT levels in rat nucleus accumbens as assessed by in vivo microdialysis ([10]; Lazenka MF, Suyama JA, Banks ML, Negus SS, unpublished results).

2.2 *Abuse-Related Behavioral Effects*

These in vitro and in vivo neurochemical effects of amphetamine, MDMA, and fenfluramine also correspond to expression of abuse-related behavioral effects by these drugs. Drug self-administration procedures are the most widely used preclinical procedures to assess abuse potential [11–13], and in these procedures, laboratory animals emit an operant response (e.g., pressing a lever) to receive a dose of drug (e.g., by intravenous infusion). Thus, animals in drug self-administration procedures engage in drug-taking behaviors that are analogous to the drug-taking behaviors displayed by human drug abusers. A drug is considered to produce "reinforcing effects" and to function as a "reinforcer" in a drug self-administration procedure if subjects respond at higher rates for delivery of some dose of drug than they respond for delivery of vehicle, and drugs that function as reinforcers in animals often function as drugs of abuse in humans. Evidence from drug self-administration procedures indicates that amphetamine produces stronger

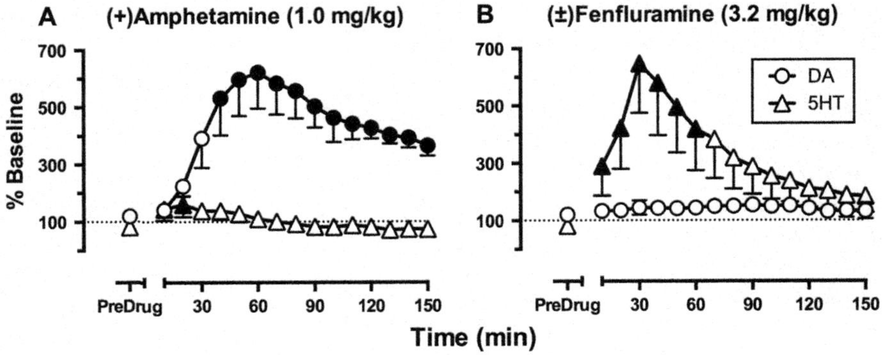

Fig. 1 (+)Amphetamine selectively increases DA > 5HT levels (**a**), and (±)fenfluramine significantly increases 5HT > DA levels (**b**), in rat nucleus accumbens as measured by in vivo microdialysis. *Abscissae*: Time relative to IP drug administration in min. *Ordinates*: Percent baseline levels of DA and 5HT. Points show mean ± SEM for 5–7 rats, and filled points show a significant difference from the "PreDrug" point ($p < 0.05$). Adapted from Suyama et al. [9]

reinforcing effects than MDMA, and fenfluramine does not produce reinforcing effects [14–16].

A related preclinical procedure, known as intracranial self-stimulation (ICSS), will be referenced extensively in this chapter [17]. As in drug self-administration, laboratory animals in ICSS procedures emit an operant response to receive a reinforcer; however, in ICSS, the reinforcer is not drug delivery, but instead is the delivery of electrical stimulation to a brain reward area via a surgically implanted microelectrode. In one common type of the ICSS procedure, the amount of electrical brain stimulation is varied during each behavioral session by manipulating the frequency of electrical pulses, and increasing frequencies of brain stimulation maintain increasing rates of ICSS responding. Figure 2a shows a photograph of a rat in an ICSS procedure, and Fig. 2b shows the sigmoidal plot that relates brain stimulation frequency to ICSS rate. Thus, low frequencies of brain stimulation maintain low rates of ICSS, whereas higher frequencies maintain high rates of ICSS. Once subjects are trained in this procedure, drugs can be administered before daily behavioral sessions, and abuse potential can be inferred from the profile of drug effects on the ICSS frequency-rate curve. For example, Fig. 3 shows the effects of (+)amphetamine, (+)MDMA, and (±)fenfluramine on ICSS in rats [18]. (+)Amphetamine produces leftward and upward shifts in the ICSS frequency-rate curve (Fig. 3a) and a dose-dependent increase in the total number of stimulations delivered across all brain stimulation frequencies (Fig. 3b). This drug-induced increase in responding is described as "facilitation of ICSS," and drugs that facilitate ICSS also usually function as reinforcers in preclinical drug self-administration procedures and display high abuse liability in humans. Accordingly, facilitation of ICSS can be viewed as a behavioral index of a drug's abuse potential. In contrast to (+)amphetamine, (±)fenfluramine produces only dose-dependent decreases in ICSS (Fig. 3e, f), and drugs that only depress ICSS usually fail to

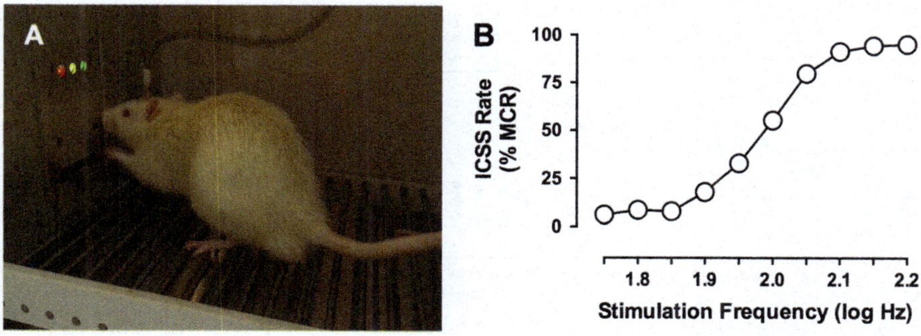

Fig. 2 Photograph of a rat engaged in intracranial self-stimulation (ICSS) (**a**), and example of a baseline frequency-rate curve from the ICSS procedure (**b**). In this ICSS procedure, responding on a lever results in the delivery of electrical brain stimulation delivered via a microelectrode surgically implanted into a brain reward area. In (**a**), a cable connects the electrode mounted on the subject's skull to a stimulator located outside the picture. In (**b**), the abscissa shows the frequency in log Hz of the electrical pulses delivered during each stimulation delivery, and the ordinate shows the ICSS rate expressed as percent maximum control rate (%MCR), which normalizes ICSS rate measurements within each subject. Low frequencies of brain stimulation maintain low ICSS rates, whereas high ICSS rates maintain high ICSS rates. Adapted from Negus and Miller [17]

function as reinforcers in preclinical drug self-administration procedures and lack abuse liability in humans. Lastly, (+)MDMA produces a mixed profile of effects that includes both facilitation of low ICSS rates maintained by low brain stimulation frequencies and depression of high ICSS rates maintained by high brain stimulation frequencies (Fig. 3c). As a result of this mixed-effect profile, MDMA produces a lower maximal stimulation of total ICSS than amphetamine (Fig. 3d). Drugs that produce this mixed profile of ICSS facilitation and depression often function as relatively weak or unreliable reinforcers in preclinical drug self-administration procedures and display relatively modest abuse liability in humans.

2.3 *Correlation Between Neurochemical and Behavioral Effects*

Figure 4a shows a correlation between maximal ICSS facilitation (defined as the maximum increase in total ICSS as in Fig. 3d–f) and DAT vs. SERT selectivity (defined as shown in Table 1) for (+)amphetamine, (+)MDMA, (±)fenfluramine, and 7 other monoamine releasers [18]. Figure 4b shows a correlation between maximal ICSS facilitation in rats and maximal reinforcing effects in a nonhuman primate drug self-administration procedure for most of these same drugs [18]. These significant correlations provide one source of evidence to suggest that ICSS can be useful both (1) as a behavioral correlate to neurochemical drug effects and (2) as a complement to drug self-administration procedures for preclinical

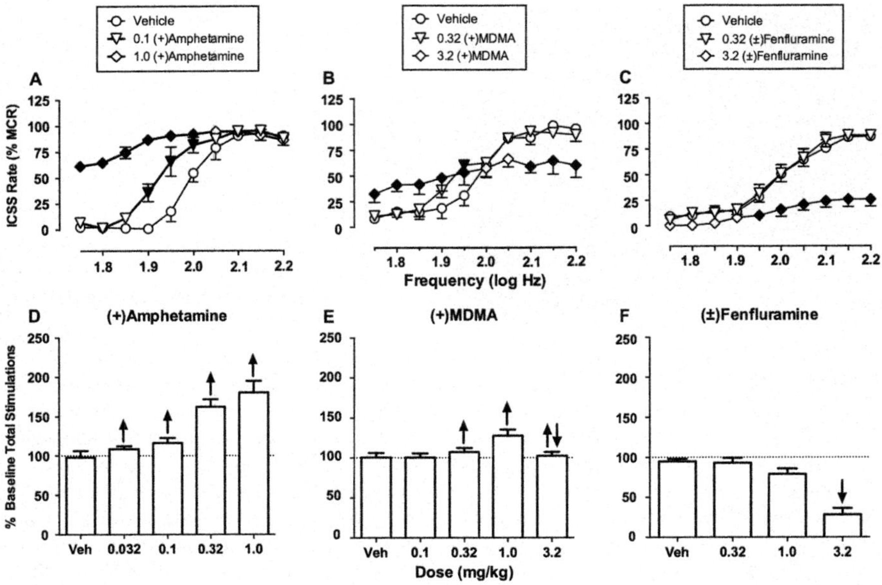

Fig. 3 (+)Amphetamine, (+)MDMA, and (±)fenfluramine produce qualitatively different effects on ICSS in rats. Top panels **a–c** show effects of selected drug doses on full frequency-rate curves. *Abscissae*: brain stimulation frequency in log Hz. *Ordinates*: ICSS rate expressed as %MCR. Filled points indicated a significant difference from "Vehicle" ($p < 0.05$). Bottom panels **e** and **f** show a summary measure of total ICSS across all 10 frequencies of brain stimulation. *Abscissae*: Drug dose in mg/kg. *Ordinates*: Total ICSS expressed as a percentage of the baseline number of total stimulations delivered in the absence of any treatment. *Upward/downward arrows* indicate a significant increase/decrease in ICSS for at least one brain stimulation frequency in the full frequency-rate curves as shown in Panels **a–c**. The maximum increase in total ICSS produced by any drug dose was used for correlations shown in Fig. 4. Adapted from Bauer et al. [18]

assessment of the abuse potential of monoamine releasers. Moreover, these results also provide evidence to suggest that drug selectivity to act at DAT vs. SERT is a significant determinant of abuse-related behavioral effects for monoamine releasers. Of course, one ultimate goal of these preclinical neurochemical and behavioral studies is to predict abuse potential of novel drugs in humans. The risk of abuse by humans is a difficult endpoint to quantify, in part because definitions of abuse include not only the extent of drug use, but also the degree of harm caused by that use [17, 19]. However, with these caveats in mind, abuse liability is generally considered highest for amphetamine and lower for MDMA, and fenfluramine is considered to have little or no abuse liability.

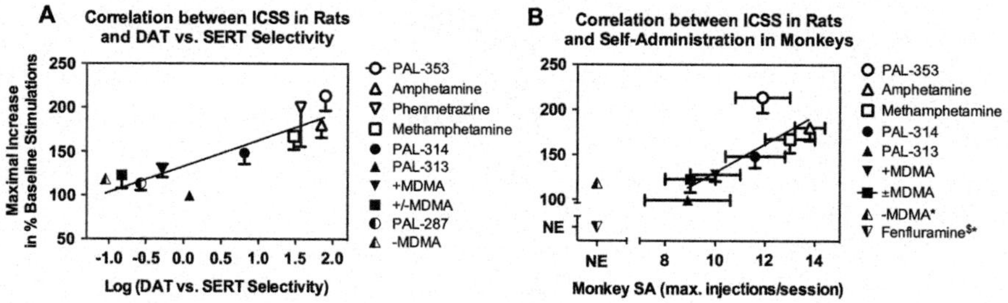

Fig. 4 Drug-induced facilitation of ICSS in rats correlates with both (**a**) DAT vs. SERT selectivity as determined from in vitro studies of monoamine release in rat brain synaptosomes as shown in Table 1 ($r = 0.89$, $p < 0.001$), or (**b**) maximum self-administration produced by any dose of each drug in a progressive-ratio assay of drug self-administration in rhesus monkeys ($r = 0.80$, $p = 0.032$). Error bars show SEM. Adapted from Bauer et al. [18]

3 Quantitative Structure–Activity Relationships for Para-Substituted Methcathinone Analogs

The results summarized above suggest a strong relationship for monoamine releasers between:

(1) in vitro neurochemical effects determined by measures of selectivity to promote monoamine release via DAT vs. SERT in rat brain synaptosomes,
(2) in vivo neurochemical effects determined by microdialysis measures of selectivity to release DA vs. 5HT in nucleus accumbens, and
(3) abuse-related behavioral effects in an ICSS procedure

These results also provide a framework for assessment of new psychoactive substances, and as one example, we conducted quantitative structure–activity relationship (QSAR) analysis for a series of seven racemic methcathinone analogs with different substitutions at the *para* (or 4-) position on the phenyl ring (Fig. 5) [9, 20, 21]. For the purposes of these studies, drugs were named using the convention "4-R MCAT," and the series included the parent compound methcathinone (MCAT) as well as the recently scheduled analogs flephedrone (4-F MCAT) and mephedrone (4-OCH$_3$ MCAT) and the other halogenated analogs brephedrone (4-Br-MCAT) and clephedrone (4-Cl-MCAT). Substituents were selected with respect to the three structural attributes as shown in Table 2: (1) steric bulk of the substituent in three-dimensional space, quantified here by volume (Vol); (2) electron-withdrawing capacity of the substituent (σ_p); and (3) lipophilicity of the substituent (π_p). A goal of the study was to evaluate the correlation between the structural attributes of these substituents and the functional effects of the associated drugs (also shown in Table 2) to produce neurochemical effects in in vitro and in vivo assays of monoamine release and abuse-related behavioral effects in the ICSS procedure.

Figure 6 shows the results of these QSAR analyses. There were two main findings. First, as discussed above, there were significant positive correlations

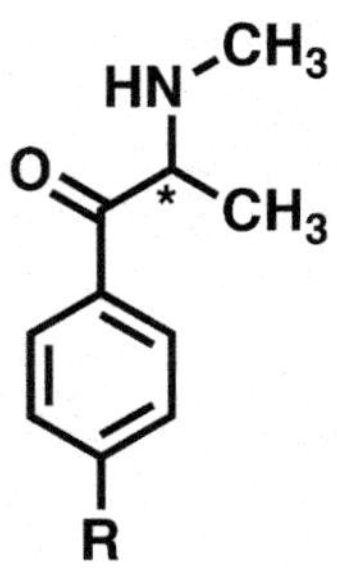

Fig. 5 Structure of 4-R methcathinone analogs used for QSAR analysis. Seven compounds were synthesized and evaluated with different 4-R substituents to vary structural parameters as shown in Table 2. *Asterisk* indicates position of the chiral carbon

Table 2 Structural and functional attributes of 4-substituted methcathinone (4-R MCAT) analogs used in quantitative structure–activity response (QSAR) analysis

		Structural attributes[a]			Neurochemical selectivity[b]		Behavior[c]
Drug	R	Vol	σ_p	π_p	In vitro	In vivo	Maximal ICSS
MCAT (methcathinone)	-H	150.36	0	0	309	12.56	191.9
4-F MCAT (flephedrone)	-F	153.78	0.06	0.14	15.4	1.24	156.3
4-Cl MCAT (clephedrone)	-Cl	164.43	0.23	0.71	3.40	1.23	114.9
4-CH_3 MCAT (mephedrone)	-CH_3	166.89	−0.17	0.56	2.41	0.62	102.5
4-Br MCAT (brephedrone)	-Br	169.43	0.23	0.86	1.01	0.89	118
4-OCH_3 MCAT (methedrone)	-OCH_3	175.01	−0.27	−0.02	0.24	0.32	110.9
4-CF_3 MCAT	-CF_3	178.40	0.54	0.88	0.07	Not determined	90.9

Drugs are listed in order of increasing volume of the 4-substituent
[a]Reported in Bonano et al. [20]; Sakloth et al. [21]
[b]In vitro selectivity calculated as effective concentration to produce a 50% increase (EC_{50}) in monoamine release via SERT ÷ EC_{50} to increase monoamine release via DAT from rat brain synaptosomes [20]. In vivo selectivity calculated as effective dose to produce a 250% increase (ED_{250}) to increase 5HT levels ÷ ED_{250} to increase DA levels in rat nucleus accumbens as assessed by in vivo microdialysis [9]
[c]Maximal facilitation of ICSS as determined in a behavioral assay of ICSS [20]

for all functional measures (Fig. 6d). Specifically, the in vitro and in vivo measures of drug selectivity to promote monoamine release via DAT vs. SERT correlated with each other and with the measure of abuse-related behavioral effects in the ICSS procedure. These correlations support the propositions that (a) in vitro measures of neurochemical selectivity at DAT vs. SERT in rat brain synaptosomes are

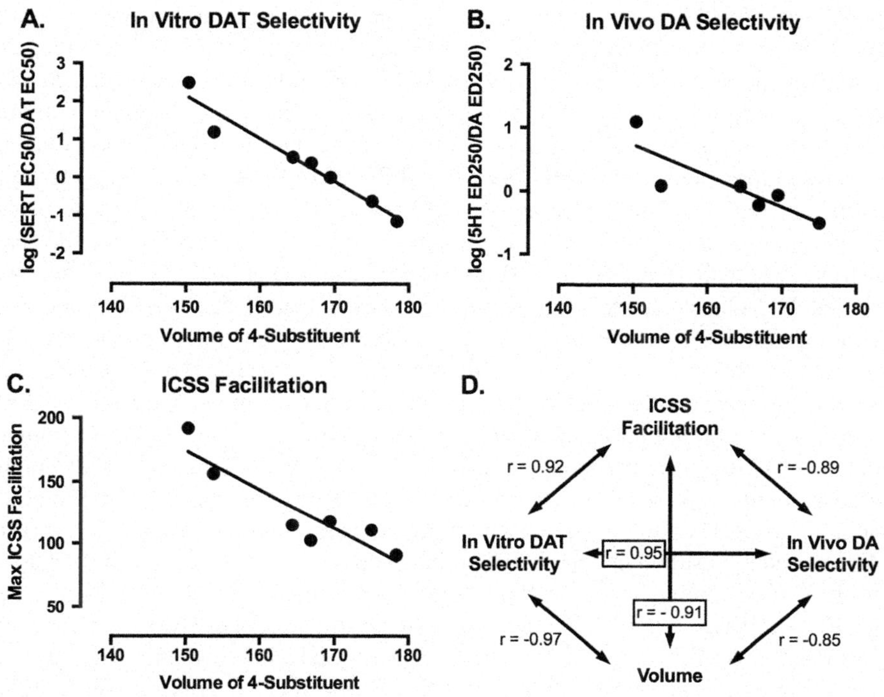

Fig. 6 Correlations between volume of the 4-substituent and (**a**) in vitro selectivity to promote monoamine release via DAT vs. SERT in rat brain synaptosomes, (**b**) in vivo selectivity to increase extracellular DA vs. 5HT levels in rat nucleus accumbens, and (**c**) in vivo effectiveness to produce abuse-related facilitation of ICSS. (**d**) Matrix of correlations between 4-substituent volume and each of the three functional endpoints. Volume correlated negatively with all functional measures, and all functional measures correlated positively with each other. All correlations were significant ($p < 0.05$)

predictive of in vivo neurochemical selectivity to promote DA vs. 5HT release, and (b) these measures of neurochemical selectivity are predictive of abuse-related behavioral effects. It is also important to note that drug effects on ICSS did not correlate reliably with potency of drugs to act at DAT alone in vitro or to release DA alone in vivo (data not shown). This indicates that expression of abuse-related behavioral effects results from an integration of DAT- and SERT-mediated effects, and it provides a rationale for QSAR studies that consider structural determinants of drugs at both transporters rather than at DAT alone.

The second main finding of the QSAR studies was that each of the three functional measures (in vitro DAT selectivity, in vivo DA selectivity, and ICSS effects) correlated negatively with volume of the 4-position substituent (Fig. 6), but none of the functional measures correlated with either the electronic or lipophilic attributes of the 4-substituent (data not shown). These results suggest that steric bulk of the 4-substituent plays a more important role than either electronic or lipophilic attributes in governing each drug's interaction with DAT and SERT.

More specifically, larger 4-substituent volumes were associated with declining DAT potencies but increasing SERT potencies, suggesting that DAT has limited tolerance for bulk at the 4-position, whereas SERT prefers larger substituents at this location, yielding a net loss in DAT vs. SERT selectivity as 4-substituent volume increases. On the basis of these observations, molecular modeling was conducted with homology models of human DAT and SERT (hDAT and hSERT, respectively) based on the *Drosophila melanogaster* DAT (dDAT) to identify the characteristics of substrate-binding pockets that might account for the differential selectivities of 4-R MCAT analogs at DAT and SERT. These results suggested two determinants of 4-R MCAT selectivity. First, docking studies indicated that hDAT contains a relatively large serine residue (S149) in the substrate-binding pocket at the site that interacts with the 4-substituent of MCAT analogs, whereas hSERT contains a smaller alanine residue (A169) at the homologous location. The larger S149 amino acid in hDAT limits the volume of the 4-substituent that can be accommodated, resulting in a preference by hDAT for 4-R MCAT analogs with small 4-substituents (e.g., 4-H for MCAT itself). Conversely, the smaller A169 amino acid in hSERT allows more space in the substrate-binding pocket for larger 4-substituents. Although the A169 amino acid in the docking pocket renders hSERT more tolerant than DAT of larger 4-substituents, it did not explain why hSERT displays a preference for larger 4-substituents. To address this issue, Hydropathic INTeraction (HINT) analysis was conducted, and this suggested a second determinant of 4-R MCAT selectivity. Specifically, HINT analysis indicated that the substrate-binding pocket of hSERT displayed a preference for relatively larger 4-substituents due in part to hydrophobic interactions between transporter and substrate. Overall, these studies indicated that hDAT prefers smaller 4-substituents, whereas SERT prefers larger 4-substituents. Figure 7 shows a simplified diagram to summarize these conclusions and their implications for abuse potential.

4 Stereoselective Effects of Methcathinone and Mephedrone

The QSAR studies summarized above were conducted with racemic compounds, but more recent studies have identified an additional role for stereoselectivity as a determinant both of 4-R MCAT interactions with transporters and of ultimate expression of abuse-related effects [22, 23]. Specifically, methcathinone, methamphetamine, and many of their analogs possess a single chiral carbon atom (the α carbon signified by the asterisk in Fig. 5), and the *S* enantiomer of these compounds is typically more potent and/or effective than the *R* enantiomer to promote DA release via DAT [7, 24] or to produce abuse-related behavioral effects in assays of drug self-administration, drug discrimination, or ICSS [18, 25–27]. However, recent studies suggest a potentially more nuanced role for stereochemistry in

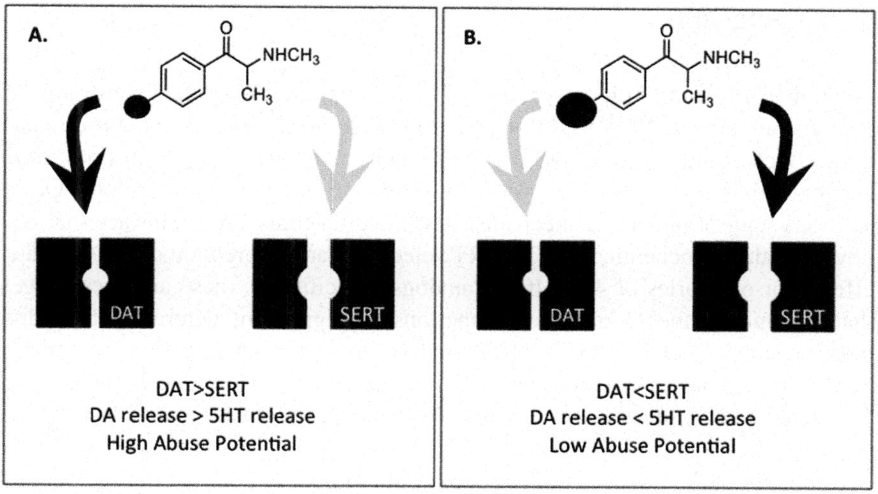

Fig. 7 QSAR and modeling studies suggest that DAT prefers small 4-substituents of 4-R MCAT analogs, whereas SERT prefers larger 4-substituents. (**a**) As a result of these structural differences in the transporters, 4-R MCAT analogs with small 4-substituents (e.g., MCAT) are more potent as substrates at DAT than SERT, leading to preferential DA release and strong abuse-related behavioral effects in vivo. (**b**) Conversely, 4-R MCAT analogs with larger 4-substituents are more potent as substrates for SERT than DAT, leading to preferential 5HT release and weak abuse-related behavioral effects in vivo

abuse-related effects of mephedrone (4-CH_3 MCAT) [22]. Specifically, the *R*(+) enantiomer of mephedrone is more effective than the *S*(−) enantiomer to produce locomotor activation, conditioned place preference, and facilitation of ICSS in rats [22]. Neurochemical evidence suggested that this apparent inversion of stereochemistry results from an unusual stereoselectivity not only in potency, but also in selectivity as a substrate at DAT vs. SERT. Thus, *R*(+)mephedrone was slightly more potent than its *S*(−) enantiomer to promote monoamine release via DAT but much less potent at SERT. As a result, the *R*(+) enantiomer displays a 50-fold greater selectivity than the *S*(−) enantiomer to promote monoamine release via DAT vs. SERT, and this stereoselectivity in neurochemical effects contributed to stereoselectivity in expression of abuse-related behavioral effects. It is unknown whether this stereoselectivity would also be apparent for other 4-R MCAT analogs, but a similar impact of stereochemistry was observed for isomers of 4-CH_3 cathinone [23]. Importantly, these results suggest that stereoselectivity at the chiral carbon at one end of the 4-R MCAT molecule can influence interactions of the 4-substituent at the other end of the molecule with its own portion of the DAT and SERT substrate-binding pockets.

5 Conclusions

Preclinical research with a wide range of monoamine transporter substrates has demonstrated that DAT > SERT selectivity is a strong determinant of abuse-related drug effects. Studies summarized in this chapter support this general proposition and extend it to a series of synthetic cathinone analogs. Furthermore, QSAR analyses suggest molecular mechanisms at the drug-transporter interface that may govern both neurochemical DAT/SERT selectivity and expression of abuse-related effects for one series of 4-R MCAT analogs. Specifically, these analyses suggest that volume of the 4-substituent functions as significant determinant of drug potency and selectivity, with DAT preferring smaller 4-substituents, whereas SERT prefers larger 4-substituents. These studies illustrate one application of QSAR analysis to investigate structural determinants of abuse-related drug effects.

Acknowledgements This work was supported by R01 DA033930.

References

1. Baumann MH, Partilla JS, Lehner KR (2013) Psychoactive "bath salts": not so soothing. Eur J Pharmacol 698:1–5
2. De Felice LJ, Glennon RA, Negus SS (2014) Synthetic cathinones: chemical phylogeny, physiology, and neuropharmacology. Life Sci 97:20–26
3. Reith ME, Blough BE, Hong WC, Jones KT, Schmitt KC, Baumann MH, Partilla JS, Rothman RB, Katz JL (2015) Behavioral, biological, and chemical perspectives on atypical agents targeting the dopamine transporter. Drug Alcohol Depend 147:1–19
4. Simmler LD, Liechti ME (2016) Interactions of cathinone NPS with human transporters and receptors in transfected cells. Curr Top Behav Neurosci
5. Sitte HH, Freissmuth M (2015) Amphetamines, new psychoactive drugs and the monoamine transporter cycle. Trends Pharmacol Sci 36:41–50
6. Solis E (2016) Electrophysiological actions of synthetic cathinones on monoamine transporters. Curr Top Behav Neurosci
7. Rothman RB, Baumann MH, Dersch CM, Romero DV, Rice KC, Carroll FI, Partilla JS (2001) Amphetamine-type central nervous system stimulants release norepinephrine more potently than they release dopamine and serotonin. Synapse 39:32–41
8. Setola V, Hufeisen SJ, Grande-Allen KJ, Vesely I, Glennon RA, Blough B, Rothman RB, Roth BL (2003) 3,4-Methylenedioxymethamphetamine (MDMA, "Ecstasy") induces fenfluramine-like proliferative actions on human cardiac valvular interstitial cells in vitro. Mol Pharmacol 63:1223–1229
9. Suyama JA, Sakloth F, Kolanos R, Glennon RA, Lazenka MF, Negus SS, Banks ML (2016) Abuse-related neurochemical effects of para-substituted methcathinone analogs in rats: microdialysis studies of nucleus accumbens dopamine and serotonin. J Pharmacol Exp Ther 356:182–190
10. Baumann MH, Clark RD, Rothman RB (2008) Locomotor stimulation produced by 3,4-methylenedioxymethamphetamine (MDMA) is correlated with dialysate levels of serotonin and dopamine in rat brain. Pharmacol Biochem Behav 90:208–217

11. Aarde SM, Taffe MA (2016) Predicting the abuse liability of entactogen-class new and emerging psychoactive substances via preclinical models of drug self-administration. Curr Top Behav Neurosci
12. Carter LP, Griffiths RR (2009) Principles of laboratory assessment of drug abuse liability and implications for clinical development. Drug Alcohol Depend 105(Suppl 1):S14–S25
13. Olive MF, Watterson L (2016) Reinforcing effects of cathinone NPS in the intravenous drug self-administration paradigm. Curr Top Behav Neurosci
14. Wang Z, Woolverton WL (2007) Estimating the relative reinforcing strength of (+/-)-3,4-methylenedioxymethamphetamine (MDMA) and its isomers in rhesus monkeys: comparison to (+)-methamphetamine. Psychopharmacology (Berl) 189:483–488
15. Wee S, Anderson KG, Baumann MH, Rothman RB, Blough BE, Woolverton WL (2005) Relationship between the serotonergic activity and reinforcing effects of a series of amphetamine analogs. J Pharmacol Exp Ther 313:848–854
16. Woods JH, Tessel RE (1974) Fenfluramine: amphetamine congener that fails to maintain drug-taking behavior in the rhesus monkey. Science 185:1067–1069
17. Negus SS, Miller LL (2014) Intracranial self-stimulation to evaluate abuse potential of drugs. Pharmacol Rev 66:869–917
18. Bauer CT, Banks ML, Blough BE, Negus SS (2013) Use of intracranial self-stimulation to evaluate abuse-related and abuse-limiting effects of monoamine releasers in rats. Br J Pharmacol 168:850–862
19. Nutt D, King LA, Saulsbury W, Blakemore C (2007) Development of a rational scale to assess the harm of drugs of potential misuse. Lancet 369:1047–1053
20. Bonano JS, Banks ML, Kolanos R, Sakloth F, Barnier ML, Glennon RA, Cozzi NV, Partilla JS, Baumann MH, Negus SS (2015) Quantitative structure-activity relationship analysis of the pharmacology of para-substituted methcathinone analogues. Br J Pharmacol 172:2433–2444
21. Sakloth F, Kolanos R, Mosier PD, Bonano JS, Banks ML, Partilla JS, Baumann MH, Negus SS, Glennon RA (2015) Steric parameters, molecular modeling and hydropathic interaction analysis of the pharmacology of para-substituted methcathinone analogues. Br J Pharmacol 172:2210–2218
22. Gregg RA, Baumann MH, Partilla JS, Bonano JS, Vouga A, Tallarida CS, Velvadapu V, Smith GR, Peet MM, Reitz AB, Negus SS, Rawls SM (2015) Stereochemistry of mephedrone neuropharmacology: enantiomer-specific behavioural and neurochemical effects in rats. Br J Pharmacol 172:883–894
23. Hutsell B, Baumann MH, Partilla J, Banks ML, Verkariya R, Glennon RA, Negus SS (2016) Abuse-related neurochemical and behavioral effects of cathinone and 4-methylcathinone stereoisomers in rats. Eur J Neuropsychopharm 26:288–297
24. Rothman RB, Vu N, Partilla JS, Roth BL, Hufeisen SJ, Compton-Toth BA, Birkes J, Young R, Glennon RA (2003) In vitro characterization of ephedrine-related stereoisomers at biogenic amine transporters and the receptorome reveals selective actions as norepinephrine transporter substrates. J Pharmacol Exp Ther 307:138–145
25. Balster RL, Schuster CR (1973) A comparison of d-amphetamine, l-amphetamine, and methamphetamine self-administration in rhesus monkeys. Pharmacol Biochem Behav 1:67–71
26. Glennon RA, Young R, Hauck AE, McKenney JD (1984) Structure-activity studies on amphetamine analogs using drug discrimination methodology. Pharmacol Biochem Behav 21:895–901
27. Johanson CE, Schuster CR (1981) A comparison of the behavioral effects of l- and dl-cathinone and d-amphetamine. J Pharmacol Exp Ther 219:355–362

Reinforcing Effects of Cathinone NPS in the Intravenous Drug Self-Administration Paradigm

Lucas R. Watterson and M. Foster Olive

Abstract Since the mid- to late 2000s, there has been a dramatic rise in the use and abuse of synthetic derivatives of cathinone, a stimulant alkaloid found in the African shrub *Catha edulis*. Cathinone novel psychoactive substances (NPS), also referred to as synthetic cathinones or "bath salt"-type drugs, have gained popularity among drug users due to their potency, low cost, ease of procurement, and diverse array of evolving chemical structures. While the ability of cathinone NPS to produce psychotomimetic effects, multiple organ system toxicity, and death in humans is well documented, there has been limited scientific investigation into the reinforcing effects and abuse liability of these drugs. In this chapter, we will summarize the existing literature on the reinforcing effects of cathinone NPS in rodents using the intravenous self-administration (IVSA) paradigm. We will also compare the ability of cathinone NPS to serve as reinforcers to that of classical psychostimulants such as cocaine, methamphetamine, and methylenedioxymethamphetamine (MDMA). The chapter will conclude with a summary and indications for future avenues of research on cathinone NPS.

Keywords Cathinone • Reinforcement • Intravenous self-administration • Animal model • Operant conditioning

---

L.R. Watterson and M.F. Olive (✉)
Department of Psychology, Arizona State University, PO Box 871104, Tempe, AZ 85287, USA
e-mail: foster.olive@asu.edu

Curr Topics Behav Neurosci (2017) 32: 133–144
DOI 10.1007/7854_2016_33
Published Online: 15 July 2016

Contents

1 Cathinone NPS Abuse and Pharmacology

Across numerous continents, the late 2000s witnessed a dramatic surge in the synthesis, marketing, and abuse of novel psychoactive substances (NPS). Many of the NPS that emerged during this time were chemical derivatives of the alkaloid cathinone, a naturally occurring amphetamine-like chemical found in the *Catha edulis* (Khat) shrub. Cathinone NPS are often referred to in the scientific literature as synthetic cathinones and are colloquially referred to as "bath salts". This latter term is a result of initial marketing tactics to disguise them as false retail bath products, in order to evade law enforcement and regulatory agencies.

In the USA, the rise in the use of cathinone NPS was alarmingly rapid, with poison control centers receiving 0, 304, and 6,156 calls reporting cathinone-related toxicity in the years 2009–2011, respectively [1]. During this time, approximately 98% of cathinone NPS revealed in toxicological investigations were identified as 4-methylmethcathinone (4-MMC, mephedrone), 3,4-methylenedioxypyrovalerone (MDPV), and 3,4-methylenedioxymethcathinone (methylone) [2–6]. While these three cathinone NPS (sometimes referred to as the "3Ms" [3]), as well as many derivatives, have since been placed into Schedule I or other illegal status in the USA and elsewhere, newer cathinone-related NPS continue to surface. Thus, cathinone NPS represent a constantly evolving class of synthetic psychostimulants with the potential for abuse.

Cathinone NPS are primarily used for their desired psychological effects, which include euphoria and increases in energy, libido, and alertness [2–6]. However, serious adverse psychological and behavioral effects are associated with the use of cathinone NPS, including agitated delirium and paranoia, persistent hallucinations and delusions, aggression, and violence. In addition, cathinone NPS pose a significant public health hazard, as their use is significantly associated with clinical toxicity of multiple physiological systems [7–10]. Despite this high risk of adverse effects, cathinone users frequently report a persistent desire to continue using the drugs, and prolonged periods of misuse have been reported [10–17]. Collectively, these observations suggest that some cathinone NPS possess a high potential for abuse and dependence.

2 The Self-Administration Model of Drug Reinforcement

The intravenous self-administration (IVSA) paradigm is generally considered to be the "gold standard" of animal models designed to assess the abuse liability of psychoactive substances [18]. Most often performed in rodents, this paradigm involves surgical implantation of an indwelling intravenous catheter into the jugular or other major vein, while the other end is tunneled under the skin and connected to a vascular access port implanted on the dorsum. Following recovery from surgery, the animal is placed in an operant conditioning chamber (see Fig. 1) equipped with response manipulanda (e.g., levers or nose poke detectors) that are interfaced to a computer. A sterile drug solution is placed in a syringe and delivered by a computer-controlled syringe pump located outside the apparatus. This solution is delivered to the animal via a single-channel liquid swivel in order to allow free rotation of the animal while maintaining a continuous flow of the solution. Responding on one of the manipulanda designated as "active" results in a computer-controlled drug infusion and simultaneous presentation of auditory and/or visual cues. Responding on the other manipulanda designated as "inactive" serves as a control for nonspecific behavior and generally has no programmed consequences at any time during the experiment. To avoid overdose or toxicity due to multiple drug infusions in close temporal proximity, each drug infusion is often followed immediately by a "timeout" period (e.g., 20 s) whereby additional active responses do not result in additional drug infusions. Self-administration sessions are typically 1–6 or more hours in length and are typically conducted 5–7 days per week.

The IVSA paradigm offers numerous advantages over other animal models of human drug-taking behavior. Such advantages include, but are not limited to:

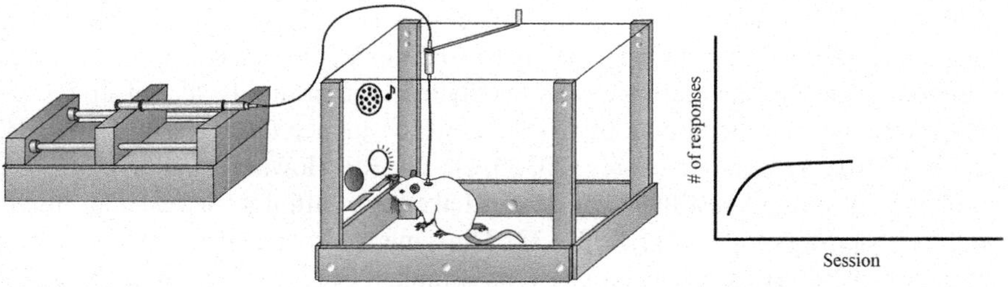

Fig. 1 Typical experimental apparatus utilized in the rodent intravenous self-administration (IVSA) paradigm. Upon pressing one of two levers, a computer-controlled syringe pump (*left*) delivers a solution containing an abused drug into an indwelling venous catheter via tubing connected to a liquid swivel. Each drug infusion is accompanied by simultaneous presentation of tone and/or illumination of a stimulus light located above the lever. The graph on the *right* represents typical response patterns across daily experimental sessions during the acquisition of drug self-administration

1. the drug is administered voluntarily by the animal at numerous points during the session, rather than as a single bolus injection administered passively by an experimenter;
2. the drug is administered directly into the bloodstream, which simulates intravenous drug use in humans;
3. additional experimental variables such as frequency and duration of drug access can be manipulated;
4. the effects of alternative reinforcers or response-contingent punishment can be assessed;
5. extinction and reinstatement procedures can be incorporated to model drug-seeking behavior and relapse;
6. the behavioral demand required to deliver each infusion can be varied to assess the efficacy of a particular drug to serve as a reinforcer. This latter phenomenon is most often integrated into the progressive ratio paradigm, where the number of responses required for each successive drug infusion can be increased via a linear or exponential function until the animal "gives up" and ceases responding.

3 Reinforcing Efficacy of Cathinone NPS and Comparison to Classical Psychostimulants

Cathinone NPS exert their stimulant and sympathomimetic effects via neurochemical mechanisms that are strikingly similar to those of classical psychostimulants such as cocaine or amphetamines. Specifically, it has been demonstrated that MDPV inhibits the activity of presynaptic transporters for dopamine (DAT) and norepinephrine (NET), but has little affinity for presynaptic serotonin transporters (SERT). As a result, MPDV produces lasting increases in synaptic levels of dopamine DA and NE, but not 5-HT [19–27]. While the neurochemical actions of MDPV are similar to those of cocaine, its effects appear to be much more potent and longer lasting. In contrast, similar to traditional illicit amphetamines, other cathinones such as mephedrone and methylone act primarily as substrates for presynaptic plasma membrane transporters, which induce the release of DA, NE, and 5-HT from presynaptic stores [20, 23, 24, 28–30]. However, unlike traditional amphetamines, ring-substituted cathinones have lower affinity for vesicular monoamine transporters (VMAT) [29, 31, 32] Thus, mephedrone and methylone appear to act more similarly to amphetamine-type stimulants, including methamphetamine and MDMA. However, it is becoming increasingly apparent that many cathinone derivatives have diverse mechanisms of action, each with differing actions on and balances of affinities for DAT, NET, and SERT [24], which appears to influence their reinforcing and entactogenic effects (see below). As a result, many studies examining the abuse potential of newer cathinone NPS often include comparisons to traditional psychostimulants.

Studies conducted up to three decades ago have reported that animals will intravenously self-administer either cathinone itself or its methylated derivative methcathinone [33–35]. However, studies examining the reinforcing efficacy of newer cathinone NPS have just recently begun to emerge [18, 36]. The first report of the ability of newer cathinone derivatives to serve as an intravenous reinforcer was published by Hadlock and colleagues [37]. In this study, male Sprague–Dawley rats were first trained to lever press for food reinforcement in an operant conditioning paradigm. Next, rats underwent catheter implantation and were randomly assigned to self-administer mephedrone or methamphetamine at a dose of 0.24 mg per 10 μl infusion, or saline in 4-hour daily sessions conducted for 8 days. Following initial training for food reinforcement, rats assigned to receive intravenous saline failed to maintain operant responding. However, rats allowed to self-administer methamphetamine or mephedrone rapidly acquired and maintained responding. Interestingly, rats self-administering mephedrone displayed more robust increases in active lever pressing across daily sessions than rats self-administering the same dose of methamphetamine. The authors speculated that these differences were attributable to the differential pharmacokinetic and monoamine-releasing properties of these stimulants.

The findings of Hadlock and colleagues represented an important first demonstration of the reinforcing effects of cathinone NPS and paved the way for more detailed examinations of patterns of self-administration across a range of doses and other experimental conditions. A subsequent study by Aarde and colleagues [38] demonstrated that mephedrone supported intravenous self-administration in male Wistar and Sprague–Dawley rats. Such effects were observed under fixed ratio conditions at several doses, most reliably at doses of 0.5 and 1 mg/kg per infusion. These doses of mephedrone were approximately an order of magnitude higher than doses of methamphetamine that support self-administration in this study and others [39], suggesting that mephedrone is approximately ten times less potent as a reinforcer than methamphetamine. Aarde et al. [38] also demonstrated that during the initial phases of acquisition of drug self-administration, male Sprague–Dawley rats exhibited lower mephedrone intake under the same dose conditions as Wistar rats, although no strain differences in pharmacokinetic parameters were noted. Finally, under progressive ratio conditions, the most robust responding for mephedrone was observed when the per-infusion dose was increased to 1.5 mg/kg following initial training on a 0.5 mg/kg dose. A similar study by Motbey and colleagues [40] also demonstrated that male Sprague–Dawley rats would self-administer mephedrone at doses ranging from 0.03 to 1 mg/kg in a typical inverted U-shaped dose–response fashion. These animals also showed increased overall numbers of mephedrone infusions compared to animals self-administering methamphetamine and less robust hyperlocomotion, supporting the notion of a lower potency of mephedrone relative to methamphetamine. Finally, two studies by Taffe and colleagues have demonstrated that mephedrone is a more efficacious reinforcer in both male and female rats than the entactogenic drug MDMA during the initial phases of drug self-administration [41, 42].

With regard to the cathinone derivative MDPV, Aarde and colleagues demonstrated that MDPV was readily self-administered in male Wistar rats at doses ranging from 0.01 to 0.5 mg/kg per infusion and at rates and amounts similar to those of rats self-administering methamphetamine [43]. These investigators also noted a dose-dependent increase in breakpoints for MDPV reinforcement on a progressive ratio schedule that were higher than those observed for similar doses of methamphetamine, suggesting that MPDV is a more potent reinforcer than this traditional psychostimulant. Subsequent studies by these investigators showed that ~60% of rats acquiring MDPV self-administration did so in "binge-like" manner, operationally defined as eight infusions in a 5 min interval, but this was not influenced by the availability of a running wheel as a nondrug reinforcer [44]. Studies by our laboratory [45] have provided similar results, where male Sprague–Dawley rats readily acquired MDPV self-administration under limited access conditions (2 h daily sessions) at all doses tested (0.05, 0.1, and 0.2 mg/kg per infusion). In addition, a positive relationship between MDPV dose and breakpoints for drug reinforcement under progressive ratio conditions was observed in this study, similar to results reported by Aarde and colleagues [43], although in contrast we observed similar breakpoints for the same (0.05 mg/kg/infusion) dose of MDPV and methamphetamine. When self-administration sessions were increased in duration to 6 h/day, we observed an escalation of drug intake over time at the 0.1 and 0.2 mg/kg per-infusion doses of MDPV.

As mentioned previously, MDPV appears to act as a long-lasting monoamine reuptake inhibitor with preferential affinity for presynaptic DA and NE transporters, as opposed to mephedrone which acts as a monoamine-releasing agent. Given the cocaine-like pharmacological action of MDPV, it is important to compare its reinforcing efficacy to that of cocaine itself, which was recently reported by Schindler and colleagues [46]. In this study it was demonstrated MDPV was readily self-administered by male Sprague–Dawley rats at doses of 0.003–0.03 mg/kg per infusion. In addition, rates of MDPV self-administration at the 0.03 mg/kg dose were similar to those of rats trained to self-administer cocaine at a dose of 0.5 mg/kg per infusion, indicating that MDPV is a more potent reinforcer than cocaine, consistent with its binding profile at monoamine transporters [19, 20, 22–27].

In contrast to the apparent robust reinforcing effects of MDPV, investigations into the ability of the monoamine-releasing cathinone NPS methylone have produced less consistent results. In 2012, we reported that male Sprague–Dawley rats did not display robust self-administration of a low dose of methylone (0.05 mg/kg per infusion) tested under limited daily access conditions (2 h/day), but more reliable self-administration was observed at higher doses (0.1, 0.2, and 0.5 mg/kg per infusion) [47]. In addition, when daily sessions were extended to 6 h in length, an escalation of methylone intake was not observed at any dose tested, unlike the escalation of intake that we observed with extended access to MDPV [45]. Other investigators have reported that intravenous methylone possesses only weak reinforcing properties in male and female Wistar rats at doses of 0.3–0.5 mg/kg per infusion (similar to that observed with the classical entactogen MDMA) [41, 42, 46], but the reinforcing effects were potentiated in rats initially trained to self-administer mephedrone [41, 42]. While the mechanism for these effects is currently unknown, these and other investigators [46] have speculated that the lower reinforcing efficacy of methylone is due to its ability to facilitate 5-HT release at similar potencies as MDMA, which is also not reliably self-administered in the IVSA paradigm.

Finally, it should be noted the aforementioned studies have focused primarily on assessing the reinforcing effects of "first-generation" cathinone NPS (mephedrone, MDPV, and methylone). However, there are dozens of other "second-generation" cathinone NPS that are already on current drug markets or will be in the coming years. Such second-generation cathinone NPS include naphthylpyrovalerone (naphyrone), 4-methoxymethcathinone (methedrone), β-keto-*N*-methylbenzodioxolylbutanamine (butylone), β-keto-methylbenzodioxolylpentanamine (pentylone), 4-methylpyrrolidinopropiophenone (4-MePPP), 2-methylamino-1-phenylpentan-1-one (pentedrone), 4-fluoro-*N*-methylcathinone (4-FMC, flephedrone), 4-methyl-*N*-ethylcathinone (4-MEC), and α-pyrrolidinopentiophenone (α-PVP). We recently reported that both α-PVP and 4-MEC facilitate brain stimulation reward [48], and today we are only aware of one study that has examined the potential reinforcing effects of second-generation cathinone NPS. In this study, Aarde and colleagues demonstrated that male Wistar rats readily self-administered relatively low doses of α-PVP (0.025–0.25 mg/kg per infusion) under fixed ratio conditions, similar to those observed in rat self-administering MDPV and displayed typical inverted U-shaped dose-dependent breakpoints for α-PVP under progressive ratio conditions [49]. Thus, α-PVP appears to have reinforcing properties similar in potency to MDPV.

4 Summary and Avenues for Future Research

In this chapter, we have reviewed the small but growing body of preclinical literature demonstrating that various cathinone NPS are dose-dependently self-administered by laboratory rodents via the intravenous route. Such patterns of self-administration are similar in nature to those observed with traditional psychostimulants such as cocaine and amphetamines and collectively suggest that many cathinone NPS possess significant abuse liability and potential for addiction. Thus, cathinone NPS should continue to be considered an emerging class of abused drugs that warrant appropriate regulatory control, as well as adequate interventions for detoxification and treatment of dependence.

Not surprisingly, however, the structure-activity relationships of different cathinone NPS at monoamine transporters influence their potential abuse liability. From the studies reviewed here, the potency of these cathinone derivatives to serve as intravenous reinforcers can be rank ordered as MDPV ~ α-PVP > mephedrone > methylone. In comparison to traditional psychostimulants, the dose-effect function of these cathinones suggests that α-PVP and MDPV are roughly equipotent with methamphetamine, whereas mephedrone appears to be roughly equipotent to cocaine. However, these potencies do not necessarily reflect their affinity or mode of action at monoamine transporters. For example, while both MPDV and α-PVP inhibit monoamine reuptake and appear to be equipotent in their reinforcing efficacy, α-PVP has a higher selectivity for DAT vs. SERT than MDPV [50]. Previous studies suggest that higher DAT/SERT inhibition ratios are more predictive of reinforcer efficacy than DAT affinity per se [51–53]. While MDPV appears to be approximately equipotent with methamphetamine as a reinforcer, MDPV acts as a long-lasting DAT inhibitor, while methamphetamine is primarily a DA/NE releasing agent. Clearly, it is not feasible to test the ability of every single cathinone NPS, whether currently in existence or predicted based on structure-activity relationships, to support IVSA and thus serve as a behavioral reinforcer. It is therefore likely that bioinformatics, computational chemistry, and in silico molecular modeling are needed to predict the abuse potential of cathinone NPS as they continue to evolve.

There are numerous other unanswered questions regarding cathinone NPS and their abuse potential. For example, what other nonhuman model organisms can be used to predict the abuse liability of cathinone NPS? Do cathinone NPS exhibit physiological affinity for any molecular entities other than monoamine transporters, which might contribute to their neurobiological or behavioral effects? Specifically, do cathinone NPS interact with trace amine-associated receptors or elements of glutamate transmission, as has been shown to be the case for traditional psychostimulants [54, 55]? One immediate direction for future studies should employ the use of selective antagonists to assess these and other potential molecular targets for their role in cathinone reinforcement. Finally, what are the effects of cathinone NPS on neuronal plasticity and function, gliotransmission, cerebrovascular function, cell viability, gene expression, and epigenetic processes?

It is clear that the field of cathinone NPS research is in its infancy. While some legislative efforts have attempted to curb the availability and abuse of these drugs, it is clear that many more NPS of this drug class (and others) will continue to evolve likely steps ahead of policymakers, scientists, educators, and treatment professionals.

Acknowledgments The authors wish to acknowledge Public Health Service grant R01DA025606 from the National Institute on Drug Abuse for supporting our initial studies on cathinone NPS abuse potential.

References

1. Association of Poison Control Centers (2013) Bath salts data. 30 April 2013
2. Baumeister D, Tojo LM, Tracy DK (2015) Legal highs: staying on top of the flood of novel psychoactive substances. Ther Adv Psychopharmacol 5:97–132
3. Karch SB (2015) Cathinone neurotoxicity ("the 3Ms"). Curr Neuropharmacol 13:21–25
4. Underwood E (2015) A new drug war. Science 347:469–473
5. Weaver MF, Hopper JA, Gunderson EW (2015) Designer drugs 2015: assessment and management. Addict Sci Clin Pract 10:8
6. Zawilska JB (2015) "Legal highs" - an emerging epidemic of novel psychoactive substances. Int Rev Neurobiol 120:273–300

7. Banks ML, Worst TJ, Rusyniak DE, Sprague JE (2014) Synthetic cathinones ("bath salts"). J Emerg Med 46:632–642
8. Froberg BA, Levine M, Beuhler MC, Judge BS, Moore PW, Engebretsen KM, McKeown NJ, Rosenbaum CD, Young AC, Rusyniak DE (2015) Acute methylenedioxypyrovalerone toxicity. J Med Toxicol 11:185–194
9. Hall C, Heyd C, Butler C, Yarema M (2014) "Bath salts" intoxication: a new recreational drug that presents with a familiar toxidrome. CJEM 16:171–176
10. Karila L, Megarbane B, Cottencin O, Lejoyeux M (2015) Synthetic cathinones: a new public health problem. Curr Neuropharmacol 13:12–20
11. Batisse A, Fortias M, Bourgogne E, Gregoire M, Sec I, Djezzar S (2014) Case series of 21 synthetic cathinones abuse. J Clin Psychopharmacol 34:411–413
12. Baumann MH (2014) Awash in a sea of 'bath salts': implications for biomedical research and public health. Addiction 109:1577–1579
13. Capriola M (2013) Synthetic cathinone abuse. Clin Pharmacol Adv Appl 5:109–115
14. Fass JA, Fass AD, Garcia AS (2012) Synthetic cathinones (bath salts): legal status and patterns of abuse. Ann Pharmacother 46:436–441
15. Paillet-Loilier M, Cesbron A, Le Boisselier R, Bourgine J, Debruyne D (2014) Emerging drugs of abuse: current perspectives on substituted cathinones. Subst Abuse Rehabil 5:37–52
16. Sadeg N, Darie A, Vilamot B, Passamar M, Frances B, Belhadj-Tahar H (2014) Case report of cathinone-like designer drug intoxication psychosis and addiction with serum identification. Addict Disord Treat 13:38–43
17. Winder GS, Stern N, Hosanagar A (2013) Are "bath salts" the next generation of stimulant abuse? J Subst Abuse Treat 44:42–45
18. Watterson LR, Olive MF (2014) Synthetic cathinones and their rewarding and reinforcing effects in rodents. Adv Neurosci 2014:209875
19. Baumann MH, Partilla JS, Lehner KR, Thorndike EB, Hoffman AF, Holy M, Rothman RB, Goldberg SR, Lupica CR, Sitte HH, Brandt SD, Tella SR, Cozzi NV, Schindler CW (2013) Powerful cocaine-like actions of 3,4-methylenedioxypyrovalerone (MDPV), a principal constituent of psychoactive 'bath salts' products. Neuropsychopharmacology 38:552–562
20. Cameron K, Kolanos R, Verkariya R, De Felice L, Glennon RA (2013) Mephedrone and methylenedioxypyrovalerone (MDPV), major constituents of "bath salts," produce opposite effects at the human dopamine transporter. Psychopharmacology (Berl) 227:493–499
21. Cameron KN, Kolanos R, Solis E, Glennon RA, De Felice LJ (2013) Bath salts components mephedrone and methylenedioxypyrovalerone (MDPV) act synergistically at the human dopamine transporter. Br J Pharmacol 168:1750–1757
22. Coppola M, Mondola R (2012) 3,4-methylenedioxypyrovalerone (MDPV): chemistry, pharmacology and toxicology of a new designer drug of abuse marketed online. Toxicol Lett 208: 12–15
23. De Felice LJ, Glennon RA, Negus SS (2014) Synthetic cathinones: chemical phylogeny, physiology, and neuropharmacology. Life Sci 97:20–26
24. Glennon RA (2014) Bath salts, mephedrone, and methylenedioxypyrovalerone as emerging illicit drugs that will need targeted therapeutic intervention. Adv Pharmacol 69:581–620
25. Kolanos R, Partilla JS, Baumann MH, Hutsell BA, Banks ML, Negus SS, Glennon RA (2015) Stereoselective actions of methylenedioxypyrovalerone (MDPV) to inhibit dopamine and norepinephrine transporters and facilitate intracranial self-stimulation in rats. ACS Chem Neurosci 6:771–777
26. Kolanos R, Solis EJ, Sakloth F, De Felice LJ, Glennon RA (2013) "Deconstruction" of the abused synthetic cathinone methylenedioxypyrovalerone (MDPV) and an examination of effects at the human dopamine transporter. ACS Chem Neurosci 4:1524–1529
27. Simmler L, Buser T, Donzelli M, Schramm Y, Dieu LH, Huwyler J, Chaboz S, Hoener M, Liechti M (2013) Pharmacological characterization of designer cathinones in vitro. J Comp Neurol 168:458–470

28. Baumann MH, Ayestas MA Jr, Partilla JS, Sink JR, Shulgin AT, Daley PF, Brandt SD, Rothman RB, Ruoho AE, Cozzi NV (2012) The designer methcathinone analogs, mephedrone and methylone, are substrates for monoamine transporters in brain tissue. Neuropsychopharmacology 37:1192–1203
29. Eshleman AJ, Wolfrum KM, Hatfield MG, Johnson RA, Murphy KV, Janowsky A (2013) Substituted methcathinones differ in transporter and receptor interactions. Biochem Pharmacol 85:1803–1815
30. Lopez-Arnau R, Martinez-Clemente J, Carbo M, Pubill D, Escubedo E, Camarasa J (2013) An integrated pharmacokinetic and pharmacodynamic study of a new drug of abuse, methylone, a synthetic cathinone sold as "bath salts". Prog Neuropsychopharmacol Biol Psychiatry 45: 64–72
31. Cozzi NV, Sievert MK, Shulgin AT, Jacob P III, Ruoho AE (1999) Inhibition of plasma membrane monoamine transporters by Beta-keto amphetamines. Eur J Pharmacol 381:63–69
32. Pifl C, Reither H, Hornykiewicz O (2015) The profile of mephedrone on human monoamine transporters differs from 3,4-methylenedioxymethamphetamine primarily by lower potency at the vesicular monoamine transporter. Eur J Pharmacol 755:119–126
33. Gosnell BA, Yracheta JM, Bell SM, Lane KE (1996) Intravenous self-administration of cathinone by rats. Behav Pharmacol 7:526–531
34. Kaminski BJ, Griffiths RR (1994) Intravenous self-injection of methcathinone in the baboon. Pharmacol Biochem Behav 47:981–983
35. Woolverton WL, Johanson CE (1984) Preference in rhesus monkeys given a choice between cocaine and d, l-cathinone. J Exp Anal Behav 41:35–43
36. Watterson LR, Watterson E, Olive MF (2013) Abuse liability of novel 'legal high' designer stimulants: evidence from animal models. Behav Pharmacol 24:341–355
37. Hadlock GC, Webb KM, McFadden LM, Chu PW, Ellis JD, Allen SC, Andrenyak DM, Vieira-Brock PL, German CL, Conrad KM, Hoonakker AJ, Gibb JW, Wilkins DG, Hanson GR, Fleckenstein AE (2011) 4-Methylmethcathinone (mephedrone): neuropharmacological effects of a designer stimulant of abuse. J Pharmacol Exp Ther 339:530–536
38. Aarde SM, Angrish D, Barlow DJ, Wright MJJ, Vandewater SA, Creehan KM, Houseknecht KL, Dickerson TJ, Taffe MA (2013) Mephedrone (4-methylmethcathinone) supports intravenous self-administration in Sprague-Dawley and Wistar rats. Addict Biol 18:768–799
39. Kitamura O, Wee S, Specio SE, Koob GF, Pulvirenti L (2006) Escalation of methamphetamine self-administration in rats: a dose-effect function. Psychopharmacology (Berl) 186:48–53
40. Motbey CP, Clemens KJ, Apetz N, Winstock AR, Ramsey J, Li KM, Wyatt N, Callaghan PD, Bowen MT, Cornish JL, McGregor IS (2013) High levels of intravenous mephedrone (4-methylmethcathinone) self-administration in rats: Neural consequences and comparison with methamphetamine. J Psychopharmacol 27:823–836
41. Creehan KM, Vandewater SA, Taffe MA (2015) Intravenous self-administration of mephedrone, methylone and MDMA in female rats. Neuropharmacology 92:90–97
42. Vandewater SA, Creehan KM, Taffe MA (2015) Intravenous self-administration of entactogen-class stimulants in male rats. Neuropharmacology 99:538–545
43. Aarde SM, Huang PK, Creehan KM, Dickerson TJ, Taffe MA (2013) The novel recreational drug 3,4-methylenedioxypyrovalerone (MDPV) is a potent psychomotor stimulant: self-administration and locomotor activity in rats. Neuropharmacology 71:130–140
44. Aarde SM, Huang PK, Dickerson TJ, Taffe MA (2015) Binge-like acquisition of 3,4-methylenedioxypyrovalerone (MDPV) self-administration and wheel activity in rats. Psychopharmacology (Berl) 232:1867–1877
45. Watterson LR, Kufahl PR, Nemirovsky NE, Sewalia K, Grabenauer M, Thomas BF, Marusich JA, Wegner S, Olive MF (2014) Potent rewarding and reinforcing effects of the synthetic cathinone 3,4-methylenedioxypyrovalerone (MDPV). Addict Biol 19:165–174
46. Schindler CW, Thorndike EB, Goldberg SR, Lehner KR, Cozzi NV, Brandt SD, Baumann MH (2016) Reinforcing and neurochemical effects of the "bath salts" constituents 3,4-methylenedioxypyrovalerone (MDPV) and 3,4-methylenedioxy-N-methylcathinone (methylone) in male rats. Psychopharmacology (Berl) 233:1981–1990

47. Watterson LR, Hood LE, Sewalia K, Johnson CT, Wegner S, Grabenauer M, Thomas BF, Marusich JA, Olive MF (2012) The reinforcing effects of the methylone, a synthetic cathinone commonly found in "bath salts". J Addict Res Ther S9:002
48. Watterson LR, Burrows BT, Hernandez RD, Moore KN, Grabenauer M, Marusich JA, Olive MF (2014) Effects of α-pyrrolidinopentiophenone and 4-methyl-N-ethylcathinone, two synthetic cathinones commonly found in second-generation "bath salts," on intracranial self-stimulation thresholds in rats. Int J Neuropsychopharmacol 18:1–9
49. Aarde SM, Creehan KM, Vandewater SA, Dickerson TJ, Taffe MA (2015) In vivo potency and efficacy of the novel cathinone α-pyrrolidinopentiophenone and 3,4-methylenedioxypyrovalerone: self-administration and locomotor stimulation in male rats. Psychopharmacology (Berl) 232: 3045–3055
50. Marusich JA, Antonazzo KR, Wiley JL, Blough BE, Partilla JS, Baumann MH (2014) Pharmacology of novel synthetic stimulants structurally related to the "bath salts" constituent 3,4-methylenedioxypyrovalerone (MDPV). Neuropharmacology 87:206–213
51. Roberts DCS, Phelan R, Hodges LM, Hodges MM, Bennett B, Childers S, Davies H (1999) Self-administration of cocaine analogs by rats. Psychopharmacology (Berl) 144:389–397
52. Wee S, Anderson KG, Baumann MH, Rothman RB, Blough BE, Woolverton WL (2005) Relationship between the serotonergic activity and reinforcing effects of a series of amphetamine analogs. J Pharmacol Exp Ther 313:848–854
53. Wee S, Carroll FI, Woolverton WL (2006) A reduced rate of in vivo dopamine transporter binding is associated with lower relative reinforcing efficacy of stimulants. Neuropsychopharmacology 31:351–362
54. Gass JT, Olive MF (2008) Glutamatergic substrates of drug addiction and alcoholism. Biochem Pharmacol 75:218–265
55. Liberles SD (2015) Trace amine-associated receptors: ligands, neural circuits, and behaviors. Curr Opin Neurobiol 34:1–7

Predicting the Abuse Liability of Entactogen-Class, New and Emerging Psychoactive Substances via Preclinical Models of Drug Self-administration

Shawn M. Aarde and Michael A. Taffe

Abstract Animal models of drug self-administration are currently the gold standard for making predictions regarding the relative likelihood that a recreational drug substance will lead to continued use and addiction. Such models have been found to have high predictive accuracy and discriminative validity for a number of drug classes including ethanol, nicotine, opioids, and psychostimulants such as cocaine and methamphetamine. Members of the entactogen class of psychostimulants (drugs that produce an "open mind state" including feelings of interpersonal closeness, intimacy and empathy) have been less frequently studied in self-administration models. The prototypical entactogen 3,4-methylenedioxymethamphetamine (MDMA; "Ecstasy") supports self-administration but not with the same consistency nor with the same efficacy as structurally related drugs amphetamine or methamphetamine. Consistent with these observations, MDMA use is more episodic in the majority of those who use it frequently. Nevertheless, substantial numbers of MDMA users will meet the criteria for substance dependence at some point in their use history. This review examines the currently available evidence from rodent self-administration studies of MDMA and two of the new and emerging psychoactive substances (NPS) that produce entactogen type neuropharmacological responses – mephedrone (4-methylmethcathinone; 4MMC; "meow meow") and methylone (3,4-methylenedioxymethcathinone). Overall, the current evidence predicts that these NPS entactogens have enhanced abuse liability compared with MDMA.

Keywords Addiction • Drug abuse • Empathogen • Entactogen • MDMA • Mephedrone • Methylone • Self-administration

---

S.M. Aarde and M.A. Taffe (✉)
Committee on the Neurobiology of Addictive Disorders Mailcode SP30-2400, The Scripps Research Institute, 10550 North Torrey Pines Road, La Jolla, CA 92037, USA
e-mail: mtaffe@scripps.edu

Curr Topics Behav Neurosci (2017) 32: 145–164
DOI 10.1007/7854_2016_54
Published Online: 8 November 2016

Contents

1 Current Epidemiology of Entactogen Use

The term "entactogen" is used to delineate recreational drugs that "produce experiences of emotional communion, oneness, relatedness, emotional openness – that is, empathy or sympathy" [1] or, as originally coined by Nichols, a drug that "powerfully enhances emotions and empathy" [2]. Although (±)3,4-methylenedioxyamphetamine (MDA) enjoyed some popularity through the mid-1980s, (±)3,4-methylenedioxymethamphetamine (MDMA) became the class-defining entactogen and has been the most common constituent of illicit "Ecstasy" [3] over recent decades. Epidemiological data from the Monitoring the Future (MtF) survey show that use of Ecstasy in high school students [4] has been stable in the past decade while annual prevalence rates for college students have gradually increased [5]. Annual prevalence of MDMA/Ecstasy in individuals in their 20s is equivalent to that for non-crack cocaine, slightly lower than for all amphetamines (including ADHD medications) and is at least threefold higher than prevalence of heroin, crack (smokable cocaine), ice (smokable methamphetamine), or phencyclidine. The MtF data also report that 17% of 29- to 30-year-olds and 10–14% of 21- to 28-year-olds have used MDMA at least once in their lifetime [5]. Thus, overall lifetime rates of recreational exposure to MDMA are substantial and will continue to be so for some time as these cohorts age. Further exposure may result from the initiation of multiple Phase I clinical trials to establish MDMA as an adjunctive treatment for psychotherapy nearly a decade ago [6–9], although results remain controversial [10–12]. Overall, the impact of MDMA on health continues to be a pressing issue for scientific investigation.

Greer, arguing for the use of MDMA (50–200 mg) in psychotherapy [13], claimed that MDMA has low abuse liability because its use is self-limiting (reduction in desirable effects and an increase in adverse effects with continued use) and some subsequent studies have indicated spontaneous disuse with time [14, 15], a transition that has been speculated to reflect lasting changes in brain serotonergic function. Nevertheless, significant proportions of heavy Ecstasy users meet criteria for dependence at some point in their use history [16–18] and there are case reports of Ecstasy use patterns that are daily or at least several times per week [19–21]. These latter examples are highly consistent with the repetitive use patterns that are common to reference drugs of abuse such as methamphetamine (METH), cocaine, and heroin. Similarly, close examination of Ecstasy use statistics in human cognitive/toxicity investigations identifies occasional (rare) individuals who use Ecstasy at least several times per week [22–28], which stands in contrast to studies of METH-related cognitive impairment in which *most* individuals are using 3 times per week or more [29–31]. It is also clear that Ecstasy consumers abuse a very wide range of drugs with some frequency [32, 33], suggesting a level of generalized substance dependence that may be elevated in those that use MDMA [34]. The evidence for human Ecstasy dependence has grown to the point where efforts are underway to establish criteria for a new MDMA-specific DSM diagnostic category [35, 36].

Recreational use of *cathinone* derivative stimulant drugs ("bath salts") is new but has increased substantially since 2009 and continues to expand worldwide. Some of the earliest appearing and most popular entities such as 4-methylmethcathinone (4MMC; *mephedrone*) and 3,4-methylenedioxymethcathinone (*methylone*; beta-keto-MDMA) were explicitly marketed as MDMA substitutes and are reported to have MDMA-like, entactogen-characteristic subjective effects [37, 38]. Use of mephedrone expanded rapidly in the UK from 2009 to 2010 [38, 39] during a reported European shortage of MDMA [40, 41]. Mephedrone and methylone appear to have sustained popularity despite legal controls [42] and have joined drugs such as MDMA or cocaine, rather than replacing them, in user populations [43]. Just as up to 40% of Ecstasy users may meet criteria for dependence [44], there is initial evidence of dependence on mephedrone [45, 46]. The 2013 and 2014 midyear reports of the US National Forensic Laboratory Information System [47, 48] show that methylone is now more common than MDMA in this database and case reports of fatalities involving mephedrone or methylone are highly reminiscent of similar deaths attributed to MDMA [49–53]. The emergence of these new, MDMA-like recreational drugs has prompted controlled, laboratory studies to determine the potential similarities and differences among these entactogens, including studies of relative abuse liability.

2 Entactogen Pharmacology and Predicted Addiction Liability

It is perhaps obvious that a recreational drug such as MDMA that exhibits distinct subjective properties compared with the structurally related prototypical psychostimulant METH would have different pharmacological and neurochemical properties. For example, in vitro investigations by several groups [54–58] show that methylone and mephedrone are monoamine transporter substrates, which act to enhance transporter-mediated release of monoamines as do both MDMA and METH, in vivo. However, mephedrone [54, 59, 60] and methylone [54, 61] each produce neuropharmacological profiles of enhanced release of serotonin (5-HT) compared to dopamine (DA) in the nucleus accumbens, similar to MDMA but dissimilar to METH. Thus, the key distinction in overall subjective effects between METH and the entactogens appears to be the relative effects on 5-HT transporters (SERT) versus DA transporters (DAT), and ensuing monoaminergic signaling.

The in vitro pharmacological data likewise show that the entactogens are dissimilar to a typical stimulant such as METH. Simmler and colleagues report a DAT/SERT ratio on transporter inhibition of 0.08 for MDMA, 1.4 for mephedrone, and 3.3 for methylone versus >10 for METH, using human transporters expressed in cells. DAT inhibition potency (IC_{50}) was 17 μM for MDMA, 4.82 μM for methylone, and 3.31 μM for mephedrone, compared with 1.05 μM for METH. Finally, an assay of monoamine release mediated by the DAT illustrates an effective concentration (EC_{50}) of 3.75 μM for mephedrone, 22 μM for MDMA, >100 μM for methylone, and 1.56 μM for METH. MDMA and mephedrone were about equipotent at inhibiting 5-HT release mediated by the SERT (5.63 μM and 5.98 μM, respectively); however, the EC_{50} for methylone was >10 μM compared to >33 μM for METH. An analysis of transporter-mediated monoamine release using rat brain synaptosomes [54] reported that the DAT/SERT ratios of the entactogens (mephedrone 2.41; methylone 1.82; MDMA 0.97) were much lower than that of METH (152.0). In summary, the in vitro pharmacological data and the in vivo neuropharmacological data predict that the entactogens would exhibit similar abuse liability, but these drugs would exhibit *less* propensity for repetitive use compared with a traditional stimulant drug like METH.

When it comes to controlled laboratory models of abuse liability such as the intravenous self-administration (IVSA) procedure, the relative DA/5-HT effects have been thought most critical. In short, relatively enhanced 5-HT effects tend to reduce the degree to which rats or monkeys will self-administer a given drug [62–64]. These findings are reinforced by demonstrations that drugs which function as 5-HT indirect agonists suppress the rate of cocaine or amphetamine self-administration in monkeys [62, 65, 66] and that exposure to a 5-HT depleting regimen of MDMA in rats enhances acquisition of the self-administration of cocaine [67] and enhances reinstatement of D-amphetamine seeking primed by either D-amphetamine or MDMA [68]. The correlation of DA/5-HT potency ratios with reinforcer efficacy agrees with a finding that prior treatment of rats with the serotonergic neurotoxin 5,7-dihydroxy-tryptamine [69] or genetic deletion of the SERT [70] enhances the acquisition of MDMA IVSA. Nevertheless, the idea of relative DAT/SERT selectivity as a major determinant of abuse liability is a dogma that was established before the emergence of the cathinones, with MDMA as the lone example of an entactogen that was popular with human users. The emergence of the two cathinone-class entactogen drugs mephedrone and methylone has provided a key opportunity to further determine abuse liability of designer stimulant drugs.

3 Self-administration of Entactogens

3.1 Overview of Methods

Drug self-administration is a powerful preclinical/non-clinical method for predicting the abuse liability of that drug [71]. Various self-administration procedures provide methods to quantify the rewarding capacity of different recreational drugs. If delivery of a drug can increase the frequency or probability of a behavioral response (such as a lever press) from a laboratory animal, it is considered to act as a reinforcer of that behavior; this capacity features prominently in the processes theorized to lead to addiction [72]. In general, the greater the efficacy of a drug as a reinforcer in laboratory animal self-administration procedures, the higher its potential abuse liability is predicted to be for humans.

Self-administration can therefore be used to evaluate the reinforcer efficacy and potency of new and emerging psychoactive substances (NPS) for which little, if any, human epidemiological data are available. Moreover, once a drug has been shown to be efficacious as a reinforcer, self-administration procedures can be used to determine the neurobiological, behavioral, or environmental determinants of drug taking behavior. For drugs that have highly variable use patterns and/or inconsistent/conflicting reports of abuse, the self-administration paradigm can be used to determine the conditions that increase reinforcer potency and efficacy.

Importantly, the most basic rodent self-administration procedure (i.e., lever pressing under a fixed-ratio schedule for the intravenous delivery of drug over a short 1–2 h access interval) is not a model of drug abuse or addiction – rather, it models drug reward and reinforcement [73]. It is clear, however, that reward or reinforcement processes contribute to, or are a component of, the phenomena outlined as being diagnostic of drug abuse/addiction [i.e., as defined by either the DSM5 (USA) or ICD10 (Europe)]. Under certain conditions, self-administration in an animal model may result in behavior that would meet the human diagnostic criteria (see below). Even if an animal's self-administration of a drug under the typically short periods of drug access does not constitute the expression of disordered use or addiction per se, the capacity of drugs to support self-administration has valuable and well-validated predictive utility [71, 74, 75].

There are two primary methods to determine the potential of a given drug to reinforce self-administration behavior. First and most simply, assessment of the initial acquisition of drug taking behavior can be used to quantify how rapidly and readily drug availability comes to support behavioral responding. This can be used to determine the threshold dose that is necessary to produce a behaviorally reinforcing effect, providing an estimate of drug potency. Comparison of different patterns of responding across drugs can provide additional insight; for example, the inter-session and intra-session variability in MDMA intake is relatively high [76] compared with METH. Similarly, comparison of the initial acquisition of responding for drug infusions can illustrate inter-individual variation in reinforcing value; for example, only about half of subjects meet criteria for MDMA self-administration, whereas virtually all acquire cocaine self-administration [77]. Drug acquisition can also vary across environmental and other contexts; for example, when alternative reinforcers are available (e.g., palatable food, wheel access), this may differentially alter responding for intravenous drug infusions [78].

A second way to assess the reinforcing capability of different drugs is by comparing dose–response curves under varied schedules of reinforcement, following initial acquisition. This involves varying the available per-infusion dose of the drug from session to session to determine how the animal alters its response pattern. In the simplest version, a fixed ratio (FR; each successive infusion within a session requires a fixed number of responses) schedule of reinforcement is typically used to determine potency (i.e., minimum amount of drug required to maintain responding). A more complex approach is the use of a *progressive* ratio (PR) schedule of reinforcement in which the number of responses required to obtain each successive infusion increases within the session. The PR procedure is typically used to determine efficacy (i.e., the maximum amount of responding the drug can support at a given dose) [79–81]. Alternatively, systematic within-session reductions in the available per-infusion dose to zero while under a fixed-ratio schedule can also permit calculation of the maximum amount of responding that the drug will support [82]. Both this within-session thresholding procedure and the PR approach are in essence a protracted transition to an extinction condition and as such it should be the case that the response rate drops to zero as either the response number becomes too high (PR) or the dose drops too low (within-session thresholding) [83, 84].

Additional modifications of the basic self-administration procedure that extend the model to address various DSM5/ICD-10 diagnostic criteria for substance abuse and/or addiction include:

1. The extended access model which usually compares self-administration sessions that are relatively short to ones that are long (e.g., 1–2 h vs. 6 h – typical for cocaine or methamphetamine) to study the phenomena of intake escalation [85–87].
2. The inclusion of adverse consequences or punishments – for example, foot shock [88–90] or histamine administration [91–93] – to study the phenomena of drug taking despite incurring negative outcomes.
3. Reinstatement of responding following its extinction by re-exposure to the drug, presentation of drug-paired cues, or exposure to a stressor to study the phenomena of relapse [94–97].
4. Making alternative "natural" reinforcers available to study the phenomena of devaluation of non-drug rewards and preoccupation with drug seeking [98–100].

3.2 MDMA

It has been repeatedly demonstrated that METH or amphetamine will readily support intravenous self-administration (IVSA) in rats [101–105], nonhuman primates (NHPs) [106–110], and cats [111]. Laboratory studies of the abuse liability of MDA or MDMA have been curiously sporadic in comparison with the typical amphetamines, perhaps because the "average" non-compulsive human use pattern does not appear to fit well with the usual animal models. While it is clear that MDA [112] or MDMA [107, 113–116] will substitute for cocaine in baboons and rhesus monkeys trained for intravenous self-administration, it has yet to be established that drug-naïve nonhuman primates will acquire MDA or MDMA self-administration by any route of administration, or that oral administration of either drug will function as a reinforce, even in drug-experienced NHPs.

Interpretation of entactogen IVSA in rats is complicated by a broad range of individual differences in drug preference compared with the IVSA of other stimulants, a relative dearth of studies and the methodological and analytical choices of authors in conducting their studies. One available study showed that rats will self-administer MDA (~0.3 mg/kg/inf) under fixed-ratio 1 (FR1) training conditions, but behavior was extinguished under progressive-ratio (PR) conditions [117]. Additional studies show that MDMA generates consistent levels (2–5 mg/kg/session) of self-administration in rats [118–123], although one laboratory report intakes several fold higher [76, 124, 125]. It is clear from the available evidence that rat IVSA of MDMA differs from the IVSA of typical psychostimulants such as cocaine, amphetamine, or METH. Dalley and colleagues [102] reported highly variable MDMA (50 μg freebase per infusion; ~0.15 mg MDMA HCl/kg/inf; 32–50 infusions over session) IVSA compared with either amphetamine or METH (same dose, 60–75 infusions with greater day to day stability) in rats that were food restricted. Schenk and colleagues have shown that only about 55–60% of individuals will reach acquisition criteria for MDMA IVSA within about 10–14 training sessions using a protocol which starts at 1 mg/kg/inf and then is reduced to 0.5 mg/kg/inf [70, 77], suggesting that subject inclusion criteria may partially contribute to differential results between laboratories. The critical effect of subject inclusion criteria was highlighted by work in which a median-split was presented for acquisition data instead of excluding subjects on arbitrary acquisition criteria [120, 123]. In particular, when rats were subjected to dose-substitution studies under an FR-1 response contingency, the less-preferring half of the distributions of male and female rats were probably not self-administering MDMA since they were insensitive to changes in the available dose.

Serotonergic dysfunction produced by either SERT deletion or a 5-HT selective neurotoxin [69, 70] dramatically increases the percent of animals reaching acquisition criteria for MDMA self-administration. In essence, these manipulations make the neuropharmacological response to MDMA more similar to that of METH which would provide one possible mechanistic explanation for the enhanced reinforcer efficacy of MDMA in such studies. In a similar vein, the ~50% of animals that met acquisition criteria for IVSA of MDMA are insensitive to antagonists of the 5-HT 1A, 1B or 2A receptor subtypes [126], possibly indicating that the half of the distribution of rats that acquire MDMA IVSA are constitutively less sensitive to the serotonergic effects of MDMA.

Lack of a systematic approach to methodological variables in reports published so far has further complicated the comparison of MDMA IVSA data with data from the broader number of studies involving METH or cocaine self-administration. For example, while studies hint that increases in MDMA intake, and the number of subjects who meet acquisition criteria, may be achieved through exposure to many sessions of access [102, 121, 127, 128], this issue has not been comprehensively explored. Other methodological variables also may be critical. For example, one report [129] demonstrates that acute increases in ambient temperature (30°C) significantly enhance self-administration of MDMA, though consistent training under elevated temperature conditions may not have any effect [130]. Additional variables which can affect self-administration include: rat strain, time of day in which sessions are conducted [131], the speed of the intravenous infusion [121], housing enrichment [132], and food restriction [102].

Relatively few studies have specifically examined MDMA IVSA under short vs long daily access conditions. This latter approach has been proposed to better reflect the transition to dependence [85, 86]. Vandewater and colleagues found that 6 h MDMA (0.5 mg/kg/inf) IVSA led to increased intake compared with 2 h sessions in male rats [123] which is consistent with the "escalated" intake reported for cocaine or METH. This finding was inconsistent with a prior report that 6 h access led to no difference in *total session MDMA intake* relative to rats trained in 2 h sessions [76], but that outcome may have depended on a relatively high per-infusion dose (1.0 mg/kg/inf) and/or training animals in the inactive (light) cycle. Additional targeted study would be required to resolve conditions under which MDMA self-administration does, or does not, escalate with longer daily access.

It has long been established that significant sex-differences exist in rat models of stimulant drug abuse. For example, female rats will self-administer more cocaine [133, 134] and more METH [135, 136] than males, and these sex differences can be more pronounced under long-access escalation and/or progressive ratio procedures. In the single study of MDMA IVSA in *female* rats that is currently available [120], the authors found that female rats showed only a slightly more consistent MDMA intake when directly compared with male rats under 2 h daily access conditions (see the Supplemental Materials of [123]). Escalation of self-administration of MDMA in females under longer access conditions has not yet been examined.

3.3 *Mephedrone and Methylone*

The original report of mephedrone IVSA in rats indicated that it can readily support IVSA [137]; however, the study was limited to a single per-infusion dose and access conditions that likely increased drug intake. Specifically, access to drug was relatively long (4 h) and under high (29°C) ambient temperature (T_A). As noted above, escalation of METH intake occurs when daily drug access is 6 h vs 2 h [138] and intake of cocaine, METH, and MDMA increase under relatively high (30°C vs. ~22°C) T_A conditions [129, 139]. Hadlock and colleagues also compared the

IVSA of mephedrone with IVSA of a similar per-infusion dose of METH and found greater intake of mephedrone. However, equipotent per-infusion doses are essential for comparison of numbers of infusions, since higher doses generate fewer infusions [138]. Subsequent work showed mephedrone to be less potent than METH in other assays [54, 60, 140]; thus, the initial report may have overestimated mephedrone IVSA relative to METH IVSA. Aarde and colleagues [101] showed that mephedrone supports IVSA similarly in both Wistar and Sprague–Dawley rats and that a training dose of 0.5 mg/kg per infusion supported approximately as many reinforcer deliveries as METH at a dose of 0.05 mg/kg per infusion. Motbey and colleagues [141] found that adolescent male Sprague–Dawley rats obtained a mean of about 60 infusions of mephedrone (0.3 mg/kg/inf) after 10 sessions, also suggesting it is a highly efficacious reinforcer in IVSA. Unfortunately, there are no other entactogen IVSA data in adolescent animals to place these data in context. Two additional studies directly compared the IVSA of mephedrone and MDMA and found that the mephedrone was a more efficacious reinforcer than MDMA, resulting in higher daily intakes of drug [120, 123].

The first examination of methylone IVSA showed robust acquisition of self-administration, with all rats reaching criteria at the 0.5 mg/kg/infusion training dose and breakpoints under a PR procedure similar to those reached by METH trained rats in 2-h access sessions [142]. Subsequent studies indicated a less-efficacious profile for methylone. Direct comparisons of methylone-trained male or female rats with those trained on MDMA or mephedrone showed the greatest intakes for mephedrone, the lowest for MDMA, and an intermediate profile for methylone [120, 123, 143]. Schindler and colleagues [61] found that male Sprague–Dawley rats showed methylone IVSA to about the same extent as Vandewater et al. [123], again in 2-h access sessions. A potential bridge across these studies was provided by an indication that training male rats in 6 h daily access sessions resulted in greater escalation of drug intake for methylone than for MDMA [123, 143]. These results suggest that as yet undetermined methodological differences may explain the differences in outcome between the study of 2 h access IVSA [61, 123, 142]. Nevertheless, the evidence at present suggests that methylone is most likely a more efficacious reinforcer than MDMA.

With respect to sex-differences, Creehan et al. [120] showed that female rats readily acquired IVSA of methylone; however, a follow-up study from the same group [123] demonstrated a similar relative abuse liability in male rats. Rats of both sexes reached a mean of about 12–15 infusions of methylone after 10 sessions of acquisition training, as compared to about 7–10 infusions of MDMA (0.5 mg/kg/inf) and 20–25 infusions of mephedrone (0.5 mg/kg/inf) in separate groups of male and female rats.

In summary, initial reports from rat IVSA studies indicate mephedrone and methylone exhibit readily established and relatively *consistent* self-administration profiles [137, 141, 142] that appear to contrast with the less reliable profile that has been established for MDMA. This conclusion was underlined most directly in studies that compared the three entactogens within a single model [120, 123, 143].

3.3.1 Future Directions

As this review has shown, there are relatively few investigations of the abuse liability of entactogen psychostimulants. Much is still to be identified about their properties, but there are nevertheless a few key issues which are of the most pressing importance for the field. The first three are relatively general: (1) study of sex-differences, (2) ability to rapidly screen novel cathinone derivatives, and (3) non-rat models of abuse liability. The remaining issues are specific to understanding why the entactogen cathinones appear to have enhanced abuse liability relative to MDMA (pharmacokinetics, speed of monoamine responses and non-monoamine pharmacological properties). These issues are discussed in further detail below.

3.3.2 Sex-Differences

The US National Institutes of Health (NIH) has recently issued a policy position which reinforces the critical importance of conducting sex-difference comparisons across biomedical domains [144]. The importance of this new initiative is certainly apparent for substance abuse research since substance use increases more quickly in women and treatment outcomes are poorer compared with men [145, 146]. METH dependence starts earlier in women [147, 148], cocaine treatment outcomes are less successful [145], and MDMA dependence occurs more frequently in women [46, 149]. It has been shown that female rats will self-administer more cocaine [133, 150] and more METH [135, 136] than male rats but, as mentioned briefly above, there is only one study of the self-administration of any entactogen in female rats that has been reported to date [120]. While that study found minimal sex-differences, it must be emphasized that this single study only begins to address the scope of work necessary for firm conclusions about potential sex-based differences.

3.3.3 Generalization or Rapid Screening

One of the clearest challenges with the cathinone derivative drugs is the diversity of the molecules which have entered recreational use markets. Driven by efforts to stay one step ahead of legal controls, the suppliers for recreational users have proven agile in providing a number of different compounds across place and time. This diversity poses a significant challenge for the research efforts to determine relative risks, and this concern is particularly acute for the relatively labor-intensive IVSA procedure. Dose-substitution tests of multiple drugs in a single group of rats is possible [120, 123], albeit this can be limited due to concerns over sequence effects and useful catheter life. Such attempts are very rare in rats for any drug class, but are much more common in nonhuman primate models, as reviewed

[151]. Additional participation in evaluating novel and emerging psychoactive substances from the many laboratories which use IVSA models to study the reinforcing properties of psychostimulants would be helpful in this regard.

3.3.4 Self-administration in Nonhuman Primates and Mice

No data are available on the self-administration of the entactogen cathinones in nonhuman primate or mouse models at present. Extension of research on the cathinones to these other relatively common self-administration models would facilitate deployment of the powerful genetic tools available with mouse models and would allow identification of any order differences in NHP models which might complicate translation to the human condition.

3.3.5 Pharmacokinetics

One key drug property which may differentiate mephedrone, methylone, and MDMA is the relative speed of brain entry, metabolism, and elimination. Rapid onset and offset of drugs is associated with more frequent re-dosing and transitions to habitual behavior. One study suggested that mephedrone may cross the blood–brain barrier more quickly than does MDMA [57], which may partially explain its enhanced efficacy as a reinforcer. A study conducted in humans found a 3.6-fold slower elimination half-life for MDMA compared with mephedrone after oral dosing [152] and methylone exhibits a 2.5-fold slower half-life compared with mephedrone after i.v. administration in rats [153, 154]. Humans variously self-administer entactogens orally, by insufflation, by intravenous injection, with e-cigarette "vape" inhalation and occasionally via rectal or vaginal routes so further understanding of how route of administration may affect pharmacokinetic profiles may be important.

3.3.6 Neuropharmacological Differences

Differences in the rapidity of the accumulation of extracellular DA versus 5-HT following injection of mephedrone and MDMA have been identified using intra-cerebral microdialysis techniques [59] although the 20 min sampling of that study gives poor temporal resolution. The reason for this difference is unknown at present and further investigation could help to clarify the predictive value of the relative temporal character of DA versus 5-HT responses. Since humans variously use entactogens by intravenous injection, insufflation, the oral route, and other methods, additional questions arise about the relative DA/5-HT responses in the context of different routes of administration.

3.3.7 Non-monoamine Targets

Although clearly the DA and 5-HT responses to the entactogens confer the majority of the signal when it comes to reinforcing and rewarding properties, it is always possible that subtle non-monoamine properties may differentiate these compounds. For example, a sigma1 affinity of mephedrone has been reported [155] and an agonist at this site would be predicted to enhance self-administration [156]. Additional study of the potential modulatory effects of non-monoamine neuropharmacological responses to various cathinone-derivative NPS would significantly advance understanding.

4 Discussion

It is clear that the relative neuropharmacological responses of DA and 5-HT in the nucleus accumbens of the rat are insufficient to completely predict the abuse liability of entactogen-class psychostimulants. Similarly, other preclinical models of abuse liability may not generate the best predictions. For example, Bonano and colleagues concluded that the lower potency of mephedrone to decrease intracranial self-stimulation reward (ICSS) thresholds [157] compared with methylone suggested that it would have lower abuse liability. Relatedly, mephedrone and methylone produce similar dose–response functions for locomotor stimulation in mice and identical potencies in a drug-discrimination assay when rats are trained on cocaine; however, methylone was less potent in METH-trained rats [158]. Thus, the most precise preclinical predictor of relative abuse liability for the entactogens will continue to be the self-administration procedure.

Acknowledgments Preparation of this review was supported in part by grants provided by the US National Institutes of Health (DA024105; DA024705) which had no further input on the content or decision to publish. The authors are grateful for the efforts of research technical staff Sophia A. Vandewater, Kevin M. Creehan, and Yanabel Grant who have assisted tremendously over the years with our efforts to understand the effects of entactogen and non-entactogen psychostimulant drugs.

References

1. Wikipedia (2015) Empathogen-entactogen 2015 [cited]. https://en.wikipedia.org/wiki/Empathogen-entactogen
2. Nichols DE (1986) Differences between the mechanism of action of MDMA, MBDB, and the classic hallucinogens. Identification of a new therapeutic class: entactogens. J Psychoactive Drugs 18:305–313
3. EcstasyData.org (2006) EcstasyData.org: ecstasy lab testing and analysis results – ecstasy pill reports 2006 [cited]. www.ecstasydata.org

4. Johnston LD, O'Malley PM, Bachman JG, Scheulenberg JE, Miech RA (2015) Monitoring the future national survey results on drug use, 1975-2014. Volume I, Secondary school students. University of Michigan, Ann Arbor, p. 599
5. Johnston LD, O'Malley PM, Bachman JG, Schulenberg JE, Miech RA (2015) Monitoring the future national survey results on drug use, 1975-2014. Volume II, College students and adults ages 19-55. University of Michigan, Ann Arbor, p. 439
6. Doblin R (2006) MAPS-sponsored cancer anxiety research. MAPS Bull 16:11
7. Mithoefer M (2006) MDMA-assisted psychotherapy in the treatment of posttraumatic stress disorder (PTSD): seventh update on study progress. MAPS Bull 16:7–8
8. Mojeiko V (2006) Israel MDMA/PTSD research project. MAPS Bull 16:10
9. Oehen P (2006) MDMA-assisted psychotherapy pilot study in Switzerland. MAPS Bull 16:9
10. Amoroso T (2015) The psychopharmacology of +/-3,4 methylenedioxymethamphetamine and its role in the treatment of posttraumatic stress disorder. J Psychoactive Drugs:1–8
11. Parrott AC (2014) The potential dangers of using MDMA for psychotherapy. J Psychoactive Drugs 46:37–43
12. White CM (2014) 3,4-Methylenedioxymethamphetamine's (MDMA's) Impact on Post-traumatic Stress Disorder. Ann Pharmacother 48:908–915
13. Greer G, Tolbert R (1986) Subjective reports of the effects of MDMA in a clinical setting. J Psychoactive Drugs 18:319–327
14. Parrott AC (2005) Chronic tolerance to recreational MDMA (3,4-methylenedioxymethamphetamine) or ecstasy. J Psychopharmacol 19:71–83
15. von Sydow K, Lieb R, Pfister H, Hofler M, Wittchen HU (2002) Use, abuse and dependence of ecstasy and related drugs in adolescents and young adults-a transient phenomenon? Results from a longitudinal community study. Drug Alcohol Depend 66:147–159
16. Cottler LB, Womack SB, Compton WM, Ben-Abdallah A (2001) Ecstasy abuse and dependence among adolescents and young adults: applicability and reliability of DSM-IV criteria. Hum Psychopharmacol 16:599–606
17. Schuster P, Lieb R, Lamertz C, Wittchen HU (1998) Is the use of ecstasy and hallucinogens increasing? Results from a community study. Eur Addict Res 4:75–82
18. Thomasius R, Petersen KU, Zapletalova P, Wartberg L, Zeichner D, Schmoldt A (2005) Mental disorders in current and former heavy ecstasy (MDMA) users. Addiction 100: 1310–1319
19. Hurault de Ligny B, El Haggan W, Comoz F, Lobbedez T, Pujo M, Griveau AM, Bottet P, Bensadoun H, Ryckelynck JP (2005) Early loss of two renal grafts obtained from the same donor: role of ecstasy? Transplantation 80:153–156
20. Jansen KL (1999) Ecstasy (MDMA) dependence. Drug Alcohol Depend 53:121–124
21. Kouimtsidis C, Schifano F, Sharp T, Ford L, Robinson J, Magee C (2006) Neurological and psychopathological sequelae associated with a lifetime intake of 40,000 ecstasy tablets. Psychosomatics 47:86–87
22. Fox HC, Parrott AC, Turner JJ (2001) Ecstasy use: cognitive deficits related to dosage rather than self-reported problematic use of the drug. J Psychopharmacol 15:273–281
23. Hanson KL, Luciana M (2004) Neurocognitive function in users of MDMA: the importance of clinically significant patterns of use. Psychol Med 34:229–246
24. MacInnes N, Handley SL, Harding GF (2001) Former chronic methylenedioxymethamphetamine (MDMA or ecstasy) users report mild depressive symptoms. J Psychopharmacol 15:181–186
25. Morgan MJ (1999) Memory deficits associated with recreational use of "ecstasy" (MDMA). Psychopharmacology (Berl) 141:30–36
26. Quednow BB, Kuhn KU, Hoppe C, Westheide J, Maier W, Daum I, Wagner M (2007) Elevated impulsivity and impaired decision-making cognition in heavy users of MDMA ("Ecstasy"). Psychopharmacology (Berl) 189:517–530
27. Ward J, Hall K, Haslam C (2006) Patterns of memory dysfunction in current and 2-year abstinent MDMA users. J Clin Exp Neuropsychol 28:306–324

28. White B, Day C, Degenhardt L, Kinner S, Fry C, Bruno R, Johnston J (2006) Prevalence of injecting drug use and associated risk behavior among regular ecstasy users in Australia. Drug Alcohol Depend 83:210–217
29. Johanson CE, Frey KA, Lundahl LH, Keenan P, Lockhart N, Roll J, Galloway GP, Koeppe RA, Kilbourn MR, Robbins T, Schuster CR (2006) Cognitive function and nigrostriatal markers in abstinent methamphetamine abusers. Psychopharmacology (Berl) 185:327–338
30. Johnson BA, Roache JD, Ait-Daoud N, Wells LT, Wallace CL, Dawes MA, Liu L, Wang XQ (2007) Effects of topiramate on methamphetamine-induced changes in attentional and perceptual-motor skills of cognition in recently abstinent methamphetamine-dependent individuals. Prog Neuropsychopharmacol Biol Psychiatry 31:123–130
31. Rippeth JD, Heaton RK, Carey CL, Marcotte TD, Moore DJ, Gonzalez R, Wolfson T, Grant I (2004) Methamphetamine dependence increases risk of neuropsychological impairment in HIV infected persons. J Int Neuropsychol Soc 10:1–14
32. Gouzoulis-Mayfrank E, Daumann J (2006) The confounding problem of polydrug use in recreational ecstasy/MDMA users: a brief overview. J Psychopharmacol 20:188–193
33. Wish ED, Fitzelle DB, O'Grady KE, Hsu MH, Arria AM (2006) Evidence for significant polydrug use among ecstasy-using college students. J Am Coll Health 55:99–104
34. Wu LT, Ringwalt CL, Weiss RD, Blazer DG (2009) Hallucinogen-related disorders in a national sample of adolescents: the influence of ecstasy/MDMA use. Drug Alcohol Depend 104:156–166
35. Cottler LB, Leung KS, Abdallah AB (2009) Test-re-test reliability of DSM-IV adopted criteria for 3,4-methylenedioxymethamphetamine (MDMA) abuse and dependence: a cross-national study. Addiction 104:1679–1690
36. Degenhardt L, Bruno R, Topp L (2009) Is ecstasy a drug of dependence? Drug Alcohol Depend 107:1–10
37. Winstock A, Mitcheson L, Ramsey J, Davies S, Puchnarewicz M, Marsden J (2011) Mephedrone: use, subjective effects and health risks. Addiction 106:1991–1996
38. Winstock AR, Mitcheson LR, Deluca P, Davey Z, Corazza O, Schifano F (2011) Mephedrone, new kid for the chop? Addiction 106:154–161
39. Winstock A, Mitcheson L, Marsden J (2010) Mephedrone: still available and twice the price. Lancet 376:1537–1537
40. Brunt TM, Niesink RJ, van den Brink W (2012) Impact of a transient instability of the ecstasy market on health concerns and drug use patterns in The Netherlands. Int J Drug Policy 23:134–140
41. Brunt TM, Poortman A, Niesink RJ, van den Brink W (2011) Instability of the ecstasy market and a new kid on the block: mephedrone. J Psychopharmacol 25(11):1543–1547
42. DEA (2011) Schedules of controlled substances: temporary placement of three synthetic cathinones in Schedule I. Final order. Fed Regist 76:65371–65375
43. Moore K, Dargan PI, Wood DM, Measham F (2013) Do novel psychoactive substances displace established club drugs, supplement them or act as drugs of initiation? The relationship between mephedrone, ecstasy and cocaine. Eur Addict Res 19:276–282
44. Leung KS, Ben Abdallah A, Copeland J, Cottler LB (2010) Modifiable risk factors of ecstasy use: risk perception, current dependence, perceived control, and depression. Addict Behav 35:201–208
45. Ding Y, He N, Shoptaw S, Gao M, Detels R (2014) Severity of club drug dependence and perceived need for treatment among a sample of adult club drug users in Shanghai, China. Soc Psychiatry Psychiatr Epidemiol 49:395–404
46. Uosukainen H, Tacke U, Winstock AR (2015) Self-reported prevalence of dependence of MDMA compared to cocaine, mephedrone and ketamine among a sample of recreational poly-drug users. Int J Drug Policy 26:78–83
47. DEA (2014) National Forensic Laboratory Information System: midyear report 2013. U.S. Drug Enforcement Administration, Springfield, VA

48. DEA (2015) National Forensic Laboratory Information System: midyear report 2014. U.S. Drug Enforcement Administration, Springfield, VA
49. Lusthof KJ, Oosting R, Maes A, Verschraagen M, Dijkhuizen A, Sprong AG (2011) A case of extreme agitation and death after the use of mephedrone in The Netherlands. Forensic Sci Int 206:e93–e95
50. Patel MM, Wright DW, Ratcliff JJ, Miller MA (2004) Shedding new light on the "safe" club drug: methylenedioxymethamphetamine (ecstasy)-related fatalities. Acad Emerg Med 11: 208–210
51. Pearson JM, Hargraves TL, Hair LS, Massucci CJ, Frazee CC, Garg U, Pietak BR (2012) Three fatal intoxications due to methylone. J Anal Toxicol 36:444–451
52. Schifano F (2004) A bitter pill. Overview of ecstasy (MDMA, MDA) related fatalities. Psychopharmacology (Berl) 173:242–248
53. Torrance H, Cooper G (2010) The detection of mephedrone (4-methylmethcathinone) in 4 fatalities in Scotland. Forensic Sci Int 202:e62–e63
54. Baumann MH, Ayestas MA Jr, Partilla JS, Sink JR, Shulgin AT, Daley PF, Brandt SD, Rothman RB, Ruoho AE, Cozzi NV (2012) The designer methcathinone analogs, mephedrone and methylone, are substrates for monoamine transporters in brain tissue. Neuropsychopharmacology 37:1192–1203
55. Baumann MH, Partilla JS, Lehner KR, Thorndike EB, Hoffman AF, Holy M, Rothman RB, Goldberg SR, Lupica CR, Sitte HH, Brandt SD, Tella SR, Cozzi NV, Schindler CW (2013) Powerful cocaine-like actions of 3,4-Methylenedioxypyrovalerone (MDPV), a principal constituent of psychoactive 'bath salts' products. Neuropsychopharmacology 38:552–562
56. Lopez-Arnau R, Martinez-Clemente J, Pubill D, Escubedo E, Camarasa J (2012) Comparative neuropharmacology of three psychostimulant cathinone derivatives: butylone, mephedrone and methylone. Br J Pharmacol 167:407–420
57. Simmler L, Buser T, Donzelli M, Schramm Y, Dieu LH, Huwyler J, Chaboz S, Hoener M, Liechti M (2013) Pharmacological characterization of designer cathinones in vitro. Br J Pharmacol 168:458–470
58. Simmler LD, Rickli A, Hoener MC, Liechti ME (2014) Monoamine transporter and receptor interaction profiles of a new series of designer cathinones. Neuropharmacology 79:152–160
59. Kehr J, Ichinose F, Yoshitake S, Goiny M, Sievertsson T, Nyberg F, Yoshitake T (2011) Mephedrone, compared with MDMA (ecstasy) and amphetamine, rapidly increases both dopamine and 5-HT levels in nucleus accumbens of awake rats. Br J Pharmacol 164: 1949–1958
60. Wright MJ Jr, Angrish D, Aarde SM, Barlow DJ, Buczynski MW, Creehan KM, Vandewater SA, Parsons LH, Houseknecht KL, Dickerson TJ, Taffe MA (2012) Effect of ambient temperature on the thermoregulatory and locomotor stimulant effects of 4-methylmethcathinone in Wistar and Sprague-Dawley rats. PLoS One 7:e44652
61. Schindler CW, Thorndike EB, Goldberg SR, Lehner KR, Cozzi NV, Brandt SD, Baumann MH (2016) Reinforcing and neurochemical effects of the "bath salts" constituents 3,4-methylenedioxypyrovalerone (MDPV) and 3,4-methylenedioxy-N-methylcathinone (methylone) in male rats. Psychopharmacology (Berl) 233:1981–1990
62. Negus SS, Mello NK, Blough BE, Baumann MH, Rothman RB (2007) Monoamine releasers with varying selectivity for dopamine/norepinephrine vs. serotonin release as candidate "agonist" medications for cocaine dependence: studies in assays of cocaine discrimination and cocaine self-administration in rhesus monkeys. J Pharmacol Exp Ther 320:627–636
63. Roberts DC, Phelan R, Hodges LM, Hodges MM, Bennett B, Childers S, Davies H (1999) Self-administration of cocaine analogs by rats. Psychopharmacology (Berl) 144:389–397
64. Wee S, Anderson KG, Baumann MH, Rothman RB, Blough BE, Woolverton WL (2005) Relationship between the serotonergic activity and reinforcing effects of a series of amphetamine analogs. J Pharmacol Exp Ther 313:848–854

65. Howell LL, Carroll FI, Votaw JR, Goodman MM, Kimmel HL (2007) Effects of combined dopamine and serotonin transporter inhibitors on cocaine self-administration in rhesus monkeys. J Pharmacol Exp Ther 320(2):757–765
66. Wee S, Woolverton WL (2006) Self-administration of mixtures of fenfluramine and amphetamine by rhesus monkeys. Pharmacol Biochem Behav 84:337–343
67. Fletcher PJ, Robinson SR, Slippoy DL (2001) Pre-exposure to (+/-)3,4-methylenedioxymethamphetamine (MDMA) facilitates acquisition of intravenous cocaine self-administration in rats. Neuropsychopharmacology 25:195–203
68. Morley KC, Cornish JL, Li KM, McGregor IS (2004) Preexposure to MDMA ("Ecstasy") delays acquisition but facilitates MDMA-induced reinstatement of amphetamine self-administration behavior in rats. Pharmacol Biochem Behav 79:331–342
69. Bradbury S, Bird J, Colussi-Mas J, Mueller M, Ricaurte G, Schenk S (2014) Acquisition of MDMA self-administration: pharmacokinetic factors and MDMA-induced serotonin release. Addict Biol 19:874–884
70. Oakly AC, Brox BW, Schenk S, Ellenbroek BA (2014) A genetic deletion of the serotonin transporter greatly enhances the reinforcing properties of MDMA in rats. Mol Psychiatry 19:534–535
71. O'Connor EC, Chapman K, Butler P, Mead AN (2011) The predictive validity of the rat self-administration model for abuse liability. Neurosci Biobehav Rev 35:912–938
72. Wise RA, Koob GF (2014) The development and maintenance of drug addiction. Neuropsychopharmacology 39:254–262
73. Robinson TE (2004) Neuroscience. Addicted rats. Science 305:951–953
74. Gardner EL (2000) What we have learned about addiction from animal models of drug self-administration. Am J Addict 9:285–313
75. Panlilio LV, Goldberg SR (2007) Self-administration of drugs in animals and humans as a model and an investigative tool. Addiction 102:1863–1870
76. Schenk S, Gittings D, Johnstone M, Daniela E (2003) Development, maintenance and temporal pattern of self-administration maintained by ecstasy (MDMA) in rats. Psychopharmacology (Berl) 169:21–27
77. Schenk S, Hely L, Lake B, Daniela E, Gittings D, Mash DC (2007) MDMA self-administration in rats: acquisition, progressive ratio responding and serotonin transporter binding. Eur J Neurosci 26:3229–3236
78. Ahmed SH (2010) Validation crisis in animal models of drug addiction: beyond non-disordered drug use toward drug addiction. Neurosci Biobehav Rev 35:172–184
79. Hodos W (1961) Progressive ratio as a measure of reward strength. Science (New York, NY) 134:943–944
80. Hodos W, Kalman G (1963) Effects of increment size and reinforcer volume on progressive ratio performance. J Exp Anal Behav 6:387–392
81. Richardson NR, Roberts DC (1996) Progressive ratio schedules in drug self-administration studies in rats: a method to evaluate reinforcing efficacy. J Neurosci Methods 66:1–11
82. Oleson EB, Richardson JM, Roberts DC (2011) A novel IV cocaine self-administration procedure in rats: differential effects of dopamine, serotonin, and GABA drug pre-treatments on cocaine consumption and maximal price paid. Psychopharmacology (Berl) 214:567–577
83. Bradshaw CM, Killeen PR (2012) A theory of behaviour on progressive ratio schedules, with applications in behavioural pharmacology. Psychopharmacology (Berl) 222:549–564
84. Killeen PR, Sanabria F, Dolgov I (2009) The dynamics of conditioning and extinction. J Exp Psychol Anim Behav Process 35:447–472
85. Ahmed SH, Koob GF (1999) Long-lasting increase in the set point for cocaine self-administration after escalation in rats. Psychopharmacology (Berl) 146:303–312
86. Anker JJ, Baron TR, Zlebnik NE, Carroll ME (2012) Escalation of methamphetamine self-administration in adolescent and adult rats. Drug Alcohol Depend 124:149–153
87. Edwards S, Koob GF (2013) Escalation of drug self-administration as a hallmark of persistent addiction liability. Behav Pharmacol 24:356–362

88. Deroche-Gamonet V, Belin D, Piazza PV (2004) Evidence for addiction-like behavior in the rat. Science 305:1014–1017
89. Kearns DN, Weiss SJ, Panlilio LV (2002) Conditioned suppression of behavior maintained by cocaine self-administration. Drug Alcohol Depend 65:253–261
90. Vanderschuren LJ, Everitt BJ (2004) Drug seeking becomes compulsive after prolonged cocaine self-administration. Science 305:1017–1019
91. Gancarz-Kausch AM, Adank DN, Dietz DM (2014) Prolonged withdrawal following cocaine self-administration increases resistance to punishment in a cocaine binge. Sci Rep 4:6876
92. Holtz NA, Carroll ME (2015) Cocaine self-administration punished by intravenous histamine in adolescent and adult rats. Behav Pharmacol 26:393–397
93. Woolverton WL, Freeman KB, Myerson J, Green L (2012) Suppression of cocaine self-administration in monkeys: effects of delayed punishment. Psychopharmacology (Berl) 220: 509–517
94. Comer SD, Lac ST, Curtis LK, Carroll ME (1993) Effects of buprenorphine and naltrexone on reinstatement of cocaine-reinforced responding in rats. J Pharmacol Exp Ther 267: 1470–1477
95. Lenoir M, Ahmed SH (2007) Heroin-induced reinstatement is specific to compulsive heroin use and dissociable from heroin reward and sensitization. Neuropsychopharmacology 32: 616–624
96. Lynch WJ, Carroll ME (2000) Reinstatement of cocaine self-administration in rats: sex differences. Psychopharmacology (Berl) 148:196–200
97. Martin-Fardon R, Maurice T, Aujla H, Bowen WD, Weiss F (2007) Differential effects of sigma1 receptor blockade on self-administration and conditioned reinstatement motivated by cocaine vs natural reward. Neuropsychopharmacology 32:1967–1973
98. Ahmed SH, Lenoir M, Guillem K (2013) Neurobiology of addiction versus drug use driven by lack of choice. Curr Opin Neurobiol 23:581–587
99. Cosgrove KP, Hunter RG, Carroll ME (2002) Wheel-running attenuates intravenous cocaine self-administration in rats: sex differences. Pharmacol Biochem Behav 73:663–671
100. Miller ML, Vaillancourt BD, Wright MJ Jr, Aarde SM, Vandewater SA, Creehan KM, Taffe MA (2012) Reciprocal inhibitory effects of intravenous d-methamphetamine self-administration and wheel activity in rats. Drug Alcohol Depend 121:90–96
101. Aarde SM, Angrish D, Barlow DJ, Wright MJ Jr, Vandewater SA, Creehan KM, Houseknecht KL, Dickerson TJ, Taffe MA (2013) Mephedrone (4-methylmethcathinone) supports intravenous self-administration in Sprague-Dawley and Wistar rats. Addict Biol 18:786–799
102. Dalley JW, Lääne K, Theobald DEH, Peña Y, Bruce CC, Huszar AC, Wojcieszek M, Everitt BJ, Robbins TW (2007) Enduring deficits in sustained visual attention during withdrawal of intravenous methylenedioxymethamphetamine self-administration in rats: results from a comparative study with d-amphetamine and methamphetamine. Neuropsychopharmacology 32:1195–1206
103. Miller ML, Aarde SM, Moreno AY, Creehan KM, Janda KD, Taffe MA (2015) Effects of active anti-methamphetamine vaccination on intravenous self-administration in rats. Drug Alcohol Depend 153:29–36
104. Ranaldi R, Poeggel K (2002) Baclofen decreases methamphetamine self-administration in rats. Neuroreport 13:1107–1110
105. Yokel RA, Pickens R (1973) Self-administration of optical isomers of amphetamine and methylamphetamine by rats. J Pharmacol Exp Ther 187:27–33
106. Balster RL, Schuster CR (1973) A comparison of d-amphetamine, l-amphetamine, and methamphetamine self-administration in rhesus monkeys. Pharmacol Biochem Behav 1: 67–71
107. Beardsley PM, Balster RL, Harris LS (1986) Self-administration of methylenedioxymethamphetamine (MDMA) by rhesus monkeys. Drug Alcohol Depend 18:149–157
108. Kangas BD, Bergman J (2015) Effects of self-administered methamphetamine on discrimination learning and reversal in nonhuman primates. Psychopharmacology (Berl) 233:373–380

109. Newman JL, Carroll ME (2006) Reinforcing effects of smoked methamphetamine in rhesus monkeys. Psychopharmacology (Berl) 188:193–200
110. Yasar S, Gaal J, Panlilio LV, Justinova Z, Molnar SV, Redhi GH, Schindler CW (2006) A comparison of drug-seeking behavior maintained by D-amphetamine, L-deprenyl (selegiline), and D-deprenyl under a second-order schedule in squirrel monkeys. Psychopharmacology (Berl) 183:413–421
111. Balster RL, Kilbey MM, Ellinwood EH Jr (1976) Methamphetamine self-administration in the cat. Psychopharmacologia 46:229–233
112. Griffiths RR, Winger G, Brady JV, Snell JD (1976) Comparison of behavior maintained by infusions of eight phenylethylamines in baboons. Psychopharmacology (Berl) 50:251–258
113. Fantegrossi WE, Ullrich T, Rice KC, Woods JH, Winger G (2002) 3,4-Methylenedioxymethamphetamine (MDMA, "ecstasy") and its stereoisomers as reinforcers in rhesus monkeys: serotonergic involvement. Psychopharmacology (Berl) 161:356–364
114. Fantegrossi WE, Woolverton WL, Kilbourn M, Sherman P, Yuan J, Hatzidimitriou G, Ricaurte GA, Woods JH, Winger G (2004) Behavioral and neurochemical consequences of long-term intravenous self-administration of MDMA and its enantiomers by rhesus monkeys. Neuropsychopharmacology 29:1270–1281
115. Lamb RJ, Griffiths RR (1987) Self-injection of d,1-3,4-methylenedioxymethamphetamine (MDMA) in the baboon. Psychopharmacology (Berl) 91:268–272
116. Lile JA, Ross JT, Nader MA (2005) A comparison of the reinforcing efficacy of 3,4-methylenedioxymethamphetamine (MDMA, "ecstasy") with cocaine in rhesus monkeys. Drug Alcohol Depend 78:135–140
117. Markert LE, Roberts DC (1991) 3,4-Methylenedioxyamphetamine (MDA) self-administration and neurotoxicity. Pharmacol Biochem Behav 39:569–574
118. Ball KT, Jarsocrak H, Hyacinthe J, Lambert J, Lockowitz J, Schrock J (2015) Yohimbine reinstates extinguished 3,4-methylenedioxymethamphetamine (MDMA; ecstasy) seeking in rats with prior exposure to chronic yohimbine. Behav Brain Res 294:1–6
119. Ball KT, Walsh KM, Rebec GV (2007) Reinstatement of MDMA (ecstasy) seeking by exposure to discrete drug-conditioned cues. Pharmacol Biochem Behav 87:420–425
120. Creehan KM, Vandewater SA, Taffe MA (2015) Intravenous self-administration of mephedrone, methylone and MDMA in female rats. Neuropharmacology 92:90–97
121. De La Garza R 2nd, Fabrizio KR, Gupta A (2007) Relevance of rodent models of intravenous MDMA self-administration to human MDMA consumption patterns. Psychopharmacology (Berl) 189:425–434
122. Ratzenboeck E, Saria A, Kriechbaum N, Zernig G (2001) Reinforcing effects of MDMA ("ecstasy") in drug-naive and cocaine-trained rats. Pharmacology 62:138–144
123. Vandewater SA, Creehan KM, Taffe MA (2015) Intravenous self-administration of entactogen-class stimulants in male rats. Neuropharmacology 99:538–545
124. Daniela E, Brennan K, Gittings D, Hely L, Schenk S (2004) Effect of SCH 23390 on (+/-)-3,4-methylenedioxymethamphetamine hyperactivity and self-administration in rats. Pharmacol Biochem Behav 77:745–750
125. Daniela E, Gittings D, Schenk S (2006) Conditioning following repeated exposure to MDMA in rats: role in the maintenance of MDMA self-administration. Behav Neurosci 120: 1144–1150
126. Schenk S, Foote J, Aronsen D, Bukholt N, Highgate Q, Van de Wetering R, Webster J (2016) Serotonin antagonists fail to alter MDMA self-administration in rats. Pharmacol Biochem Behav 148:38–45
127. Reveron ME, Maier EY, Duvauchelle CL (2006) Experience-dependent changes in temperature and behavioral activity induced by MDMA. Physiol Behav 89:358–363
128. Schenk S, Colussi-Mas J, Do J, Bird J (2012) Profile of MDMA self-administration from a large cohort of rats: MDMA develops a profile of dependence with extended testing. J Drug Alcohol Res 1:1–6

129. Cornish JL, Shahnawaz Z, Thompson MR, Wong S, Morley KC, Hunt GE, McGregor IS (2003) Heat increases 3,4-methylenedioxymethamphetamine self-administration and social effects in rats. Eur J Pharmacol 482:339–341
130. Feduccia AA, Kongovi N, Duvauchelle CL (2010) Heat increases MDMA-enhanced NAcc 5-HT and body temperature, but not MDMA self-administration. Eur Neuropsychopharmacol 20:884–894
131. Roberts DC, Brebner K, Vincler M, Lynch WJ (2002) Patterns of cocaine self-administration in rats produced by various access conditions under a discrete trials procedure. Drug Alcohol Depend 67:291–299
132. Aarde SM, Miller ML, Creehan KM, Vandewater SA, Taffe MA (2015) One day access to a running wheel reduces self-administration of d-methamphetamine, MDMA and methylone. Drug Alcohol Depend 151:151–158
133. Roth ME, Carroll ME (2004) Sex differences in the escalation of intravenous cocaine intake following long- or short-access to cocaine self-administration. Pharmacol Biochem Behav 78:199–207
134. Smith MA, Walker KL, Cole KT, Lang KC (2011) The effects of aerobic exercise on cocaine self-administration in male and female rats. Psychopharmacology (Berl) 218:357–369
135. Reichel CM, Chan CH, Ghee SM, See RE (2012) Sex differences in escalation of methamphetamine self-administration: cognitive and motivational consequences in rats. Psychopharmacology (Berl) 223:371–380
136. Roth ME, Carroll ME (2004) Sex differences in the acquisition of IV methamphetamine self-administration and subsequent maintenance under a progressive ratio schedule in rats. Psychopharmacology (Berl) 172:443–449
137. Hadlock GC, Webb KM, McFadden LM, Chu PW, Ellis JD, Allen SC, Andrenyak DM, Vieira-Brock PL, German CL, Conrad KM, Hoonakker AJ, Gibb JW, Wilkins DG, Hanson GR, Fleckenstein AE (2011) 4-Methylmethcathinone (mephedrone): neuropharmacological effects of a designer stimulant of abuse. J Pharmacol Exp Ther 339:530–536
138. Kitamura O, Wee S, Specio SE, Koob GF, Pulvirenti L (2006) Escalation of methamphetamine self-administration in rats: a dose-effect function. Psychopharmacology (Berl) 186: 48–53
139. Cornish JL, Clemens KJ, Thompson MR, Callaghan PD, Dawson B, McGregor IS (2008) High ambient temperature increases intravenous methamphetamine self-administration on fixed and progressive ratio schedules in rats. J Psychopharmacol 22:100–110
140. Huang PK, Aarde SM, Angrish D, Houseknecht KL, Dickerson TJ, Taffe MA (2012) Contrasting effects of d-methamphetamine, 3,4-methylenedioxymethamphetamine, 3,4-methylenedioxypyrovalerone, and 4-methylmethcathinone on wheel activity in rats. Drug Alcohol Depend 126:168–175
141. Motbey CP, Clemens KJ, Apetz N, Winstock AR, Ramsey J, Li KM, Wyatt N, Callaghan PD, Bowen MT, Cornish JL, McGregor IS (2013) High levels of intravenous mephedrone (4-methylmethcathinone) self-administration in rats: neural consequences and comparison with methamphetamine. J Psychopharmacol 27:823–836
142. Watterson LR, Hood L, Sewalia K, Tomek SE, Yahn S, Johnson CT, Wegner S, Blough BE, Marusich JA, Olive MF (2012) The Reinforcing and Rewarding effects of methylone, a synthetic cathinone commonly found in "bath salts". J Addict Res Ther S9(002):1–8
143. Nguyen JD, Grant Y, Creehan KM, Vandewater SA, Taffe MA (2016) Escalation of intravenous self-administration of methylone and mephedrone under extended access conditions. Addict Biol. doi:10.1111/adb.12398
144. Clayton JA, Collins FS (2014) Policy: NIH to balance sex in cell and animal studies. Nature 509:282–283
145. DeVito EE, Babuscio TA, Nich C, Ball SA, Carroll KM (2014) Gender differences in clinical outcomes for cocaine dependence: randomized clinical trials of behavioral therapy and disulfiram. Drug Alcohol Depend 145:156–167

146. Westermeyer J, Boedicker AE (2000) Course, severity, and treatment of substance abuse among women versus men. Am J Drug Alcohol Abuse 26:523–535
147. Kim JY, Fendrich M (2002) Gender differences in juvenile arrestees' drug use, self-reported dependence, and perceived need for treatment. Psychiatr Serv 53:70–75
148. Rawson RA, Gonzales R, Obert JL, McCann MJ, Brethen P (2005) Methamphetamine use among treatment-seeking adolescents in Southern California: participant characteristics and treatment response. J Subst Abuse Treat 29:67–74
149. Bruno R, Matthews AJ, Topp L, Degenhardt L, Gomez R, Dunn M (2009) Can the severity of dependence scale be usefully applied to 'ecstasy'? Neuropsychobiology 60:137–147
150. Smith MA, Pennock MM, Walker KL, Lang KC (2012) Access to a running wheel decreases cocaine-primed and cue-induced reinstatement in male and female rats. Drug Alcohol Depend 121:54–61
151. Huskinson SL, Naylor JE, Rowlett JK, Freeman KB (2014) Predicting abuse potential of stimulants and other dopaminergic drugs: overview and recommendations. Neuropharmacology 87C:66–80
152. Papaseit E, Perez-Mana C, Mateus JA, Pujadas M, Fonseca F, Torrens M, Olesti E, de la Torre R, Farre M (2016) Human pharmacology of mephedrone in comparison with MDMA. Neuropsychopharmacology 41(11):2704–2713
153. Lopez-Arnau R, Martinez-Clemente J, Carbo M, Pubill D, Escubedo E, Camarasa J (2013) An integrated pharmacokinetic and pharmacodynamic study of a new drug of abuse, methylone, a synthetic cathinone sold as "bath salts". Prog Neuropsychopharmacol Biol Psychiatry 45:64–72
154. Martinez-Clemente J, Lopez-Arnau R, Carbo M, Pubill D, Camarasa J, Escubedo E (2013) Mephedrone pharmacokinetics after intravenous and oral administration in rats: relation to pharmacodynamics. Psychopharmacology (Berl) 229:295–306
155. Iversen L, Gibbons S, Treble R, Setola V, Huang XP, Roth BL (2013) Neurochemical profiles of some novel psychoactive substances. Eur J Pharmacol 700:147–151
156. Katz JL, T-P S, Hiranita T, Hayashi T, Tanda G, Kopajtic T, Tsai S-Y (2011) A role for sigma receptors in stimulant self administration and addiction. Pharmaceuticals (Basel) 4: 880–914
157. Bonano JS, Glennon RA, De Felice LJ, Banks ML, Negus SS (2014) Abuse-related and abuse-limiting effects of methcathinone and the synthetic "bath salts" cathinone analogs methylenedioxypyrovalerone (MDPV), methylone and mephedrone on intracranial self-stimulation in rats. Psychopharmacology (Berl) 231(1):199–207
158. Gatch MB, Taylor CM, Forster MJ (2013) Locomotor stimulant and discriminative stimulus effects of 'bath salt' cathinones. Behav Pharmacol 24(5–6):437–447

The Affective Properties of Synthetic Cathinones: Role of Reward and Aversion in Their Abuse

Heather E. King and Anthony L. Riley

Abstract The drug class known as synthetic cathinones has gained significant attention in the last few years as a result of increased use and abuse. These compounds have been shown to possess reinforcing efficacy in that they are abused in human populations and are self-administered in animal models. The present chapter outlines the affective properties of synthetic cathinones that are thought to impact drug self-administration in general and presents research confirming that these drugs have both rewarding and aversive effects in standalone and concurrent assessments. The implications of these affective properties for the overall abuse potential of these compounds are discussed along with directions for future research.

Keywords Affective properties • Avoidance • Cathinones • Reward

Contents

H.E. King (✉) and A.L. Riley
Psychopharmacology Laboratory, Department of Psychology, American University, 4400 Massachusetts Ave, NW, Washington, DC 20016, USA
e-mail: heather.elise39@gmail.com

Curr Topics Behav Neurosci (2017) 32: 165–182
DOI 10.1007/7854_2016_32
Published Online: 15 July 2016

1 Introduction

In recent years, synthetic cathinones have made it to the forefront of both public health concerns and drug abuse research [1–4]. Often associated with severe hallucinations, paranoia, violent behaviors, tachycardia, and even death [5, 6], these drugs pose a significant threat to those who use them, and yet their pharmacological profile is still relatively uncharacterized compared to other abused compounds. In the present review, we discuss the affective properties of synthetic cathinones and address how these properties might contribute to their abuse.

2 Animal Models of Use and Abuse

Drugs of abuse have historically been examined in animal models, given the obvious limitations on such experimentation in humans. The model with the most face validity for human drug taking is drug self-administration [7] (see Watterson and Olive, this volume). In this model, animals are allowed to respond, e.g., press a lever, for the delivery of a specific drug, and the rate or level of responding is generally concluded to be a function of the rewarding effects of that drug [8, 9]. Such a procedure has been widely used since its initial demonstration (see [10]), and a wide range of compounds used and abused by humans support such behavior [11–13]. Traditionally, drug taking in animals (as assessed in self-administration models) has been assumed to reflect the drug's positive (rewarding) effects, an assumption supported by findings that these same drugs appear to be rewarding (see [14]) in other models of drug reward, e.g., intracranial self-stimulation and conditioned place preference.

Although the rewarding effects of a drug are certainly important to the initiation and maintenance of drug-taking behavior, many drugs of abuse also produce aversive effects, as indexed by their ability to induce a conditioned taste avoidance (see [15]). Specifically, when a psychoactive drug is paired with access to a novel-tasting solution, animals will come to avoid the drug-associated taste on subsequent exposures. This avoidance is referred to as a conditioned taste avoidance (CTA) and has now been reported with a wide variety of compounds, including many drugs of abuse (see [16, 17]). These aversive effects also may be important to drug intake, basically by limiting intake. In fact, drug intake may be conceptualized as a function of the balance between these two affective states. These dual effects have been seen with a range of compounds and under a variety of experimental conditions, demonstrating a reliable co-occurrence of both rewarding and aversive effects for a variety of drugs of abuse, including amphetamine [18, 19], caffeine [20], morphine [21], and nicotine [22].

If drug use and abuse are a function of the balance of the drug's rewarding and aversive effects, it is important to determine these effects for specific drugs and the various factors that are known to affect these properties. Such factors include, but are not limited to, sex, concurrent drug use, age, drug history, species, strain, drug dose, drug duration, and route of administration (see [23]). As we begin to characterize new synthetic compounds, examination of both their rewarding and aversive effects, and the range of factors that may influence them, is important in the determination of overall abuse liability.

3 Affective Properties of Synthetic Cathinones

As reviewed elsewhere in this volume, synthetic cathinones are self-administered (see Watterson and Olive, this volume) and produce rewarding effects in intracranial self-stimulation (see Bonano and Negus, this volume). However, given that self-administration is a function of the balance between the rewarding and aversive effects of a drug, evaluation of their affective properties and the factors that may influence these properties is important. To this end, we review the effects of synthetic cathinones in two major animal models of drug reward (conditioned place preference) and aversion (conditioned taste avoidance).

3.1 Assessments of the Rewarding Effects of Cathinones

The conditioned place preference (CPP) procedure is a well-established method for directly assessing the rewarding effects of drugs of abuse (for reviews, see [8, 24, 25]). These assessments typically involve a two-chambered place-conditioning apparatus, containing distinct environmental and tactile cues in each chamber. A baseline test determines each subject's initial chamber preference, allowing any subsequent drug-induced changes in preference to be measured. During conditioning, an injection of a drug is paired with one chamber of the place-conditioning apparatus, and injections of the drug's vehicle are paired with the opposite chamber (for control subjects, the drug vehicle is typically paired with both chambers). If a drug is rewarding, the animal will typically increase its time spent in the drug-paired chamber. This procedure has been used extensively to evaluate the rewarding effects of a wide range of drugs of abuse, including those of a number of psychostimulants (see [26–29]).

Several studies have used this procedure to investigate the rewarding effects of synthetic cathinones. In the first such assessment, Lisek et al. [30] gave male Sprague–Dawley rats an injection of 3-, 10-, or 30-mg/kg mephedrone and confined them to one side of the place-conditioning apparatus for 30 min in one of two daily conditioning sessions. The second session consisted of an injection of vehicle and placement in the opposite conditioning chamber for 30 min. After 4 days of two-session place conditioning, animals were given free access to both chambers and tested for any acquired preference for the drug-paired chamber. Rats conditioned with 30-mg/kg mephedrone showed a greater shift in their preference for the

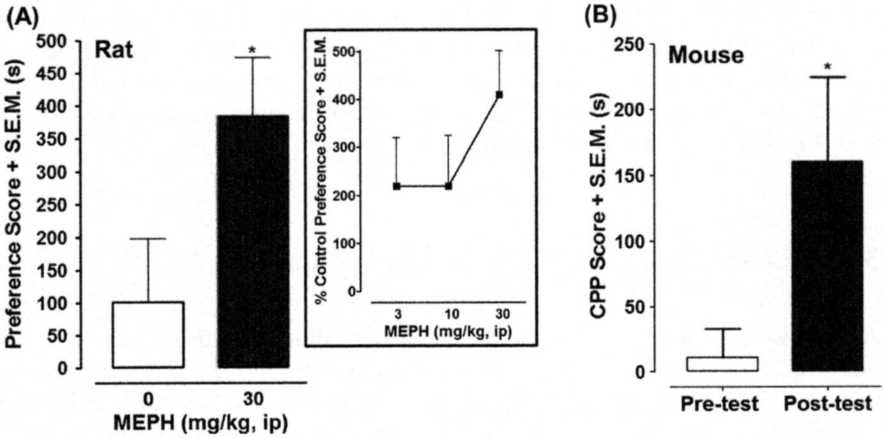

Fig. 1 Effects of mephedrone (MEPH) in the CPP assay in rats and mice. Rat CPP: data are presented as preference score (time in seconds spent on the drug-paired side posttest minus pretest) in rats conditioned with saline (0 mg/kg) ($n = 12$) or MEPH (30 mg/kg) ($n = 10$). $^*p < 0.05$ compared to saline control. *Box*: data from rat CPP experiments are presented as percentage of control (saline-treated) preference score for different doses of MEPH (3, 10, 30 mg/kg; $n = 10$–12 rats per group). Panel **B**, mouse CPP: data obtained from 23 mice are presented as time spent in 30-mg/kg MEPH-paired compartment minus the time spent in saline-paired compartment pre- and post-conditioning. $^*p < 0.05$ compared to the pretest. From Lisek et al. [30]; figure reprinted with permission from Elsevier

drug-paired side compared to that of vehicle-treated controls (see Fig. 1; Panel A), with lower doses causing shifts in side preference of lesser magnitudes. An additional assessment using CD-1 mice showed that 30-mg/kg mephedrone also induced a significant place preference (see Fig. 1; Panel B).

In a related evaluation of the cathinone compounds in place preference conditioning, Karlsson et al. [31] utilized C57BL/6 mice to assess the ability of MDPV, mephedrone, methylone, and amphetamine to support CPP at a range of doses (0.5, 2, 5, 10, and 20 mg/kg of each compound). After baseline assessments of chamber preference, mice were given two 15-min conditioning sessions per day for 4 days; one with saline paired with a distinct chamber in the morning and the other with an injection of drug paired with the opposite chamber 4 h later. Results showed that all of the synthetic cathinone compounds produced reliable dose-dependent place preferences, but effective doses varied across compounds. Specifically, methylone induced a significant place preference at 5, 10, and 20 mg/kg, whereas mephedrone produced place preferences at 5 and 20 mg/kg. The strongest rewarding effect was seen in the MDPV-treated animals, in which all doses produced significant place preferences. In addition, only MDPV-treated animals showed significantly stronger preferences than their counterparts treated with the positive control, amphetamine.

These results certainly indicate that the synthetic cathinones possess rewarding effects and that these effects are comparable to those of other psychostimulants; however, this work utilized C57/BL6 mice, an inbred strain that is particularly susceptible to the reinforcing effects of ethanol and other drugs [32]. Given that

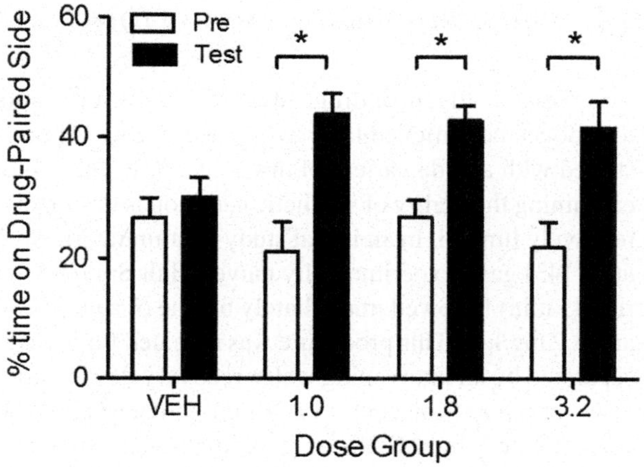

Fig. 2 Change in percent time spent on the MDPV-paired side from pretest to posttest for subjects injected with 0-, 1-, 1.8-, and 3.2-mg/kg MDPV ($n = 12$ for all groups). From King et al. [36, 37]; figure reprinted with permission from Elsevier

both strain and species have been shown to influence a host of effects induced by other abused drugs [33–35], generalization of results between strains and across species may be difficult. To facilitate comparison with the earlier Lisek et al. [30] study, King et al. [36, 37] assessed the ability of MDPV (1, 1.8, or 3.2 mg/kg) to induce place preferences in adult male Sprague–Dawley rats. After an initial baseline assessment, rats were injected with MDPV or vehicle once daily on alternate days and confined to the drug- or vehicle-associated chamber, respectively. Rats received a total of eight conditioning sessions, four with MDPV and four with vehicle. On the final test, all doses of MDPV produced significant shifts in preference for the drug-paired environment from their initial preconditioning baselines (see Fig. 2). Although all MDPV-injected subjects increased the amount of time spent on the drug-paired side, it should be noted that actual preferences were not produced, i.e., animals still spent a majority of time on the non-drug-paired side, which may have been a function of initial preferences for that side.

Additionally, these shifts in preference were of similar magnitude (i.e., not dose dependent), contradictory with results seen in the prior assessment of MDPV-induced CPP in mice [31] and with other stimulants [38, 39]. For example, a meta-analysis of place-conditioning studies with amphetamine and cocaine reported a significant effect of amphetamine dose on the magnitude of place conditioning and a trend toward this same relationship with cocaine [40]. It is possible that the doses used in the work by King et al. [36, 37] produced a ceiling effect and that dose-dependent differences may have emerged at lower doses.

In summary, the synthetic cathinones clearly possess rewarding effects. These effects have been demonstrated in a variety of preparations, in male and female rodents and across several species/strains. However, given that overall drug intake (i.e., self-administration) is a function of the balance of the drug's rewarding and aversive effects, assessment of the aversive effects of these compounds is important before making conclusions regarding their abuse potential. In that context, our laboratory has focused on such assessments with MDPV, as reviewed in the next section.

3.2 Assessments of the Aversive Effects of Cathinones

The aversive effects of drugs of abuse are most commonly assessed by their ability to produce conditioned taste avoidance (CTA). Although CTAs have been demonstrated with a wide variety of drugs of abuse (for reviews, see [16, 41, 42]), studies examining the ability of synthetic cathinones to support taste avoidance learning are relatively limited. In an initial study examining MDPV's aversive effects, Merluzzi et al. [43] gave experimentally naïve, adult Sprague–Dawley rats access to saccharin (45 min) followed immediately by one of a number of doses of MDPV (0, 1, 1.8, and 3.2 mg/kg). This procedure was repeated for a total of 5 conditioning days, each separated by an intervening water-recovery day in which animals were given access to water but not injected. In addition, to determine whether core body temperature was affected by MDPV, the animals' temperatures were taken via scans of implanted telemetry probes immediately prior to drug administration, as well as at 30-, 60-, 90-, and 120-min postinjection.

Results from this study demonstrated a clear dose-dependent MDPV-induced CTA that developed over the course of the 5-day conditioning period (see Fig. 3; Panel A). A final two-bottle assessment (where animals were given simultaneous access to both saccharin and water, and percent saccharin out of total fluid consumption was measured as an index of the aversive effects of MDPV) revealed that all MDPV-treated subjects showed significantly stronger avoidance than vehicle-treated animals and the mid- and high-dose groups showed stronger avoidance than the low-dose group (Fig. 3; Panel B). Further, while MDPV also produced a hyperthermic effect, no clear dose-dependent relationship between the degree of hyperthermia and strength of taste avoidance emerged, suggesting that this physiological effect was not involved in the aversive effects of the drug (for a discussion of the role of hyperthermia in ethanol-induced taste avoidance, see [44]).

Because taste avoidance is age related with adolescent animals tending to show weaker avoidance than adults (for reviews, see [45, 46]), Merluzzi et al. [43] extended their work on MDPV-induced avoidance in adults to adolescents. The procedure used with adolescents was identical to that described above for adults. Results revealed that adolescent animals acquired significant MDPV-induced taste avoidance. As illustrated in Fig. 4 (Panel A), drug-treated groups drank significantly less saccharin than the vehicle-treated group on Trials 3 and 4, although no differences between drug-treated groups were observed. A two-bottle test showed that the mid- and high-dose groups drank a lower percentage of saccharin than the vehicle-treated group and that the mid-dose group drank less saccharin than the low-dose group (see Fig. 4; Panel B). Hence, while adolescents acquired avoidance of the MDPV-paired saccharin solution, avoidance was weaker and acquired more slowly in adolescents (compared to adults), suggesting that adolescents may be relatively insensitive to MDPV's aversive effects and particularly susceptible to use and abuse of MDPV. Little is known about possible age differences in the rewarding effects of MDPV. The examination of adolescent and adult animals will be important in assessing its overall abuse vulnerability.

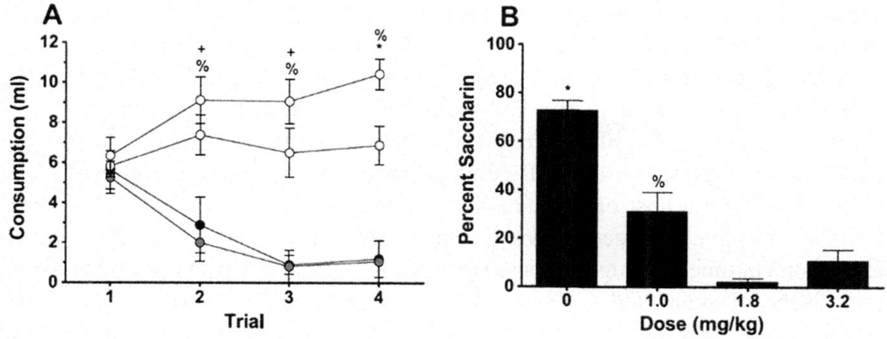

Fig. 3 Mean (± SEM) saccharin consumption (ml) by adult rats during taste avoidance acquisition (**A**) and mean (± SEM) percent saccharin consumed on a final two-bottle test (**B**), $n = 8$–9 for all groups. + denotes a significant difference between Group 0 and Groups 1.8 and 3.2. % denotes a significant difference between Group 1.0 and Groups 1.8 and 3.2. * denotes a significant difference between Group 0 and all drug-treated groups. From Merluzzi et al. [43]; figure reprinted with permission from Wiley

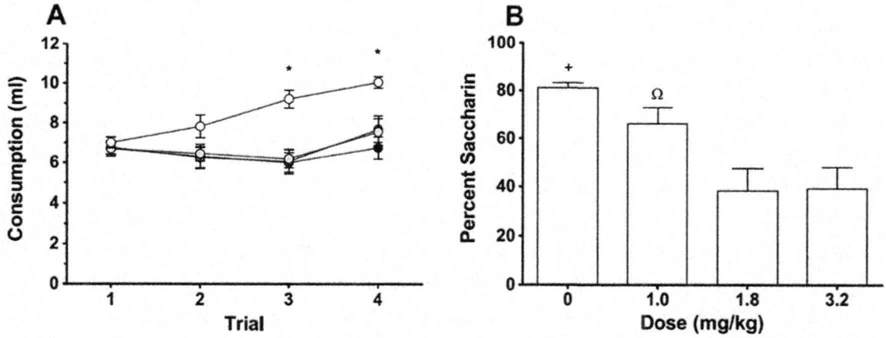

Fig. 4 Mean (± SEM) saccharin consumption (ml) by adolescent rats during taste avoidance acquisition (**A**) and mean (± SEM) percent saccharin consumed on a final two-bottle test (**B**), $n = 8$–9 for all groups. * denotes a significant difference between Group 0 and all drug-treated groups. + denotes a significant difference between Group 0 and Groups 1.8 and 3.2. Ω denotes a significant difference between Groups 1.0 and 1.8. From Merluzzi et al. [43]; figure reprinted with permission from Wiley

Interestingly, while MDPV produced a hyperthermic effect in adult animals, the same doses of MDPV (and under the same parametric procedures) produced a hypothermic effect in the adolescent animals. As with adults, these changes in the thermic response following MDPV were unrelated to the strength of the MDPV-induced taste avoidance. The MDPV changes in body temperature are not surprising given that such effects are common with the administration of various stimulants (see [47–49]) and that Fantegrossi et al. [50] reported MDPV-induced changes in body temperature in mice, although it is interesting that adults and adolescents display opposite effects in response to an injection of MDPV. The fact that temperature has been related to the taste avoidance induced by other drugs, e.g., alcohol (see [44]), suggests that such a relationship may be drug specific.

Subsequently, King et al. [51] examined potential strain differences in MDPV-induced taste avoidance in the F344 and LEW rat strains. Although primarily characterized for their differences in the rewarding effects of drugs, where the LEW strain displays greater self-administration and stronger conditioned place preferences with a host of compounds ([52–56]; though see [57]), the F344 and LEW strains have also been shown to display differential sensitivity to the aversive effects of a number of drugs of abuse (for reviews, see [16, 58]). For example, LEW rats displayed attenuated taste avoidance compared to F344 rats for morphine, nicotine, and ethanol [57, 59–61]; however, they acquire stronger cocaine-induced taste avoidance ([62, 63]; for other drug comparisons, see [16]). These differences suggest that there may be a genetic component in the relative sensitivity to these affective properties, with the direction of the difference being drug dependent. In the King et al. [51] study, male F344 and LEW rats were tested with vehicle or one of three doses of MDPV (1.0, 1.8, or 3.2 mg/kg) in a taste avoidance procedure similar to that used by Merluzzi et al. [43] and described above.

Because differential consumption at baseline between the two strains was observed, consumption for each dose group in each strain was calculated as a percentage of its respective control group on the final conditioning trial. Similar to our work in outbred adult Sprague–Dawley rats, MDPV induced significant dose-dependent taste avoidance in both rat strains. Further, strain differences in these percentage shifts were not observed at any dose. The lack of strain difference in MDPV-induced avoidance is somewhat surprising given prior data showing stronger cocaine-induced avoidance in LEW animals (see [62]); however, the degree of avoidance was strong for both groups at the two highest doses (between 50 and 75 % reductions in consumption), with some animals in both groups displaying complete suppression on this trial. Such strong avoidance may have produced a floor effect that precluded seeing differences among groups. As above with our work with outbred adult rats [43], the temperature assessment revealed that at 30-, 60-, and 90-min postinjection, all groups in both strains injected with MDPV displayed a hyperthermic response. The fact that MDPV produced dose-dependent avoidance but no dose-dependent effects on hyperthermia again suggests that the taste avoidance was independent of the changes in body temperature.

In summary, MDPV reliably produces taste avoidance in several strains and in adolescents and adults, supporting the position that MDPV has aversive effects. Unfortunately, studies have not yet examined the ability of mephedrone or methylone to produce CTA. This significant gap in the literature needs to be addressed in order to assess abuse potential of this class of drugs.

4 Concurrent Assessments of the Affective Properties of Cathinones

Synthetic cathinones are self-administered and have both aversive and rewarding effects (presumably mediating/moderating drug taking). However, the work discussed here involved studies looking exclusively at either the aversive or rewarding effects of these drugs. Therefore, the presence of either reward or avoidance could simply reflect the different parametric conditions (e.g., dose of drug, route of administration, drug duration, subjects) under which the rewarding and aversive effects were assessed.

To address the rewarding and aversive effects of a drug concurrently, many studies have utilized a combined CTA/CPP design in which animals are given access to a taste, injected with the drug and then placed in a distinct chamber of a place preference apparatus. Under these conditions, decreases in consumption of the drug-associated taste index the drug's aversive effects, whereas shifts in preferences for the drug-associated side index the rewarding effects of the drug. As noted earlier, this procedure has reliably demonstrated concurrent place preferences (reward) and taste avoidance (aversion) using a range of compounds and under a variety of experimental conditions (see [20–22, 64]). In this context, we recently examined the affective properties of MDPV (see [36, 37]). Given that females display both stronger taste avoidance and place preferences induced by stimulants such as cocaine, amphetamine, and nicotine [65–67], both sexes were included in this assessment.

Specifically, adult Sprague–Dawley male and female rats were habituated to daily 20-min water access to ensure stable consumption. On the day before conditioning began, all animals were given 15-min access to the place-conditioning apparatus to assess initial side preferences. On the first conditioning day, animals were given 20-min access to a novel saccharin solution during their daily fluid-access period and then immediately transported to a room adjacent to the CPP chambers and assigned to a dose group. Subjects were given an injection of either drug (1.0-, 1.8-, and 3.2-mg/kg MDPV) or vehicle and confined to the initially non-preferred chamber for 30 min. The following day, they were given 20-min access to water, followed immediately by a saline injection and then confinement to the opposite (originally preferred) chamber of the previous day. This 2-day cycle was repeated for a total of four consecutive cycles over 8 days, followed by a final place preference test and a two-bottle taste avoidance test.

Under these conditions, MDPV induced taste avoidance in all animals, but males and females differed in the speed of acquisition and degree of this suppression (see Fig. 5). Specifically, males in the 1.8- and 3.2-mg/kg groups showed significant decreases in saccharin consumption from control subjects following only a single pairing of saccharin and MDPV, whereas avoidance was evident on this trial for females only in the 3.2-mg/kg group. In later trials, all male drug-treated groups drank less than controls, whereas only females treated with the 1.8- and 3.2-mg/kg doses displayed avoidance. While males treated with 3.2-mg/kg MDPV showed a

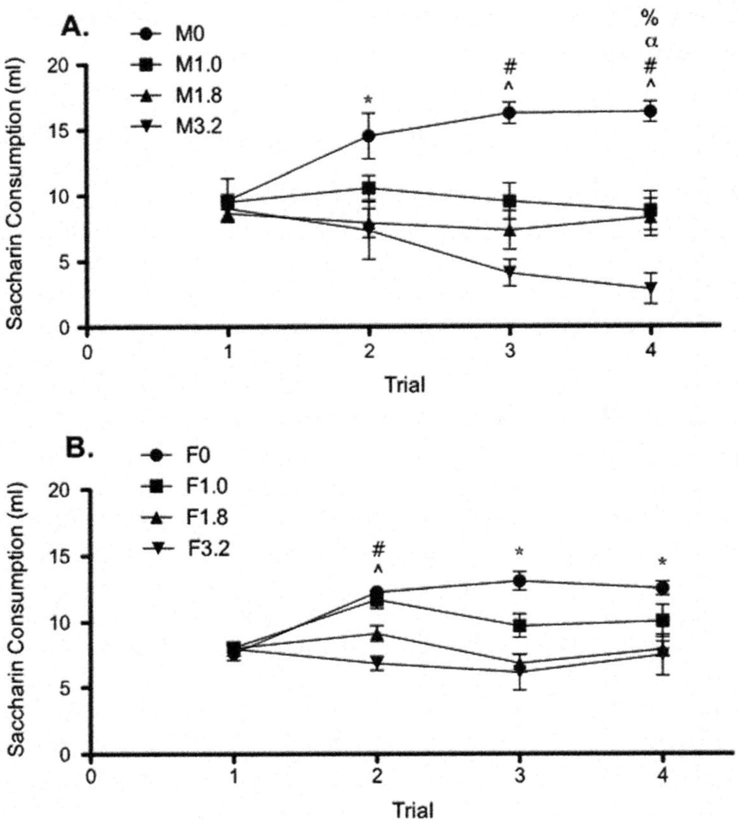

Fig. 5 Mean (± SEM) saccharin consumption in ml over all conditioning trials for male and female subjects injected with vehicle or 1-, 1.8-, and 3.2-mg/kg MDPV ($n = 8$ for each group). Panel **A** (males): *M0 significantly greater than M1.8 and M3.2; ^M0 significantly greater than all drug-treated groups; %M1.0 and M1.8 significantly greater than M3.2; #M3.2 significant decrease from Trial 1; αM3.2 significant decrease from Trial 2. Panel **B** (females): *F0 significantly greater than F1.8 and F3.2; ^F0 and F1.0 significantly greater than F3.2; #F1.0 significant increase from Trial 1. From King et al. [36, 37]; figure reprinted with permission from Elsevier

significant decrease in consumption from baseline on Trials 3–4 and a significant decrease from Trial 2 to Trial 4, none of the female dose groups displayed a significant decrease from their own baseline consumption level. Further, in a direct comparison between males and females, males injected with 3.2-mg/kg MDPV drank significantly less than females on Trial 4. It is important to note that the sex differences in acquisition of MDPV-induced taste avoidance were not evident on the final two-bottle assessment; all drug-treated groups drank a smaller percentage of saccharin than control subjects. This finding is likely a reflection of the sensitivity of the two-bottle test relative to forced-choice consumption, i.e., when animals are given access to both the drug-paired taste and water in the two-bottle assessment, aversions tend to be stronger (with no forced drinking), and differences among groups are not always evident in this more sensitive index of the drug's aversive effects (see [68, 69]). Although taste avoidance was induced by MDPV in both males and females, MDPV-induced avoidance was weaker in females compared to males, an effect that is consistent with results in other work on sex differences in taste avoidance with a variety of drugs [65, 67, 70].

In the concomitant place preference conditioning, all groups significantly increased time on the drug-paired side, independent of drug treatment. Because these preferences did not vary as a function of sex, the data were collapsed across sex for analysis (see Fig. 6). Despite significant increases in the vehicle-treated animals, the 1.8- and 3.2-mg/kg groups spent significantly more time on the drug-paired side at the final preference test than did the vehicle animals, indicating that MDPV was rewarding.

However, these preferences did not vary as a function of sex. Given the pharmacological similarities to cocaine, it might have been predicted that MDPV would have produced stronger place preferences in females compared to males [66, 71]. However, prior work showing enhanced sensitivity of females in place preference conditioning with cocaine utilized different strains [66] and initiated conditioning at different ages and under a different experimental design [71], suggesting that such effects may be dependent upon a variety of experiential and subject variables. Although sex differences in place preference conditioning with MDPV were not seen, the fact that females showed a weaker avoidance response compared to males suggests that sex may be an important factor in determining susceptibility to use and abuse of MDPV and argues for the importance of combined assessments of reward and aversion in predicting abuse liability.

Although MDPV is both rewarding and aversive in the same animal (as indexed by changes in the combined CTA/CPP design), little is known about the extent (if any) to which these two affective properties are related. To assess whether there was a relationship between avoidance and preferences, we ran a correlational analysis on the change in the amount consumed over conditioning (Trial 1 to Trial 4) and the change in percent time on the drug-paired side (pretest to posttest) within each sex and dose group tested in the combined CTA/CPP procedure. This analysis revealed minimal correlations between taste avoidance and place preferences, suggesting that the strength of preference is unrelated to the strength of avoidance (see [19] for similar findings with morphine and amphetamine).

The fact that MDPV induced concurrent taste avoidance and place preference, but that sex differences were only evident in taste avoidance, argues for a dissociation between these two affective properties, i.e., these effects likely function independently. This position is further supported by the correlational analyses showing no consistent significant relationship between strength of taste avoidance and strength of place preference within sex and dose groups. These data parallel those by Verendeev and Riley [19] who evaluated the relationship between strength of avoidance and preferences, using both morphine and amphetamine in a combined CTA/CPP procedure. While both avoidance and preferences were seen in drug-treated animals, there was no consistent relationship between these factors in individual animals (see also [72]), although strength of amphetamine-induced

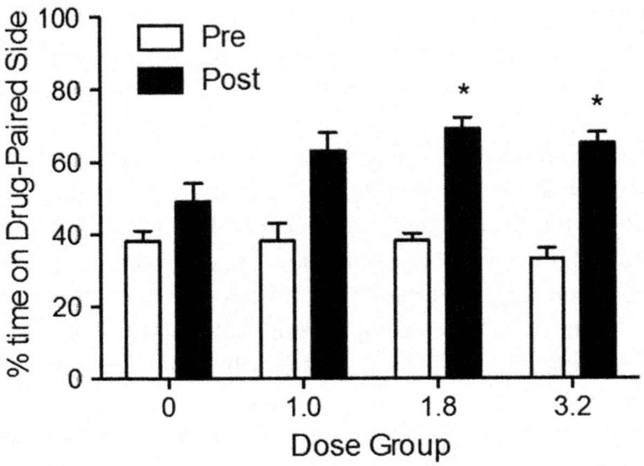

Fig. 6 Mean percent time spent on the drug-paired side (± SEM) for all groups at pre- and posttest, collapsed across sex. All groups (including vehicle) showed significant increases from pretest to posttest. *Significant difference from Group 0. Animals in the two higher doses spent more time in the MDPV-paired side than did control subjects. From King et al. [36, 37]; figure reprinted with permission from Elsevier

avoidance was positively related to strength of amphetamine-induced place preferences in a similar assessment that was based upon group averages and serial assessments in CTA and CPP (see [73]). The dissociation between reward and avoidance suggests that these effects are independent and differentially associated with specific stimuli (i.e., rewarding effects with environmental cues and aversive effects with gustatory stimuli) that condition opposite behaviors (approach and avoidance) (see also [21]).

5 Drug Use and Abuse

In summary, a variety of animal models clearly indicate that the synthetic cathinones have both aversive and rewarding effects and are self-administered (see Watterson and Olive chapter in this volume). The ultimate goal underlying exploration of the aversive and rewarding properties of these drugs is to assess the relationship of these affective properties to self-administration. While the aversive and rewarding effects of the cathinones are becoming well characterized, their relationship to self-administration has not been established. As our knowledge about factors that impact aversion and reward grows, examination of the ways in which their relative balance contributes to eventual drug taking (i.e., self-administration) is crucial for understanding abuse vulnerability of these drugs as well as potential targets for treatment.

In fact, few such studies exist for any drug. One major issue in these demonstrations concerns the fact that the procedures needed to assess this relationship are generally serial in nature. That is, one first assesses reward and aversion (using designs such as place preference conditioning and taste avoidance; see above) and then examines the ability of the drug to support self-administration (see [74]). The difficulty with this procedure is that exposure to the drug during the assessments of reward and aversion could impact the drug's self-administration, precluding an unconfounded assessment of drug taking. For example, if the initial exposures to the drug during taste avoidance and/or place preference conditioning sensitized (or adapted) the animal to the drug effect, this could impact the likelihood or degree of self-administration and affect any relationships being examined [75–77]. Limited attempts have been made to examine these relationships by using concurrent procedures, that is, with the assessments of reward and/or aversion and self-administration occurring at the same time. This has been nicely done in a series of recent papers by Grigson and her colleagues. In one of the first such assessments (see [78]), animals were given access to a novel saccharin solution to drink and then allowed to self-administer cocaine. Results showed that animals that displayed the greater suppression of saccharin consumption (i.e., displaying the strongest taste avoidance) also displayed the greatest cocaine self-administration, suggesting that taste avoidance was directly associated with self-administration of cocaine. Although the work of Grigson and Twinning demonstrates a relationship between the aversive effects of the drug as indexed by taste avoidance and the drug's self-administration, it is difficult to conclude from these demonstrations that there is any causal relationship between the two, i.e., that the drug's aversive effects are responsible for the greater self-administration. It is equally possible that the degree of self-administration which by definition increases the amount of the drug that was paired with the saccharin solution that preceded drug access may have simply conditioned a stronger taste avoidance response. In this view, self-administration was responsible for the greater taste avoidance. Although a relationship can be seen in such concurrent assessments, conclusions are less clear (see also [79–81]).

The most interpretable manner by which associations and relationships may be evaluated between the drug's rewarding and aversive effects and its self-administration may require independent assessments in which both affective properties of the drug and its self-administration are established (in independent groups of animals), and various manipulations are made and assessed for their effects on these behavioral models. In such assessments, the extent and direction of effects from these manipulations can then be evaluated to determine if and to what degree the behavioral models are similarly or differently affected, providing insight into the possible relationships of reward and aversion to drug taking. While less subject to interpretation, this procedure requires systematic evaluation of a wide variety of subject and experiential factors in a host of different models of reward, aversion, and intake. At the present time, the questions remaining for the rapidly growing list of synthetic cathinones indicate that more work is required to answer many of the questions that would be posed by this approach. In combination with the serial and concurrent procedures outlined above, characterizing these compounds for their abuse potential as well as the factors that may impact it will be crucial to understand abuse vulnerability of these drugs as well as potential targets for treatment.

References

1. Baumann MH (2014) Awash in a sea of 'bath salts': implications for biomedical research and public health. Addiction 109(10):1577–1579
2. Baumann MH, Partilla JS, Lehner KR (2013) Psychoactive "bath salts": not so soothing. Eur J Pharmacol 698(1):1–5
3. Katz DP, Bhattacharya D, Bhattacharya S, Deruiter J, Clark CR, Suppiramaniam V, Dhanasekaran M (2014) Synthetic cathinones: "a khat and mouse game". Toxicol Lett 229 (2):349–356
4. Kyle PB, Iverson RB, Gajagowni RG, Spencer L (2011) Illicit bath salts: not for bathing. J Miss State Med Assoc 52(12):375–377
5. Baumann MH, Solis E, Watterson LR, Marusich JA, Fantegrossi WE, Wiley JL (2014) Baths salts, spice, and related designer drugs: the science behind the headlines. J Neurosci 34 (46):15150–15158
6. Spiller HA, Ryan ML, Weston RG, Jansen J (2011) Clinical experience with and analytical confirmation of "bath salts" and "legal highs"(synthetic cathinones) in the United States. Clin Toxicol 49(6):499–505
7. O'Connor EC, Chapman K, Butler P, Mead AN (2011) The predictive validity of the rat self-administration model for abuse liability. Neurosci Biobehav Rev 35(3):912–938
8. Sanchis-Segura C, Spanagel R (2006) REVIEW: behavioural assessment of drug reinforcement and addictive features in rodents: an overview. Addict Biol 11(1):2–38
9. Zernig G, Ahmed SH, Cardinal RN, Morgan D, Acquas E, Foltin RW, Stöckl P (2007) Explaining the escalation of drug use in substance dependence: models and appropriate animal laboratory tests. Pharmacology 80(2–3):65–119
10. Weeks JR (1962) Experimental morphine addiction: method for automatic intravenous injections in unrestrained rats. Science 138(3537):143–144
11. Ator NA, Griffiths RR (1987) Self-administration of barbiturates and benzodiazepines: a review. Pharmacol Biochem Behav 27(2):391–398
12. Balster RL, Lukas SE (1985) Review of self-administration. Drug Alcohol Depend 14 (3):249–261
13. Fischman M (1988) Behavioral pharmacology of cocaine. J Clin Psychiatry 49:7–10
14. Koob G, Le Moal M (2006) Neurobiology of addiction. Elsevier, Amsterdam
15. Freeman KB, Riley AL (2008) The origins of conditioned taste aversion learning: a historical analysis. In: Rcilly S, Schachtman TR (eds) Conditioned taste aversions: neural and behavioral processes. Oxford university press, New York, pp 9–36
16. Davis CM, Riley AL (2010) Conditioned taste aversion learning. Ann N Y Acad Sci 1187 (1):247–275
17. Riley AL, Davis CM, Roma PG (2009) Strain differences in taste aversion learning: implications for animal models of drug abuse. Ann N Y Acad Sci 1187:247–275
18. Sherman JE, Roberts T, Roskam SE, Holman EW (1980) Temporal properties of the rewarding and aversive effects of amphetamine in rats. Pharmacol Biochem Behav 13(4):597–599
19. Verendeev A, Riley AL (2011) Relationship between the rewarding and aversive effects of morphine and amphetamine in individual subjects. Learn Behav 39(4):399–408
20. Brockwell NT, Eikelboom R, Beninger RJ (1991) Caffeine-induced place and taste conditioning: production of dose-dependent preference and aversion. Pharmacol Biochem Behav 38 (3):513–517

21. King HE, Riley AL (2013) A history of morphine-induced taste aversion learning fails to affect morphine-induced place preference conditioning in rats. Learn Behav 41(4):433–442
22. Pomfrey RL, Bostwick TA, Wetzell BB, Riley AL (2015) Adolescent nicotine exposure fails to impact cocaine reward, aversion and self-administration in adult male rats. Pharmacol Biochem Behav 137:30–37
23. Schramm-Sapyta NL, Francis R, MacDonald A, Keistler C, O'Neill L, Kuhn CM (2014) Effect of sex on ethanol consumption and conditioned taste aversion in adolescent and adult rats. Psychopharmacology 231(8):1831–1839
24. Tzschentke TM (1998) Measuring reward with the conditioned place preference paradigm: a comprehensive review of drug effects, recent progress and new issues. Prog Neurobiol 56 (6):613–672
25. Tzschentke TM (2007) Review on CPP: measuring reward with the conditioned place preference (CPP) paradigm: update of the last decade. Addict Biol 12(3-4):227–462
26. Bardo M, Valone J, Bevins RA (1999) Locomotion and conditioned place preference produced by acute intravenous amphetamine: role of dopamine receptors and individual differences in amphetamine self-administration. Psychopharmacology 143(1):39–46
27. Bedingfield JB, King DA, Holloway FA (1998) Cocaine and caffeine: conditioned place preference, locomotor activity, and additivity. Pharmacol Biochem Behav 61(3):291–296
28. Bilsky EJ, Hui Y, Hubbell CL, Reid LD (1990) Methylenedioxymethamphetamine's capacity to establish place preferences and modify intake of an alcoholic beverage. Pharmacol Biochem Behav 37(4):633–638
29. Yates JR, Beckmann JS, Meyer AC, Bardo MT (2013) Concurrent choice for social interaction and amphetamine using conditioned place preference in rats: effects of age and housing condition. Drug Alcohol Depend 129(3):240–246
30. Lisek R, Xu W, Yuvasheva E, Chiu Y-T, Reitz AB, Liu-Chen L-Y, Rawls SM (2012) Mephedrone ('bath salt') elicits conditioned place preference and dopamine-sensitive motor activation. Drug Alcohol Depend 126(1):257–262
31. Karlsson L, Andersson M, Kronstrand R, Kugelberg FC (2014) Mephedrone, methylone and 3, 4-methylenedioxypyrovalerone (MDPV) induce conditioned place preference in mice. Basic Clin Pharmacol Toxicol 115(5):411–416
32. Cunningham CL, Gremel CM, Groblewski PA, Reilly S, Schachtman T (2009) Genetic influences on conditioned taste aversion. Oxford University Press, New York
33. Broadbent J, Muccino KJ, Cunningham CL (2002) Ethanol-induced conditioned taste aversion in 15 inbred mouse strains. Behav Neurosci 116(1):138
34. Ralph RJ, Caine SB (2005) Dopamine D1 and D2 agonist effects on prepulse inhibition and locomotion: comparison of Sprague-Dawley rats to Swiss-Webster, 129X1/SvJ, C57BL/6J, and DBA/2J mice. J Pharmacol Exp Ther 312(2):733–741
35. Thomsen M, Caine SB (2011) Psychomotor stimulant effects of cocaine in rats and 15 mouse strains. Exp Clin Psychopharmacol 19(5):321
36. King HE, Wakeford A, Taylor W, Wetzell B, Rice KC, Riley AL (2015) Sex differences in 3, 4-methylenedioxypyrovalerone (MDPV)-induced taste avoidance and place preferences. Pharmacol Biochem Behav 137:16–22
37. King HE, Wetzell B, Rice KC, Riley AL (2015) An assessment of MDPV-induced place preference in adult Sprague-Dawley rats. Drug Alcohol Depend 146:116–119
38. Costello N, Carlson J, Glick S, Bryda M (1989) Dose-dependent and baseline-dependent conditioning with d-amphetamine in the place conditioning paradigm. Psychopharmacology 99(2):244–247
39. Durazzo TC, Gauvin DV, Goulden KL, Briscoe RJ, Holloway FA (1994) Cocaine-induced conditioned place approach in rats: the role of dose and route of administration. Pharmacol Biochem Behav 49(4):1001–1005
40. Bardo M, Rowlett J, Harris M (1995) Conditioned place preference using opiate and stimulant drugs: a meta-analysis. Neurosci Biobehav Rev 19(1):39–51

41. Hunt T, Amit Z (1987) Conditioned taste aversion induced by self-administered drugs: paradox revisited. Neurosci Biobehav Rev 11(1):107–130
42. Verendeev A, Riley AL (2012) Conditioned taste aversion and drugs of abuse: history and interpretation. Neurosci Biobehav Rev 36(10):2193–2205
43. Merluzzi AP, Hurwitz ZE, Briscione MA, Cobuzzi JL, Wetzell B, Rice KC, Riley AL (2014) Age-dependent MDPV-induced taste aversions and thermoregulation in adolescent and adult rats. Dev Psychobiol 56(5):943–954
44. Cunningham CL, Niehus JS, Bachtold JF (1992) Ambient temperature effects on taste aversion conditioned by ethanol: contribution of ethanol-induced hypothermia. Alcohol Clin Exp Res 16(6):1117–1124
45. Doremus-Fitzwater TL, Varlinskaya EI, Spear LP (2010) Motivational systems in adolescence: possible implications for age differences in substance abuse and other risk-taking behaviors. Brain Cogn 72(1):114–123
46. Spear LP (2011) Rewards, aversions and affect in adolescence: emerging convergences across laboratory animal and human data. Dev Cogn Neurosci 1(4):390–403
47. Cappon GD, Morford LL, Vorhees CV (1998) Enhancement of cocaine-induced hyperthermia fails to elicit neurotoxicity. Neurotoxicol Teratol 20(5):531–535
48. Dafters RI, Lynch E (1998) Persistent loss of thermoregulation in the rat induced by 3, -4-methylenedioxymethamphetamine (MDMA or "Ecstasy") but not by fenfluramine. Psychopharmacology 138(2):207–212
49. Fukumura M, Cappon GD, Pu C, Broening HW, Vorhees CV (1998) A single dose model of methamphetamine-induced neurotoxicity in rats: effects on neostriatal monoamines and glial fibrillary acidic protein. Brain Res 806(1):1–7
50. Fantegrossi WE, Gannon BM, Zimmerman SM, Rice KC (2013) In vivo effects of abused 'bath salt' constituent 3, 4-methylenedioxypyrovalerone (MDPV) in mice: drug discrimination, thermoregulation, and locomotor activity. Neuropsychopharmacology 38(4):563–573
51. King HE, Wetzell B, Rice KC, Riley AL (2014) 3, 4-Methylenedioxypyrovalerone (MDPV)-induced conditioned taste avoidance in the F344/N and LEW rat strains. Pharmacol Biochem Behav 126:163–169
52. Horan B, Smith M, Gardner EL, Lepore M, Ashby CR (1997) (—)-nicotine produces conditioned place preference in Lewis, but not Fischer 344 rats. Synapse 26(1):93–94
53. Kosten TA, Miserendino M, Chi S, Nestler EJ (1994) Fischer and Lewis rat strains show differential cocaine effects in conditioned place preference and behavioral sensitization but not in locomotor activity or conditioned taste aversion. J Pharmacol Exp Ther 269(1):137–144
54. Picetti R, Caccavo JA, Ho A, Kreek MJ (2012) Dose escalation and dose preference in extended-access heroin self-administration in Lewis and Fischer rats. Psychopharmacology 220(1):163–172
55. Picetti R, Ho A, Butelman ER, Kreek MJ (2010) Dose preference and dose escalation in extended-access cocaine self-administration in Fischer and Lewis rats. Psychopharmacology 211(3):313–323
56. Stöhr T, Wermeling DS, Weiner I, Feldon J (1998) Rat strain differences in open-field behavior and the locomotor stimulating and rewarding effects of amphetamine. Pharmacol Biochem Behav 59(4):813–818
57. Davis CM, Roma PG, Dominguez JM, Riley AL (2007) Morphine-induced place conditioning in Fischer and Lewis rats: acquisition and dose-response in a fully biased procedure. Pharmacol Biochem Behav 86(3):516–523
58. Riley AL (2011) The paradox of drug taking: the role of the aversive effects of drugs. Physiol Behav 103(1):69–78
59. Lancellotti D, Bayer BM, Glowa JR, Houghtling RA, Riley AL (2001) Morphine-induced conditioned taste aversions in the LEW/N and F344/N rat strains. Pharmacol Biochem Behav 68(3):603–610
60. Pescatore KA, Glowa JR, Riley AL (2005) Strain differences in the acquisition of nicotine-induced conditioned taste aversion. Pharmacol Biochem Behav 82(4):751–757

61. Roma PG, Flint WW, Higley JD, Riley AL (2006) Assessment of the aversive and rewarding effects of alcohol in Fischer and Lewis rats. Psychopharmacology 189(2):187–199
62. Glowa JR, Shaw AE, Riley AL (1994) Cocaine-induced conditioned taste aversions: comparisons between effects in LEW/N and F344/N rat strains. Psychopharmacology 114(2):229–232
63. Grigson PS, Freet CS (2000) The suppressive effects of sucrose and cocaine, but not lithium chloride, are greater in Lewis than in Fischer rats: evidence for the reward comparison hypothesis. Behav Neurosci 114(2):353
64. Simpson GR, Riley AL (2005) Morphine preexposure facilitates morphine place preference and attenuates morphine taste aversion. Pharmacol Biochem Behav 80(3):471–479
65. Roma PG, Davis CM, Kohut SJ, Huntsberry ME, Riley AL (2008) Early maternal separation and sex differences in the aversive effects of amphetamine in adult rats. Physiol Behav 93 (4):897–904
66. Russo SJ, Jenab S, Fabian SJ, Festa ED, Kemen LM, Quinones-Jenab V (2003) Sex differences in the conditioned rewarding effects of cocaine. Brain Res 970(1):214–220
67. Torres OV, Natividad LA, Tejeda HA, Van Weelden SA, O'Dell LE (2009) Female rats display dose-dependent differences to the rewarding and aversive effects of nicotine in an age-, hormone-, and sex-dependent manner. Psychopharmacology 206(2):303–312
68. Dragoin W, Mccleary GE, McCleary P (1971) A comparison of two methods of measuring conditioned taste aversions. Behav Res Methods Instrum 3(6):309–310
69. Grote FW, Brown RT (1971) Conditioned taste aversions: two-stimulus tests are more sensitive than one-stimulus tests. Behav Res Methods Instrum 3(6):311–312
70. Chambers KC, Sengstake CB (1976) Sexually dimorphic extinction of a conditioned taste aversion in rats. Anim Learn Behav 4(2):181–185
71. Zakharova E, Wade D, Izenwasser S (2009) Sensitivity to cocaine conditioned reward depends on sex and age. Pharmacol Biochem Behav 92(1):131–134
72. Cunningham CL, Tull LE, Rindal KE, Meyer PJ (2002) Distal and proximal pre-exposure to ethanol in the place conditioning task: tolerance to aversive effect, sensitization to activating effect, but no change in rewarding effect. Psychopharmacology 160(4):414–424
73. Turenne SD, Miles C, Parker LA, Siegel S (1996) Individual differences in reactivity to the rewarding/aversive properties of drugs: assessment by taste and place conditioning. Pharmacol Biochem Behav 53(3):511–516
74. Ettenberg A, Fomenko V, Kaganovsky K, Shelton K, Wenzel JM (2015) On the positive and negative affective responses to cocaine and their relation to drug self-administration in rats. Psychopharmacology 232(13):2363–2375
75. Horger BA, Giles MK, Schenk S (1992) Preexposure to amphetamine and nicotine predisposes rats to self-administer a low dose of cocaine. Psychopharmacology 107(2-3):271–276
76. Horger BA, Shelton K, Schenk S (1990) Preexposure sensitizes rats to the rewarding effects of cocaine. Pharmacol Biochem Behav 37(4):707–711
77. Schenk S, Partridge B (1997) Sensitization and tolerance in psychostimulant self-administration. Pharmacol Biochem Behav 57(3):543–550
78. Grigson PS, Twining RC (2002) Cocaine-induced suppression of saccharin intake: a model of drug-induced devaluation of natural rewards. Behav Neurosci 116(2):321
79. Cason AM, Grigson PS (2013) Prior access to a sweet is more protective against cocaine self-administration in female rats than in male rats. Physiol Behav 112:96–103
80. Imperio CG, Grigson PS (2015) Greater avoidance of a heroin-paired taste cue is associated with greater escalation of heroin self-administration in rats. Behav Neurosci 129(4):380
81. Twining RC, Bolan M, Grigson PS (2009) Yoked delivery of cocaine is aversive and protects against the motivation for drug in rats. Behav Neurosci 123(4):913

MDMA, Methylone, and MDPV: Drug-Induced Brain Hyperthermia and Its Modulation by Activity State and Environment

Eugene A. Kiyatkin and Suelynn E. Ren

Abstract Psychomotor stimulants are frequently used by humans to intensify the subjective experience of different types of social interactions. Since psychomotor stimulants enhance metabolism and increase body temperatures, their use under conditions of physiological activation and in warm humid environments could result in pathological hyperthermia, a life-threatening symptom of acute drug intoxication. Here, we will describe the brain hyperthermic effects of MDMA, MDPV, and methylone, three structurally related recreational drugs commonly used by young adults during raves and other forms of social gatherings. After a short introduction on brain temperature and basic mechanisms underlying its physiological fluctuations, we will consider how MDMA, MDPV, and methylone affect brain and body temperatures in awake freely moving rats. Here, we will discuss the role of drug-induced heat production in the brain due to metabolic brain activation and diminished heat dissipation due to peripheral vasoconstriction as two primary contributors to the hyperthermic effects of these drugs. Then, we will consider how the hyperthermic effects of these drugs are modulated under conditions that model human drug use (social interaction and warm ambient temperature). Since social interaction results in brain and body heat production, coupled with skin vasoconstriction that impairs heat loss to the external environment, these physiological changes interact with drug-induced changes in heat production and loss, resulting in distinct changes in the hyperthermic effects of each tested drug. Finally, we present our recent data, in which we compared the efficacy of different pharmacological strategies for reversing MDMA-induced hyperthermia in both the brain and body. Specifically, we demonstrate increased efficacy of the centrally acting atypical neuroleptic compound clozapine over the peripherally acting

E.A. Kiyatkin (✉) and S.E. Ren
Behavioral Neuroscience Branch, National Institute on Drug Abuse – Intramural Research Program, NIH, 333 Cassell Drive, Baltimore, MD 21224, USA
e-mail: ekiyatki@intra.nida.nih.gov

Curr Topics Behav Neurosci (2017) 32: 183–208
DOI 10.1007/7854_2016_35
Published Online: 20 July 2016

vasodilator drug, carvedilol. These data could be important for understanding the potential dangers of MDMA in humans and the development of pharmacological tools to alleviate drug-induced hyperthermia – potentially saving the lives of highly intoxicated individuals.

Keywords Active ingredients of "bath salts" • Brain metabolism • Cerebral heat production • Drug-induced intoxication • Drugs of abuse • MDMA (Ecstasy) • Psychomotor stimulants • Rave parties • Vasoconstriction

Contents

Abbreviations

ip	Intraperitoneal
iv	Intravenous
MDMA	3,4-Methylenedioxymethamphetamine (Ecstasy)
MDPV	3,4-Methylenedioxypyrovalerone
METH	Methamphetamine
methylone	3,4-Methylenedioxymethcathinone
NAc	Nucleus accumbens
sc	Subcutaneous

1 Introduction: Psychoactive Drugs of Abuse

In contrast to therapeutic psychoactive drugs, which are taken voluntarily or given by medical professionals to alleviate specific pathological symptoms, individuals take drugs recreationally to induce pleasurable, novel, or unusual psycho-emotional effects. Each psychoactive drug induces specific behavioral, physiological, and psycho-emotional effects, which vary significantly depending upon the dosage, individual activity state, and environmental conditions in which the drug is taken. While the physiological effects of such drugs may mimic physiological responses induced by naturally arousing stimuli at lower doses, drug-induced responses at higher doses may reach pathological levels, resulting in acute intoxication and posing a significant risk to human health. It is generally believed that drug dosage is the primary parameter in predicting acute intoxication, but many other factors play important roles in determining the severity of drug-induced responses, including individual responsiveness, previous drug experience(s), concurrent poly-drug use, preexisting pathologies, and finally, the specific circumstances or context in which the drug is taken.

In this review, we will be focused on the latter variable, discussing how activity state and the environmental conditions of drug use modulate the physiological and behavioral effects of MDMA (3,4-methylenedioxymethamphetamine or "Ecstasy," "Molly"), methylone (3,4-methylenedioxymethcathinone), and MDPV (3,4-methylenedioxypyrovalerone). While the psychoactive effects of MDMA have been known for decades, methylone and MDPV are synthetic cathinones that became popular among young adults in the last few years with the emergence of "bath salts." Although MDMA, MDPV, and methylone have distinct behavioral and physiological effects, they are structurally similar psychostimulants that all induce sympathetic activation and body hyperthermia. As such, our focus here will be on drug-induced perturbations in temperature homeostasis, with a special emphasis on brain temperature as an important parameter that not only reflects the metabolic aspects of brain activity, but also affects neural activity and neural functioning [1].

2 Brain Temperature

While it is traditionally believed that brain temperature in healthy homeothermic organisms is stable and close to 37°C, abundant data obtained in different animal models suggest that relatively large fluctuations in brain temperature occur during different types of natural motivated behavior and following exposure to various environmental challenges [2–6]. It is difficult to quantify the range of physiological fluctuations in brain temperature in humans, but, as shown by direct monitoring with chronically implanted thermocouple sensors, hypothalamic temperature recorded from freely moving rats could fall to ~35°C during deep sleep [7, 8] and phasically peak at ~39.5°C at the time of ejaculation during copulatory behavior [9].

While the direct monitoring of brain temperature in rats is a relatively simple procedure, human data are limited and often restricted to neurological patients [10, 11] or indirect measurements [12–14] that are questionable with respect to their validity and accuracy. Therefore, it has not been definitely proven that similar, relatively large brain temperature fluctuations could occur in healthy humans. However, several observations suggest that this could be the case. First, monkeys show robust physiological changes in brain temperature within a range comparable to that observed in rats [3, 15]. Second, humans show ~1.0–2.0°C diurnal fluctuations in body temperature as well as clear pathological hyperthermia (>40–41°C) during acute intoxication by METH or MDMA [16–19], similar to the temperature changes seen in rats [20]. Finally, direct measurements of venous outflow from healthy human volunteers wearing water-impermeable clothing which impaired normal heat dissipation to the external environment revealed that brain temperatures could reach 39.5–40.0°C during a 30-min bicycle exercise [21]. Importantly, even at such high brain temperatures, the physical and mental states of these volunteers remained normal, suggesting that the brain can tolerate relatively large, but transient temperature increases.

3 Basic Mechanisms Underlying Brain Temperature Homeostasis

Brain temperature is determined by the balance of two opposing forces: (1) metabolism-related intra-brain heat production and (2) heat loss via cerebral blood outflow to the rest of the body and then to the external environment. Although the brain represents only ~2% of body weight in humans, it accounts for ~20% of an organism's total energy consumption [22, 23], suggesting intense intra-cerebral heat production. This heat is removed from brain tissue by the cerebral circulation due to arrival from the lungs of cooler arterial blood [24–26].

Although mechanistic, the cooling of an internal combustion engine is a good analogy when considering brain temperature exchange. Similar to circulating coolant that continuously removes heat from a working engine, cool, oxygenated arterial blood removes heat from the brain via heat exchange. The now warmed venous blood then returns to the heart to be cooled and re-oxygenated in the lungs. Such an arrangement determines the critical role of cerebral blood flow in brain temperature homeostasis and the essential interdependence between temperature in the brain and the rest of the body. While brain temperature tends to increase due to metabolism-related intra-brain heat production, it also rises when brain-generated heat cannot be properly dissipated to the body and then to the external environment. Similarly, a decrease in cerebral metabolism tends to lower brain temperature; however, this effect could be strongly enhanced by a peripheral vasodilation that promotes heat loss to a cooler environment [27, 28].

In our experiments, we routinely used a three-point thermorecording paradigm. In addition to a brain representative site, usually the nucleus accumbens (a critical structure in brain motivation-reinforcement circuitry [29–31]), temperatures were also recorded from two peripheral locations: the temporal muscle and skin. Since the brain and temporal muscle receive arterial blood from the same common carotid artery and are equally exposed to blood-delivered heat from the body, temperature difference between these locations (NAc-Muscle temperature differential) shows the source of heat production, providing a measure of drug-induced metabolic brain activation. As a result, increases in NAc-Muscle differential reflect increased metabolic brain activation. Skin temperature is primarily determined by the state of peripheral vessels, but it also depends on the temperature of arterial blood inflow. Therefore, Skin-Muscle temperature differential serves as an accurate measure of

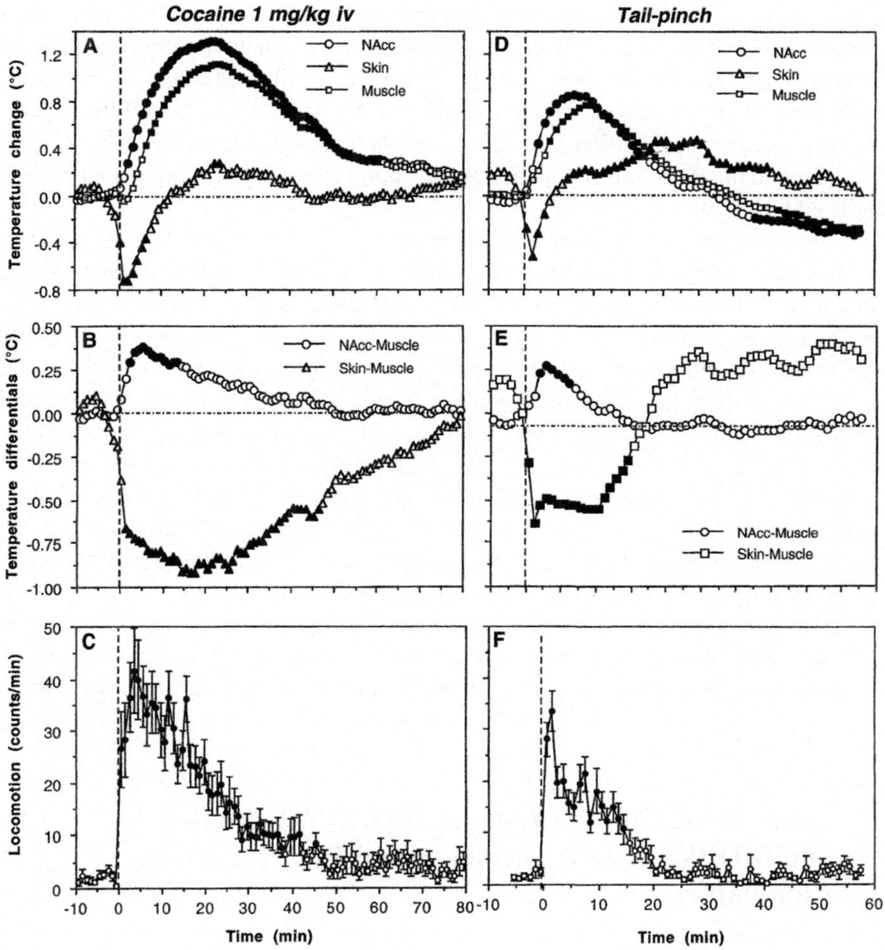

Fig. 1 Changes in brain (NAc), temporal muscle, and skin temperatures induced by iv injection of cocaine (1 mg/kg) and 3-min tail-pinch in freely moving rats under quiet resting conditions. *Top graphs* (**a**, **d**) show relative changes in temperatures, *middle graphs* show changes in NAc-Muscle and Skin-Muscle temperature differentials (**b**, **e**), and *bottom graphs* (*c–f*) show changes in locomotor activity. *Filled symbols* mark values significantly different from baseline. Original data shown in this graph were reported in Kiyatkin [1]

peripheral vascular tone, another important contributor to changes in brain temperature [1]. Increases in Skin-Muscle differential reflect vasodilation, whereas decreases reflect vasoconstriction.

Figure 1 shows temperature and locomotor effects induced by two different stimuli: a three-min tail-pinch and intravenous (iv) cocaine injected to freely moving rats at a typical self-administration dose (1 mg/kg). Despite the differing natures of these stimuli, they induced similar locomotor activation and moderate increases in brain and muscle temperatures. Importantly, NAc-Muscle differentials significantly increased in both cases, suggesting metabolic brain activation as a common factor elicited by both pharmacological and natural arousing stimuli. Similarly, both stimuli significantly decreased Skin-Muscle differentials, suggesting another common factor – prolonged skin vasoconstriction. While qualitatively similar, cocaine-induced effects on both temperature differentials were stronger and more prolonged, obviously indicating a larger elevation in temperature induced by cocaine.

This example, as well as the results of our other studies (see [1] for review), demonstrates that brain temperature increases, induced by both natural arousing stimuli and drugs, are determined by two basic mechanisms: (1) increased intrabrain heat production due to metabolic brain activation and (2) decreased heat loss due to skin vasoconstriction. Since psychostimulant drugs increase metabolism and locomotor activity [32–36] and induce peripheral vasoconstriction [34, 37], it could be assumed that drug-induced brain temperature responses will depend upon the ongoing state of the organism and the environmental conditions associated with drug use. A drug at a certain dose could induce minimal temperature effects in an environment where the adaptive mechanisms of heat loss are fully effective, but the same drug at the exact same dose could induce powerful hyperthermic effects in an environment where heat dissipation mechanisms are significantly impaired. Since peripheral vasodilation and perspiration are powerful means for heat loss in humans, drug-induced impairment of these adaptive mechanisms may be a very important determinant of drug-induced increases in brain and body temperatures.

4 Brain Hyperthermia Induced by MDMA, Methylone, and MDPV: State Dependency and Environmental Modulation

MDMA is a typical "club drug" that is often used by young adults under conditions of physical and emotional activation, often in a warm and humid environment. Similar to other psychostimulants, MDMA increases metabolism and induces hyperactivity coupled with hyperthermia [32–36]. The influence of environmental conditions and specific activity states could be especially important for MDMA because, in addition to metabolic activation, it also induces peripheral vasoconstriction [34, 37], thus diminishing heat dissipation from body surfaces and enhancing heat accumulation in the brain and body.

During recent years, there has been a rapid increase in the abuse of synthetic cathinone analogs, which are sold with innocuous names such as "bath salts" or "plant food" [38, 39]. Such products were designed to circumvent regulations controlling the sale and use of psychoactive substances. Two very popular synthetic cathinones are methylone and MDPV [40]. While low recreational doses of synthetic cathinones enhance mood and increase energy, high doses or chronic use can cause serious medical complications, including agitation, psychosis, tachycardia, hyperthermia, and even death [41, 42]. Due to these risks, methylone and MDPV have been classified as Schedule I controlled substances in the USA.

Methylone and MDPV are structurally similar to MDMA. Like MDMA, methylone and MDPV exert their major effects by interacting with monoamine transporter proteins in the central and peripheral nervous systems [43]. Methylone is a non-selective transporter substrate that evokes the release of dopamine, norepinephrine, and serotonin, analogous to the effects of MDMA [44–46]. By contrast, MDPV is a potent transporter blocker that inhibits the uptake of dopamine and norepinephrine, with minimal effects on serotonin uptake [43, 47].

Although robust increases in body temperature have been reported in humans as the result of acute intoxication with both methylone [48] and MDPV [49–51], data on temperature effects of methylone and MDPV in laboratory animals are limited and controversial, varying according to species, dose, and experimental conditions [44, 52, 53]. MDPV is reported to cause hyperthermia in mice (3–30 mg/kg, intraperitoneal or ip administration) but only under elevated ambient temperature [53]. In rats, MDPV (1.0–5.6 mg/kg, subcutaneous or sc administration) had no evident effect on core body temperature [52].

To elucidate how MDMA, MDPV, and methylone affect brain and body temperature, and which mechanisms underlie these changes, we first examined the effects of these drugs on NAc, muscle, and skin temperatures under standard laboratory conditions (quiet rest at 22°C ambient temperatures). We also examined the effects of these drugs on locomotion. In these initial experiments, drugs were delivered using sc injections at a wide range of recreational extremes. MDMA and methylone were tested within a dosage range of 1–9 mg/kg and the more potent MDPV was delivered at a range of 0.1–1.0 mg/kg. The data for each individual drug were described in detail elsewhere [54, 55]. While these conditions are usually used in most preclinical studies in animals, humans typically use these drugs during psychophysiological activation and often in warm environments that impair normal heat dissipation from the body to the external environment. To model conditions relevant to humans, the effects of MDMA, MDPV, and methylone were examined in rats during social interaction between two animals. Under these conditions, brain and body temperatures moderately increased but skin temperature rapidly decreased, reflecting an adaptive response elicited by arousal stimulation. Finally, the effects of each of the three drugs were tested in quietly resting rats maintained in a moderately warm environment (29°C). While this temperature is ~7°C higher than standard housing conditions (22–23°C), it corresponds to the thermoneutral zone in rats, where endogenous heat production is minimal and balanced with heat loss [56].

Since our focus was on pathological hyperthermia that develops in some drug users and could result in serious health complications (including lethality), in these experiments we used relatively large drug doses (MDMA and methylone –9 mg/kg, sc; MDPV –1 mg/kg, sc), which induce relatively large effects, both behaviorally as well as on temperature. While exceeding the typical doses consumed by humans, all drugs at these doses are well tolerated by freely moving rats tested under standard environmental conditions. For example, a 9 mg/kg dose of MDMA is ~1/5 of the

LD50 (50 mg/kg; [57]), but and with sc injection this dose never results in lethality. Since the psychoactive effects of MDMA possess significant latencies after ingestion, sometimes people consume one or two additional MDMA tablets, shifting the dose close to 9 mg/kg. Similar to MDMA, methylone and MDPV are usually ingested, with typical active doses of 100–300 mg for methylone and 5–20 mg for MDPV (see www.erowid.org for human self-reports). However, these drugs can be taken at much higher doses (up to 1,000 and 200 mg, respectively) and via other administration routes.

Figure 2 shows that MDMA, MDPV, and methylone each increased brain and muscle temperatures, rapidly decreased skin temperature, and induced locomotor activation. These changes differed from a transient temperature increase induced by the saline injection, which did not result in an evident motor response (panels a–c). While brain temperature increases induced by all three drugs were approximately equal in their magnitude (1.5–1.8°C), their time-course and duration differed in each case. Methylone and MDMA induced prolonged temperature increases, but the effect of MDPV was shorter and more rapid. Injections of each drug induced rapid increases in NAc-Muscle differentials, but this effect, suggesting metabolic brain activation and enhanced intra-brain heat production, was clearly larger vs. saline injection only with MDMA (panel k). Each drug also rapidly decreased

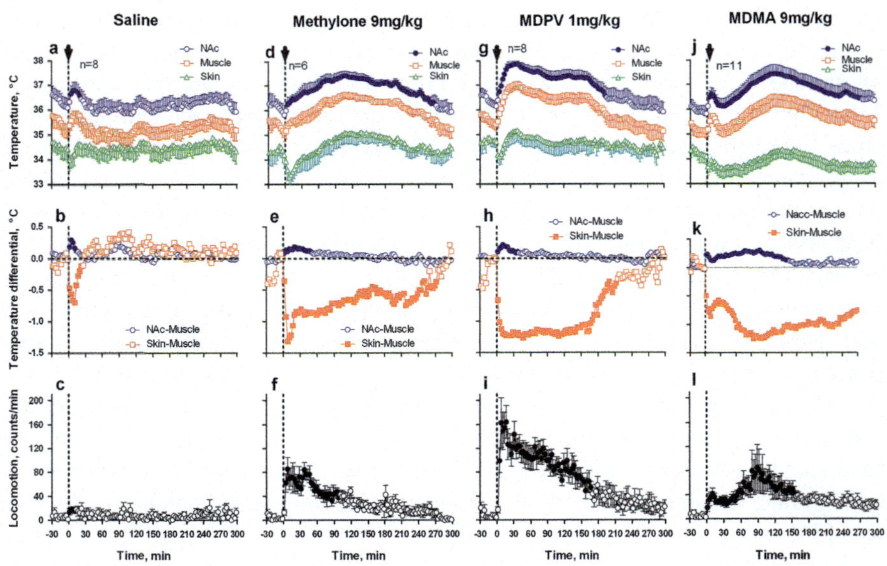

Fig. 2 Mean changes in brain (NAc), temporal muscle, and skin temperatures and locomotor responses induced by sc injections of methylone (9 mg/kg), MDPV (1 mg/kg), MDMA (9 mg/kg), and saline in freely moving rats under quiet resting conditions. *Top graphs* show mean (±SEM) values of absolute temperature changes; *middle graphs* show changes in NAc-Muscle and Skin-Muscle differentials; and *bottom graphs* show mean (±SEM) changes in locomotor activity. *Filled symbols* mark values significantly different from pre-injection baselines. *Bold arrows* mark the moment of injection. Original data shown in this graph were reported in Kiyatkin et al. [54, 55]

Skin-Muscle differential, suggesting cutaneous vasoconstriction; this effect was about the same for each drug. Finally, all three drugs induced locomotor activation; this effect was clearly the greatest for MDPV.

Consistent with our previous work [1], the introduction of a novel "guest" rat into the chamber occupied by the experimental rat undergoing recording induced locomotor activation, a rapid, strong rise in NAc and muscle temperatures (~1.5°C), and a brief decrease in skin temperature that was rapidly transformed into a more tonic, rebound-like increase (Fig. 3a–c). While the changes in NAc and muscle temperatures were generally parallel, the rise was stronger and more rapid in the brain. This resulted in a significant increase in NAc-Muscle differentials that indicated metabolic brain activation. Social interaction was also accompanied by a strong decrease in Skin-Muscle temperature differentials, indicating stimulus-induced cutaneous vasoconstriction. While NAc-Muscle differentials rapidly returned to baseline after the end of the social interaction period, Skin-Muscle differentials increased above baseline, suggesting a rebound vasodilation. These physiological parameters and locomotion showed consistent changes at the start and end of the social interaction. A saline injection during social interaction (+10 min after its onset; black arrow in Fig. 3a) had no effect on the parameters above.

When rats received methylone instead of saline, temperature differences were minimal, but the decreases in NAc and muscle temperatures and Skin-Muscle differentials after the end of social interaction were more prolonged (Fig. 3d–f). Interestingly, the combination of two hyperthermic effects (methylone + social interaction) did not result in their summation or potentiation. Mean values of all parameters did not differ statistically vs. those seen in rats that received methylone under quiet resting conditions. The mean increase in NAc temperature was calculated for the entire 5-h post-drug interval as the area under the curve was maximal with methylone used under quiet resting conditions (1.12 ± 0.23°C) and even slightly lower when methylone was used during social interaction (0.88 ± 0.18°C, no significant change).

Similar to methylone, injection of MDPV delayed brain and muscle temperature decreases after social interaction (Fig. 3g–i). However, in contrast to methylone, changes were more robust and MDPV-induced NAc temperature increases significantly potentiated during social interaction (MDPV + social interaction 1.91 ± 0.18°C vs. MDPV, quiet resting conditions 0.79 ± 0.25°C; $p < 0.01$). The "pure" effect of MDPV (difference vs. saline control for each condition) was also clearly amplified when drug was administered during social interaction (1.38°C vs. 0.74°C). When taken in an activated physiological state, MDPV also enhanced both increases in NAc-Muscle differentials and decreases in Skin-Muscle differentials, suggesting a weak potentiation of MDPV-induced metabolic and vasoconstrictive effects.

The most robust potentiation of hyperthermic effects was found with MDMA used during social interaction (Fig. 3j–l). In this case, mean NAc temperatures exceed 39°C at their peak and the overall response was the most prolonged, extending the 5-h observation window. These changes, however, were highly

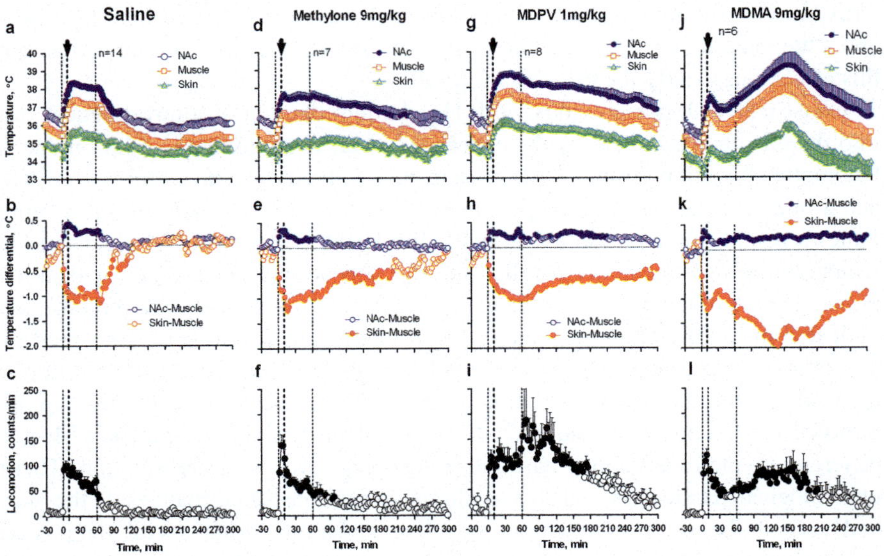

Fig. 3 Mean changes in brain (NAc), temporal muscle, and skin temperatures and locomotor responses induced by sc injections of methylone (9 mg/kg), MDPV (1 mg/kg), MDMA (9 mg/kg), and saline during social interaction. *Top graphs* show mean (±SEM) values of absolute temperature changes; *middle graphs* show changes in NAc-Muscle and Skin-Muscle differentials; and *bottom graphs* show mean (±SEM) changes in locomotor activity. *Filled symbols* mark values significantly different from the pre-injection baseline. The first and third vertical hatched lines in each graph show onset and offset of social interaction (60 min) and *black arrows* at the second hatched lines mark the moment of drug administration. Original data shown in this graph were reported in Kiyatkin et al. [54, 55]

variable in individual rats, with peak brain temperatures ranging from 37.1 to 42.1°C. The mean temperature elevations were maximal in this case (2.39 ± 0.27°C), significantly exceeding modest temperature increases seen under quiet resting conditions (0.95 ± 0.25°C; $p < 0.01$). The pure effects of MDMA on brain temperature were also significantly stronger during social interaction than in quiet resting conditions (1.86°C vs. 0.89°C), suggesting a super-additive interaction. Despite the use of a moderate, non-lethal dose of MDMA (9 mg/kg, 1/5 LD50; [57]), one of the seven rats tested with MDMA died overnight after the recording session.

The potentiation of MDMA-induced hyperthermia during social interaction could be explained by two primary contributors: a stronger rise in NAc-Muscle differentials and a strong, tonic drop in Skin-Muscle differentials. The increase in the NAc-Muscle differential was very prolonged (>5-h) although its amplitude did not differ from that seen with the drug used in quiet resting conditions. In contrast, MDMA-induced decrease in Skin-Muscle differential (−1.31 ± 0.29°C) was larger and more prolonged than with either methylone (−0.56 ± 0.13°C; $p < 0.05$) or MDPV (−0.66 ± 0.27°C; $p < 0.05$) and it did not recover at the recording end (5 h). Interestingly, robust potentiation of MDMA's hyperthermic effect profile

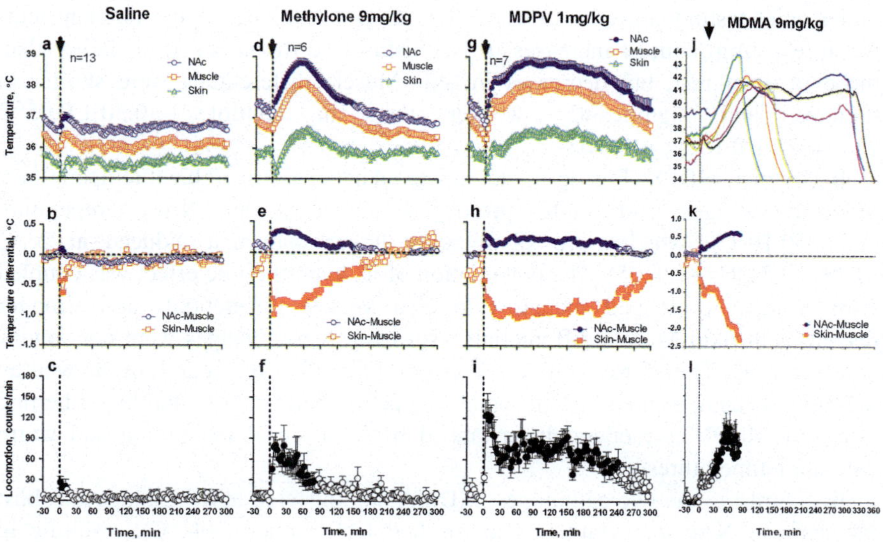

Fig. 4 Mean changes in brain (NAc), temporal muscle, and skin temperatures and locomotor responses induced by sc injections of methylone (9 mg/kg), MDPV (1 mg/kg), MDMA (9 mg/kg), and saline in freely moving rats at warm ambient temperatures (29°C). *Top graphs* show mean (±SEM) values of absolute temperature changes; *middle graphs* show changes in NAc-Muscle and Skin-Muscle differentials; and *bottom graphs* show mean (±SEM) changes in locomotor activity. *Filled symbols* mark values significantly different from the pre-injection baseline. *Black arrows* at the hatched lines mark the moment of drug administration. Since all rats exposed to MDMA died within 6 h post-injection, MDMA data are shown as individual changes (*j*) and mean values of NAc-Muscle, Skin-Muscle differentials and locomotion (*k* and *l*) for the first 80 min post-injection when all rats were still alive. Original data shown in this graph were reported in Kiyatkin et al. [54, 55]

by social interaction was not accompanied by significant changes in its locomotor effects, which remained similar in both conditions (compare Figs. 2 and 3).

The profile of temperature effects of MDMA, MDPV, and methylone was robustly modulated by warm ambient temperatures; the type of modulation was also unique to each compound (Fig. 4). Rats exposed to warm ambient temperatures (28–29°C) maintained stable but slightly higher internal temperatures (mean: 36.7 ± 0.2, 36.2 ± 0.2, and 35.2 ± 0.4°C for NAc, muscle, and skin, respectively) than rats housed under standard laboratory conditions at 22–23°C (mean: 35.9 ± 0.2, 35.2 ± 0.3, and 34.0 ± 0.26°C, respectively). The difference was minimal for the brain ($\Delta = 0.76$°C) and maximal for skin ($\Delta = 1.17$°C), suggesting weak tonic vasodilation as a way to promote dissipation of metabolic heat into the warmer environment. Saline injections under these conditions induced transient, weak temperature responses similar to that seen with saline injection at standard room temperature (Fig. 4a–c).

Although methylone (9 mg/kg) injected at 29°C increased NAc temperature (Fig. 4d–f), this change was more transient than at 23°C and its mean value assessed as the area under the curve (0.47 ± 0.15°C) was significantly lower than that seen under quiet resting conditions (1.12 ± 0.23°C, $p < 0.01$). At higher ambient temperatures, drug-induced increases in NAc-Muscle differentials were larger than those in quiet rest, but decreases in Skin-Muscle differentials were shorter in duration and significantly weaker than in control (−0.30 ± 0.05°C vs. −0.72 ± 0.15°C; $p < 0.05$).

In contrast, MDPV (Fig. 4g–i), administered under warm ambient temperatures, induced stronger and more prolonged elevations in NAc temperature (1.51 ± 0.16°C), which were significantly larger than that induced at 23°C (0.79 ± 0.25°C; $p < 0.05$). The potentiation of its hyperthermic effect was coupled with stronger prolonged increases in NAc-Muscle differentials and stronger decreases in Skin-Muscle differentials. However, these changes were not significant vs. 23°C. Therefore, while the hyperthermic effect of methylone showed an unusual weakening, both during social interaction and at 29°C, the hyperthermic effect of MDPV is moderately enhanced by both social interaction and warm external temperatures.

In contrast to the two cathinones, MDMA administered at 29°C induced robust increases in NAc temperatures that reached fatal values (>41°C), resulting in lethality in all 6 tested rats within 6-h post-injection (Fig. 4j–l). Peak values of NAc temperatures after MDMA administration ranged from 41.4°C to 43.8°C, and their means were more than 4°C larger than in the control condition (42.2 ± 0.4°C vs. 37.7 ± 0.4°C; $p < 0.01$). When calculated as the pre-lethal temperature peak, the mean increase in NAc temperature was 2.29 ± 0.22°C, significantly larger than mean MDMA-induced NAc temperature elevation in 23°C environment (0.95 ± 0.25°C; $p < 0.01$). By this parameter, the degree of potentiation for MDMA was twice that for MDPV (1.34°C vs. 0.72°C). Additionally, while significant increases in NAc-Muscle differentials were evident with MDMA at standard ambient temperatures (mean = 0.17 ± 0.07°C), NAc-Muscle differentials more than doubled at 29°C (0.36 ± 0.04°C; $p < 0.05$). When used at warm temperatures, MDMA also induced much stronger and more prolonged vasoconstriction as evidenced by Skin-Muscle differentials, which decreased on average more than 2°C below pre-injection baseline (see Fig. 3k).

Drug-specific differences in modulation are especially evident when we compare the hyperthermic effects of each drug under different experimental conditions (Fig. 5). The upper panel in Fig. 5a shows temperature change (as an area under the curve for the duration of significant drug effect) with respect to quiet resting baseline and the bottom panel shows "pure" or "net" effects of drugs, when changes occurring in saline control being subtracted. As can be seen, each of the three drugs induced an approximately equal brain hyperthermic effect when tested at standard laboratory conditions. However, the effects of these drugs showed distinct differences when they were administered during social interaction and at warm ambient temperatures. The effect of methylone (blue bars) administered during social interaction did not change and, surprisingly, decreased when the drug was administered at 29°C. This pattern of interaction could suggest that methylone and social interaction share common effector mechanisms (e.g., sympathetic activation) to induce brain hyperthermia. When these mechanisms are naturally activated during

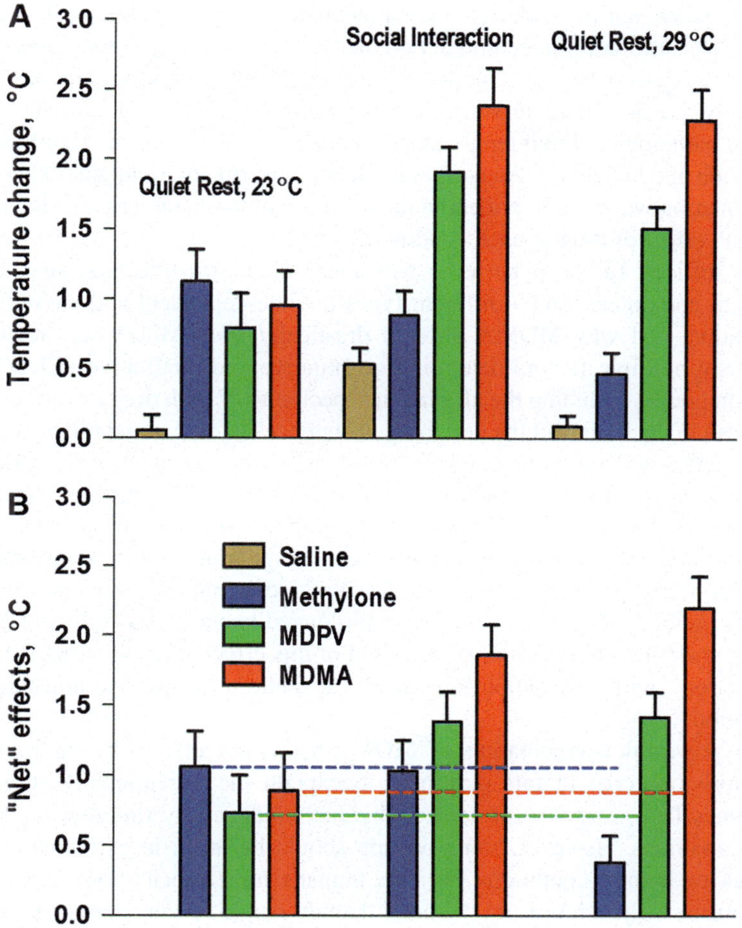

Fig. 5 Mean values of brain (NAc) temperature increases (area under curve for 5-h post-injection) induced by sc injections of methylone (9 mg/kg), MDPV (1 mg/kg), MDMA (9 mg/kg), and saline in rats under quiet resting conditions, during social interaction, and at warm ambient temperatures. *Top graphs* (**a**) show mean (±SEM) values and *bottom graphs* (**b**) show "net" or "pure" drug effects (drug–saline). *Horizontal hatched lines* in (**a**) show values in control (saline) group. *Horizontal hatched lines* in (**b**) show values induced by each drug under quiet resting conditions. Original data shown in this graph were reported in Kiyatkin et al. [54, 55]

social interaction (i.e., when brain metabolic activity is increased and/or cutaneous vessels are physiologically constricted), the effects of the drug *per se* become weaker but the overall hyperthermic response does not change.

However, a different pattern of interaction occurred with MDPV (Fig. 5, green bars). Brain temperature increases induced by this drug during social interaction and at 29°C become significantly stronger (A) and the net effect of MDPV was larger (B), suggesting an additive interaction and the involvement of different mechanisms in mediating the enhanced brain temperature response to MDPV.

While the mechanisms underlying this potentiation remain unclear, it is likely that MDPV, in addition to its central action, acts directly on peripheral vessels potentiating skin vasoconstriction and increasing intra-brain heat accumulation.

Finally, potentiation was supra-additive with MDMA (Fig. 5, red bars), which induced pathological brain temperature increases when administered during social interaction and lethality when used in moderately warm environments. This pattern of modulation was clearly evident in mean values of temperature elevation (A) and the "net" effect of drug in each condition (B).

It is difficult to speculate as to why these three structurally similar psychostimulant compounds show different types of state-dependent and environmental modulation, and why MDMA differed drastically from MDPV and methylone. Obviously multiple factors determine quantitative and qualitative differences in each compound, including the affinity and specificity of each drug to various neural substrates in the brain and the periphery, differences in drug metabolism, and drug-specific entry in brain tissue via the blood–brain barrier. Despite the importance of all these factors – many of which are still unknown and require further work – at the physiological level, MDMA, compared to cathinones, has strong, sustained effects on intra-brain heat production coupled with very strong, tonic peripheral vasoconstriction that prevents normal heat loss to the external environment. These two features facilitate the stronger and more prolonged hyperthermic effect of MDMA and the profound, often lethal, potentiation of this effect when the drug is taken in conjunction with psychophysiological activation and/or diminished heat dissipation.

The powerful augmentation of MDMA's thermal effects by environmental conditions observed in rats may help to explain the exceptionally strong, and sometimes fatal, response of some individuals induced by this drug under rave party conditions. However, some caution should be taken in extrapolating these findings to human conditions because humans have much more sophisticated mechanisms of heat loss from the body than do rats [58], thus making them more resistant to high environmental temperatures and thermogenic effects of psychomotor stimulants. In contrast to rats, humans have a well-developed ability to sweat and have a very high dynamic range of flow rates in the skin, thus allowing them to lose more metabolic heat (1 kW) than could be maximally produced in the body [59]. These differences in the effector mechanisms of heat loss could explain weaker MDMA-induced body temperature increases and their lesser dependence on ambient temperatures found in monkeys [60–62] and humans [33, 63]. Despite their high efficiency, the compensatory mechanisms of heat loss in humans could be greatly impaired under specific conditions, resulting in progressive heat accumulation in the organism. Even a simple bicycle exercise that produces ~1°C brain temperature elevation under normal conditions produced strong hyperthermia (39.0–39.5°C) when the exercise is conducted in a special water-impermeable cloth that prevents heat dissipation to the external environment [64]. Therefore, pathological brain hyperthermia induced by overdose of psychomotor stimulants under rave conditions results not only from excessive heat production due to drug-induced and associated psychophysiological activation, but also from the powerful drug-induced peripheral vasoconstriction and impaired ability to dissipate metabolic heat due to warm, humid environment. Since partygoers are often engaged in intense social/physical interactions with other individuals (dancing, sexual activity, etc.), drug effects are additionally potentiated by psychophysiological activation typical to these conditions.

5 Pharmacological Strategies to Reverse Pathological Hyperthermia Induced by MDMA

Current emergency therapeutic options to counteract MDMA-induced pathological hyperthermia mainly focus on whole body cooling, water substitution, and sedative therapy. Indeed, body cooling should have a hypothermic effect, but the effectiveness of this treatment is limited due to strong, sustained MDMA-induced vasoconstriction and the natural vasoconstrictive effect of skin cooling. Similarly, water consumption and/or saline infusions are minimally effective in counteracting MDMA-induced hyperthermia [65].

Since our previous work established a critical role of intra-brain heat production and peripheral vasoconstriction in potentiating brain and body hyperthermic effects of MDMA, we explored two alternative pharmacological strategies for possible reversal of this effect in order to decrease brain and body temperature [55]. We assessed the effects of clozapine, an atypical neuroleptic, and carvedilol and labetalol, mixed alpha/beta adrenoceptor blockers. These medications are routinely used to treat chronic health problems in humans, and were chosen because of their preclinical success in attenuating MDMA-induced body hyperthermia [66, 67]. Clozapine acts on multiple neural receptors and glial cells in the brain, presumably inhibiting MDMA-induced metabolic neural activation, sympathetic tone, and centrally mediated vasoconstriction [68, 69]. Carvedilol and labetalol act peripherally to dilate skin vessels by blocking alpha- and beta-adrenoceptors [70, 71]. Similar to our previous studies, we used three-point thermorecording paradigm (brain site, temporal muscle, and skin) and determined changes in NAc-Muscle and Skin-Muscle differentials to determine the basic physiological mechanisms underlying brain temperature responses. This three-point recording technique allowed us to evaluate the effects of the drugs on intra-brain heat production due to metabolic neural activation and heat loss due to changes in peripheral vascular tone [1].

Our experimental protocol has three important features. First, we exposed rats to MDMA at a relatively modest, non-toxic dose [9 mg/kg or ~ 1/5 of LD50; [72]] and delivered the drug subcutaneously (in 0.3 ml of saline), providing the slowest pharmacokinetics, analogous to oral consumption in humans [87]. Second, in contrast to most studies that utilized drug administration in quietly resting laboratory animals, we administered MDMA during social interaction with another rat at the height of “psychophysiological activation,” when brain and body temperatures are significantly increased [54, 73]. This protocol is more relevant for human conditions because MDMA is recreationally used in social settings associated with high arousal (e.g., rave parties and music festivals). Finally, in contrast to most studies, where a treatment drug was administered before or at the same time as MDMA [67, 74–77], we injected each of the three test drugs – clozapine, carvedilol, and labetalol – after the MDMA injection, when brain and body temperatures were already significantly increased (>38°C). This dosing regimen closely mimics the clinical situation, in which MDMA-intoxicated patients are treated for pathological hyperthermia in hospital emergency rooms.

Although we strived for a fully factorial, within-subjects design, the variability associated with MDMA temperature response made this difficult. However, generally we exposed all rats to a 1-week-long experimental protocol where they randomly received MDMA–Saline (control), MDMA–Treatment Drug, Saline, or Treatment Drug Alone on alternating days. In rats from the first two groups, we injected MDMA 10 min after the onset of the 1-h social interaction followed by a counterbalanced injection of either a treatment drug (clozapine, carvedilol, and labetalol) or saline. In rats from two other groups, we injected either a treatment drug (clozapine, carvedilol, and labetalol) or saline under quiet resting conditions. Each rat received only two MDMA injections, either alone or with a treatment drug. We injected all "treatment" drugs ip at the same dose (5 mg/kg); this dose was chosen based on previous studies [66, 67].

As shown in Fig. 6e–f, after clozapine injection, NAc temperature rapidly decreased, resulting in a large difference vs. saline control (saline). The clozapine-induced temperature decrease was rapid and profound; the final temperature values in the clozapine treatment group were lower than the initial baseline and significantly lower than in the control group that received MDMA with saline. These post-treatment temperature values, however, remain within the physiological range; similarly, low or even lower values occur in well-habituated rats during day-time recording [1]. Muscle and skin temperatures also decreased after clozapine injection, and the difference vs. control was also significant.

In contrast to the stable increase in NAc-Muscle differentials present in the control group (Fig. 6c), this parameter decreased after clozapine injection (Fig. 6g). The decrease developed with ~10–15 min onset latency, producing a significant difference vs. control from ~30 min post-injection. However, the most rapid and strong effects of clozapine were found for the Skin-Muscle differentials, which reflect the vasomotor tone of skin vessels. While this parameter further decreased after saline injection (Fig. 6c), suggesting MDMA-induced sustained skin vasoconstriction, the Skin-Muscle differential began to rapidly increase immediately after clozapine injection (Fig. 6g), reflecting a full blockade of drug-induced vasoconstriction. The difference between clozapine and saline appeared from the second data point (i.e., within 3–6 min) and this difference continued to increase throughout the entire post-treatment interval. Finally, clozapine inhibited MDMA-induced hyperlocomotion (Fig. 6h) while maintaining a normal level of locomotion similar to that seen before MDMA injections.

Carvedilol also had a strong attenuating effect on MDMA-induced increases in NAc, temporal muscle, and skin temperatures, returning these parameters to near-

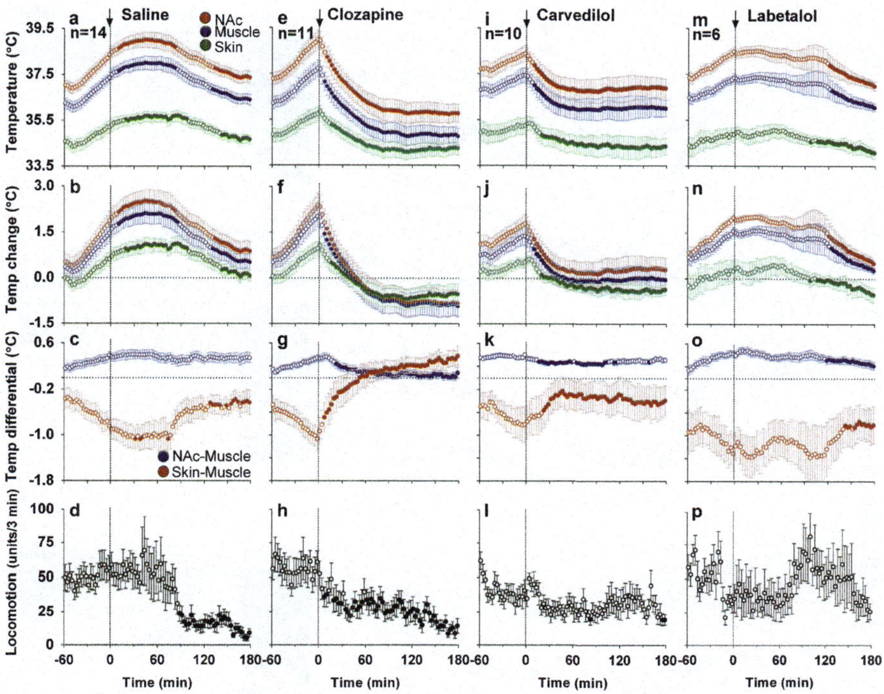

Fig. 6 The effects of clozapine, carvedilol, and labetalol on MDMA-induced temperature and locomotor responses. (**a–d**) MDMA–saline: absolute temperatures, relative temperatures, NAc-Muscle and Skin-Muscle differentials, and locomotion. (**e–h**) MDMA–clozapine, (**i–l**) MDMA–carvedilol, and (**m–p**) MDMA–labetalol. The graphs show changes in different parameters before and after administration of each testing drug and saline (0 min). In each group, the rats received a single injection of MDMA and the testing drug/saline was injected at the time of gradual NAc temperature increase in the range of 38.5°C (mean 83 min, range 57–140 min). *Filled symbols* show values significantly different from the last pre-treatment value; the absence of *filled symbols* indicates the absence of a significant effect on a specified parameter evaluated with one-way ANOVA, *n*: the number of rats (original data of this study were published in [78])

baseline levels by the end of the session (Fig. 6i–l). Carvedilol also attenuated the vasoconstrictive effects of MDMA as evidenced by a moderate increase in Skin-Muscle differentials, but this effect was short-lived and lasted only ~100 min post-treatment. Carvedilol minimally affected NAc-Muscle differential (Fig. 6k) and the effect of treatment vs. control was not significant. Lastly, carvedilol inhibited MDMA-induced locomotor activation, although locomotor activity was maintained at normal levels throughout the recording session (Fig. 6l). Labetalol had no evident effects on any of the temperature responses caused by MDMA plus social interaction (Fig. 6m–p).

To further compare the efficacy of each drug in attenuating MDMA-induced hyperthermia, we calculated the mean differences between the effects of MDMA + Treatment drug and MDMA + Vehicle (Fig. 7, left panels). Additionally, we approximated temperature change, as analyzed by the integral of the difference

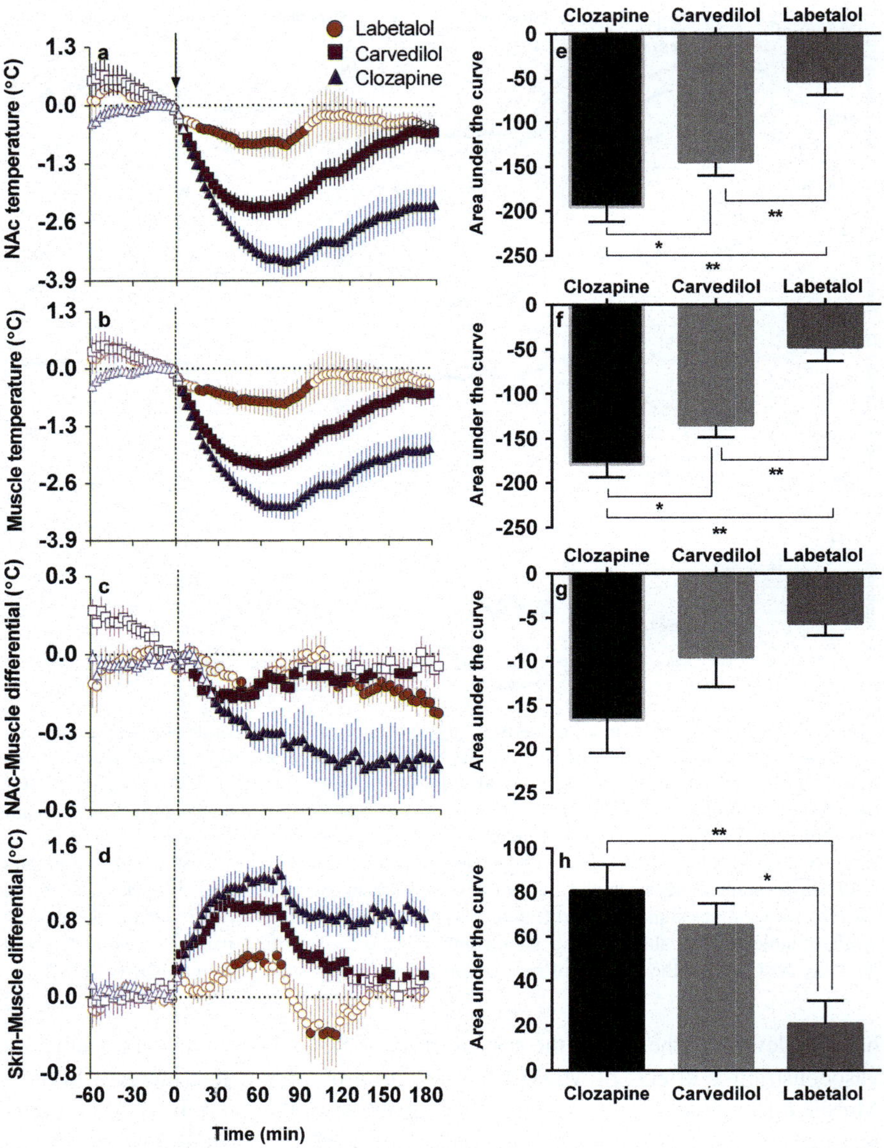

Fig. 7 Comparative effectiveness of different drugs in reversing MDMA-induced temperature responses. *Left panel* shows the time-course of the effects of each drug (difference vs. saline) on (**a**) brain and (**b**) muscle temperatures as well as (**c**) NAc-Muscle and (**d**) Skin-Muscle temperature differentials. *Right panel* shows the mean effects (area under curve for 80 min post-injection) for each testing drug (**e–h**). Time-course data were analyzed using one-way ANOVA with repeated measures; *filled symbols* show the values significantly different from the last pre-treatment value (Fisher PLSD test). *Bar graph* data were analyzed using one-way ANOVA; *asterisks* show significant between-drug differences (*$p < 0.05$; **$p < 0.01$). Original data shown in this graph were reported in Kiyatkin et al. [78]

between saline and drug treatment groups (i.e., area under the curve) for the time of maximal effect (80 min; Fig. 7, right panels).

As can be seen in Fig. 7, the attenuating effect of clozapine on NAc and muscle temperatures appeared with the shortest latencies (~6 min), and displayed the greatest magnitude and longest duration. Accordingly, the effect of clozapine on NAc and Muscle temperatures was significantly stronger ($p < 0.05$) than the effects of carvedilol and labetalol.

Clozapine also has the greatest attenuating effects on MDMA-induced intra-brain heat production and skin vasoconstriction, showing the largest decrease in NAc-Muscle differential and a strong, sustained increase in Skin-Muscle differential, respectively. In contrast, carvedilol had a much weaker effect on NAc-Muscle differentials and a less pronounced, short-lived effect on Skin-Muscle differentials. Finally, labetalol had minimal effects on MDMA-induced changes in all temperature measures.

6 Targets and Pathways to Alleviate MDMA-Induced Hyperthermia

The downstream pharmacological and biochemical mechanisms underlying MDMA-induced hyperthermia are complex and involve multiple factors, including excessive sympathetic activation, dopamine-mediated hyperactivity, decoupling of mitochondrial ATP, and heat production in both the brain and periphery (see [79] for review). As such, rather than focusing on the specific downstream neurotransmitter, receptor, or signaling mechanisms, we chose to focus on the basic physiological mechanisms underlying MDMA-induced hyperthermia and its reversal by the treatment drugs.

Clozapine (1–5 mg/kg) has been previously shown to decrease high-dose MDMA-induced body hyperthermia via vasodilation, as assessed by an ear pinna artery Doppler signal in rabbits, and blood flow Doppler signal in the tails of rats [66]. Consistent with these data, we found that a low ip dose of clozapine (5 mg/kg or ~2% of LD50; [80]) completely reversed MDMA-induced brain and body hyperthermia. The temperature decrease was exceptionally rapid, appearing within ~6 min, and both brain and muscle temperatures returned to the initial, quiet resting baseline within ~30 min post-injection. Consistent with its central, antipsychotic action, clozapine strongly reduced MDMA-induced increases in NAc-Muscle temperature differentials, suggesting a gradual decrease in intra-brain heat production due to blockade of drug-induced metabolic brain activation. Within the first 3 min post-treatment, clozapine also reversed the decreases in Skin-Muscle temperature differentials, suggesting a rapid blockade of skin vasoconstriction. In addition, clozapine decreased MDMA-induced locomotor activation while maintaining normal activity levels and showing no evident signs of sedation. Interestingly, when used in intact rats, clozapine at the same dose (5 mg/kg, ip) slightly decreased NAc and muscle temperatures, with no effects on locomotion and NAc-Muscle temperature differentials [55]. This pattern of action suggests the absence of a true sedative effect.

Carvedilol and labetalol are mixed alpha–beta adrenoceptor blockers that directly dilate blood vessels [70, 71], thus increasing heat dissipation from skin surfaces [81, 82]. At a relatively low dose (5 mg/kg or 0.6% of LD50; [83]) carvedilol also reversed MDMA-induced brain and body hyperthermia but its effects were slower, weaker, and more transient than those of clozapine. In contrast to the strong effects of clozapine on MDMA-induced increases in NAc-Muscle differentials, peripherally acting carvedilol had virtually no effects on this metabolism-related parameter. Surprisingly, the effects of carvedilol on MDMA-induced changes in Skin-Muscle differentials, an index of cutaneous vascular tone, were weaker and incomplete compared to the effects of clozapine. Carvedilol also had minimal effects on MDMA-induced hyperlocomotion, and the rats' activity remained at normal levels, suggesting lack of sedative effects. When used in intact rats, carvedilol slightly decreased brain and muscle temperatures, and this effect was observed in conjunction with the rise of Skin-Muscle temperature differentials [55], suggesting peripheral vasodilation. Consistent with its peripheral mechanism of action, carvedilol had no effects on either NAc-Muscle differentials or animal locomotion, suggesting a slight drop in brain and body temperature, resulting from transient dilation of skin vessels and increased heat dissipation.

Although human reports suggest that labetalol taken before MDMA exposure decreases MDMA-induced hyperthermia [84], in our hands this drug was ineffective and had minimal, if any, effects on all temperature parameters. The reasons for the differences between carvedilol and labetalol are unclear and may be due to their different affinities for the various subgroups of alpha- and beta-adrenoceptors [85]. As a vasodilator, labetalol also appears to be less potent than carvedilol [86]; therefore, the different effects of the drugs in our study may be due to the choice of a uniform drug dose. However, our exploratory tests with a higher dose of labetalol (20 mg/kg) did not reveal clear effects in attenuating MDMA-induced hyperthermia and also led to adverse side-effects.

In conclusion, the data indicate that carvedilol, by acting directly on blood vessels, is modestly effective in attenuating MDMA-induced brain and body hyperthermia. In contrast, clozapine induces much more rapid and powerful hypothermic effects by both decreasing MDMA-induced brain activation and diminishing the sympathetic outflow to peripheral vessels. A therapeutic agent such as clozapine that not only mitigates, but reverses, MDMA-induced hyperthermia could be indispensable for emergency situations and could save the lives of highly intoxicated individuals.

7 Conclusions

While usually underappreciated, brain temperature is an important physiological parameter that reflects metabolic neural activity and affects all types of neural function. Despite widely held beliefs that brain temperature is a stable homeostatic parameter, the data presented in this review clearly demonstrate that brain temperature can vary within relatively large limits during normal physiological activities and after exposure to different psychoactive drugs. Our focus in this review was on psychomotor stimulants, which in some drug users are known to induce extreme hyperthermia – the most dangerous health complication produced by these drugs. While a large dose of injected or ingested drug is usually viewed as the most important parameter for inducing pathological hyperthermia, this review demonstrates that the temperature effects of drugs are strongly modulated when they are used during physiological activation and in moderately warm environments. Social interaction, a procedure used in rats to model human psychophysiological activation moderately enhances intra-brain heat production and limits heat dissipation due to peripheral vasoconstriction; its combination with the similar but more sustained effects induced by psychomotor stimulants results in a strong potentiation of drug-induced hyperthermic responses. Similar potentiation could occur when the drug is used in a warm environment where proper heat dissipation is important to maintain brain and body temperature homeostasis. Therefore, a certain drug used under adverse environmental conditions could induce unexpectedly strong effects and even lethality at doses that are usually viewed as "safe."

Among the drugs considered in this review, MDMA showed the most powerful potentiation both during social interaction and in a warm environment. While the exact molecular and receptor mechanisms underlying this potentiation remain unclear, at the physiological level large doses of MDMA induce prolonged increase in metabolic neural activity coupled with strong and sustained peripheral vasoconstriction. Although slightly less but significant vasoconstriction was also induced by MDPV and methylone, these two drugs have minimal effects on intra-brain heat production possibly reducing the effects of the cathinone drugs on state-dependent and environmental potentiation of their hyperthermic effects. Finally, we tried to demonstrate how the knowledge of physiological mechanisms underlying hyperthermia could help in finding pharmacological tools for its reversal. Using MDMA, we showed that centrally acting clozapine, an atypical neuroleptic, is more efficient in reversing pathological hyperthermia than peripherally acting vasodilators. Since there are no effective drugs that could be used during MDMA-induced overdose in emergency-room environment, these data could have translational importance, helping to save the lives of highly intoxicated individuals.

Acknowledgements This work was supported by the National Institute on Drug Abuse, Intramural Research Program (DA000445-14).

Disclosure of Potential Conflicts of Interest No potential conflicts of interest are disclosed.

References

1. Kiyatkin EA (2010) Brain temperature homeostasis: physiological fluctuations and pathological shifts. Front Biosci 15:73–92
2. Abrams R, Hammel HT (1964) Hypothalamic temperature in unanesthetized albino rats during feeding and sleeping. Am J Physiol 206:641–646
3. Baker MA, Frye FM, Millet VE (1973) Origin of temperature changes evoked in the brain by sensory stimulation. Exp Neurol 38:502–519
4. Delgado JMR, Hanai T (1966) Intracerebral temperatures in free-moving cats. Am J Physiol 211:755–769
5. McElligott JC, Melzack R (1967) Localized thermal changes evoked in the brain by visual and auditory stimulation. Exp Neurol 17:293–312
6. Serota HM, Gerard RM (1938) Localized thermal changes in cat's brain. J Neurophysiol 1: 115–124
7. Kovalson VM (1972) Brain temperature variations during natural sleep and arousal in white rats. Physiol Behav 10:667–670
8. Smirnov MS, Kiyatkin EA (2008) Fluctuations in central and peripheral temperatures associated with feeding behavior in rats. Am J Physiol 295:R1414–R1424
9. Kiyatkin EA, Mitchum R (2003) Fluctuations in brain temperatures during sexual behavior in male rats: an approach for evaluating neural activity underlying motivated behavior. Neuroscience 119:1169–1183
10. Mariak Z, Lyso T, Peikarski P, Lewko J, Jadeszko M, Szydlik P (2000) Brain temperature in patients with central nervous system lesions. Neurol Neurochir Pol 34:509–522
11. Rumana CS, Gopinath SP, Uzura M, Valadka AB, Robertson CS (1998) Brain temperatures exceed systemic temperatures in head-injured patients. Crit Care Med 26:562–567
12. Rango M, Arighi A, Bonifati C, Bresolin N (2012) Increased brain temperature in Parkinson's disease. Neuroreport 23:129–133
13. Rango M, Arighi A, Bonifati C, Del Bo R, Comi G, Bresolin N (2014) The brain is hypothermic in patients with mitochondrial diseases. J Cereb Blood Flow Metab 34:915–920
14. Sukstanskii AL, Yablonskiy DA (2006) Theoretical model of temperature regulation in the brain during changes in functional activity. Proc Natl Acad Sci U S A 103:12144–12149
15. Fuller CA, Baker MA (1983) Selective regulation of brain and body temperatures in the squirrel monkey. Am J Physiol 245:R293–R297
16. Armenian P, Mamantov TM, Tsutaoka BT, Gerona RR, Silman EF, Wu AH, Olson KR (2013) Multiple MDMA (Ecstasy) overdoses at a rave event: a case series. J Intensive Care Med 28: 252–258
17. Dafters RI (1995) Hyperthermia following MDMA administration in rats: effects of ambient temperature, water consumption, and chronic dosing. Physiol Behav 58:877–882
18. Kalant H (2001) The pharmacology and toxicology of "ecstasy" (MDMA) and related drugs. CMAJ 165:917–928
19. Nimmo SM, Kennedy BW, Tullett WM, Blyth AS, Dougall JR (1993) Drug-induced hyperthermia. Anesthesia 48:892–895
20. Sei H, Furano N, Morita Y (1997) Diurnal changes in blood pressure, heart rate and body temperature during sleep in the rat. J Sleep Res 6:113–119
21. Nybo L (2008) Hyperthermia and fatigue. J Appl Physiol 104:871–877
22. Schmidt-Nielsen K (1997) Animal physiology. Adaptation and environment, 5th edn. Cambridge University Press, Cambridge
23. Siesjo B (1978) Brain energy metabolism. Wiley, New York

24. Feitelberg S, Lampl H (1935) Warmetonung der Grosshirnrinde bei Erregung und Ruhe. Functionshemmung. Arch Exp Path Pharmak 177:726–736 (in German)
25. Hayward JN, Baker MA (1968) Role of cerebral blood flow in the regulation of brain temperature in the monkey. Am J Physiol 215:389–403
26. Kiyatkin EA, Brown PL, Wise RA (2002) Brain temperature fluctuation: a reflection of functional neural activation. Eur J Neurosci 16:164–168
27. Bola RA, Kiyatkin EA (2016) Robust brain hyperglycemia during general anesthesia: relationships with metabolic brain inhibition and vasodilation. Front Physiol 7:39. doi:10.3389/fphys.2016.00039
28. Kiyatkin EA, Brown PL (2005) Brain and body temperature homeostasis during sodium pentobarbital anesthesia with and without body warming in rats. Physiol Behav 84:563–570
29. Di Chiara G (2002) Nucleus accumbens shell and core dopamine: differential role in behavior and addiction. Behav Brain Res 137:75–114
30. Mogenson GJ, Jones DL, Yim CY (1980) From motivation to action: functional interface between the limbic system and the motor system. Prog Neurobiol 14:69–97
31. Wise RA, Bozarth MA (1987) A psychomotor stimulant theory of addiction. Psychol Rev 94: 469–492
32. Alberts DS, Sonsalla PK (1995) Methamphetamine-induced hyperthermia and dopaminergic neurotoxicity in mice: pharmacological profile of protective and nonprotective agents. J Pharmacol Exp Ther 275:1104–1114
33. Freedman RR, Johanson C-E, Tancer ME (2005) Thermoregulatory effects of 3,4-methylenedioxymethamphetamine (MDMA) in humans. Psychopharmacology 183: 248–256
34. Gordon CJ, Watkinson WP, O'Callaghan PP, Miller DB (1991) Effects of 3,4-methylenedioxymetamphetamine on autonomic thermoregulatory responses of the rat. Pharmacol Biochem Behav 38:339–344
35. Green AR, Mechan AO, Elliott JM, O'Shea E, Colado MI (2003) The pharmacology and clinical pharmacology of 3,4-methylenedioxymethamphetamine (MDMA, "Ecstasy"). Pharmacol Rev 55:463–508
36. Mechan AO, Esteban B, O'Shea E, Elliott JM, Colado MI, Green AR (2002) The pharmacology of the acute hyperthermic response that follows administration of 3,4-methylenediomethamphetamine (MDMA, "ecstasy") to rats. Br J Pharmacol 135:170–180
37. Pederson NP, Blessing WW (2001) Cutaneous vasoconstriction contributes to hyperthermia induced by 3,4-methylenedioxymethamphetamine (ecstasy) on conscious rabbits. J Neurosci 21:8648–8654
38. Baumann MH, Partilla JS, Lehner KR, Thorndike EB, Hoffman AF, Holy M et al (2013) Powerful cocaine-like actions of 3,4-methylenedioxypyrovalerone (MDPV), a principal constituent of psychoactive 'bath salts' products. Neuropsychopharmacology 38:552–562
39. German CL, Fleckenstein AE, Hanson GR (2014) Bath salts and synthetic cathinones: an emerging designer drug phenomenon. Life Sci 97:2–8
40. Spiller HA, Ryan ML, Weston RG, Jansen J (2011) Clinical experience with and analytical confirmation of "bath salts" and "legal highs" (synthetic cathinones) in the United States. Clin Toxicol 49:499–505
41. Prosser JM, Nelson LS (2012) The toxicology of bath salts: a review of synthetic cathinones. J Med Toxicol 8:33–42
42. Ross EA, Reisfield GM, Watson MC, Chronister CW, Goldberger BA (2012) Psychoactive "bath salts" intoxication with methylenedioxypyrovalerone. Am J Med 125:854–858
43. Baumann MH, Partilla JS, Lehner KR (2013) Psychoactive "bath salts": not so soothing. Eur J Pharmacol 698:1–5
44. Baumann MH, Ayestas MA Jr, Partilla JS, Sink JR, Shulgin AT, Daley PF et al (2012) The designer methcathinone analogs, mephedrone and methylone, are substrates for monoamine transporters in brain tissue. Neuropsychopharmacology 37:1192–1203

45. Eshleman AJ, Wolfrum KM, Hatfield MG, Johnson RA, Murphy KV, Janowsky A (2013) Substituted methcathinones differ in transporter and receptor interactions. Biochem Pharmacol 85:1803–1815
46. Simmler LD, Buser TA, Donzelli M, Schramm Y, Dieu LH, Huwyler J et al (2013) Pharmacological characterization of designer cathinones in vitro. Br J Pharmacol 168:458–470
47. Cameron KN, Kolanos R, Solis E Jr, Glennon RA, De Felice LJ (2013) Bath salts components mephedrone and methylenedioxypyrovalerone (MDPV) act synergistically at the human dopamine transporter. Br J Pharmacol 168:1750–1757
48. Pearson JM, Hargraves TL, Hair LS, Massucci CJ, Frazee CC 3rd, Garg U et al (2001) Three fatal intoxications due to methylone. J Anal Toxicol 36:444–451
49. Borek HA, Holstege CP (2012) Hyperthermia and multiorgan failure after abuse of "bath salts" containing 3,4-methylenedioxypyrovalerone. Ann Emerg Med 60:103–105
50. Kesha K, Boggs CL, Ripple MG, Allan CH, Levine B, Jufer-Phipps R et al (2013) Methylenedioxypyrovalerone ("bath salts")-related death: case report and review of the literature. J Forensic Sci 58:1654–1659
51. Murray BL, Murphy CM, Beuhler MC (2012) Death following recreational use of designer drug "bath salts" containing 3,4-methylenedioxypyrovalerone (MDPV). J Med Toxicol 8:69–75
52. Aarde SM, Huang PK, Creehan KM, Dickerson TJ, Taffe MA (2013) The novel recreational drug 3,4-methylenedioxypyrovalerone (MDPV) is a potent psychomotor stimulant: self-administration and locomotor activity in rats. Neuropharmacology 71:130–140
53. Fantegrossi WE, Gannon BM, Zimmerman SM, Rice KC (2013) In vivo effects of abused 'bath salt' constituent 3,4-methylenedioxypyrovalerone (MDPV) in mice: drug discrimination, thermoregulation, and locomotor activity. Neuropsychopharmacology 38:563–573
54. Kiyatkin EA, Kim AH, Wakabayashi KT, Baumann MH, Shaham Y (2014) Critical role of peripheral vasoconstriction in fatal brain hyperthermia induced by MDMA (Ecstasy) under conditions that mimic human drug use. J Neurosci 34:7754–7762
55. Kiyatkin EA, Kim AH, Wakabayashi KT, Baumann MH, Shaham Y (2015) Effects of social interaction and warm ambient temperature on brain hyperthermia induced by the designer drugs methylone and MDPV. Neuropsychopharmacology 40:436–445
56. Romanovsky AA, Ivanov AI, Shimansky YP (2002) Ambient temperature for experiments in rats: a new method for determining the zone of thermal neutrality. J Appl Physiol 92: 2667–2679
57. Davis WM, Hatoum HT, Waters IW (1987) Toxicity of MDA (3,4-methylenedioxyamphetamine) considered for relevance to hazards of MDMA (Ecstasy) abuse. Alcohol Drug Res 7:123–134
58. Gordon CJ (1990) Thermal biology of the laboratory rat. Physiol Behav 47:963–991
59. Rowell LB (1983) Cardiovascular aspects of human thermoregulation. Circ Res 52:367–379
60. Banks ML, Sprague JE, Kisor DF, Czoty PW, Nichols DE, Nader MA (2007) Ambient temperature effects on 3,4-methylenedioxymethamphetamine-induced theremodysregulation and pharmacokinetics in male monkeys. Drug Metab Dispos 35:1840–1845
61. Taffe MA, Lay CC, Von Huben SN, Davis SA, Crean RD, Katner SN (2006) Hyperthermia induced by 3,4-methylenedioxymethamphetamine in unrestrained rhesus monkeys. Drug Alcohol Depend 20:276–281
62. Von Huben RD, Lay CC, Crean RD, Davis SA, Katner SN, Taffe MA (2007) Impact of ambient temperature on hyperthermia induced by (+/-)3,4- methylenediomethamphetamine in rhesus macaques. Neuropsychopharmacology 32:673–681
63. Parrott AC (2014) The potential dangers of using MDMA for psychotherapy. J Psychoactive Drugs 46:37–43
64. Nybo L, Secher NH, Nielson B (2002) Inadequate heat release from the human brain during prolonged exercise with hyperthermia. J Physiol 545:697–704
65. Proulx CI, Ducharme MB, Kenny GP (2003) Effect of water temperature on cooling efficiency during hyperthermia in humans. J Appl Physiol 94:1317–1323

66. Blessing W, Seaman B, Pedersen N, Ootsuka Y (2003) Clozapine reverses hyperthermia and sympathetically mediated cutaneous vasoconstriction induced by 3,4-methylenedioxymethamphetamine (ecstasy) in rabbits and rats. J Neurosci 15:6385–6391
67. Sprague JE, Moze P, Caden D, Rusyniak DE, Holmes C, Goldstein DS et al (2005) Carvedilol reverses hyperthermia and attenuates rhabdomyolysis induced by 3,4-methylenedioxymethamphetamine (MDMA, Ecstasy) in an animal model. Crit Care Med 33:1311–1316
68. Baldessarini RJ, Frankenburg FR (1991) Clozapine. N Engl J Med 11:746–754
69. Breier A, Buchanan RW, Waltrip Ii RW, Listwak S, Holmes C, Goldstein DS (1994) The effect of clozapine on plasma norepinephrine: relationship to clinical efficacy. Neuropsychopharmacology 10:1–7
70. Bakris G (2009) An in-depth analysis of vasodilation in the management of hypertension: focus on adrenergic blockade. J Cardiovasc Pharmacol 53:379–387
71. Sponer G, Strein K, Bartsch W, Muller-Beckmann B (1992) Vasodilatory action of carvedilol. J Cardiovasc Pharmacol 19(Suppl 1):S5–S11
72. Hardman HF, Haavik CO, Seevers MH (1973) Relationship of the structure of mescaline and seven analogs to toxicity and behavior in five species of laboratory animals. Toxicol Appl Pharmacol 25:299–309
73. Brown PL, Kiyatkin EA (2004) Brain hyperthermia induced by MDMA ("ecstasy"): modulation by environmental conditions. Eur J Neurosci 20:51–58
74. Hysek C, Schmid Y, Rickli A, Simmler L, Donzelli M, Grouzmann E et al (2012) Carvedilol inhibits the cardiostimulant and thermogenic effects of MDMA in humans. Br J Pharmacol 166:2277–2288
75. Shioda K, Nisijima K, Yoshino T, Kuboshima K, Iwamura T, Yui K et al (2008) Risperidone attenuates and reverses hyperthermia induced by 3,4-methylenedioxymethamphetamine (MDMA) in rats. Neurotoxicology 29:1030–1036
76. Taffe MA (2012) Delta9-Tetrahydrocannabinol attenuates MDMA-induced hyperthermia in rhesus monkeys. Neuroscience 201:125–133
77. Yeh SY (1997) Effects of salicylate on 3,4-methylenedioxymethamphetamine (MDMA)-induced neurotoxicity in rats. Pharmacol Biochem Behav 58:701–708
78. Kiyatkin EA, Kim AH, Wakabayashi KT, Baumann MH, Shaham Y (2016) Clinically relevant pharmacological strategies that reverse MDMA-induced brain hyperthermia potentiated by social interaction. Neuropsychopharmacology 41:549–559
79. Dao CK, Nowinski SM, Mills EM (2014) The heat is on: molecular mechanisms of drug-induced hyperthermia. Temperature 1:183–191
80. Lindt S, Lauener H, Eichenberger E (1971) The toxicology of 8-chloro-11-(4-methyl-1-piperazinyl)-5H-dibenzo(b, e)(1,4)diazepine (clozapine). Farmaco Prat 26:585–602
81. Charkoudian N (2010) Mechanisms and modifiers of reflex induced cutaneous vasodilation and vasoconstriction in humans. J Appl Physiol 109:1221–1228
82. Kellogg DL Jr (2006) In vivo mechanisms of cutaneous vasodilation and vasoconstriction in humans during thermoregulatory challenges. J Appl Physiol 100(5):1709–1718
83. Carvedilol (2012) HSDB [Internet] Bethesda (MD): National Library of Medicine (US). Hazardous Substances Databank Number: 7044. Available from: http://toxnet.nlm.nih.gov/cgi-bin/sis/search2012/r?dbs+hsdb:@term+@DOCNO+7044
84. Liechti ME (2014) Effects of MDMA on body temperature in humans. Temperature 1:192–200
85. Ruffolo RR Jr, Gellai M, Hieble JP, Willette RN, Nichols AJ (1990) The pharmacology of carvedilol. Eur J Clin Pharmacol 38:S82–S88
86. Tomlinson B, Cronin CJ, Graham BR, Prichard BN (1987) Haemodynamics of carvedilol in normal subjects compared with propranolol, pindolol, and labetalol. J Cardiovasc Pharmacol 10(Suppl 11):S69–S75
87. Baumann MH, Zolkowska D, Kim I, Scheidweiler KB, Rothman RB, Huestis MA (2009) Effects of dose and route of administration on pharmacokinetics of (+/-)3.4-methylenedioxymethamphetamine in the rat. Drug Metab Dispos 37:2163–2170

Neurotoxicology of Synthetic Cathinone Analogs

Mariana Angoa-Pérez, John H. Anneken, and Donald M. Kuhn

Abstract The present review briefly explores the neurotoxic properties of methcathinone, mephedrone, methylone, and methylenedioxypyrovalerone (MDPV), four synthetic cathinones most commonly found in "bath salts." Cathinones are β-keto analogs of the commonly abused amphetamines and display pharmacological effects resembling cocaine and amphetamines, but despite their commonalities in chemical structures, synthetic cathinones possess distinct neuropharmacological profiles and produce unique effects. Among the similarities of synthetic cathinones with their non-keto analogs are their targeting of monoamine systems, the release of neurotransmitters, and their stimulant properties. Most of the literature on synthetic cathinones has focused on describing their properties as psychostimulants, their behavioral effects on locomotion, memory, and potential for abuse, whereas descriptions of their neurotoxic properties are not abundant. The biochemical gauges of neurotoxicity induced by non-keto analogs are well studied in humans and experimental animals and include their ability to induce neuroinflammation, oxidative stress, excitotoxicity, temperature alterations as well as dysregulation of neurotransmitter systems and induce changes in monoamine transporters and receptors. These neurotoxicity gauges will serve as parameters to discuss the effects of the four previously mentioned synthetic cathinones alone or in combination with either another cathinone or with some of their non-keto analogs. Bath salts are not a defined combination of drugs and may consist of one synthetic cathinone compound or combinations of more cathinones. Furthermore, this review also presents some of the mechanisms that are thought to

M. Angoa-Pérez (✉), J.H. Anneken, and D.M. Kuhn
Research & Development Service, John D. Dingell VA Medical Center, Detroit, MI 48201, USA

Department of Psychiatry and Behavioral Neurosciences, Wayne State University School of Medicine, Detroit, MI 48201, USA
e-mail: maperez@med.wayne.edu

Curr Topics Behav Neurosci (2017) 32: 209–230
DOI 10.1007/7854_2016_21
Published Online: 17 June 2016

underlie this toxicity. A better understanding of the cellular and molecular mechanisms involved in the synthetic cathinones-induced neurotoxicity should contribute to generate modern therapeutic approaches to prevent or attenuate the adverse consequences of use of these drugs in humans.

Keywords MDPV • Mephedrone • Methcathinone • Methylone • Neurotoxicity • Synthetic cathinones

Contents

1 Neuroinflammation

One of the most relevant issues related to non-ketoamphetamines-induced neurotoxicity is that they can trigger inflammatory processes in those brain areas that exhibit terminal degeneration [1]. Studies have demonstrated that glial activation participates in the events that induce neuronal damage, since chronic neuroinflammation elevates the levels of glia-derived cytokines that exert neurotoxic effects on vulnerable neurons [2]. Microglia and astrocytes are the primary modulators of inflammation in the CNS and have been associated with the toxicity induced by administration of methamphetamine [3], amphetamine and parachloroamphetamine [4], and 3,4-methylenedioxy-methamphetamine (MDMA) [5]. Studies of the effects of β-ketoamphetamines on neuroinflammation are summarized in Table 1.

To the best of our knowledge, no reports have been made on the neuroinflammatory effects of methcathinone.

With regard to mephedrone, in vivo studies reported that there were no signs of striatal [17] or cortical [25] astroglial activation after administration of mephedrone in a binge paradigm. Similarly, no signs of microglial activation were observed in the striatum at 2 or 7 days after administration of mephedrone [17]. Lopez-Arnau and colleagues measured [^{3}H]PK11195-specific binding to investigate the microglial activation after neuronal injury in rats killed 24 h post-treatment with mephedrone. PK11195 is an isoquinoline carboxamide that purportedly binds to microglia in conditions of brain injury. In these animals, no increase in the density of [^{3}H]PK11195 binding sites was detected, indicating a lack of microglial

Table 1 Summary of studies that have investigated the neurotoxic effects of β-ketoamphetamines in vivo and in vitro

	Inflammation	Thermoregulation	Neurotransmitter depletion	Oxidative stress	Neurotransmitter transporters and receptors	Trans-endothelial dysfunction	Combinations
Methcathinone	n/a	[6, 7]	[8–11]	n/a	[8, 9, 12–15]	[15]	[16]
Mephedrone	[17–22]	[17–20, 23–27]	[7, 17–25, 28, 29]	[20, 30, 31]	[15, 24, 26, 30, 32–36]	[15]	[18, 19, 23, 30, 32, 37]
Methylone	[19, 38, 39]	[19, 24, 38–41]	[19, 24, 28, 38, 39]	n/a	[13, 15, 24, 33, 34, 42, 43]	[15]	[19, 43]
MDPV	[19]	[19, 40, 44–48]	[19, 45]	n/a	[15, 32, 33, 36, 49]	[15, 50]	[19, 32, 46, 51]

activation [20]. However, Martinez-Clemente reported an increase in reactive astrocytes in the dentate gyrus of the hippocampus of mice treated with mephedrone at 7 days after a binge of 3 doses of 25 mg/kg for 2 days [25].

There were no signs of striatal, hippocampal, or cortical microgliosis in methylone-treated rats using a regimen of 4 doses of 20 mg/kg every 3 h [39]. However, a significant increase of reactive astrocytes was reported in the frontal cortex of methylone-treated rats. By contrast, no significant differences were found in the expression of the astroglial marker GFAP in striatum or any subregion of the hippocampus (dentate gyrus, CA1 or CA3) [39]. MDPV was not found to elicit GFAP increases when administered in a binge regimen of 4 doses of 30 mg/kg every 2 h in mice [19].

None of the β-ketoamphetamines mephedrone, methylone, or MDPV in combination with each other resulted in changes in the levels of GFAP in the striatum of mice [19]. However, methylone was able to enhance by approximately 50% the expression of GFAP induced by administration of methamphetamine [19]. MDPV, on the contrary, prevented the striatal increases in GFAP observed after administration of methamphetamine and MDMA [19].

2 Thermoregulation

A common adverse effect of non-ketoamphetamines such as methamphetamine [52], amphetamine, and MDMA [53] is an increase in body temperature. This hyperthermia is dependent on the frequency of exposure, dosing, age, ambient temperature, and route of exposure [52]. MDMA and methamphetamine produce hyperthermia following acute and repeated exposure at ambient temperature and elevated ambient temperature [14, 54–58]. Hyperthermia is an important factor known to exacerbate the deleterious effects of amphetamine-type drugs. Studies of the effects of β-ketoamphetamines on thermoregulation are summarized in Table 1.

A single dose of 10 mg/kg of methcathinone caused a sustained increase in rectal temperature that was not accompanied by any significant concomitant change in tail temperature in individually housed rats [7]. Similarly, an acute intoxication with a methcathinone infusion (5 mg/kg/min; 100 mg/mL) caused hyperthermia in rats [6]. By contrast, acute exposure to mephedrone produces hypothermia in rats [7, 59]. This hypothermic response at ambient and elevated temperatures is rat-strain specific, with the reduction in body temperature detected in Wistar but not Sprague-Dawley rats [27]. Alpha-1 adrenoreceptor and dopamine D1 receptor blockade seem to enhance the hypothermic response induced by mephedrone [7]. On the other hand, when mephedrone is administered repeatedly in a binge paradigm, it produces hyperthermia in both mice [17, 19, 25] and rats [24]. These studies indicate that mephedrone differs from MDMA and methamphetamine in its thermoregulatory effects despite their neuropharmacological similarity. Lopez-Arnau and colleagues showed that the effect of mephedrone changes with the dose using a 2-day binge paradigm of 3 doses of 25 mg/kg a day every 2 h

[20]. On day 1, after receiving the first dose of mephedrone, the treated animals showed a significant transient reduction in body temperature; after the second dose, temperature increased over the saline values but this hyperthermic response was significant only after the third dose. On day 2, no significant hyperthermic response was evidenced. Consistent with that report, other studies have suggested that the mephedrone-induced hypothermia is attenuated with repeated dosing and that this response can be attenuated by 6-hydroxydopamine or abolished by 5,7-dihydroxytryptamine [26]. Taken together, mephedrone produces hypothermia following acute exposure, while producing hyperthermia following binge models of dosing [60].

Studies of self-administration of mephedrone in rats evidenced that its thermoregulatory effects also differed between rat strains, with the Sprague-Dawley rats being most sensitive. While in Sprague-Dawley rats the administration of mephedrone produced a dose-dependently decreased body temperature, in Wistar rats, the response was biphasic, starting with a decrease during the first 15 min, followed by an increase during the next 25–30 min [61]. Another study with these two strains of rats monitored the effects of the subcutaneous administration of mephedrone under conditions of low (23°C) and high (27°C) ambient temperature. A reliable reduction of body temperature was produced by mephedrone in Wistar rats at low and high temperatures with only minimal effect in Sprague-Dawley rats. Furthermore, hypothermia produced by serotonin (5-HT) 1A/7 receptor agonists was similar in each strain [27].

Similar to mephedrone, methylone produces hyperthermia following binge dosing [60]. In mice, 4 injections of methylone (20 mg/kg) every 3 h produced a robust hyperthermic response that reached a peak between 25 and 35 min after each drug administration. This effect of methylone increased significantly with the dose, so that the last dose induced a greater increase in body temperature than the first one [38]. In rats, methylone treatment (3 injections at 3 and 10 mg/kg every 2 h) caused significant hyperthermia from 2 h through 6 h post-injection [24]. The thermoregulatory effects caused by a single administration of methylone did not differ from the outcomes observed after repeated administration of the drug. Piao and colleagues evaluated the acute effects of methylone using 5-HT transporter (SERT) and dopamine (DA) transporter (DAT) knockout (KO) mice and observed a slight diminution in the hyperthermic effects of methylone in DAT KO mice, whereas a slight enhancement of these effects was seen in SERT KO mice. Administration of selective D1 and D2 receptor antagonists reduced methylone-induced hyperthermia, but these drugs also had hypothermic effects of their own in saline-treated mice, which complicates interpretation of the findings [41].

MDPV exhibits a few important differences in altering body temperature by comparison to other β-ketoamphetamines. Acute exposure to MDPV produces hyperthermia [47] at elevated temperatures but not at normal ambient temperatures, which contrasts with what is observed for MDMA and methamphetamine [60]. MDPV has been shown to elevate body temperature in some cases of human medical emergency and fatal overdose deaths [62]. In rats, treatment with 1.0, 5.6, and 10.0 mg/kg of MDPV elicited a significant hypothermia when compared with the vehicle condition [44]. The effects were dose dependent and lasted up to 3 h after dosing [44]. However, an unexpected lack of dose dependence was characterized when MDPV was administered in the 20°C ambient environment. At that cool ambient temperature, MDPV doses from 1 to 30 mg/kg each induced a rise in core temperature of approximately 1.5°C, which was not different from that observed following saline administration, and the time course of this effect was also similar across all tested doses [46]. Besides affecting body temperature, MDPV has been found to induce brain hyperthermia through an increase in peripheral vasoconstriction [40]. Furthermore, an age-dependent effect of MDPV on thermoregulation was documented in rats. While adolescent rats increased their body temperature following MDPV administration, adults showed a decrease in temperature [48].

The β-ketoamphetamines mephedrone, methylone, and MDPV differentially affect the temperature effects of their non-β-ketoamphetamine counterparts [19]. In vivo studies demonstrated that the mephedrone-induced hyperthermia is not enhanced by concomitant administration of methamphetamine [23]. When given in two-drug combinations, mephedrone, methylone and MDPV caused significant increases in body temperature [19]. When methylone or MDPV are given with mephedrone, the initial hypothermic effect of mephedrone was retained and slightly exaggerated. Combined treatment with MDPV and methylone results in a steady 1–2°C increase in core body temperature that becomes evident within 15 min of treatment and persists for at least 8–9 h [19]. Neither MDPV nor methylone attenuated the hyperthermic effects of methamphetamine in mice [19].

3 Neurotransmitters

Alterations in monoaminergic systems are one of the hallmarks of the most studied non-keto-amphetamines. Methamphetamine is perhaps best known for its toxic effects on DA nerve terminals of the striatum [63, 64]. MDMA has also been shown to cause long-term deficits in DA and 5-HT nerve ending in both laboratory animals and humans [65–67]. Even amphetamine has been linked to nerve terminal damage [66]. Amphetamine neurotoxicity manifests as long-term depletion of DA and 5-HT, inhibition of their biosynthetic enzymes tyrosine hydroxylase (TH) and tryptophan hydroxylase 2 (TPH-2), inactivation of DAT and SERT, reduction in function of the vesicle monoamine transporter (VMAT), degeneration of fine, unmyelinated axons, and apoptosis [3]. Reductions in DA, TH, and DAT have been documented in the postmortem striatum of chronic methamphetamine users [68]. Studies of the effects of β-ketoamphetamines on neurotransmitter systems are summarized in Table 1.

Methcathinone is a potent releaser of DA but not 5-HT, similar to amphetamine and methamphetamine [15]. Methcathinone has previously been shown to release radiolabeled DA [16] and 5-HT from rat brain preparations with similar DA versus 5-HT selectivity to amphetamine and methamphetamine, but with two- to threefold lower potency [15]. Multiple administrations of methcathinone caused persistent deficits in monoaminergic systems [11], reflected by decreases in DA and 5-HT uptake capacity, tissue content and associated TH and TPH-2 activities in frontal cortex, hippocampus, and neostriatum after four doses of 30 mg/kg in rats [8, 9]. However, the effects of this drug seemed to be more accentuated in rats compared to mice. While in mice methcathinone produced long lasting depletions of striatal DA, in rats it caused significant depletions of both DA and 5-HT [11]. A single high-dose administration of methcathinone increased striatal DA release, as measured by microdialysis in conscious rats [8]. Methcathinone was also found to increase the metabolites homovanillic acid (HVA) and 5-hydroxyindoleacetic acid (5-HIAA) levels in striatum [7].

Mephedrone alone did not cause persistent reductions in the levels of DA, 5-HT, or TPH-2 [18, 19, 22, 28, 69] aside from a small decrease in the DA metabolite HVA in the mouse striatum [28]. Mephedrone is considered a more potent releaser of DA than MDMA [15, 70]. An in vivo microdialysis study in rats showed that mephedrone produced a rapid and pronounced increase in DA levels in the nucleus accumbens that was comparable with amphetamine and greater than MDMA, which only elevates DA levels moderately [71]. While both mephedrone and MDMA also produced strong increases in extracellular 5-HT, amphetamine had only a moderate effect on 5-HT levels [71]. Self-administration of mephedrone was shown to decrease the levels of striatal 5-HIAA in rats [21]. Mephedrone administered in a binge of 3 doses of 10 mg/kg every 2 h showed that the extracellular increase in striatal DA seen after the first mephedrone injection was similar in magnitude and time course to those following the second and third injections. However, the extracellular overflow of striatal 5-HT was more variable but was enhanced when second and third injections were given when compared with the first response [26]. Other microdialysis studies in the rat nucleus accumbens showed that mephedrone elevated extracellular DA and 5-HT levels, with relatively higher effects on 5-HT levels [24, 27], similar to MDMA and unlike methamphetamine, which preferentially increases DA [24]. Thus, mephedrone shares some of the DA-releasing properties of amphetamine and methamphetamine and the 5-HT-releasing property of MDMA [15]. Repeated administration of mephedrone in rats showed no significant effect on tissue concentrations of DA, 5-HT, or their metabolites in the striatum or frontal cortex and hippocampus 7 days post-administration although the concentration of DOPAC was significantly increased in this region following mephedrone [29]. The same study further evaluated the acute effects of mephedrone in comparison with MDMA, and reported reductions in hippocampal 5-HT and 5-HIAA 2 h after a single injection of MDMA but not following acute mephedrone [29]. The expression of TH, a biochemical marker of neuronal integrity in DA neurons, was found to be decreased in frontal cortex but not in the striatum after a binge regimen of 25 mg/kg of mephedrone [20]. The overall effects of mephedrone do not involve long-lasting depletions of DA but they seem to affect 5-HT. Repeated administration of mephedrone in rats caused persistent decreases in hippocampal 5-HT levels but no changes were observed in striatal DA after 7 days of treatment [70]. Another study reported decreases in TH and TPH-2 after a binge of mephedrone for 2 consecutive days [25]. The fact that mephedrone has DA-releasing capability resembling methamphetamine and yet does not cause DA deficits is of significant interest for studying the differential mechanisms of damage induced by stimulants.

Methylone administration does not result in damage to DA nerve endings in mice [19]. Binge administration of methylone to single-housed rats (3 or 10 mg/kg, 3 doses) has no long-lasting effects on brain tissue monoamines [24] but produced significant elevations on extracellular levels of DA and 5-HT [24, 34, 42]. There seem to be species differences in the sensitivity to long-term neurochemical effects of methylone [28]. The effects of drug treatments on mouse brain monoamine levels in the frontal cortex, striatum, and hippocampus indicate that methylone did not cause any significant changes in neurotransmitter levels. However, in the rat brain methylone had a profound impact on 5-HT levels, causing a decrease in 5-HT levels in the frontal cortex, striatum, and hippocampus comparable to that induced by amphetamine. Additionally, 5-HIAA levels were reduced in the striatum and hippocampus [28].

In vivo microdialysis studies indicate that MDPV increases extracellular concentrations of DA in the nucleus accumbens in a dose-related manner similar to cocaine. However, MDPV was tenfold more potent than cocaine in its ability to increase extracellular dopamine [49]. This robust stimulation of DA transmission by MDPV predicts serious potential for abuse [49]. Despite the high potency to block the DAT, MDPV did not produce DA efflux. Thus, this cathinone is thought to be a pure transporter uptake inhibitor [15].

Studies in mice indicate that combined treatment with mephedrone and methamphetamine or MDMA did not change the status of 5-HT nerve endings to an extent that was different from either drug alone [18]. Methamphetamine and MDMA alone caused mild reductions in 5-HT but did not change SERT and TPH2 levels [23]. While mephedrone did not produce changes in the 5-HT system, it enhanced the DA and TH depletions induced by methamphetamine, amphetamine, and MDMA in striatum [23].

In mice, none of the β-ketoamphetamines mephedrone, methylone, or MDPV in combination with each other resulted in changes in striatal DA or TH, but mephedrone and methylone potentiated the depletions of DA and TH induced by administration of methamphetamine [19, 23]. On the other hand, MDPV was able to prevent the striatal decreases in DA and TH observed after administering methamphetamine, MDMA, and MPTP [19]. Consistent with this study, in vitro data showed that MDPV blocked methamphetamine-induced DA release with high potency reflecting its elevated efficiency as an uptake inhibitor. The finding suggests that the more potently a drug antagonizes the DA release produced by methamphetamine, the more potently it also blocks DA uptake [51].

4 Biochemical Mechanisms: Oxidative Stress and Cytotoxicity

Evidence indicates that reactive oxygen species (ROS) are responsible for amphetamine-related damage but neither the manner by which these drugs cause oxidative stress nor the cellular source of the reactant species is known [4, 68]. Oxidative stress is believed to be a prominent factor in methamphetamine-induced cellular toxicity [72]. By increasing DA release, amphetamines increase the available DA for oxidation and its metabolism into ROS [73]. Methamphetamine's ability to flood the intracellular medium with DA is thought to be the first step in a cascade that leads to mitochondrial dysfunction, enhanced excitatory neurotransmission, increases in oxidative stress, nerve ending damage, and apoptosis [74]. Similar to the other amphetamines, metabolism of MDMA also results in the formation of ROS, which ultimately induce long-term neurotoxic effects [2]. However, none of the β-ketoamphetamines (methcathinone, mephedrone, methylone, and MDPV) showed cytotoxicity at the highest concentrations tested in functional assays [36]. Studies of the effects of β-ketoamphetamines on oxidative stress and cytotoxicity are summarized in Table 1.

Methcathinone is manufactured by a clandestine process that involves potassium permanganate oxidation of ephedrine and pseudoephedrine contained in readily available pharmaceuticals. Intravenous injections of such methcathinone preparations expose users to a high manganese load because the resultant methcathinone is not purified [75]. Although studies of methcathinone abusers have described movement disorders similar to Parkinson's disease attributed to the manganese toxicity, the syndrome lacks typical features of this condition such as resting tremor and gait initiation failure [75]. The accumulation of manganese can lead to the development of encephalopathy and might trigger secondary pathogenic mechanisms, such as mitochondrial dysfunction and oxidative stress [76].

Animal studies showed that mephedrone induced an increase in the expression of the antioxidant enzymes superoxide dismutase, catalase, and glutathione peroxidase in the hippocampus, striatum, and frontal cortex in rats. Along with these enzyme protein increases, treatment with mephedrone caused a significant increase in the levels of lipid peroxidation in the frontal cortex [20]. In mice, a single injection of mephedrone at 2.5 mg/kg caused both an increase in lipid peroxidation levels and a decrease of catalase activity in the hippocampus and prefrontal cortex [37] whereas at a dose of 25 mg/kg, mephedrone induced a significant increase of glutathione peroxidase in striatum [30].

The exposure of mouse cultured cortical cells to various concentrations of mephedrone for 24 h or 48 h caused a concentration-dependent decrease in metabolically active cells. The calculated LD50 value for mephedrone after 24 h of incubation was significantly higher to that obtained after 48 h of drug exposure [25]. In addition, neuroblastoma cells exposed to mephedrone showed an increase in cytotoxic damage only at high concentrations [31]. Cell culture studies documented the cytotoxicity of methylone and methamphetamine in CHO cells stably expressing the rat transporters DAT, NET, and SERT. Methylone was not cytotoxic to any cell line except that expressing the SERT [43], indicating higher specificity for the 5-HT system.

The cytotoxic effects of MDPV have only been evaluated in a developing brain mouse model, where a single administration of the drug caused a prominent increase in the number of apoptotic cells in the piriform cortex, retrosplenial area, hippocampus CA1, and nucleus accumbens, without an overall change in the density of cells. The neurons of the nucleus accumbens appeared to be especially sensitive to MDPV as they showed an increase of apoptotic cells in the core and shell regions of the accumbens. However, this effect of MDPV was not observed in the brain of adult mice [45]. While methylone alone was reported to be cytotoxic in cell lines expressing the rat SERT, the combination of methamphetamine with methylone caused a significant increase in the toxicity in the cells stably expressing the rat monoamine transporters DAT, NET, and SERT but not in the control CHO cells [43].

5 Neurotransmitter Transporters and Receptors

The non-keto-amphetamines including amphetamine, methamphetamine, and MDMA cause neurotoxic effects to monoaminergic systems in part through alterations in DA and 5-HT transporters and receptors [66]. Methamphetamine causes acute increases in both DA and 5-HT release that result from the direct and indirect actions on the DAT and SERT. Amphetamines can disrupt the vesicle proton gradient to cause an increase in cytoplasmic DA and 5-HT from vesicular compartments by altering the function of the vesicular monoamine transporter-2 (VMAT-2). Both methamphetamine and MDMA also increase 5-HT release through similar transporter mediated mechanisms, though MDMA has a preferential affinity for SERT over DAT and consequently more pronounced effects on the 5-HT system [74]. Studies of the effects of β-ketoamphetamines on neurotransmitter transporters and receptors are summarized in Table 1.

In animals, repeated administration of methcathinone was shown to reduce the content of DA and 5-HT, the number of transporter sites, as well as the activity of TPH-2 and TH [9]. In humans, persistent reductions of DAT density have been reported using PET in methcathinone users and are suggestive of loss of DAT or loss of DA terminals [10, 77]. Methcathinone exhibited a monoamine transporter inhibition profile that was very similar to that of the non-keto analogs amphetamine and methamphetamine, with high inhibitory potencies at the DAT and low potencies at the SERT [13, 15, 78]. It is believed that the deficits in DAT and SERT produced by methcathinone may reflect potential long-term damage to DA and 5-HT neurons. Nevertheless, to become evident, these neural deficits require massive, multiple doses of methcathinone over several days. Such doses are 10 to 100 times higher than behaviorally active doses [12]. Deficits in DA function induced by methcathinone were prevented by pretreatment with dopamine D1 or

D2 receptor antagonists, whereas 5-HT changes were prevented with a depletion of striatal DA by lesioning with 6-hydroxydopamine [8]. Apparently the serotonergic neurotoxicity of methcathinone is promoted by the presence of the *N*-methyl group on the drug molecule as it was earlier reported that no long-term changes in 5-HT levels were observed with repeated high doses of the *des*methyl parent compound, cathinone [12]. In a cell study, methcathinone was less efficacious in releasing preloaded radiolabeled neurotransmitter via VMAT-2 than methamphetamine [33]. Methcathinone also exhibited low μM potency at 5-HT_{2A} receptor binding [15].

Mephedrone administration alone did not cause persistent reductions in the levels of SERT [18]. Uptake inhibition studies using rat synaptosomes found that mephedrone potently inhibits DAT and SERT [35, 70], and the drug is a substrate for DAT, SERT, and the norepinephrine transporter (NET) [24]. Similar to methcathinone, mephedrone can bind to 5-HT_{2A} receptors and stimulation of these receptors has also been shown to enhance DA release potentially increasing abuse liability [79]. Mephedrone and methcathinone also exhibited affinity for α1A adrenoceptors, which have been implicated in stimulant-induced vasoconstriction, hyperthermia, and euphoria [15] and methcathinone has been found to be a low potency partial agonist at the 5-HT_{1A} receptors [33].

In mice, repeated administration of mephedrone induced a significant loss in DA and 5-HT reuptake sites in striatum, hippocampus, and frontal cortex [25]. In addition, mephedrone decreased the number of D2 receptors in striatum and the number of 5-HT_{2A} receptors in frontal cortex and hippocampus of treated mice [25]. In adolescent mice, mephedrone elicited an increase in D3 receptors in the striatum [30]. In rats, a binge of mephedrone induced a significant loss in DAT in frontal cortex and a decrease in the density of SERT in striatum, cortex, and hippocampus. This effect was accompanied by decreased TPH-2 expression in all the three brain areas and a moderate decrease in the concentration of D2 receptors in the striatum [20]. The effects of mephedrone on the human monoamine transporters were studied using cell lines stably expressing the human NET, DAT, and SERT. These data indicate that mephedrone and MDMA were equally potent in inhibiting noradrenaline uptake at NET. Compared to their NET inhibition potency, both drugs were weaker uptake inhibitors at DAT and SERT, with mephedrone being more potent than MDMA at DAT and less potent than MDMA at SERT. Nonetheless, mephedrone and MDMA differed most in their inhibition of DA uptake by synaptic vesicles isolated from human striatum, with MDMA being tenfold more potent than mephedrone, and their ability to release DA from human VMAT expressing cells [80]. In general, the in vitro releasing capabilities of mephedrone resemble those of MDMA. With regard to selectivity ratios, mephedrone displayed NET/DAT ratios and DAT/SERT ratios close to unity, similar to MDMA [24]. Interestingly, a recent report suggests that the para ring-substitution of the methyl group in mephedrone left-shifted the SERT inhibition curves over the DAT inhibition curves (DAT:SERT inhibition ratios <1), resulting in monoamine transporter inhibition profiles that were more similar to MDMA and less similar to the parent compound methcathinone [36]. Similarly, in vitro and in vivo studies have shown that methcathinone para ring-substituted analogs increase monoamine release via SERT relative to DAT, and that this shift in selectivity markedly reduces the abuse-related effects of the drugs as assessed by intracranial self-stimulation [81, 82].

Methylone was reported to act somewhat more potently in inhibiting DAT than SERT at the human transporter [15], but equally potent for DAT and SERT inhibition in rat synaptosomes [83]. Methylone was a substrate for NET and DAT, with slightly lower potency at SERT, displaying a selectivity profile similar to mephedrone but about half as potent. In general, the in vitro releasing capabilities of methylone resembled those of MDMA. With regard to selectivity ratios, methylone displayed NET/DAT ratios and DAT/SERT ratios close to unity, similar to MDMA [24]. In cells expressing the VMAT-2, methylone elicited less than 35% of methamphetamine maximal efficacy to stimulate release of neurotransmitter via the VMAT-2 [33]. Similar to mephedrone, methylone is also a low potency partial agonist at the 5-HT_{1A} receptors, and an antagonist with very low potency at the 5-HT_{2A} receptor [33]. While methamphetamine and MDMA are likely to be substrates for VMAT-2, methcathinone and methylone are not. Therefore, the behavioral effects of methcathinone and methylone arise largely from the drugs' effects at the plasma membrane transporters, not VMAT-2. In summary, due to the large decrease in potency at VMAT-2, methcathinone and methylone are highly selective for the plasma membrane catecholamine transporters and moderately selective for SERT. As a result of its greater potency at the SERT, methylone is somewhat less discriminating than methcathinone at the plasma membrane [13].

MDPV exhibited very high affinity for the DAT and NET in the low nanomolar range ($<$10 nM) in vitro, consistent with its high potency as a DAT and NET inhibitor [15, 33, 49]. MDPV exhibited the most potent DAT inhibition [15], being at least tenfold more potent than cocaine and methamphetamine [15]. In contrast, MDPV is a weak inhibitor of the SERT, resulting in high DAT selectivity, with DAT/SERT inhibition ratios $>$100. MDPV is also one of the most potent NET inhibitors [15]. Studies using fast-scan cyclic voltammetry in mouse striatal slices indicate that MDPV is more potent than cocaine at inhibiting DA clearance [49]. In contrast to mephedrone, MDPV is a very potent NET and DAT inhibitor with very low 5-HT activity, reflected by high DAT:SERT inhibition ratios. The 3,4-methylenedioxy ring substitution that is shared by MDMA and MDPV would be predicted to increase serotonergic activity compared with the non-substituted compound methamphetamine. However, in the case of MDPV, the SERT inhibition potency is very low despite the presence of this substitution [36]. In this regard, data have shown that the carbonyl and the extended alpha alkyl groups in MDPV have greater contributions to this drug's affinity for DAT than the methylenedioxy group [84]. In addition, in vitro findings revealed that the presence of a pyrrolidine ring in any cathinone-like compound such as MDPV confers potent blocking properties at DAT and NET [85].

An examination of methylone's ability to influence the reverse transport of substrates through DAT, NET, and SERT was done in comparison with methamphetamine, since unlike cocaine, methamphetamine induces the release of monoamines via a reversal of transport. Similar to methamphetamine, methylone elicited the release of radiolabeled DA, NE, and 5-HT from CHO cells expressing the rat DAT, NET, and SERT. In addition, the combination of methylone with methamphetamine did not cause a further increase in the release of substrates [43]. None of the β-ketoamphetamines mephedrone, methylone, or MDPV in combination with each other resulted in changes in striatal DAT [19]. In combination with methamphetamine, mephedrone and methylone enhanced the reductions in DAT observed in mouse striatum [19, 23]. In contrast, administration of MDPV prevented the depletions in DAT observed after methamphetamine, amphetamine, MDMA, and MPTP [19].

6 Transendothelial Blood–Brain Barrier Dysfunction

The rate at which drugs reach the brain parenchyma depends not only on their route of administration but also on their ability to cross the cerebral endothelium, also called the blood–brain barrier (BBB), which constitutes the main brain interface modulating the exchange of compounds between the brain and blood [50]. Alterations in BBB function are likely involved in drug abuse neurotoxicity [1, 86]. Both MDMA and METH have been shown to produce disruption of the BBB as reflected by IgG extravasation and Evans Blue leakage [5, 87]. In fact, it was previously shown that METH compromises BBB function and its capacity to protect the brain against infection by the human immunodeficiency virus [88]. Studies of the effects of β-ketoamphetamines on BBB dysfunction are summarized in Table 1.

Methcathinone exhibited a brain permeability ratio ≥ 3, indicating high permeability. However, the apical to basolateral transport of methcathinone was not consistent with active transport by one of the blood-to-brain influx carriers [15]. No studies on the compromise of the BBB by methcathinone have been reported to date. The permeability ratio for mephedrone was >10, suggesting very high BBB permeability [15], and confirming that mephedrone readily enters the brain [15, 70]. Although highly permeable into the brain, mephedrone administration has not been linked to any BBB dysfunction. It is well recognized that compounds with a brain/plasma concentration ratio greater than 1 freely cross the blood–brain barrier and the obtained brain/plasma ratio for methylone of 1.42 demonstrates access to central nervous system [89]. As a reference, methamphetamine, amphetamine, and MDMA brain/plasma ratios are >3 [15].

Similar to mephedrone, the permeability ratio for MDPV was >10, suggesting a very high permeability [15]. The apical to basolateral transport of MDPV was significantly greater than basolateral to apical transport, consistent with active transport by one of the blood-to-brain influx carriers [15]. MDPV is a monoamine uptake inhibitor that is more lipophilic and potent than other cathinone derivatives. The high lipophilicity of this substance is caused by the pyrrolidine ring and the tertiary amino group creating a less polar molecule more able to cross the blood–brain barrier [90]. No combinations of the β-ketoamphetamines mephedrone, methylone, or MDPV with each other or any other amphetamine compound have been evaluated on BBB dysfunction to date.

7 Mechanisms of Action

Drugs that target monoamine transporters can be classified generally as either substrates, such as methamphetamine, or blockers, such as cocaine [83]. Both types of compounds elicit profound psychostimulant effects that render them liable for recreational abuse [91]. Substrates or blockers increase monoamine neurotransmitter concentrations in the synaptic cleft but this action can be the result of at least two distinct mechanisms [13]. One mechanism is through drug inhibition of plasma membrane transporter-mediated uptake of released neurotransmitters (i.e., for transporter blockers). The inhibition is believed to arise from competition by drugs for substrate binding sites in the monoamine uptake transporters, thereby reducing the effectiveness with which DA, 5-HT, and NE are cleared from the synapse following release. Typical DAT blockers are expected to fully inhibit DA uptake and to fully inhibit binding of another blocker, as well as release of substrate by reverse transport [92]. A second mechanism is through the drug-evoked release of the monoamine neurotransmitters, apparently by transporter-mediated exchange (i.e., for transporter substrates). The drug-evoked neurotransmitter release arises from two intracellular compartments. Methamphetamine and MDMA induce the release of newly synthesized, cytosolic pools of monoamines and also release monoamines from synaptic vesicle stores [13]. Typical DAT releasers are expected to fully release another substrate accumulated in cell or synaptosomes [92]. This mechanistic distinction is important to consider because transporter substrates and blockers display critical differences in their acute and long-term effects. Only substrates are translocated into cells where they could disrupt vesicular storage and stimulate non-exocytotic release of neurotransmitters by reversing the normal direction of transporter flux, and could produce persistent deficits in monoamine neurons, including depletion of neurotransmitters and loss of functional transporters [83]. The flux-coupled channel model suggests that whereas some cathinones, such as mephedrone, behave as DA-releasing agents (depolarizing current), some others such as MDPV act as DA-reuptake inhibitors (hyperpolarizing current) [93]. An "excitatory substrate" implies that in addition to the proposed transporter-mediated chemical effects of methamphetamine, mephedrone, and related cathinones, these substances have a depolarizing action that could itself promote exocytotic neurotransmitter release [32]. Structurally analogous MDPV, however, induces an outward hyperpolarizing current under similar conditions and therefore acting as an "inhibitory," non-substrate blocker [93]. Results from release assays reveal that mephedrone and methylone function as substrates at monoamine transporters [33], thereby stimulating the release of radiolabeled substrates via DAT, NET, and SERT [83]. Mephedrone, methylone, and MDMA are non-selective transporter substrates (i.e., non-selective releasers), while methcathinone and amphetamine are selective substrates at DAT and NET. Mephedrone displays similar releasing potency at all three transporters and is about twice as potent as methylone [83]. While mephedrone, methylone, MDMA, and amphetamine are fully efficacious in the release assays, MDPV and cocaine are inactive as releasers [15]. MDPV displays a novel pharmacological profile when compared to other synthetic cathinones as it is a potent uptake blocker at DAT and NET with no measurable substrate activity [83]. When compared to the prototypical transporter blocker cocaine, MDPV was 50-fold more potent at DAT, 10-fold more potent at NET, and 10-fold less potent at SERT [94]. Taken together, the in vitro results indicate that mephedrone and methylone are non-selective transporter substrates, whereas MDPV is a pure catecholamine-selective transporter blocker [94]. This dichotomy of interaction with the DAT by mephedrone and methylone on one hand and by MDPV on the other can explain their opposing effects on methamphetamine-induced neurotoxicity [19]. While mephedrone and methylone enhanced the neurotoxic effects of methamphetamine, MDPV protected. It has previously shown that treatments resulting in an increase in the releasable pool of DA significantly accentuate methamphetamine-induced damage in DA nerve endings [95]. MDPV has an effect that is similar to more classical DAT blockers, including amphonelic acid and nomifensine, which also provide protection against methamphetamine-induced neurotoxicity [19]. By blocking DAT-mediated transport (inward or outward), MDPV blocks methamphetamine-induced efflux of DA [15]. Therefore, these properties as substrates or blockers represent an important mechanism by which synthetic cathinones influence the synaptic levels of monoamines but they do not explain why they lack neurotoxic properties on their own or how they enhance the neurotoxicity of the amphetamines.

While the principal targets of amphetamines are plasmalemmal transporters, these drugs have concerted actions on other two important elements of the monoamine nerve ending: vesicular transporters and the degrading enzymes monoamine oxidase (MAO) and catechol-*O*-methyl transferase [91], both of which may contribute to their toxic properties. Amphetamine interactions with these three targets are the core tenet of the so-called weak base hypothesis [96]. Amphetamines enter nerve terminals via plasmalemmal transporters and disrupt vesicular storage as weak bases by dissipating the proton gradient across the membrane [96]. A reduction of the vesicular pH gradient promotes the reverse transport of DA into the cytosol. DA is then released into the synaptic space via reverse transport through the DAT. This flooding of the cytoplasm and synaptic space with the oxidatively labile DA is thought to be a critical first step in the neurotoxic cascade of the amphetamines [73]. These conditions of elevated concentrations of cytosolic monoamines could be further aggravated by inhibition of MAO [91]. Unlike amphetamines, mephedrone and methylone have little if any affinity for VMAT-2 [33]. Therefore, their lack of neurotoxicity could derive from an inability to promote the release of DA from storage vesicles into the cytoplasm.

If not toxic on their own, how can mephedrone and methylone increase the neurotoxicity of methamphetamine, MDMA, and amphetamine? We hypothesize that the enhancement of neurotoxicity elicited by the combination of methamphetamine plus either mephedrone or methylone could be explained by a reversal of greater numbers of DAT molecules than caused by either drug alone, resulting in heightened DA efflux into the cytoplasm (i.e., via methamphetamine actions on the VMAT) and synapse (i.e., via combined methamphetamine plus mephedrone actions on the DAT). This possibility is supported by the observation that amphetamine-induced DA release is greater when originating from both synaptic vesicles and cytoplasmic stores than from cytoplasmic stores only [97]. In addition, a possible inhibition of MAO could be speculated for mephedrone and methylone as it has been shown that MAO inhibitors increase significantly the DA depletion induced by methamphetamine [95]. However, a well-established mechanism to explain why some β-ketoamphetamines such as mephedrone and methylone are not neurotoxic on their own but are capable of potentiating the damage induced by amphetamines remains to be elucidated.

8 Conclusion

As β-keto analogs of amphetamines, synthetic cathinones may be expected to have amphetamine-like effects because of their structural similarity. However, β-ketoamphetamines are a diverse class of chemical compounds with differential neurotoxic properties on monoaminergic neurons. Some of the benchmarks used to gauge the neurotoxicity induced by amphetamines include inflammation, disruption of monoaminergic neurotransmitters, their transporters and receptors, alterations in thermoregulation, oxidative stress, and cytotoxicity. Compared to the effects induced by amphetamines on these parameters, the effects described for β-ketoamphetamines seem to be more moderate. Administration of synthetic cathinones is not consistently associated with long-term depletions in the levels of DA and 5-HT or with inhibition of these neurotransmitters biosynthetic enzymes. While hyperthermia has been established as one of the hallmark effects of amphetamines, synthetic cathinones elicit more complex responses that involve hypothermia and oscillations between hyper and hypothermia. Neuroinflammation markers such as microglial activation have not been documented after administration of synthetic cathinones and reports of increases in GFAP have been sparse. The evaluation of the effects of these cathinones on oxidative stress and cytotoxicity are limited and mostly circumscribed to in vitro studies, where concentrations are very high. Nonetheless, some studies in animals have described increases in lipid peroxidation and in the expression of antioxidant enzymes after mephedrone. Deficits in DAT and SERT were only observed after multiple doses that are several times higher than behaviorally active doses or with exacerbation of other factors such as high ambient temperature. Although only a few studies have reported the neurotoxic effects of β-ketoamphetamines alone or in combination with other drugs of the same group or with amphetamines, the overall outcome appears to be associated with their interaction with the vesicular and plasma transporters. The fact that mephedrone and methylone cause little or no toxicity themselves on the one hand, while being capable of enhancing amphetamines toxicity on the other hand, remains a provocative and open question that requires additional research. The role of these synthetic cathinones as weak bases to collapse the vesicular pH gradient necessary for monoamine storage, their capacity to increase the releasable pool of cytosolic monoamines, their potential to inhibit monoamine degrading enzymes, their ability to increase monoamine oxidation and metabolism into ROS, as well as their additive effects in recruiting DAT molecules along with amphetamines to enhance the DA efflux into the synapse, constitute some of the possible manners in which these β-ketoamphetamines may heighten the neurotoxicity induced by amphetamines.

References

1. Yamamoto BK, Moszczynska A, Gudelsky GA (2010) Amphetamine toxicities: classical and emerging mechanisms. Ann N Y Acad Sci 1187:101–121
2. Moratalla R, Khairnar A, Simola N, Granado N, Garcia-Montes JR, Porceddu PF, Tizabi Y, Costa G, Morelli M (2015) Amphetamine-related drugs neurotoxicity in humans and in experimental animals: main mechanisms. Prog Neurobiol
3. Thomas DM, Walker PD, Benjamins JA, Geddes TJ, Kuhn DM (2004) Methamphetamine neurotoxicity in dopamine nerve endings of the striatum is associated with microglial activation. J Pharmacol Exp Ther 311:1–7
4. Thomas DM, Dowgiert J, Geddes TJ, Francescutti-Verbeem D, Liu X, Kuhn DM (2004) Microglial activation is a pharmacologically specific marker for the neurotoxic amphetamines. Neurosci Lett 367:349–354
5. Rubio-Araiz A, Perez-Hernandez M, Urrutia A, Porcu F, Borcel E, Gutierrez-Lopez MD, O'Shea E, Colado MI (2014) 3,4-Methylenedioxymethamphetamine (MDMA, ecstasy) disrupts blood-brain barrier integrity through a mechanism involving P2X7 receptors. Int J Neuropsychopharmacol 17:1243–1255
6. Rockhold RW, Carlton FB Jr, Corkern R, Derouen L, Bennett JG, Hume AS (1997) Methcathinone intoxication in the rat: abrogation by dextrorphan. Ann Emerg Med 29:383–391
7. Shortall SE, Green AR, Swift KM, Fone KC, King MV (2013) Differential effects of cathinone compounds and MDMA on body temperature in the rat, and pharmacological characterization of mephedrone-induced hypothermia. Br J Pharmacol 168:966–977
8. Gygi MP, Fleckenstein AE, Gibb JW, Hanson GR (1997) Role of endogenous dopamine in the neurochemical deficits induced by methcathinone. J Pharmacol Exp Ther 283:1350–1355
9. Gygi MP, Gibb JW, Hanson GR (1996) Methcathinone: an initial study of its effects on monoaminergic systems. J Pharmacol Exp Ther 276:1066–1072
10. McCann UD, Wong DF, Yokoi F, Villemagne V, Dannals RF, Ricaurte GA (1998) Reduced striatal dopamine transporter density in abstinent methamphetamine and methcathinone users: evidence from positron emission tomography studies with [11C]WIN-35,428. J Neurosci 18:8417–8422
11. Sparago M, Wlos J, Yuan J, Hatzidimitriou G, Tolliver J, Dal Cason TA, Katz J, Ricaurte G (1996) Neurotoxic and pharmacologic studies on enantiomers of the N-methylated analog of cathinone (methcathinone): a new drug of abuse. J Pharmacol Exp Ther 279:1043–1052
12. Cozzi NV, Foley KF (2003) Methcathinone is a substrate for the serotonin uptake transporter. Pharmacol Toxicol 93:219–225
13. Cozzi NV, Sievert MK, Shulgin AT, Jacob P 3rd, Ruoho AE (1999) Inhibition of plasma membrane monoamine transporters by beta-ketoamphetamines. Eur J Pharmacol 381:63–69

14. Metzger RR, Haughey HM, Wilkins DG, Gibb JW, Hanson GR, Fleckenstein AE (2000) Methamphetamine-induced rapid decrease in dopamine transporter function: role of dopamine and hyperthermia. J Pharmacol Exp Ther 295:1077–1085
15. Simmler LD, Buser TA, Donzelli M, Schramm Y, Dieu LH, Huwyler J, Chaboz S, Hoener MC, Liechti ME (2013) Pharmacological characterization of designer cathinones in vitro. Br J Pharmacol 168:458–470
16. Glennon RA, Yousif M, Naiman N, Kalix P (1987) Methcathinone: a new and potent amphetamine-like agent. Pharmacol Biochem Behav 26:547–551
17. Angoa-Perez M, Kane MJ, Francescutti DM, Sykes KE, Shah MM, Mohammed AM, Thomas DM, Kuhn DM (2012) Mephedrone, an abused psychoactive component of 'bath salts' and methamphetamine congener, does not cause neurotoxicity to dopamine nerve endings of the striatum. J Neurochem 120:1097–1107
18. Angoa-Perez M, Kane MJ, Herrera-Mundo N, Francescutti DM, Kuhn DM (2014) Effects of combined treatment with mephedrone and methamphetamine or 3,4-methylenedioxymethamphetamine on serotonin nerve endings of the hippocampus. Life Sci 97:31–36
19. Anneken JH, Angoa-Perez M, Kuhn DM (2015) 3,4-Methylenedioxypyrovalerone prevents while methylone enhances methamphetamine-induced damage to dopamine nerve endings: beta-ketoamphetamine modulation of neurotoxicity by the dopamine transporter. J Neurochem 133:211–222
20. Lopez-Arnau R, Martinez-Clemente J, Rodrigo T, Pubill D, Camarasa J, Escubedo E (2015) Neuronal changes and oxidative stress in adolescent rats after repeated exposure to mephedrone. Toxicol Appl Pharmacol 286:27–35
21. Motbey CP, Clemens KJ, Apetz N, Winstock AR, Ramsey J, Li KM, Wyatt N, Callaghan PD, Bowen MT, Cornish JL, McGregor IS (2013) High levels of intravenous mephedrone (4-methylmethcathinone) self-administration in rats: neural consequences and comparison with methamphetamine. J Psychopharmacol 27:823–836
22. Motbey CP, Karanges E, Li KM, Wilkinson S, Winstock AR, Ramsay J, Hicks C, Kendig MD, Wyatt N, Callaghan PD, McGregor IS (2012) Mephedrone in adolescent rats: residual memory impairment and acute but not lasting 5-HT depletion. PLoS One 7, e45473
23. Angoa-Perez M, Kane MJ, Briggs DI, Francescutti DM, Sykes CE, Shah MM, Thomas DM, Kuhn DM (2013) Mephedrone does not damage dopamine nerve endings of the striatum, but enhances the neurotoxicity of methamphetamine, amphetamine, and MDMA. J Neurochem 125:102–110
24. Baumann MH, Ayestas MA Jr, Partilla JS, Sink JR, Shulgin AT, Daley PF, Brandt SD, Rothman RB, Ruoho AE, Cozzi NV (2012) The designer methcathinone analogs, mephedrone and methylone, are substrates for monoamine transporters in brain tissue. Neuropsychopharmacology 37:1192–1203
25. Martinez-Clemente J, Lopez-Arnau R, Abad S, Pubill D, Escubedo E, Camarasa J (2014) Dose and time-dependent selective neurotoxicity induced by mephedrone in mice. PLoS One 9, e99002
26. Shortall SE, Spicer CH, Ebling FJ, Green AR, Fone KC, King MV (2015) Contribution of serotonin and dopamine to changes in core body temperature and locomotor activity in rats following repeated administration of mephedrone. Addict Biol
27. Wright MJ Jr, Angrish D, Aarde SM, Barlow DJ, Buczynski MW, Creehan KM, Vandewater SA, Parsons LH, Houseknecht KL, Dickerson TJ, Taffe MA (2012) Effect of ambient temperature on the thermoregulatory and locomotor stimulant effects of 4-methylmethcathinone in Wistar and Sprague-Dawley rats. PLoS One 7, e44652
28. den Hollander B, Rozov S, Linden AM, Uusi-Oukari M, Ojanpera I, Korpi ER (2013) Long-term cognitive and neurochemical effects of "bath salt" designer drugs methylone and mephedrone. Pharmacol Biochem Behav 103:501–509
29. Shortall SE, Macerola AE, Swaby RT, Jayson R, Korsah C, Pillidge KE, Wigmore PM, Ebling FJ, Richard Green A, Fone KC, King MV (2013) Behavioural and neurochemical comparison

of chronic intermittent cathinone, mephedrone and MDMA administration to the rat. Eur Neuropsychopharmacol 23:1085–1095
30. Ciudad-Roberts A, Camarasa J, Ciudad CJ, Pubill D, Escubedo E (2015) Alcohol enhances the psychostimulant and conditioning effects of mephedrone in adolescent mice; postulation of unique roles of D3 receptors and BDNF in place preference acquisition. Br J Pharmacol 172:4970–4984
31. den Hollander B, Sundstrom M, Pelander A, Ojanpera I, Mervaala E, Korpi ER, Kankuri E (2014) Keto amphetamine toxicity-focus on the redox reactivity of the cathinone designer drug mephedrone. Toxicol Sci 141:120–131
32. Cameron KN, Kolanos R, Solis E Jr, Glennon RA, De Felice LJ (2013) Bath salts components mephedrone and methylenedioxypyrovalerone (MDPV) act synergistically at the human dopamine transporter. Br J Pharmacol 168:1750–1757
33. Eshleman AJ, Wolfrum KM, Hatfield MG, Johnson RA, Murphy KV, Janowsky A (2013) Substituted methcathinones differ in transporter and receptor interactions. Biochem Pharmacol 85:1803–1815
34. Lopez-Arnau R, Martinez-Clemente J, Pubill D, Escubedo E, Camarasa J (2012) Comparative neuropharmacology of three psychostimulant cathinone derivatives: butylone, mephedrone and methylone. Br J Pharmacol 167:407–420
35. Martinez-Clemente J, Escubedo E, Pubill D, Camarasa J (2012) Interaction of mephedrone with dopamine and serotonin targets in rats. Eur Neuropsychopharmacol 22:231–236
36. Rickli A, Hoener MC, Liechti ME (2015) Monoamine transporter and receptor interaction profiles of novel psychoactive substances: para-halogenated amphetamines and pyrovalerone cathinones. Eur Neuropsychopharmacol 25:365–376
37. Budzynska B, Boguszewska-Czubara A, Kruk-Slomka M, Kurzepa J, Biala G (2015) Mephedrone and nicotine: oxidative stress and behavioral interactions in animal models. Neurochem Res 40:1083–1093
38. Lopez-Arnau R, Martinez-Clemente J, Abad S, Pubill D, Camarasa J, Escubedo E (2014) Repeated doses of methylone, a new drug of abuse, induce changes in serotonin and dopamine systems in the mouse. Psychopharmacology (Berl) 231:3119–3129
39. Lopez-Arnau R, Martinez-Clemente J, Pubill D, Escubedo E, Camarasa J (2014) Serotonergic impairment and memory deficits in adolescent rats after binge exposure of methylone. J Psychopharmacol 28:1053–1063
40. Kiyatkin EA, Kim AH, Wakabayashi KT, Baumann MH, Shaham Y (2015) Effects of social interaction and warm ambient temperature on brain hyperthermia induced by the designer drugs methylone and MDPV. Neuropsychopharmacology 40:436–445
41. Piao YS, Hall FS, Moriya Y, Ito M, Ohara A, Kikura-Hanajiri R, Goda Y, Lesch KP, Murphy DL, Uhl GR, Sora I (2015) Methylone-induced hyperthermia and lethal toxicity: role of the dopamine and serotonin transporters. Behav Pharmacol 26:345–352
42. Nagai F, Nonaka R, Satoh Hisashi Kamimura K (2007) The effects of non-medically used psychoactive drugs on monoamine neurotransmission in rat brain. Eur J Pharmacol 559:132–137
43. Sogawa C, Sogawa N, Ohyama K, Kikura-Hanajiri R, Goda Y, Sora I, Kitayama S (2011) Methylone and monoamine transporters: correlation with toxicity. Curr Neuropharmacol 9:58–62
44. Aarde SM, Creehan KM, Vandewater SA, Dickerson TJ, Taffe MA (2015) In vivo potency and efficacy of the novel cathinone alpha-pyrrolidinopentiophenone and 3,4-methylenedioxypyrovalerone: self-administration and locomotor stimulation in male rats. Psychopharmacology (Berl) 232:3045–3055
45. Adam A, Gerecsei LI, Lepesi N, Csillag A (2014) Apoptotic effects of the 'designer drug' methylenedioxypyrovalerone (MDPV) on the neonatal mouse brain. Neurotoxicology 44:231–236

46. Fantegrossi WE, Gannon BM, Zimmerman SM, Rice KC (2013) In vivo effects of abused 'bath salt' constituent 3,4-methylenedioxypyrovalerone (MDPV) in mice: drug discrimination, thermoregulation, and locomotor activity. Neuropsychopharmacology 38:563–573
47. King HE, Wetzell B, Rice KC, Riley AL (2014) 3,4-Methylenedioxypyrovalerone (MDPV)-induced conditioned taste avoidance in the F344/N and LEW rat strains. Pharmacol Biochem Behav 126:163–169
48. Merluzzi AP, Hurwitz ZE, Briscione MA, Cobuzzi JL, Wetzell B, Rice KC, Riley AL (2014) Age-dependent MDPV-induced taste aversions and thermoregulation in adolescent and adult rats. Dev Psychobiol 56:943–954
49. Baumann MH, Partilla JS, Lehner KR, Thorndike EB, Hoffman AF, Holy M, Rothman RB, Goldberg SR, Lupica CR, Sitte HH, Brandt SD, Tella SR, Cozzi NV, Schindler CW (2013) Powerful cocaine-like actions of 3,4-methylenedioxypyrovalerone (MDPV), a principal constituent of psychoactive 'bath salts' products. Neuropsychopharmacology 38:552–562
50. Chapy H, Smirnova M, Andre P, Schlatter J, Chiadmi F, Couraud PO, Scherrmann JM, Decleves X, Cisternino S (2015) Carrier-mediated cocaine transport at the blood-brain barrier as a putative mechanism in addiction liability. Int J Neuropsychopharmacol 18
51. Simmler LD, Wandeler R, Liechti ME (2013) Bupropion, methylphenidate, and 3,4-methylenedioxypyrovalerone antagonize methamphetamine-induced efflux of dopamine according to their potencies as dopamine uptake inhibitors: implications for the treatment of methamphetamine dependence. BMC Res Notes 6:220
52. Numachi Y, Ohara A, Yamashita M, Fukushima S, Kobayashi H, Hata H, Watanabe H, Hall FS, Lesch KP, Murphy DL, Uhl GR, Sora I (2007) Methamphetamine-induced hyperthermia and lethal toxicity: role of the dopamine and serotonin transporters. Eur J Pharmacol 572:120–128
53. Crean RD, Davis SA, Von Huben SN, Lay CC, Katner SN, Taffe MA (2006) Effects of (+/-) 3,4-methylenedioxymethamphetamine, (+/-)3,4-methylenedioxyamphetamine and methamphetamine on temperature and activity in rhesus macaques. Neuroscience 142:515–525
54. Cappon GD, Morford LL, Vorhees CV (1997) Ontogeny of methamphetamine-induced neurotoxicity and associated hyperthermic response. Brain Res Dev Brain Res 103:155–162
55. Dafters RI (1995) Hyperthermia following MDMA administration in rats: effects of ambient temperature, water consumption, and chronic dosing. Physiol Behav 58:877–882
56. Makisumi T, Yoshida K, Watanabe T, Tan N, Murakami N, Morimoto A (1998) Sympatho-adrenal involvement in methamphetamine-induced hyperthermia through skeletal muscle hypermetabolism. Eur J Pharmacol 363:107–112
57. Paulson PE, Robinson TE (1995) Amphetamine-induced time-dependent sensitization of dopamine neurotransmission in the dorsal and ventral striatum: a microdialysis study in behaving rats. Synapse 19:56–65
58. Yoshida K, Morimoto A, Makisumi T, Murakami N (1993) Cardiovascular, thermal and behavioral sensitization to methamphetamine in freely moving rats. J Pharmacol Exp Ther 267:1538–1543
59. Miller ML, Creehan KM, Angrish D, Barlow DJ, Houseknecht KL, Dickerson TJ, Taffe MA (2013) Changes in ambient temperature differentially alter the thermoregulatory, cardiac and locomotor stimulant effects of 4-methylmethcathinone (mephedrone). Drug Alcohol Depend 127:248–253
60. Gregg RA, Rawls SM (2014) Behavioral pharmacology of designer cathinones: a review of the preclinical literature. Life Sci 97:27–30
61. Aarde SM, Angrish D, Barlow DJ, Wright MJ Jr, Vandewater SA, Creehan KM, Houseknecht KL, Dickerson TJ, Taffe MA (2013) Mephedrone (4-methylmethcathinone) supports intravenous self-administration in Sprague-Dawley and Wistar rats. Addict Biol 18:786–799
62. Froberg BA, Levine M, Beuhler MC, Judge BS, Moore PW, Engebretsen KM, McKeown NJ, Rosenbaum CD, Young AC, Rusyniak DE (2015) Acute methylenedioxypyrovalerone toxicity. J Med Toxicol 11:185–194

63. Cadet JL, Jayanthi S, Deng X (2003) Speed kills: cellular and molecular bases of methamphetamine-induced nerve terminal degeneration and neuronal apoptosis. FASEB J 17:1775–1788
64. Kita T, Wagner GC, Nakashima T (2003) Current research on methamphetamine-induced neurotoxicity: animal models of monoamine disruption. J Pharmacol Sci 92:178–195
65. Baumann MH, Wang X, Rothman RB (2007) 3,4-Methylenedioxymethamphetamine (MDMA) neurotoxicity in rats: a reappraisal of past and present findings. Psychopharmacology (Berl) 189:407–424
66. Cadet JL, Krasnova IN, Jayanthi S, Lyles J (2007) Neurotoxicity of substituted amphetamines: molecular and cellular mechanisms. Neurotox Res 11:183–202
67. Capela JP, Carmo H, Remiao F, Bastos ML, Meisel A, Carvalho F (2009) Molecular and cellular mechanisms of ecstasy-induced neurotoxicity: an overview. Mol Neurobiol 39:210–271
68. Imam SZ, el-Yazal J, Newport GD, Itzhak Y, Cadet JL, Slikker W Jr, Ali SF (2001) Methamphetamine-induced dopaminergic neurotoxicity: role of peroxynitrite and neuroprotective role of antioxidants and peroxynitrite decomposition catalysts. Ann N Y Acad Sci 939:366–380
69. Motbey CP, Hunt GE, Bowen MT, Artiss S, McGregor IS (2012) Mephedrone (4-methylmethcathinone, 'meow'): acute behavioural effects and distribution of Fos expression in adolescent rats. Addict Biol 17:409–422
70. Hadlock GC, Webb KM, McFadden LM, Chu PW, Ellis JD, Allen SC, Andrenyak DM, Vieira-Brock PL, German CL, Conrad KM, Hoonakker AJ, Gibb JW, Wilkins DG, Hanson GR, Fleckenstein AE (2011) 4-Methylmethcathinone (mephedrone): neuropharmacological effects of a designer stimulant of abuse. J Pharmacol Exp Ther 339:530–536
71. Kehr J, Ichinose F, Yoshitake S, Goiny M, Sievertsson T, Nyberg F, Yoshitake T (2011) Mephedrone, compared with MDMA (ecstasy) and amphetamine, rapidly increases both dopamine and 5-HT levels in nucleus accumbens of awake rats. Br J Pharmacol 164:1949–1958
72. Toborek M, Seelbach MJ, Rashid CS, Andras IE, Chen L, Park M, Esser KA (2013) Voluntary exercise protects against methamphetamine-induced oxidative stress in brain microvasculature and disruption of the blood-brain barrier. Mol Neurodegener 8:22
73. Cadet JL, Brannock C (1998) Free radicals and the pathobiology of brain dopamine systems. Neurochem Int 32:117–131
74. Halpin LE, Collins SA, Yamamoto BK (2014) Neurotoxicity of methamphetamine and 3,4-methylenedioxymethamphetamine. Life Sci 97:37–44
75. Stepens A, Groma V, Skuja S, Platkajis A, Aldins P, Eksteina I, Martinsone I, Bricis R, Donaghy M (2014) The outcome of the movement disorder in methcathinone abusers: clinical, MRI and manganesemia changes, and neuropathology. Eur J Neurol 21:199–205
76. Levin OS (2005) "Ephedron" encephalopathy. Zh Nevrol Psikhiatr Im S S Korsakova 105:12–20
77. Kelly JP (2011) Cathinone derivatives: a review of their chemistry, pharmacology and toxicology. Drug Test Anal 3:439–453
78. Rothman RB, Vu N, Partilla JS, Roth BL, Hufeisen SJ, Compton-Toth BA, Birkes J, Young R, Glennon RA (2003) In vitro characterization of ephedrine-related stereoisomers at biogenic amine transporters and the receptorome reveals selective actions as norepinephrine transporter substrates. J Pharmacol Exp Ther 307:138–145
79. Gudelsky GA, Yamamoto BK, Nash JF (1994) Potentiation of 3,4-methylenedioxymethamphetamine-induced dopamine release and serotonin neurotoxicity by 5-HT2 receptor agonists. Eur J Pharmacol 264:325–330
80. Pifl C, Reither H, Hornykiewicz O (2015) The profile of mephedrone on human monoamine transporters differs from 3,4-methylenedioxymethamphetamine primarily by lower potency at the vesicular monoamine transporter. Eur J Pharmacol 755:119–126

81. Bonano JS, Banks ML, Kolanos R, Sakloth F, Barnier ML, Glennon RA, Cozzi NV, Partilla JS, Baumann MH, Negus SS (2015) Quantitative structure-activity relationship analysis of the pharmacology of para-substituted methcathinone analogues. Br J Pharmacol 172:2433–2444
82. Sakloth F, Kolanos R, Mosier PD, Bonano JS, Banks ML, Partilla JS, Baumann MH, Negus SS, Glennon RA (2015) Steric parameters, molecular modeling and hydropathic interaction analysis of the pharmacology of para-substituted methcathinone analogues. Br J Pharmacol 172:2210–2218
83. Baumann MH, Partilla JS, Lehner KR (2013) Psychoactive "bath salts": not so soothing. Eur J Pharmacol 698:1–5
84. Kolanos R, Solis E Jr, Sakloth F, De Felice LJ, Glennon RA (2013) Deconstruction of the abused synthetic cathinone methylenedioxypyrovalerone (MDPV) and an examination of effects at the human dopamine transporter. ACS Chem Neurosci 12:1524–1529
85. Marusich JA, Antonazzo KR, Wiley JL, Blough BE, Partilla JS, Baumann MH (2014) Pharmacology of novel synthetic stimulants structurally related to the "bath salts" constituent 3,4-methylenedioxypyrovalerone (MDPV). Neuropharmacology 87:206–213
86. Kiyatkin EA, Sharma HS (2009) Acute methamphetamine intoxication: brain hyperthermia, blood-brain barrier, brain edema, and morphological cell abnormalities. Int Rev Neurobiol 88:65–100
87. O'Shea E, Urrutia A, Green AR, Colado MI (2014) Current preclinical studies on neuroinflammation and changes in blood-brain barrier integrity by MDMA and methamphetamine. Neuropharmacology 87:125–134
88. Coelho-Santos V, Leitao RA, Cardoso FL, Palmela I, Rito M, Barbosa M, Brito MA, Fontes-Ribeiro CA, Silva AP (2015) The TNF-alpha/NF-kappaB signaling pathway has a key role in methamphetamine-induced blood-brain barrier dysfunction. J Cereb Blood Flow Metab 35:1260–1271
89. Lopez-Arnau R, Martinez-Clemente J, Carbo M, Pubill D, Escubedo E, Camarasa J (2013) An integrated pharmacokinetic and pharmacodynamic study of a new drug of abuse, methylone, a synthetic cathinone sold as "bath salts". Prog Neuropsychopharmacol Biol Psychiatry 45:64–72
90. Coppola M, Mondola R (2012) Synthetic cathinones: chemistry, pharmacology and toxicology of a new class of designer drugs of abuse marketed as "bath salts" or "plant food". Toxicol Lett 211:144–149
91. Sitte HH, Freissmuth M (2015) Amphetamines, new psychoactive drugs and the monoamine transporter cycle. Trends Pharmacol Sci 36:41–50
92. Reith ME, Blough BE, Hong WC, Jones KT, Schmitt KC, Baumann MH, Partilla JS, Rothman RB, Katz JL (2015) Behavioral, biological, and chemical perspectives on atypical agents targeting the dopamine transporter. Drug Alcohol Depend 147:1–19
93. De Felice LJ, Glennon RA, Negus SS (2014) Synthetic cathinones: chemical phylogeny, physiology, and neuropharmacology. Life Sci 97:20–26
94. Araujo AM, Valente MJ, Carvalho M, Dias da Silva D, Gaspar H, Carvalho F, de Lourdes Bastos M, Guedes de Pinho P (2015) Raising awareness of new psychoactive substances: chemical analysis and in vitro toxicity screening of 'legal high' packages containing synthetic cathinones. Arch Toxicol 89:757–771
95. Thomas DM, Francescutti-Verbeem DM, Kuhn DM (2008) The newly synthesized pool of dopamine determines the severity of methamphetamine-induced neurotoxicity. J Neurochem 105:605–616
96. Sulzer D, Maidment NT, Rayport S (1993) Amphetamine and other weak bases act to promote reverse transport of dopamine in ventral midbrain neurons. J Neurochem 60:527–535
97. Pifl C, Drobny H, Reither H, Hornykiewicz O, Singer EA (1995) Mechanism of the dopamine-releasing actions of amphetamine and cocaine: plasmalemmal dopamine transporter versus vesicular monoamine transporter. Mol Pharmacol 47:368–373

Combination Chemistry: Structure–Activity Relationships of Novel Psychoactive Cannabinoids

Jenny L. Wiley, Julie A. Marusich, and Brian F. Thomas

Abstract Originally developed as research tools for use in structure–activity relationship studies, synthetic cannabinoids contributed to significant scientific advances in the cannabinoid field. Unfortunately, a subset of these compounds was diverted for recreational use beginning in the early 2000s. As these compounds were banned, they were replaced with additional synthetic cannabinoids with increasingly diverse chemical structures. This chapter focuses on integration of recent results with those covered in previous reviews. Whereas most of the early compounds were derived from the prototypic naphthoylindole JWH-018, currently popular synthetic cannabinoids include tetramethylcyclopropyl ketones and indazole-derived cannabinoids (e.g., AB-PINACA, AB-CHMINACA). Despite their structural differences, psychoactive synthetic cannabinoids bind with high affinity to CB_1 receptors in the brain and, when tested, have been shown to activate these receptors and to produce a characteristic profile of effects, including suppression of locomotor activity, antinociception, hypothermia, and catalepsy, as well as Δ^9-tetrahydrocannabinol (THC)-like discriminative stimulus effects in mice. When they have been tested, synthetic cannabinoids are often found to be more efficacious at activation of the CB_1 receptor and more potent in vivo. Further, their chemical alteration by thermolysis during use and their uncertain stability and purity may result in exposure to degradants that differ from the parent compound contained in the original product. Consequently, while their intoxicant effects may be similar to those of THC, use of synthetic cannabinoids may be accompanied by unpredicted, and sometimes harmful, effects.

Keywords Cannabinoids • Indoles • JWH-018 • Receptor binding • Synthetic cannabinoids • THC

J.L. Wiley (✉), J.A. Marusich, and B.F. Thomas
RTI International, 3040 Cornwallis Road, Research Triangle Park, NC 27709-2194, USA
e-mail: jwiley@rti.org

Curr Topics Behav Neurosci (2017) 32: 231–248
DOI 10.1007/7854_2016_17
Published Online: 20 June 2016

Contents

1 Introduction

Synthetic cannabinoids are chemicals that interact with the endogenous system through which the psychoactive components of the cannabis plant act. Although these compounds were originally developed as tools for probing receptors and/or as part of early phase drug discovery efforts, reports from drug abuse monitoring sites beginning in the early 2000s suggested that some of these research chemicals were being diverted for recreational use. These compounds, contained in products labeled "Spice" or "herbal incense," shared the cannabimimetic subjective effects of Δ^9-tetrahydrocannabinol (THC), the primary psychoactive constituent of the cannabis plant. However, because they differed structurally from THC and its analogs, synthetic cannabinoids were legal when they were first marketed as recreational drugs. Since then, drug enforcement agencies worldwide have struggled to develop strategies to manage the continuous influx of novel synthetic cannabinoids that have increased in structural diversity as older compounds have been banned. This chapter presents an overview of the context in which synthetic cannabinoids were discovered followed by a more in-depth look at their pharmacology.

Determination of the strength of relationships between the chemical structures of compounds and their activity is fundamental to the science of pharmacology as well as to the process of drug discovery and development. Orderly structure-activity relationships (SAR) are a cardinal sign of receptor activation or blockade, processes through which many drug treatments for disorders of the central nervous system (CNS) work. Synthetic cannabinoids were originally designed and synthesized for use in SAR studies in academic and pharmaceutical laboratories. Although computational chemistry and other technological advances in recent years have led to innovative approaches to drug discovery, assessment of SAR was considered "state-of-the-science" in investigation of receptor mechanisms in the 1980s and early 1990s when the story of non-plant-based synthetic cannabinoids began. Cannabinoid receptors (CB_1 and CB_2) also were identified and cloned during this time of intensive SAR evaluation [1–3]. Furthermore, synthesis of a radiolabeled synthetic cannabinoid, [^{3}H]CP55,940, played an integral part in discovery of this receptor system [4].

The endocannabinoid system is comprised of CB_1 and CB_2 receptors, along with their endogenous ligands [e.g., anandamide and 2-arachidonoylglycerol (2-AG)] and synthetic and metabolic enzymes for these ligands (reviewed in [5]). Both cannabinoid receptors are G-protein coupled receptors, with CB_1 receptors located widely throughout the brain and CB_2 receptors found primarily in the periphery [6–8]. THC is a partial agonist at both receptor sub-types [9, 10], but produces its cannabimimetic psychoactive effects via activation of CB_1 receptors in the CNS [11], as do synthetic cannabinoids [12, 13]. While medical and legal problems associated with the manufacture and use of synthetic cannabinoids are causes for concern, these compounds were created within a research context, were not meant for human use without further development, and contributed to significant scientific advances in the cannabinoid field. For example, these pharmacological tools aided researchers in discovery of CB_1 and CB_2 receptors [1], delineation of separate functions mediated by CB_1 and CB_2 receptors [14, 15], determination of CB_1 receptor mediation of cannabis intoxication [11], and demonstration of possible roles that the endocannabinoid system may play in physiology and pathology [5].

For many years, the focus was on synthesis of compounds that directly activated or blocked cannabinoid receptors; however, separation of psychoactive and therapeutic effects proved problematic for compounds that activated CB_1 receptors directly. In addition, the considerable homology between CB_1 and CB_2 receptors presented difficulties in designing CB_2-selective compounds that did not activate CB_1 receptors. Hence, many of the earlier compounds bind to and activate both cannabinoid receptors. More recently, the scope of cannabinoid synthesis has broadened to include compounds that inhibit endocannabinoid metabolic enzymes (e.g., inhibitors of fatty acid amide hydrolase and monoacylglycerol lipase for anandamide and 2-AG, respectively) [16–18]. These new compounds, as well as selective CB_2 receptor agonists, offer promising leads for development of therapeutics to treat the many disorders or conditions that may be related directly or indirectly to dysfunction of the endocannabinoid system, including pain, neurodegeneration, substance abuse, obesity, and psychiatric disorders [19–22].

2 Diversion and Development of an Illicit Industry

Diversion of synthetic cannabinoids from their use in research was first recognized by drug enforcement agencies across the world during the early 2000s and has increased in scope since then. The concomitant rise of the worldwide web and its public availability facilitated the spread of awareness of these chemicals [23], including greater access to patent literature and to scientific papers on synthetic methods. Further, the web provided opportunities for direct marketing to consumers and organization of forums of like-minded drug users to spread information on the latest compounds. The synergistic effects of these developments arguably led to the rapid proliferation of synthetic cannabinoid use. Currently, the primary location for bulk production of the compounds is believed to be China [24].

Once synthesized, the compounds are shipped to product manufacturers who spray the cannabinoid compound(s) on plant material (e.g., marshmallow leaf) and package it for individual sale. Even though the packets often are labeled as "not for human consumption," users use the substance via the same methods employed with cannabis, such as rolling the material into smokeable joints or placing it in pipes or other devices for smoking. Two notable issues arise as a direct result of this process. The first is that spraying often leads to uneven distribution of the chemical across the sample contained in the package, creating the possibility of "hot spots" containing enhanced concentrations of active chemicals. The second is that the combustion involved in smoking the plant material can conceivably alter the composition of the chemical it contains, resulting in exposure to different chemicals [25]. These problems have not been ameliorated by the recent switch by some users to e-liquids that contain synthetic cannabinoids. Given the low solubility of cannabinoids in the e-liquid vehicle, crystallization or precipitation may occur and use of atomizers or vaporizers marketed for nicotine still involve intense heating of the chemical.

Metabolism of the synthetic cannabinoids may also result in creation of additional chemicals that may modulate and/or extend the duration of its effects [26–28].

3 Receptor Affinity and Efficacy

Like phytocannabinoid agonists, synthetic cannabinoids exhibit structurally specific receptor recognition and can affect the activation state of the receptor in a variety of signal transduction pathways. Assessment of cannabinoid receptor recognition is typically affected by measurement of the strength with which the synthetic cannabinoid displaces a radiolabeled ligand that binds to the receptor (e.g., [^{3}H]CP55,940) and is expressed as affinity (k_{i}), with lower numbers indicative of higher affinity. The ability of synthetic cannabinoids to alter the activation state of cannabinoid receptors (i.e., efficacy) has been measured through examination of their effects on signaling pathways (e.g., GTPγS, beta arrestin, and calcium influx). In general, synthetic cannabinoids tend to have greater affinities for the CB_1 receptor than does THC, which often correlates with their greater potencies in vivo. To the extent that they have been evaluated, synthetic cannabinoids also show greater efficacy for the CB_1 receptor than the partial agonist THC [13, 29, 30]. The acute in vivo correlates of greater efficacy are not fully understood, although some research suggests that efficacy differences may have implications for development of tolerance and cross-tolerance following repeated administration [31].

THC and most of the abused synthetic cannabinoids bind to and activate both CB_1 and CB_2 receptors, with variable degrees of selectivity for one or the other receptor. Cannabimimetic psychoactivity of the compounds, and their consequent abuse liability, is most closely associated with their high CB_1 receptor affinities [11]; hence, this section maintains a primary focus on CB_1 receptor SAR, with minimal attention to CB_2 and noncannabinoid receptors.

3.1 CB_1 Receptor

Until recently, the most prevalent synthetic cannabinoids identified in spice or herbal incense products were classified into seven structural groups, as depicted in Fig. 1: naphthoylindoles (e.g., JWH-018, JWH-073, and AM-2201), naphthylmethylindoles (JWH-185), naphthoylpyrroles (JWH-030), naphthylmethylindenes (JWH-176), phenylacetylindoles (JWH-250, RCS-4), cyclohexylphenols (CP47,497), and tetrahydrocannabinols (THC, HU-210). Previous publications have reviewed the in vitro and in vivo pharmacology of indole- and pyrrole-derived cannabinoids [32, 33]. Systematic legal restrictions placed on these cannabinoid classes have decreased their prevalence in recent samples and have resulted in synthesis of cannabinoids with increased structural diversity. In the

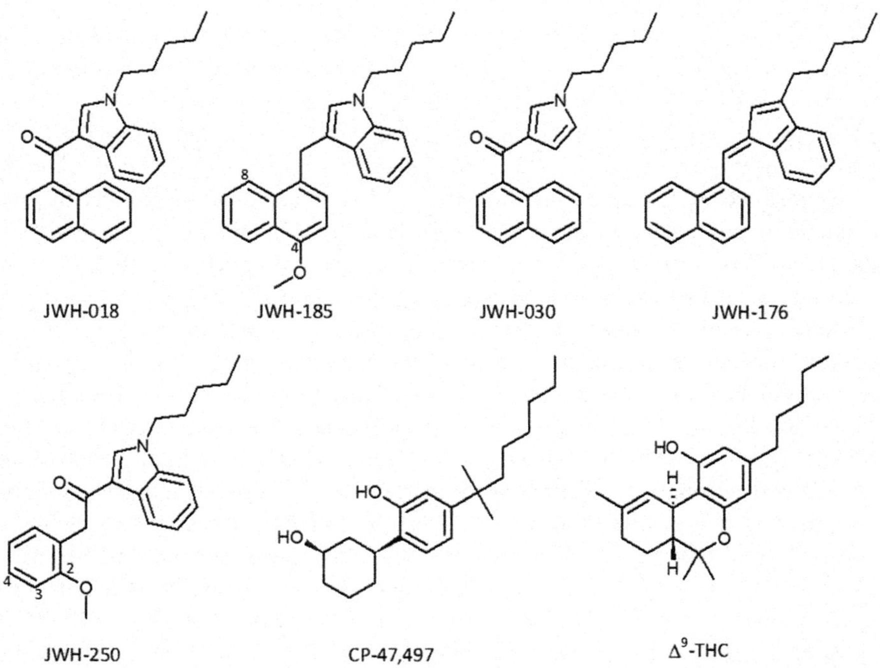

Fig. 1 Chemical structures of representative compounds from major structural groups of synthetic cannabinoids: naphthoylindoles (JWH-018), naphthylmethylindoles (JWH-185), naphthoylpyrroles (JWH-030), naphthylmethylindenes (JWH-176), phenylacetylindoles (JWH-250), cyclohexylphenols (CP-47,497), and tetrahydrocannabinols (Δ^9-THC)

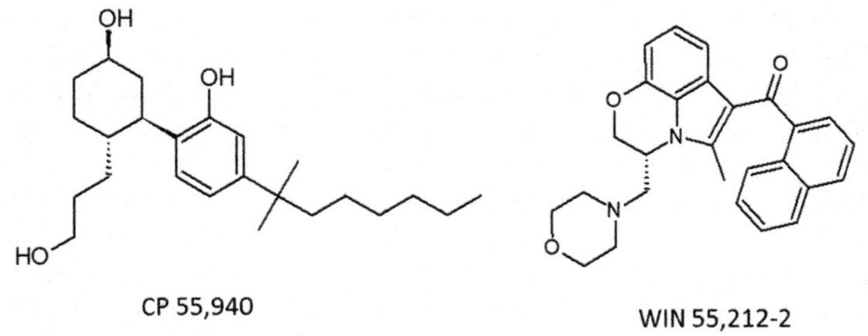

Fig. 2 Chemical structures of CP55,940 and WIN55,212-2

present chapter, a brief overview of the pharmacology of these older cannabinoids is provided, but the primary concentration is on review of the pharmacology of synthetic cannabinoids that have appeared on the market more recently.

JWH-018 (1-pentyl-3-1-naphthoylindole), a naphthoylindole, was the first cannabinoid compound identified in herbal incense products and, as such, has been referenced as the prototypic synthetic cannabinoid [33]. It is structurally similar to the aminoalkylindole WIN55,212-2 (Fig. 2), with the exception that the latter's oxazine and morpholino substituents are replaced with an *n*-pentyl group. SAR studies showed that affinity and potency varied systematically with the length of *n*-alkyl substituent, with optimal activity from *n*-butyl to *n*-hexyl and absence or reduction of receptor binding at shorter or longer carbon chains [34–36]. Replacement of JWH-018's *n*-pentyl group with *n*-fluoropentyl resulted in AM-2201, a potent psychoactive cannabinoid that appeared in confiscated samples as JWH-018 was fading in popularity [37, 38]. 2-Methylation of the indole in the alkylindole series resulted in compounds with decreased CB_1 receptor affinities and in vivo cannabimimetic potencies and a slight shift in optimal chain length. Conversion of naphthoylindoles to naphthoylpyrroles decreased CB_1 receptor affinities and reduced potencies to an even greater extent than 2-methylation [36].

While variations in the structures of early compounds focused primarily on manipulation of the alkyl group or conversion of the indole core to a pyrrole, structural innovations involving the naphthoyl group soon began to appear (e.g., JWH-185, Fig. 1). These changes included additions to and substitutions for this functional group. Additions to the naphthoyl group concentrated on alteration of the steric and electronic effects through addition of two types of substituents: electron withdrawing halogen substituents and electron donating methoxy [39, 40]. C-4 substitution of either type of substituent resulted in compounds with the best CB_1 receptor affinities and in vivo activity, as compared to substitution at other positions. Unlike substituents at other positions on the naphthoyl, the rotation of C-4 substituents is less hindered and thereby, less likely to interfere with optimal aromatic stacking, which has been shown to be important for cannabinoid receptor recognition [41, 42]. Together, these results suggest that steric effects play a stronger role in determining the nature of CB_1 receptor affinity and in vivo activity than do electronic effects.

Steric influences and aromatic stacking are also important determinants of the effects of substitutions for the naphthoyl group. For example, SAR investigation of a series of 1-pentyl-3-phenylacetylindoles (e.g., JWH-250, Fig. 1) showed that the decrease in the number of aromatic rings on the non-indole side of the carbonyl from two (naphthoyl) to one (phenylacetyl) resulted in reduction of CB_1 receptor affinities and in vivo potencies [43]. As with addition of halogen and methoxy substituents to the naphthoyl group, the position of substituents on the phenyl ring (i.e., 2-, 3- or 4-position) also affected CB_1 receptor affinities and potencies, with 2- and 3-phenylacetyl substituents showing enhanced affinities compared to 4-substituents. Hence, steric influences are also important in binding for the 1-pentyl-3-phenylacetylindole series of synthetic cannabinoids.

Tetramethylcyclopropyl ketone indoles represent another category of synthetic cannabinoid, in which the core change is substitution of a tetramethylcyclopropyl group for the naphthoyl substituent of the parent 3-naphthoylindole. Specific compounds that have been sold over the internet include UR-144, XLR-11, and A834735 (Fig. 3) [44]. These compounds resemble those synthesized by Abbott Laboratories in their effort to develop CB_2-selective agonists for pain and inflammation [45, 46]. While many of the Abbott compounds showed higher affinity for the CB_2 receptor, a number of the compounds also possessed significant affinity for the CB_1 receptor, which undoubtedly serves as the basis for their inclusion in "herbal incense" products. All three compounds have high affinities for both CB_1 and CB_2 receptors, with XLR-11 and UR-144 having similar affinities for the CB_1 receptor ($k_{\mathrm{i}} = 24$ and 29 nM, respectively) and A834735 having greater affinity ($k_{\mathrm{i}} = 4.6$ nM) than the other two compounds [13, 47]. Unlike THC, both XLR-11

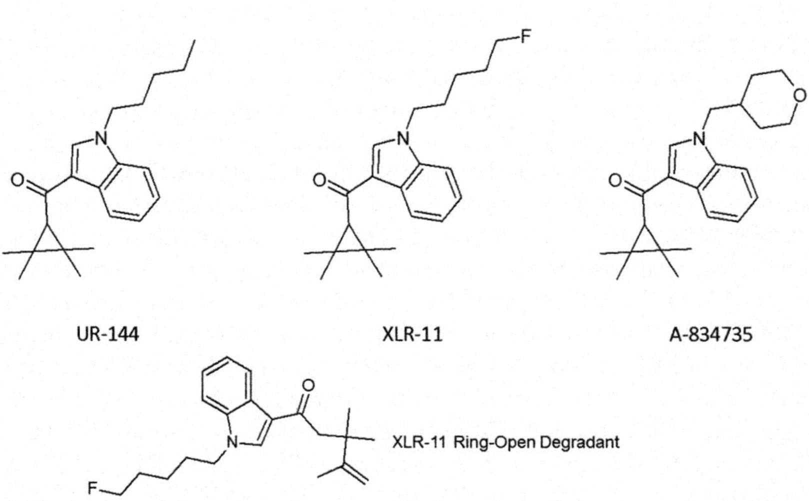

Fig. 3 Chemical structures of tetramethylcyclopropyl ketones: UR-144, XLR-11, and A-834735. Also shown is the chemical structure of the open-ring degradant of XLR-11

and UR-144 are fully efficacious CB_1 receptor agonists, as measured by GTPγS binding [13]. Further, recent data show that repeated exposure of the parent compounds to high heat (as would occur during smoking or vaping) resulted in thermolysis of the tetramethylcyclopropyl group of each compound and formation of open-ring degradants with substantially increased CB_1 receptor affinities and efficacies (Fig. 3; [48]). In human users of UR-144, analytical findings showed that the majority of urine samples contained metabolites of the pyrolysis product and only minimal amounts of the parent compound [49]. The discrepancies between chemicals contained in the product and chemicals created when the product is combusted highlight the importance of thorough analysis for accurate prediction of the effects of exposure, an idea that has been echoed in other studies [25, 44].

The tetramethylcyclopropyl substituent is only one of several novel substitutions for the naphthoyl group in products containing synthetic cannabinoids. Using a fluorometric imaging plate reader (FLIPR) assay that measures cross-membrane ion flux, Banister et al. [50–52] reported the potencies for activation of CB_1 and CB_2 receptors of several series of synthetic cannabinoids. Unfortunately, these studies did not assess binding affinities for these compounds. Since CB_1 receptor binding affinity is the single most frequent measure available across synthetic cannabinoid SAR studies [53, 54], direct integration of their results into the body of previous research is complicated. Nevertheless, several of the findings from the Australian group are of note. First, previously untested series of cannabinoids, including adamantane-derived indoles [e.g., adamantan-1-yl(1-pentyl-1H-indol-3-yl) methanone (AB-001) and N-(adamtan-1-yl)-1-pentyl-1H-indole-3-carboxamide (SDB-001); Fig. 4] and indole-3-carboxamides and -carboxylates (e.g., AB-PICA, AB-FUBICA), were shown to activate the CB_1 receptor (as measured by FLIPR) [50, 52]. Second, indole-derived compounds with a fluorine at the terminal end of the *n*-alkyl substituent were reported to exhibit more potent activation of the receptor in the FLIPR assay than those without this substitution [51]. Finally, potent in vitro activity was observed for compounds in which an indazole was substituted for the indole and in which various carboxamide and carboxylate substituents were substituted for the naphthoyl group [52]. In most cases, potencies for compounds in the indazole series exceeded those of the comparable compounds in the indole series. These results support previous work which has suggested that CB_1 receptor binding site(s) can tolerate a large degree of structural variability for agonists [55].

AB-PINACA, one of the indazoles evaluated in the Banister et al. [52] study, was assessed in additional assays in another recent paper [30]. A second indazole cannabinoid (AB-CHMINACA) and a compound with a new benzimidazole structure (FUBIMINA) were also tested (Fig. 4). Of the three compounds, FUBIMINA had the lowest CB_1 receptor affinity ($k_i = 296$ nM). While these results suggest that FUBIMINA would not be likely to show cannabinoid psychoactivity in vivo except at high doses, it has appeared in some samples confiscated in Japan [56]. Presence of the two indazole compounds in samples has also been reported [57, 58], which is not surprising given their high CB_1 receptor affinities. Of the two indazoles, AB-CHMINACA had the highest CB_1 receptor affinity ($k_i = 0.78$ nM), an affinity that was comparable to that of CP55,940 ($k_i = 0.59$ nM). Although AB-PINACA had somewhat less affinity ($k_i = 2.87$ nM), its affinity still exceeded that of the

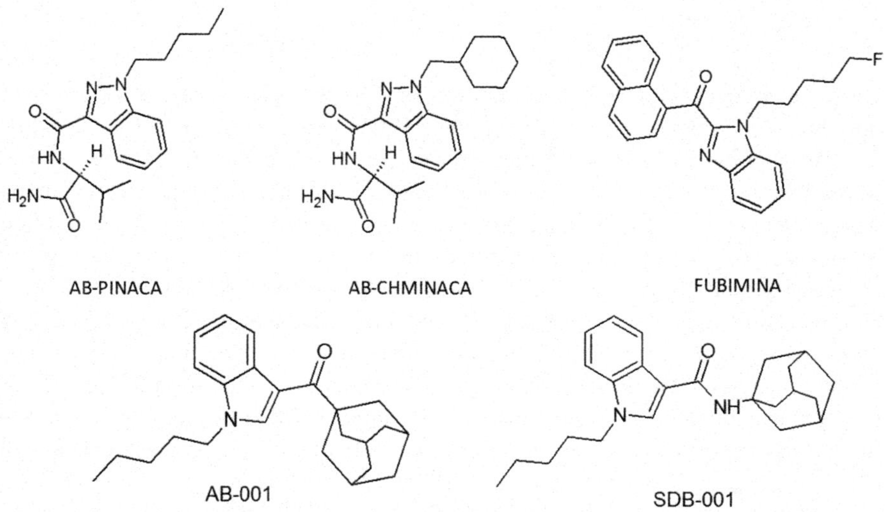

Fig. 4 Chemical structures of indazole-derived synthetic cannabinoids (AB-PINACA and AB-CHMINACA), a benzimidazole cannabinoid (FUBIMINA), and adamantane-derived cannabinoids (AB-001 and SDB-001)

prototypic phytocannabinoid THC ($k_i = 41$ nM; [59]). The sole structural difference between these two indazole cannabinoids is the substitution of a cyclohexylmethyl moiety in AB-CHMINACA for the *n*-pentyl of AB-PINACA, suggesting that receptor recognition is facilitated by the conformational restraint provided by the ring system. Results of GTPγS binding showed that both indazole and benzimidazole compounds were full agonists at CB_1 receptors [30]. Interestingly, efficacies of AB-CHMINACA and AB-PINACA for stimulation of the receptor were greater than those produced by other full agonists such as CP55,940 and WIN55,212-2 [30, 60]. In contrast, efficacies of FUBIMINA and CP55940 were comparable.

As illustrated by structural modifications present in these indazole and benzimidazole cannabinoids, structures of the most popular synthetic cannabinoids today often contain substitutions for more than one substituent. For example, compounds within the PINACA series show each of the following alterations: an indazole (vs. indole) core, lack of a cyclic structural group at the position of the naphthoyl group of the naphthoylindoles, and may contain a terminal substitution on the *n*-alkyl group [52]. This increased variety of structural modifications compared to the prototype JWH-018 is arguably the consequence of increases in the number of banned substances, highlighting the continued evolution and sophistication of synthetic cannabinoid manufacturers in response to regulation.

3.2 *CB_2 Receptor*

Once researchers realized that separation of CB_1 and CB_2 receptor affinity was possible, CB_2 receptor selectivity became a viable target for pharmaceutical industry investigation [61]. While a thorough review of SAR for the CB_2 receptor is beyond the scope of this chapter, several points are worth mentioning. As with most drug development efforts, determination of SAR for the target of interest often results in synthesis of many compounds that are off-target – in this case, many compounds that do not have high CB_2 selectivity. Many of these "off-target" compounds have high CB_1 receptor affinity, a property that has been exploited by manufacturers of synthetic cannabinoids contained in herbal incense products. Little information exists on the practical consequences of activation of CB_2 receptors for users of synthetic cannabinoids. For example, CB_2 receptor activation may be related to peripheral effects of synthetic cannabinoids, which have not been well characterized. On the other hand, the CB_2 receptor activating effects of synthetic cannabinoids may be enhanced in the CNS in users who have certain brain disorders or injuries (e.g., neuroinflammation) due to the proposed induction of CB_2 receptors by brain microglia under these conditions [62]. Ironically, the effects of CB_2 receptor activation, the property for which many of the currently abused compounds were originally synthesized, have received minimal research attention compared to the amount of attention that has been given to the effects of their activation of CB_1 receptors.

3.3 *Noncannabinoid Receptors*

Published research on synthetic cannabinoids has concentrated almost exclusively on examination of their in vitro and/or in vivo cannabinoid effects. While an occasional paper may mention lack of affinity of specific compounds for major receptor classes (e.g., [63]), for the most part, published literature on the noncannabinoid effects of synthetic cannabinoids is virtually nonexistent.

4 In Vivo Pharmacology

Although hundreds of synthetic cannabinoids have been evaluated for their CB_1 and CB_2 receptor affinities [29, 35, 39, 41, 54, 64, 65], in vivo pharmacology and toxicology studies of these compounds were rare until they were discovered in products confiscated from human users. Early studies with a limited number of compounds showed that potencies in a battery of four tests in mice (the "tetrad") were highly correlated with CB_1 receptor binding affinities [59, 66]. Psychoactive cannabinoids of various structural classes, including indole-derived cannabinoids, produce a characteristic profile of effects in the tetrad, including suppression of locomotor activity, antinociception, hypothermia, and catalepsy [66]. They also possess THC-like discriminative stimulus effects in rodents and nonhuman primates [67, 68]. This section focuses on a review of recent in vivo studies with synthetic cannabinoids, as results of earlier in vivo studies were reviewed previously [32].

As reported in a previous review [32], XLR-11 and UR-144 produced the full complement of tetrad effects and substituted for THC in drug discrimination in mice, in each case with potencies several-fold greater than THC [13]. The tetrad effects of these two compounds were attenuated by co-administration of the prototypic CB_1 receptor antagonist rimonabant, suggesting CB_1 receptor mediation of these effects. CB_1 receptor affinities were similar for these two compounds and they possessed similar in vivo potencies. In contrast, a later investigation reported that XLR-11, a fluorine-containing analog of UR-144, showed substantially enhanced potency for activation of CB_1 receptors (compared to the non-fluorinated UR-144) and decreased body temperature at a lower dose [51]. Several differences across the studies may account for their discrepant results. First, activation of the CB_1 receptor was measured in two different assays: membrane ion flux [51] or GTPγS binding [13]. The degree to which these two preparations assess the same phenomenon is uncertain, particularly given recent findings that ligands for G-protein coupled receptors, including cannabinoid receptors [69, 70], may exhibit signaling bias [71, 72]. In contrast with functional potency in assays that measure the in vitro activation of cannabinoid receptors, receptor binding affinity (as evaluated via displacement of a radiolabeled agonist) is highly correlated with the in vivo potency of cannabinoids in the tetrad and drug discrimination procedures [36, 59, 66]. Consistent with these findings, the similar binding affinities of XLR-11 and UR-144 were predictive of their similar in vivo potencies in cannabinoid-selective procedures [13]. In vivo potencies were also calculated in a different manner across the two studies, partly as a result of different procedures. Whereas in vivo potencies in the tetrad and in THC discrimination were calculated through a least squares linear regression procedure [13], potency for producing hypothermia across time (as measured via implanted telemetric devices) was defined as the lowest dose which significantly decreased body temperature [51]. The differences between these two studies highlight the difficulty of SAR research across labs in the absence of a single common measure.

More recently, open-ring degradants of the tetramethylcyclopropyl ketones (XLR-11, UR-144, and A834735), but not a degradant of PB-22 (1-pentyl-1H-indole-3-carboxylic acid 8-quinolinyl ester), were shown to produce tetrad effects in mice and to substitute for JWH-018 in mice trained to discriminate JWH-018 from vehicle [48]. These data are consistent with anecdotal reports showing that tetramethylcyclopropyl ketone cannabinoids within this class are more potent than would be expected given affinities of the parent compounds [73].

Indazole cannabinoids, AB-CHMINACA and AB-PINACA, and the benzimidazole FUBIMINA have also been tested in vivo in the tetrad and THC discrimination procedures [30]. While AB-CHMINACA and AB-PINACA produced the full profile of cannabinoid effects in the tetrad battery in mice, FUBIMINA was inactive except at a relatively high (56 mg/kg) intravenous dose. The effects of the three compounds also differed in THC discrimination in mice. AB-CHMINACA fully substituted for THC across a dose range that did not affect overall responding. These results are consistent with the compound's high CB_1 receptor affinity and resemble those obtained with other psychoactive cannabinoid agonists from a variety of structural classes [67, 68]. Consistent with its relatively low CB_1 receptor affinity, FUBIMINA only partially substituted for THC in mouse drug discrimination, which is also consistent with its modest CB_1 receptor affinity. The most puzzling results emerged with AB-PINACA. Although AB-PINACA produced full dose-dependent substitution for THC, this substitution was achieved only at a dose that was accompanied by substantial decreases in response rate. Previously, response rate decreases induced by other synthetic cannabinoids were observed only with doses that were suprathreshold for full substitution [12, 13]. This lack of separation between doses that are THC-like and those that substantially suppress responding suggest that AB-PINACA is a potent psychoactive CB_1 receptor agonist, but they also suggest that the doses that induce intoxication may be very close to doses associated with behavioral toxicity.

The brevity of this section on the in vivo pharmacology of synthetic cannabinoids reflects the sporadic nature of research in this area. Only a small number of the hundreds of compounds that have appeared on the illicit synthetic cannabinoid market has been tested in animals. For many of these cannabinoids, the first test subject has been human. A review of research on the toxic effects of synthetic cannabinoids follows.

5 Toxicology

Preclinical toxicological assessment of synthetic cannabinoids has been sparse, with forensic toxicology comprising the bulk of the research. Because much of the forensic research is related to identification and detection of synthetic cannabinoid metabolites, the reader is referred to the chapter on metabolism of synthetic cannabinoids for a review of relevant literature. Non-laboratory research in this area has consisted primarily of anecdotal and clinical reports and epidemiological studies, which have been reviewed previously [74, 75]. One of the primary problems with this research is the difficulty in identification of specific compounds that are associated with the various reported effects.

In general, the pharmacological effects of synthetic cannabinoids in humans resemble those of THC and may include subjective intoxication, tachycardia (fast pulse rate), and conjunctival injection ("red eyes") [76]. The degree of acute intoxication produced by synthetic cannabinoids may be more intense or milder than that produced by marijuana [77]; however, experienced marijuana users tended to prefer natural cannabis over synthetic cannabinoids [78]. After repeated use, dependence may occur [79, 80]. Differences in the clinical effects of THC and synthetic cannabinoids have also been reported. For example, users of synthetic cannabinoids may show increased incidence (compared to marijuana) of anxiety or agitation [81], nausea and vomiting [82], hypertension [83], seizures [83, 84], and psychiatric disturbance (e.g., suicidality, exacerbation of pre-existing psychosis, and hallucinations) [79, 83, 85, 86]. Acute kidney injury may be associated with the use of XLR-11 [87, 88] and death has been known to occur as a result of synthetic cannabinoid use [89]. Consequently, synthetic cannabinoids account for a greater proportion of cannabinoid-related emergency room visits than does marijuana [90, 91].

6 Summary

In the 1980s, cannabinoid researchers developed potent synthetic cannabinoids that were used to identify the molecular and biochemical foundations of the endogenous cannabinoid system and facilitate the development of experimental therapeutics. The transition of novel synthetic cannabinoids from research chemicals to recreational use occurred in the early 2000s, increased rapidly to a multimillion-dollar designer drug industry, and continued to evolve as a public health concern despite ongoing regulatory efforts. The recreational use of synthetic cannabinoids persists in an expanding variety of chemical forms and formulations, particularly in uninformed youth, "psychonauts," and individuals attempting to avoid drug testing (e.g., military, ex-convicts, and individuals involved in public transport). Even with the current trend towards decriminalization and legalization of cannabis use, the long elimination half-life of phytocannabinoids and their metabolites constrains its recreational use in certain populations attempting to avoid detection, such that synthetic cannabinoid use continues to be of significant interest. Very little is known about the chemical purity or stability of these new chemical entities, the exposures that occur during their use, or their in vitro or in vivo pharmacological and toxicological effects. As a result, there are frequent reports of overdose and untoward effects being attributed to their use as intoxicants. The current state of affairs creates a paradoxical situation, where the potential for abuse and harm from synthetic cannabinoids must be recognized and dealt with effectively, while simultaneously enabling the development and testing of novel synthetic cannabinoids in carefully controlled preclinical and clinical studies to further elucidate the role of the endogenous cannabinoid system in health and disease states.

Acknowledgements Preparation of this review was supported in part by National Institute on Drug Abuse grant DA-003672; NIDA had no further role in the writing of the review or in the decision to submit the paper for publication.

References

1. Devane WA, Dysarz FA, Johnson MR, Melvin LS, Howlett AC (1988) Determination and characterization of a cannabinoid receptor in rat brain. Mol Pharmacol 34:605–613
2. Matsuda LA, Lolait SJ, Brownstein MJ, Young AC, Bonner TI (1990) Structure of a cannabinoid receptor and functional expression of the cloned cDNA. Nature 346:561–564
3. Munro S, Thomas KL, Abu-Shaar M (1993) Molecular characterization of a peripheral receptor for cannabinoids. Nature 365:61–64
4. Howlett AC, Bidaut-Russell M, Devane WA, Melvin LS, Johnson MR, Herkenham M (1990) The cannabinoid receptor: biochemical, anatomical and behavioral characterization. Trends Neurosci 13:420–423
5. Pertwee RG (2008) Ligands that target cannabinoid receptors in the brain: from THC to anandamide and beyond. Addict Biol 13:147–159
6. Galiegue S, Mary S, Marchand J, Dussossoy D, Carriere D, Carayon P et al (1995) Expression of central and peripheral cannabinoid receptors in human immune tissues and leukocyte subpopulations. Eur J Biochem 232:54–61
7. Herkenham M, Lynn AB, Little MD, Johnson MR, Melvin LS, DeCosta BR et al (1990) Cannabinoid receptor localization in the brain. Proc Natl Acad Sci U S A 87:1932–1936
8. Thomas BF, Wei X, Martin BR (1992) Characterization and autoradiographic localization of the cannabinoid binding site in rat brain using [3H]11-OH-delta 9-THC-DMH. J Pharmacol Exp Ther 263:1383–1390
9. Breivogel CS, Childers SR (2000) Cannabinoid agonist signal transduction in rat brain: comparison of cannabinoid agonists in receptor binding, G-protein activation, and adenylyl cyclase inhibition. J Pharmacol Exp Ther 295:328–336
10. Pertwee RG (2008) The diverse CB1 and CB2 receptor pharmacology of three plant cannabinoids: delta9-tetrahydrocannabinol, cannabidiol and delta9-tetrahydrocannabivarin. Br J Pharmacol 153:199–215
11. Wiley JL, Lowe JA, Balster RL, Martin BR (1995) Antagonism of the discriminative stimulus effects of delta 9-tetrahydrocannabinol in rats and rhesus monkeys. J Pharmacol Exp Ther 275:1–6
12. Wiley JL, Barrett RL, Lowe J, Balster RL, Martin BR (1995) Discriminative stimulus effects of CP 55,940 and structurally dissimilar cannabinoids in rats. Neuropharmacology 34:669–676
13. Wiley JL, Marusich JA, Lefever TW, Grabenauer M, Moore KN, Thomas BF (2013) Cannabinoids in disguise: Delta-9-tetrahydrocannabinol-like effects of tetramethylcyclopropyl ketone indoles. Neuropharmacology 75:145–154
14. Pertwee RG (2006) The pharmacology of cannabinoid receptors and their ligands: an overview. Int J Obes (Lond) 30 Suppl 1:S13–S18
15. Poso A, Huffman JW (2008) Targeting the cannabinoid CB2 receptor: modelling and structural determinants of CB2 selective ligands. Br J Pharmacol 153:335–346
16. Fowler CJ (2015) The potential of inhibitors of endocannabinoid metabolism for drug development: a critical review. Handb Exp Pharmacol 231:95–128
17. Kohnz RA, Nomura DK (2014) Chemical approaches to therapeutically target the metabolism and signaling of the endocannabinoid 2-AG and eicosanoids. Chem Soc Rev 43:6859–6869
18. McKinney MK, Cravatt BF (2005) Structure and function of fatty acid amide hydrolase. Annu Rev Biochem 74:411–432
19. de Kloet AD, Woods SC (2009) Minireview: endocannabinoids and their receptors as targets for obesity therapy. Endocrinology 150:2531–2536
20. Guindon J, Hohmann AG (2009) The endocannabinoid system and pain. CNS Neurol Disord Drug Targets 8:403–421
21. Hillard CJ, Weinlander KM, Stuhr KL (2012) Contributions of endocannabinoid signaling to psychiatric disorders in humans: genetic and biochemical evidence. Neuroscience 204:207–229

22. Serrano A, Parsons LH (2011) Endocannabinoid influence in drug reinforcement, dependence and addiction-related behaviors. Pharmacol Ther 132:215–241
23. Wax PM (2002) Just a click away: recreational drug Web sites on the Internet. Pediatrics 109, e96
24. Holley P, Wan W (2015) Deadly Chinese drugs are flooding the U.S., and police can't stop them. *Washington Post*: Washington, DC
25. Bell S, Nida C (2015) Pyrolysis of drugs of abuse: a comprehensive review. Drug Test Anal 7:445–456
26. Brents LK, Reichard EE, Zimmerman SM, Moran JH, Fantegrossi WE, Prather PL (2011) Phase I hydroxylated metabolites of the K2 synthetic cannabinoid JWH-018 retain in vitro and in vivo cannabinoid 1 receptor affinity and activity. PLoS One 6, e21917
27. Brents LK, Gallus-Zawada A, Radominska-Pandya A, Vasiljevik T, Prisinzano TE, Fantegrossi WE et al (2012) Monohydroxylated metabolites of the K2 synthetic cannabinoid JWH-073 retain intermediate to high cannabinoid 1 receptor (CB1R) affinity and exhibit neutral antagonist to partial agonist activity. Biochem Pharmacol 83:952–961
28. Seely KA, Brents LK, Radominska-Pandya A, Endres GW, Keyes GS, Moran JH et al (2012) A major glucuronidated metabolite of JWH-018 is a neutral antagonist at CB1 receptors. Chem Res Toxicol 25:825–827
29. Huffman JW, Szklennik PV, Almond A, Bushell K, Selley DE, He H et al (2005) 1-Pentyl-3-phenylacetylindoles, a new class of cannabimimetic indoles. Bioorg Med Chem Lett 15:4110–4113
30. Wiley JL, Marusich JA, Lefever TW, Antonazzo KR, Wallgren MT, Cortes RA et al (2015) AB-CHMINACA, AB-PINACA, and FUBIMINA: affinity and potency of novel synthetic cannabinoids in producing delta9-tetrahydrocannabinol-like effects in mice. J Pharmacol Exp Ther 354:328–339
31. Hruba L, Ginsburg B, McMahon LR (2012) Apparent inverse relationship between cannabinoid agonist efficacy and tolerance/cross-tolerance produced by {Delta}9-tetrahydrocannabinol treatment in rhesus monkeys. J Pharmacol Exp Ther 342:843–849
32. Wiley JL, Marusich JA, Huffman JW (2014) Moving around the molecule: relationship between chemical structure and in vivo activity of synthetic cannabinoids. Life Sci 97:55–63
33. Wiley JL, Marusich JA, Huffman JW, Balster RL, Thomas BF (2011) Hijacking of basic research: the case of synthetic cannabinoids. RTI Press, Research Triangle Park, NC
34. Huffman JW, Dai D, Martin BR, Compton DR (1994) Design, synthesis and pharmacology of cannabimimetic indoles. Bioorg Med Chem Lett 4:563–566
35. Lainton JAH, Huffman JW, Martin BR, Compton DR (1995) 1-Alkyl-3-(1-naphthoyl)pyrroles: a new cannabinoid class. Tetrahedron Lett 36:1401–1404
36. Wiley JL, Compton DR, Dai D, Lainton JA, Phillips M, Huffman JW et al (1998) Structure-activity relationships of indole- and pyrrole-derived cannabinoids. J Pharmacol Exp Ther 285:995–1004
37. Denooz R, Vanheugen JC, Frederich M, de Tullio P, Charlier C (2013) Identification and structural elucidation of four cannabimimetic compounds (RCS-4, AM-2201, JWH-203 and JWH-210) in seized products. J Anal Toxicol 37:56–63
38. Logan BK, Reinhold LE, Xu A, Diamond FX (2012) Identification of synthetic cannabinoids in herbal incense blends in the United States. J Forensic Sci 57:1168–1180
39. Aung MM, Griffin G, Huffman JW, Wu M, Keel C, Yang B et al (2000) Influence of the N-1 alkyl chain length of cannabimimetic indoles upon CB(1) and CB(2) receptor binding. Drug Alcohol Depend 60:133–140
40. Wiley JL, Marusich JA, Martin BR, Huffman JW (2012) 1-Pentyl-3-phenylacetylindoles and JWH-018 share in vivo cannabinoid profiles in mice. Drug Alcohol Depend 123:148–153
41. Huffman JW, Mabon R, Wu MJ, Lu J, Hart R, Hurst DP et al (2003) 3-Indolyl-1-naphthylmethanes: new cannabimimetic indoles provide evidence for aromatic stacking interactions with the CB(1) cannabinoid receptor. Bioorg Med Chem 11:539–549

42. Reggio PH, Basu-Dutt S, Barnett-Norris J, Castro MT, Hurst DP, Seltzman HH et al (1998) The bioactive conformation of aminoalkylindoles at the cannabinoid CB1 and CB2 receptors: insights gained from (E)- and (Z)-naphthylidene indenes. J Med Chem 41:5177–5187
43. Wiley JL, Smith VJ, Chen J, Martin BR, Huffman JW (2012) Synthesis and pharmacology of 1-alkyl-3-(1-naphthoyl)indoles: steric and electronic effects of 4- and 8-halogenated naphthoyl substituents. Bioorg Med Chem 20:2067–2081
44. Kavanagh P, Grigoryev A, Savchuk S, Mikhura I, Formanovsky A (2013) UR-144 in products sold via the Internet: identification of related compounds and characterization of pyrolysis products. Drug Test Anal 5:683–692
45. Frost JM, Dart MJ, Tietje KR, Garrison TR, Grayson GK, Daza AV et al (2010) Indol-3-ylcycloalkyl ketones: effects of N1 substituted indole side chain variations on CB(2) cannabinoid receptor activity. J Med Chem 53:295–315
46. Frost JM, Dart MJ, Tietje KR, Garrison TR, Grayson GK, Daza AV et al (2008) Indol-3-yl-tetramethylcyclopropyl ketones: effects of indole ring substitution on CB2 cannabinoid receptor activity. J Med Chem 51:1904–1912
47. Chin CL, Tovcimak AE, Hradil VP, Seifert TR, Hollingsworth PR, Chandran P et al (2008) Differential effects of cannabinoid receptor agonists on regional brain activity using pharmacological MRI. Br J Pharmacol 153:367–379
48. Thomas BF, Endres GW, Wiley JL, Pollard GT, Decker AM, Gay EA, Patel PR, Kovach AL, Grabenauer M. (2015). Chemical exposures and risks associated with vaporization and inhalation of synthetic cannabinoids. Paper presented at the annual meeting of the International Cannabinoid Research Society, Wolfville, Nova Scotia, Canada
49. Grigoryev A, Kavanagh P, Melnik A, Savchuk S, Simonov A (2013) Gas and liquid chromatography-mass spectrometry detection of the urinary metabolites of UR-144 and its major pyrolysis product. J Anal Toxicol 37:265–276
50. Banister SD, Wilkinson SM, Longworth M, Stuart J, Apetz N, English K et al (2013) The synthesis and pharmacological evaluation of adamantane-derived indoles: cannabimimetic drugs of abuse. ACS Chem Neurosci 4:1081–1092
51. Banister SD, Stuart J, Kevin RC, Edington A, Longworth M, Wilkinson SM et al (2015) Effects of bioisosteric fluorine in synthetic cannabinoid designer drugs JWH-018, AM-2201, UR-144, XLR-11, PB-22, 5F-PB-22, APICA, and STS-135. ACS Chem Neurosci 6:1445–1458
52. Banister SD, Moir M, Stuart J, Kevin RC, Wood KE, Longworth M et al (2015) Pharmacology of indole and indazole synthetic cannabinoid designer drugs AB-FUBINACA, ADB-FUBINACA, AB-PINACA, ADB-PINACA, 5F-AB-PINACA, 5F-ADB-PINACA, ADBICA, and 5F-ADBICA. ACS Chem Neurosci 6:1546–1559
53. Huffman JW, Padgett LW (2005) Recent developments in the medicinal chemistry of cannabimimetic indoles, pyrroles and indenes. Curr Med Chem 12:1395–1411
54. Manera C, Tuccinardi T, Martinelli A (2008) Indoles and related compounds as cannabinoid ligands. Mini Rev Med Chem 8:370–387
55. Reggio PH (2003) Pharmacophores for ligand recognition and activation/inactivation of the cannabinoid receptors. Curr Pharm Des 9:1607–1633
56. Uchiyama N, Shimokawa Y, Matsuda S, Kawamura M, Kikura-Hanajiri R, Goda Y (2014) Two new synthetic cannabinoids, AM-2201 benzimidazole analog (FUBIMINA) and (4-methylpiperazin-1-yl)(1-pentyl-1H-indol-3-yl)methanone (MEPIRAPIM), and three phenethylamine derivatives, 25H-NBOMe 3,4,5-trimethoxybenzyl analog, 25B-NBOMe, and 2C-N-NBOMe, identified in illegal products. Forensic Toxicol 32:105–115
57. Drug Enforcement Agency (2015) Schedules of controlled substances: temporary placement of three synthetic cannabinoids into schedule I. Final order. Fed Regist 80:5042–5047
58. Peterson BL, Couper FJ (2015) Concentrations of AB-CHMINACA and AB-PINACA and driving behavior in suspected impaired driving cases. J Anal Toxicol 39:642–647

59. Compton DR, Rice KC, De Costa BR, Razdan RK, Melvin LS, Johnson MR et al (1993) Cannabinoid structure-activity relationships: correlation of receptor binding and in vivo activities. J Pharmacol Exp Ther 265:218–226
60. Griffin G, Atkinson PJ, Showalter VM, Martin BR, Abood ME (1998) Evaluation of cannabinoid receptor agonists and antagonists using the guanosine-5′-O-(3-[^{35}S]thio)-triphosphate binding assay in rat cerebellar membranes. J Pharmacol Exp Ther 285:553–560
61. Huffman JW (2000) The search for selective ligands for the CB2 receptor. Curr Pharm Des 6:1323–1337
62. Stella N (2010) Cannabinoid and cannabinoid-like receptors in microglia, astrocytes, and astrocytomas. Glia 58:1017–1030
63. Yao BB, Hsieh G, Daza AV, Fan Y, Grayson GK, Garrison TR et al (2009) Characterization of a cannabinoid CB2 receptor-selective agonist, A-836339 [2,2,3,3-tetramethylcyclopropanecarboxylic acid [3-(2-methoxy-ethyl)-4,5-dimethyl-3H-thiazol-(2Z)-ylidene]-amide], using in vitro pharmacological assays, in vivo pain models, and pharmacological magnetic resonance imaging. J Pharmacol Exp Ther 328:141–151
64. Huffman JW, Padgett LW, Isherwood ML, Wiley JL, Martin BR (2006) 1-Alkyl-2-aryl-4-(1-naphthoyl)pyrroles: new high affinity ligands for the cannabinoid CB(1) and CB(2) receptors. Bioorg Med Chem Lett 16:5432–5435
65. Huffman JW, Zengin G, Wu MJ, Lu J, Hynd G, Bushell K et al (2005) Structure-activity relationships for 1-alkyl-3-(1-naphthoyl)indoles at the cannabinoid CB(1) and CB(2) receptors: steric and electronic effects of naphthoyl substituents. New highly selective CB(2) receptor agonists. Bioorg Med Chem 13:89–112
66. Martin BR, Compton DR, Thomas BF, Prescott WR, Little PJ, Razdan RK et al (1991) Behavioral, biochemical, and molecular modeling evaluations of cannabinoid analogs. Pharmacol Biochem Behav 40:471–478
67. Jarbe TU, Gifford RS (2014) "Herbal incense": designer drug blends as cannabimimetics and their assessment by drug discrimination and other in vivo bioassays. Life Sci 97:64–71
68. Wiley JL (1999) Cannabis: discrimination of "internal bliss"? Pharmacol Biochem Behav 64:257–260
69. Atwood BK, Wager-Miller J, Haskins C, Straiker A, Mackie K (2012) Functional selectivity in CB(2) cannabinoid receptor signaling and regulation: implications for the therapeutic potential of CB(2) ligands. Mol Pharmacol 81:250–263
70. Mackie K (2008) Cannabinoid receptors: where they are and what they do. J Neuroendocrinol 20(Suppl 1):10–14
71. Kenakin T (2014) What is pharmacological 'affinity'? Relevance to biased agonism and antagonism. Trends Pharmacol Sci 35:434–441
72. Kenakin T, Christopoulos A (2013) Signalling bias in new drug discovery: detection, quantification and therapeutic impact. Nat Rev Drug Discov 12:205–216
73. Louis A, Peterson BL, Couper FJ (2014) XLR-11 and UR-144 in Washington state and state of Alaska driving cases. J Anal Toxicol 38:563–568
74. Gurney SM, Scott KS, Kacinko SL, Presley BC, Logan BK (2014) Pharmacology, toxicology, and adverse effects of synthetic cannabinoid drugs. Forensic Sci Rev 26:53–78
75. Seely KA, Lapoint J, Moran JH, Fattore L (2012) Spice drugs are more than harmless herbal blends: a review of the pharmacology and toxicology of synthetic cannabinoids. Neuropsychopharmacol Biol Psychiatry 39:234–243
76. Auwarter V, Dresen S, Weinmann W, Muller M, Putz M, Ferreiros N (2009) 'Spice' and other herbal blends: harmless incense or cannabinoid designer drugs? J Mass Spectrom 44:832–837
77. Vardakou I, Pistos C, Spiliopoulou C (2010) Spice drugs as a new trend: mode of action, identification and legislation. Toxicol Lett 197:157–162
78. Winstock AR, Barratt MJ (2013) Synthetic cannabis: a comparison of patterns of use and effect profile with natural cannabis in a large global sample. Drug Alcohol Depend 131:106–111

79. Gunderson EW, Haughey HM, Ait-Daoud N, Joshi AS, Hart CL (2012) "Spice" and "K2" herbal highs: a case series and systematic review of the clinical effects and biopsychosocial implications of synthetic cannabinoid use in humans. Am J Addict 21:320–326
80. Zimmermann US, Winkelmann PR, Pilhatsch M, Nees JA, Spanagel R, Schulz K (2009) Withdrawal phenomena and dependence syndrome after the consumption of "spice gold". Dtsch Arztebl Int 106:464–467
81. Schneir AB, Cullen J, Ly BT (2011) "Spice" girls: synthetic cannabinoid intoxication. J Emerg Med 40:296–299
82. Schifano F, Corazza O, Deluca P, Davey Z, Di Furia L, Farre M et al (2009) Psychoactive drug or mystical incense? Overview of the online available information on Spice products. Int J Cult Ment Health 2:137–144
83. Hermanns-Clausen M, Kneisel S, Szabo B, Auwarter V (2013) Acute toxicity due to the confirmed consumption of synthetic cannabinoids: clinical and laboratory findings. Addiction 108:534–544
84. Lapoint J, James LP, Moran CL, Nelson LS, Hoffman RS, Moran JH (2011) Severe toxicity following synthetic cannabinoid ingestion. Clin Toxicol 49:760–764
85. Every-Palmer S (2011) Synthetic cannabinoid JWH-018 and psychosis: an explorative study. Drug Alcohol Depend 117:152–157
86. Müller H, Sperling W, Köhrmann M, Huttner HB, Kornhuber J, Maler J-M (2010) The synthetic cannabinoid Spice as a trigger for an acute exacerbation of cannabis induced recurrent psychotic episodes. Schizophr Res 118:309–310
87. Bhanushali GK, Jain G, Fatima H, Leisch LJ, Thornley-Brown D (2013) AKI associated with synthetic cannabinoids: a case series. Clin J Am Soc Nephrol 8:523–526
88. Center for Disease Control and Prevention (2013) Acute kidney injury associated with synthetic cannabinoid use - multiple States, 2012. MMWR Morb Mortal Wkly Rep 62:93–98
89. Trecki J, Gerona RR, Schwartz MD (2015) Synthetic cannabinoid-related illnesses and deaths. N Engl J Med 373:103–107
90. Forrester MB, Kleinschmidt K, Schwarz E, Young A (2012) Synthetic cannabinoid and marijuana exposures reported to poison centers. Hum Exp Toxicol 31:1006–1011
91. Winstock A, Lynskey M, Borschmann R, Waldron J (2015) Risk of emergency medical treatment following consumption of cannabis or synthetic cannabinoids in a large global sample. J Psychopharmacol 29:698–703

Pharmacological and Toxicological Effects of Synthetic Cannabinoids and Their Metabolites

Sherrica Tai and William E. Fantegrossi

Abstract Commercial preparations containing synthetic cannabinoids (SCBs) are rapidly emerging as drugs of abuse. Although often assumed to be "safe" and "legal" alternatives to cannabis, reports indicate that SCBs induce toxicity not often associated with the primary psychoactive component of marijuana, Δ9-tetrahydrocannabinol (Δ^9-THC). This chapter will summarize the evidence that use of SCBs poses greater health risks relative to marijuana and suggest that distinct pharmacological properties and metabolism of SCBs relative to Δ^9-THC may contribute to this increased toxicity. Studies reviewed will indicate that in contrast to partial agonist properties of Δ^9-THC typically observed in vitro, SCBs act as full CB1 and CB2 receptor agonists both in cellular assays and animal studies. Furthermore, unlike Δ^9-THC metabolism, several SCB metabolites retain high affinity for and exhibit a range of intrinsic activities at CB1 and CB2 receptors. Finally, the potential for SCBs to cause adverse drug–drug interactions with other drugs of abuse, as well as with common therapeutic agents, will be discussed. Collectively, the evidence provided in this chapter indicates that SCBs should not be considered safe and legal alternatives to marijuana. Instead, the enhanced toxicity of SCBs relative to marijuana, perhaps resulting from the combined actions of a complex mixture of different SCBs present and their active metabolites that retain high affinity for CB1 and CB2 receptors, highlights the inherent danger that may accompany use of these substances.

S. Tai
Department of Pharmacology and Toxicology, University of Arkansas for Medical Sciences College of Medicine, Mail Slot 638, 4301 West Markham Street, Little Rock, AR 72207, USA

Department of Pharmacology, University of Michigan Medical School, 2301 MSRB III, 1150 W. Medical Center Drive, Ann Arbor, MI 48109, USA

W.E. Fantegrossi (✉)
Department of Pharmacology and Toxicology, University of Arkansas for Medical Sciences College of Medicine, Mail Slot 638, 4301 West Markham Street, Little Rock, AR 72207, USA
e-mail: wefantegrossi@uams.edu

Curr Topics Behav Neurosci (2017) 32: 249–262
DOI 10.1007/7854_2016_60
Published Online: 19 December 2016

Keywords Cannabinoids • Drug–drug interactions • Efficacy • Metabolism

Contents

1 Introduction

Synthetic cannabinoids (SCBs) have become popular recreational drugs among young adults in the USA. Use of these substances first emerged in Europe, where they were marketed as *Spice*, then quickly spread throughout the USA, where they were marketed as *K2* [1]. SCBs are sometimes marketed for commercial distribution in the form of capsules, tablets, and powders but are most commonly laced onto herbal mixtures (e.g., potpourri or incense) to be smoked (as reviewed by Tai and Fantegrossi [2]). Commercial SCB products are widely available on the internet and, despite regulatory efforts to curtail their availability, remain accessible at "brick and mortar" establishments such as head shops and convenience stores. Their widespread distribution can be attributed to clever marketing tactics, which usually involve colorful packaging and mislabeling designed to portray a harmless herbal blend which is "not intended for human consumption." These deceptive marketing tactics were adopted to circumvent legal ramifications for selling drugs of abuse, but it is unlikely that criminal proceedings stemming from the sale of prohibited substances would be impeded by arguments about product labels. Importantly, the marketing of these products appears to be specifically designed to give users the false assumption that these drugs are harmless, "natural," legal alternatives to cannabis.

Examining the pharmacological similarities between SCBs and Δ9-tetrahydrocannabinol (Δ^9-THC), the main psychoactive constituent in marijuana, has been a topic of great interest among scientists and lawmakers. In this regard, SCBs have been reported to exhibit higher binding affinity at both CB1 and CB2 cannabinoid receptor (CBR) subtypes, and also to display varying intrinsic activity relative to Δ^9-THC, both in cellular assays and animal studies [3–7]. Unlike Δ^9-THC, which consistently exhibits partial agonist efficacy in vitro [4, 5], SCBs are fully efficacious at both CBRs across a range of in vitro and in vivo assays [4, 5, 8]. In addition, metabolites of SCBs often retain higher CBR affinity than Δ^9-THC and may elicit pharmacological and toxicological effects distinct from those induced by Δ^9-THC. These active metabolites could potentially explain the increased morbidity and mortality associated with SCB exposure, as compared to what is typically seen with marijuana. The remainder of this chapter will discuss the pharmacological and toxicological effects related to the metabolism of SCBs with a particular emphasis on their active metabolites and show that these substances are not safe, have a greater toxicological profile than has been reported with marijuana, and should not be considered a legal alternative to cannabis.

2 Synthetic Cannabinoid Metabolism

Metabolism of xenobiotics occurs through several biotransformation pathways which are conserved across species and quite old from an evolutionary perspective. The goal of drug metabolism is to detoxify potentially harmful compounds, removing them from the circulation and ultimately excreting them from the body altogether. The liver plays a major role in this detoxification process. In some cases, metabolism may activate inert compounds (the concept of a "pro-drug") or produce metabolic intermediates which may themselves induce toxicity. In most cases, oxidative metabolism of xenobiotics first occurs via the hepatic cytochrome P450 (CYP) enzyme system at which point the metabolites are conjugated with a sugar moiety, glucuronic acid, by a class of enzymes called UDP-glucuronosyltransferases (UGTs). The resulting metabolites are then soluble enough to be removed from the body.

The metabolism of Δ^9-THC has been a reference standard for understanding cannabinoid pharmacokinetics [9]. Metabolism of Δ^9-THC by human hepatic microsomes initially occurs via oxidation by CYP subtypes 2C9 and 3A4 [10]. In brief, hydroxylation of Δ^9-THC by CPY2C9 produces 11-hydroxy-Δ^9-THC, which is the only psychoactive metabolite of Δ^9-THC [11, 12]. Further oxidation of the remaining hydroxyl groups of Δ^9-THC produces carboxylic acids at several positions along the alkyl side chain, and these metabolites are devoid of biological effects. Further oxidation of the active metabolite 11-hydroxy-Δ^9-THC abolishes pharmacological activity and leads to the production of 11-nor-9-carboxy-Δ^9-THC, which is then conjugated at the carboxyl position to form *O*-ester glucuronide, the major metabolite excreted in human urine [13].

In the early 2000s, initial reports of SCB metabolism emerged with a focus on in vitro metabolism of CBR ligands (11R)-2-methyl-11-[(morpholin-4-yl)methyl]-3-(naphthalene-1-carbonyl)-9-oxa-1-azatricyclo[6.3.1.$0^{4,12}$]dodeca-2,4(12),5,7-tetraene (WIN-55,212-2) [14], 1-[2-(morpholin-4-yl)ethyl]-2-methyl-3-(4-methoxybenzoyl)-6-iodoindole (AM-630) [15], and (2-methyl-1-propyl-1H-indol-3-yl)-1-naphthalenyl-methanone (JWH-015) [16]. The subsequent recreational use of SCBs shifted the focus more towards in vivo metabolism of the commonly abused CBR ligand naphthalen-1-yl-(1-pentylindol-3-yl)methanone (JWH-018) and its analogs. Metabolites of JWH-018 were first identified in urine specimens obtained from three people who had smoked a commercial product containing this compound [17]. Predominant phase I metabolites are formed by oxidation of the indole ring or the N-alkyl chain to form mono-hydroxylated compounds. The phase I JWH-018 metabolites are excreted in urine almost exclusively in the form of phase II glucuronide conjugates, as

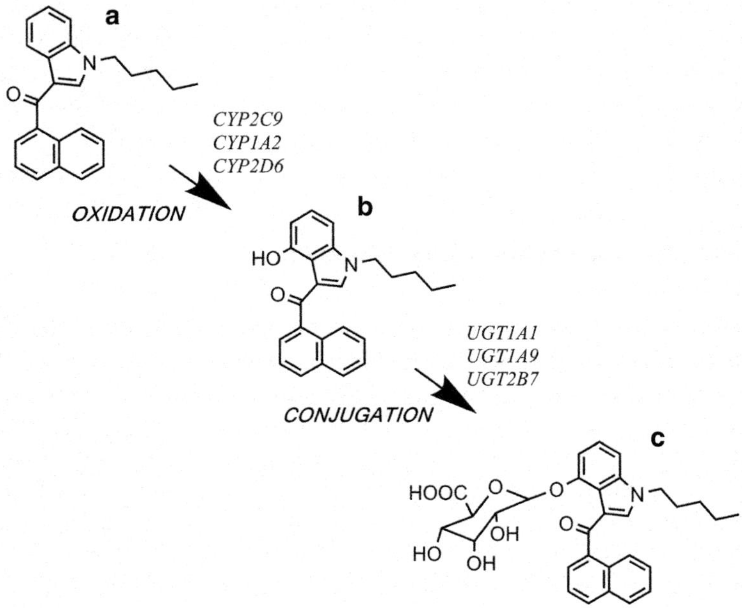

Fig. 1 Metabolism of the synthetic cannabinoid (SCB) JWH-018. The parent compound JWH-018 (**a**) undergoes phase I oxidation by cytochrome P450 (CYP) enzymes to form the bioactive JWH-018 4-hydroxyindole metabolite (**b**) [5]. Phase II conjugation by UDP-glucuronosyltransferases (UGTs) forms the corresponding glucuronide conjugate (**c**) [18]. Specific CYP enzymes and UGTs responsible for these biotransformation reactions are noted

determined by gas- and liquid-chromatography mass spectrometry (GC-MS and LC-MS/MS). Figure 1 depicts the metabolism of JWH-018 to its 4-hydroxyindole metabolite and corresponding glucuronide conjugate. Subsequent investigations demonstrated the formation of phase I mono-hydroxylated metabolites and phase II glucuronides in human hepatic microsomes incubated with JWH-018 [19], urine specimens obtained from individual users of *Spice*/*K2* products, and from rats administered JWH-018 [20, 21], naphthalen-1-yl-(1-butylindol-3-yl)methanone (JWH-073) [22], or 2-(2-methoxyphenyl)-1-(1-pentylindol-3-yl)ethanone (JWH-250) [20]. Collectively, these studies confirmed that the primary urinary metabolites of these SCBs are excreted in the form of a single hydroxylation and subsequent glucuronidation.

It was not until 2011 that reference standards for identifying specific JWH-018 and JWH-073 metabolites were developed allowing for both in vivo and in vitro quantitative measurements by LC-MS [18, 23, 24]. Human urine specimens were obtained from individuals who ingested JWH-018 or a mixture of JWH-018 and JWH-073. Analysis determined that the metabolites found in the urine were excreted in high concentrations and primarily in the form of glucuronic acid conjugates [18]. Several additional laboratories have replicated with these findings [20, 25–30].

It is well accepted that CYPs are involved in the biotransformation of SCBs in humans as recognized by the formation of hydroxylated metabolites [14, 17–19, 24, 29]. In vitro metabolism studies have determined that SCBs JWH-018 and its fluorinated analogue 1-[(5-fluoropentyl)-1H-indol-3-yl]-(naphthalen-1-yl)methanone (AM-2201) are metabolized via oxidation by hepatic CYP subtypes 2C9 and 1A2 [23]. Outside the liver, CYPs are ubiquitously expressed in the body in a tissue-specific manner. CYP2C9 is also highly expressed in the intestine [31], so it is likely that intestinal CYP2C9 is involved in the metabolism of SCBs when ingested orally. CYP1A2 is highly expressed in the lung and is likely responsible for the metabolism of smoked SCBs [32]. There is evidence of the involvement of CYP2D6 in the metabolism of JWH-018 and AM-2201 in the brain especially in brain regions that have a high expression of CB1Rs, including the cortex, hippocampus, and cerebellum. It is likely that CYP2D6 is involved in the management of brain concentrations of these SCBs and their active metabolites, more so than in the liver.

Urine specimens obtained from individuals who ingested SCBs contain high concentrations of glucuronide metabolites [17–19, 24], coupled with no traces in serum [33, 34], suggesting that glucuronic acid conjugation plays a key role in the excretion of these drugs in urine. In the liver, JWH-018 and JWH-073 metabolites are formed by major UGT isoforms UGT1A1, UGT1A9, and UGT2B7 [18] (see Fig. 1). In addition, extrahepatic UGT isoforms involved in metabolism include UGT1A7 expressed in lung, UGT1A3 and UGT2B7 expressed in brain (as well as liver), and UGT1A10. Importantly, human UGT1A3 and UGT2B7 are predominant isoforms responsible for generating the major metabolites of JWH-018 and JWH-073 found in urine [35]. Since UGT1A3 and UGT2B7 are also expressed in the brain, these UGT isoforms may play a role in regulating brain concentrations of SCBs and their active metabolites at the CB1R, similar to CYP2D6 (as previously discussed).

3 Synthetic Cannabinoid Cellular Signaling

The pharmacological profiling of Δ^9-THC has led to the characterization of metabolites produced by SCBs JWH-018 and JWH-073 [36]. Like the phytocannabinoid Δ^9-THC, both JWH-018 and JWH-073 have high affinity for the CB1 and CB2Rs [3–7]. Chemical structures for JWH-073 many of the related SBCs discussed in this chapter are shown in Fig. 2. Emerging SCBs found in commercial preparations also have high affinity for the CB1 and CB2Rs, including 2-[(1S,3R)-3-hydroxycyclohexyl]-5-(2-methyloctan-2-yl)phenol (CP-47,497) [37], (1-pentylindol-3-yl)naphthalen-1-ylmethane (JWH-175) [38], 1-([(1E)-3-pentylinden-1-ylidine]methyl)naphthalene (JWH-176) [38], (1-pentylindol-3-yl)-(2,2,3,3-tetramethylcyclopropyl)methanone (UR-144) [39], naphthalen-1-yl-(1-pentylpyrrol-3-yl)methanone (JWH-030) [40], 2-(2-methoxyphenyl)-1-(1-pentylindol-3-yl)ethanone (JWH-250) [41], N-(1-adamantyl)-1-pentylindazole-3-carboxamide (APINACA, or AKB48) [42], and 1-[(N-methylpiperidin-2-yl)methyl]-3-(adamant-1-oyl)indole (AM-1248) [43]. Furthermore, most commercial SCB products display high potency and efficacy as CB1R

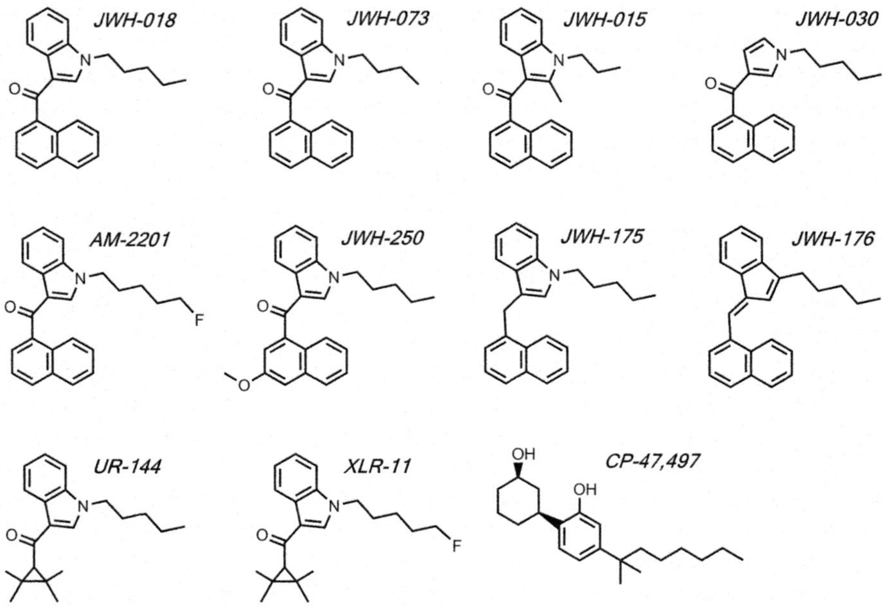

Fig. 2 Chemical structures of SCBs discussed in this chapter. Abbreviations are as follows: *JWH-018* naphthalen-1-yl-(1-pentylindol-3-yl)methanone, *JWH-073* naphthalen-1-yl-(1-butylindol-3-yl)methanone, *JWH-015* (2-methyl-1-propyl-1H-indol-3-yl)-1-naphthalenylmethanone, *JWH-030* naphthalen-1-yl-(1-pentylpyrrol-3-yl)methanone, *AM-2201* 1-[(5-fluoropentyl)-1H-indol-3-yl]-(naphthalen-1-yl)methanone, *JWH-250* 2-(2-methoxyphenyl)-1-(1-pentylindol-3-yl)ethanone, *JWH-175* (1-pentylindol-3-yl)naphthalen-1-ylmethane, *JWH-176* 1-([(1E)-3-pentylinden-1-ylidine]methyl)naphthalene, *UR-144* (1-pentylindol-3-yl)-(2,2,3,3-tetramethylcyclopropyl)methanone, *XLR-11* (1-(5-fluoropentyl)-1H-indol-3-yl)(2,2,3,3-tetramethylcyclopropyl)methanone, *CP-47,497* (2-[(1S,3R)-3-hydroxycyclohexyl]-5-(2-methyloctan-2-yl)phenol

agonists both in vitro and in vivo, including JWH-018 [5, 8], JWH-073 [4, 44], AM-1248 [39], CP-55,940 [40], WIN-55,512-2 [40], and (6aR,10aR)-9-(hydroxymethyl)-6,6-dimethyl-3-(2-methyloctan-2-yl)-6H,6aH,7H,10H,10aH-benzo[c]isochromen-1-ol (HU-210) [45].

The abuse liability of SCB products may be attributed to the presence of high affinity and fully efficacious CB1 agonists in these commercial products [46]. For instance, the in vitro efficacy of Δ^9-THC is partial relative to JWH-018 and JWH-073, which are fully efficacious [4, 8, 46]. Importantly, not all in vitro or ex vivo assessments can identify differences in efficacy between THC and SCBs. An interesting study by Hoffman et al. [47] demonstrated similar efficacy for SCBs and THC in an electrophysiological assay reflecting inhibition of transmitter release. Nevertheless, the often demonstrated low in vitro efficacy of Δ^9-THC does not necessarily translate to partial agonism in vivo, and Δ^9-THC often displays in vivo efficacy comparable to fully efficacious agonists [48]. In addition, abrupt discontinuation of chronic marijuana use or Δ^9-THC administration produces a withdrawal syndrome in humans and rodents that is accompanied by a region-specific downregulation and desensitization of CB1Rs in the brain [49–51]. Thus, it is possible that commercial SCB products that contain high-efficacy agonists may intensify the adverse effects related to tolerance, dependence, and withdrawal of SCB abuse relative to Δ^9-THC. For instance, the commercial SCB product "Spice Gold" produced a withdrawal phenomenon and dependence syndrome in humans that transpired after abrupt discontinuation of use in the form of drug craving, elevated blood pressure, nausea, tremor, profuse sweating, and nightmares [52]. The active constituents of this product were not forensically determined in the product itself or in fluids or tissue from the case subject, but contemporaneous laboratory studies determined that a mixture of the SCBs JWH-018 and CP-47,497 was present in this commercial smoking blend at the time and in the geographic area where the aforementioned case occurred [52].

Metabolism of Δ^9-THC produces a single active metabolite (11-hydroxy-Δ^9-THC) that exhibits reduced CB1R affinity compared with the parent compound [53]. On the other hand, metabolism of commercial SCBs including JWH-018, JWH-073, and AM-2201 produces numerous major mono-hydroxylated metabolites that retain nanomolar binding affinity for CB1Rs [4, 5, 23] (see Fig. 1), unlike their carboxylated metabolites, which do not bind to nor activate CB1Rs.

In addition to retaining high CB1R affinity, in vitro functional assays (G-protein activation) demonstrate that the major mono-hydroxylated metabolites of JWH-018, JWH-073, and AM-2201 exhibit partial to full efficacy at the CB1Rs similar to fully efficacious CP-55,940 [4, 5, 23]. Of further importance, the in vivo cannabimimetic effects of JWH-018 and JWH-073 mono-hydroxylated metabolites elicited profound hypothermic and locomotor depressant effects in mice [4]. These effects are attenuated by CB1R antagonist/inverse agonist 1-(2,4-dichlorophenyl)-5-(4-iodophenyl)-4-methyl-N-(1-piperidyl)pyrazole-3-carboxamide (AM-251), suggesting that these metabolites are mediating their effects through the CB1R, similar to the parent ligands. Moreover, the effects elicited by these metabolites may be associated with adverse effects of SCB use. For instance, it has been shown that several mono-hydroxylated metabolites of JWH-018, JWH-073, and AM-2201 retain high CB1R affinity and activity and may display additive or synergistic interactions with other SCBs [54]. Additional contributing factors to adverse effects of SCB use can be: (1) drug effects of non-cannabinoid-like ligands found in commercial SCB products, (2) variations in batch-to-batch preparations with differences in the concentration and content found in commercial SCB products, and (3) an exacerbation of SCB adverse effects in drug users with preexisting conditions [36, 55–58].

Although evidence suggests that mono-hydroxylated metabolites of JWH-018, JWH-073, and AM-2201 are active at CB1Rs in both in vitro and in vivo assays [54], it is also possible for some oxidized metabolites of SCBs to function as antagonists at CB1Rs. Despite the fact that the 7-hydroxyindole derivative of JWH-073 has not been detected in human urine, it has been shown to bind to CB1Rs with nanomolar affinity without eliciting any G-protein activation at concentrations that are pharmacologically relevant, up to 10 μM [4]. Furthermore, Schild analysis has shown that the 7-hydroxyindole derivative of JWH-073 competitively antagonizes G-protein activation in vitro. In mice, hypothermia induced by JWH-018 was attenuated by pretreatment with this oxidized derivative of JWH-073. Alternatively,

JWH-018 induced analgesia, catalepsy, and locomotor activity were not altered by this metabolite. Overall, these in vitro and in vivo findings along with the lack of the 7-hydroxyindole derivative of JWH-073 detected in human urine suggest that this oxidized derivative of JWH-073 may not be formed in humans or readily cross the blood–brain barrier.

In addition to a human oxidation product of JWH-018 acting as a CB1R antagonist, it has also been shown that a major human glucuronidated metabolite of JWH-018 (5-hydroxypentyl-β-D-glucuronide) retains significant affinity for CB1Rs and displays CB1R antagonism in vitro [59]. Interestingly, this study also showed that a major structurally similar glucuronidated metabolite of THC (11-nor-9-carboxy-THC-β-D-glucuronide) lacked CB1R affinity and activity. Collectively, these findings demonstrate that both hydroxylated and glucuronidated metabolites of SCBs may retain significant CB1R affinity in the absence of intrinsic activity, which suggest that some SCB metabolites can produce physiologically relevant antagonism of effects of CB1Rs.

When discussing the pharmacological and toxicological effects of SCBs, the primary focus has been geared towards understanding CB1R-induced responses. Many SCBs not only have high affinity and significant intrinsic activity at the CB1R, they very often also have comparable binding and functional activity at the second major CBR subtype, CB2R [3, 6]. CB1Rs are primarily located within the CNS, while CB2Rs are most abundantly located on immune cells in peripheral regions [60] and are associated with immune functions, inflammation, and bone formation [61]. More recent studies have demonstrated that activation of low numbers of CB2R in the CNS can modulate the abuse-related properties of alcohol [62], nicotine [63], and cocaine [64]. In addition, mono-hydroxylated metabolites of JWH-018 and JWH-073 have also been shown to retain high CB2R affinity and efficacy [7].

Interestingly, findings have implicated the involvement of endocannabinoid signaling in the modulation of the serotonin system (as reviewed by Haj-Dahmane and Shen [65]. For instance, chronic activation of CB2Rs has been shown to produce an upregulation of 5-HT_{2A} receptors in the prefrontal cortex of mice [66, 67]. CNS abnormalities in 5-HT_{2A} function can lead to mental disorders, including anxiety [68] and psychosis [69]. Furthermore, 5-HT_{2A} receptor signaling is a major site of action for hallucinogenic drugs [70]. Common adverse effects associated with SCB use are not often observed with Δ^9-THC, such as anxiety and psychosis [71]. It is possible to speculate that SCB- or SCB metabolite-induced upregulation of 5-HT_{2A} receptors, mediated via CB2R activation, might contribute to anxiety and psychosis that are observed after exposure to SCBs found in commercial abuse-ready preparations. In this regard, a case study reported on four patients hospitalized for psychosis who smoked a product containing the SCB AM-2201 while in the clinic. The authors described the appearance of new psychotic symptoms, and a marked worsening of mood and anxiety symptoms in four patients, and noted that even though they all ingested the same drug, the clinical picture differed markedly among the individual patients [72], perhaps implicating individual metabolism of AM-2201 in the diversity of effects observed. Thus, it is clearly important for future studies to explore both CB1R and CB2R signaling as it relates to the pharmacological and toxicological effects of SCBs and its metabolites. Furthermore, future studies should consider the capacity of these drugs to modulate the expression and function of other, non-CBR systems.

4 Synthetic Cannabinoid Drug–Drug Interactions

Commercial SCB preparations often contain multiple drugs in combination, and the concentrations of these specific SCBs vary widely from product to product, or even within a product from batch to batch [58, 59]. Therefore, it is possible for drug–drug interactions to exist both within and between these diverse mixtures of SCBs, which may contribute to abuse-related and adverse effects associated with the use of these drugs. As described above, mouse studies have shown that coadministration of JWH-018 and JWH-073 produced additive, synergistic, or antagonistic interactions compared to administration of either drug alone, depending on the specific endpoint examined and the drug dose ratio employed [54]. Evidence of synergistic effects with these two SCBs was demonstrated both in vivo and in vitro, with mouse assays of Δ^9-THC discrimination and analgesia, and with displacement of radioligand binding from CB1Rs in a cellular model. In addition to synergism, SCB blended mixtures can influence the relative potency of both their subjective and adverse effects. Furthermore, polysubstance abuse may lead to unpredictable effects of SCBs that may contribute to even greater abuse-related and adverse effects. Future studies are needed to understand the drug–drug interactions among SCBs and co-exposure to other drugs of abuse.

Given their shared metabolism via P450 isoforms, combined use of SCBs with various prescription medications could also potentially result in adverse drug–drug interactions. Commonly prescribed drugs such as valproic acid (an anticonvulsant and mood stabilizer) and sertraline (an antidepressant) potently inhibit CYP2C9, while drugs such as ciprofloxacin (an antibiotic) and fluvoxamine (an antidepressant) strongly inhibit CYP1A2. Additionally, CYP2C9 is a major polymorphic enzyme [31] and is responsible for the metabolism of a number of clinically important drugs such as warfarin (a blood thinner), phenytoin (an anticonvulsant), tolbutamide (an antidiabetic agent), losartan (an antihypertensive), and ibuprofen (a nonsteroidal anti-inflammatory drug). Over five allelic variants of CYP2D6 have been identified, including two "loss of function" variants (CYP2C9*4 and CYP2C9*5) [73]. Similarly, CYP1A2 is responsible for the metabolism of numerous psychiatric medications including antipsychotics (olanzapine, clozapine, haloperidol, and thioridazine), antidepressants (imipramine, clomipramine, and fluvoxamine), and cholinesterase inhibitors used in the treatment of Alzheimer's disease (tacrine) [74], but this enzyme is well conserved without common functional polymorphisms [75]. Because SCBs are also substrates for these P450 isoforms, the possibility of drug–drug interactions is a serious consideration with the use of SCBs.

5 Conclusions

Commercial SCB products are not safe alternatives to cannabis and pose significant threats to public health. It is anticipated that morbidity and mortality rates will continue to increase in correlation to SCB exposure. Recent reports have demonstrated that SCBs present a pharmacological and toxicological profile that is distinct from Δ^9-THC found in marijuana. For instance, SCBs primarily act as full CB1 and CB2R agonists both in vitro and in vivo, while Δ^9-THC is a weak partial agonist. Furthermore, several SCB metabolites bind with high affinity at CB1 and CB2Rs, while displaying a range in intrinsic activities from neutral antagonists to partial agonists to full agonists, both in cellular assays and animal studies. These findings illustrate that commercial SCBs products are not safe and should not be considered an alternate form of marijuana. Rather, they produce greater toxicity relative to marijuana, which could be attributed to the combined actions of the varying SCBs or their metabolites present in these products. Still, these findings present the supporting evidence that the pharmacological and toxicological properties of SCBs pose a severe health risk.

References

1. Wiley JL, Marusich JA, Huffman JW, Balster RL, Thomas BF (2011) Hijacking of basic research: the case of synthetic cannabinoids. Methods Rep RTI Press 2011:17971
2. Tai S, Fantegrossi W (2014) Synthetic cannabinoids: pharmacology, behavioral effects, and abuse potential. Curr Addict Rep:1–8
3. Aung MM, Griffin G, Huffman JW, Wu M, Keel C, Yang B, Showalter VM, Abood ME, Martin BR (2000) Influence of the N-1 alkyl chain length of cannabimimetic indoles upon CB (1) and CB(2) receptor binding. Drug Alcohol Depend 60:133–140
4. Brents LK, Gallus-Zawada A, Radominska-Pandya A, Vasiljevik T, Prisinzano TE, Fantegrossi WE, Moran JH, Prather PL (2012) Monohydroxylated metabolites of the K2 synthetic cannabinoid JWH-073 retain intermediate to high cannabinoid 1 receptor (CB1R) affinity and exhibit neutral antagonist to partial agonist activity. Biochem Pharmacol 83:952–961
5. Brents LK, Reichard EE, Zimmerman SM, Moran JH, Fantegrossi WE, Prather PL (2011) Phase I hydroxylated metabolites of the K2 synthetic cannabinoid JWH-018 retain in vitro and in vivo cannabinoid 1 receptor affinity and activity. PLoS One 6:e21917
6. Chin CN, Murphy JW, Huffman JW, Kendall DA (1999) The third transmembrane helix of the cannabinoid receptor plays a role in the selectivity of aminoalkylindoles for CB2, peripheral cannabinoid receptor. J Pharmacol Exp Ther 291:837–844
7. Rajasekaran M, Brents LK, Franks LN, Moran JH, Prather PL (2013) Human metabolites of synthetic cannabinoids JWH-018 and JWH-073 bind with high affinity and act as potent agonists at cannabinoid type-2 receptors. Toxicol Appl Pharmacol 269:100–108
8. Atwood BK, Huffman J, Straiker A, Mackie K (2010) JWH018, a common constituent of 'Spice' herbal blends, is a potent and efficacious cannabinoid CB receptor agonist. Br J Pharmacol 160:585–593
9. Huestis MA (2005) Pharmacokinetics and metabolism of the plant cannabinoids, delta9-tetrahydrocannabinol, cannabidiol and cannabinol. Handb Exp Pharmacol (168):657–690

10. Watanabe K, Yamaori S, Funahashi T, Kimura T, Yamamoto I (2007) Cytochrome P450 enzymes involved in the metabolism of tetrahydrocannabinols and cannabinol by human hepatic microsomes. Life Sci 80:1415–1419
11. Bornheim LM, Lasker JM, Raucy JL (1992) Human hepatic microsomal metabolism of delta 1-tetrahydrocannabinol. Drug Metab Dispos 20:241–246
12. Watanabe K, Matsunaga T, Yamamoto I, Funae Y, Yoshimura H (1995) Involvement of CYP2C in the metabolism of cannabinoids by human hepatic microsomes from an old woman. Biol Pharm Bull 18:1138–1141
13. Yamamoto I, Watanabe K, Kuzuoka K, Narimatsu S, Yoshimura H (1987) The pharmacological activity of cannabinol and its major metabolite, 11-hydroxycannabinol. Chem Pharm Bull 35:2144–2147
14. Zhang Q, Ma P, Iszard M, Cole RB, Wang W, Wang G (2002) In vitro metabolism of R(+)-[2,3-dihydro-5-methyl-3-[(morpholinyl)methyl]pyrrolo [1,2,3-de]1,4-benzoxazinyl]-(1-naphthalenyl) methanone mesylate, a cannabinoid receptor agonist. Drug Metab Dispos 30:1077–1086
15. Zhang Q, Ma P, Wang W, Cole RB, Wang G (2004) Characterization of rat liver microsomal metabolites of AM-630, a potent cannabinoid receptor antagonist, by high-performance liquid chromatography/electrospray ionization tandem mass spectrometry. J Mass Spectrom 39:672–681
16. Zhang Q, Ma P, Cole RB, Wang G (2006) Identification of in vitro metabolites of JWH-015, an aminoalkylindole agonist for the peripheral cannabinoid receptor (CB2) by HPLC-MS/MS. Anal Bioanal Chem 386:1345–1355
17. Sobolevsky T, Prasolov I, Rodchenkov G (2010) Detection of JWH-018 metabolites in smoking mixture post-administration urine. Forensic Sci Int 200:141–147
18. Chimalakonda KC, Bratton SM, Le VH, Yiew KH, Dineva A, Moran CL, James LP, Moran JH, Radominska-Pandya A (2011) Conjugation of synthetic cannabinoids JWH-018 and JWH-073, metabolites by human UDP-glucuronosyltransferases. Drug Metab Dispos 39:1967–1976
19. Wintermeyer A, Moller I, Thevis M, Jubner M, Beike J, Rothschild MA, Bender K (2010) In vitro phase I metabolism of the synthetic cannabimimetic JWH-018. Anal Bioanal Chem 398:2141–2153
20. Grigoryev A, Melnik A, Savchuk S, Simonov A, Rozhanets V (2011) Gas and liquid chromatography-mass spectrometry studies on the metabolism of the synthetic phenylacetylindole cannabimimetic JWH-250, the psychoactive component of smoking mixtures. J Chromatogr B Analyt Technol Biomed Life Sci 879:2519–2526
21. Moller I, Wintermeyer A, Bender K, Jubner M, Thomas A, Krug O, Schanzer W, Thevis M (2011) Screening for the synthetic cannabinoid JWH-018 and its major metabolites in human doping controls. Drug Test Anal 3:609–620
22. Grigoryev A, Savchuk S, Melnik A, Moskaleva N, Dzhurko J, Ershov M, Nosyrev A, Vedenin A, Izotov B, Zabirova I, Rozhanets V (2011) Chromatography-mass spectrometry studies on the metabolism of synthetic cannabinoids JWH-018 and JWH-073, psychoactive components of smoking mixtures. J Chromatogr B Analyt Technol Biomed Life Sci 879:1126–1136
23. Chimalakonda KC, Seely KA, Bratton SM, Brents LK, Moran CL, Endres GW, James LP, Hollenberg PF, Prather PL, Radominska-Pandya A, Moran JH (2012) Cytochrome P450-mediated oxidative metabolism of abused synthetic cannabinoids found in K2/Spice: identification of novel cannabinoid receptor ligands. Drug Metab Dispos 40:2174–2184
24. Moran CL, Le VH, Chimalakonda KC, Smedley AL, Lackey FD, Owen SN, Kennedy PD, Endres GW, Ciske FL, Kramer JB, Kornilov AM, Bratton LD, Dobrowolski PJ, Wessinger WD, Fantegrossi WE, Prather PL, James LP, Radominska-Pandya A, Moran JH (2011) Quantitative measurement of JWH-018 and JWH-073 metabolites excreted in human urine. Anal Chem 83:4228–4236
25. Beuck S, Moller I, Thomas A, Klose A, Schlorer N, Schanzer W, Thevis M (2011) Structure characterisation of urinary metabolites of the cannabimimetic JWH-018 using chemically synthesised reference material for the support of LC-MS/MS-based drug testing. Anal Bioanal Chem 401:493–505

26. de Jager AD, Warner JV, Henman M, Ferguson W, Hall A (2012) LC-MS/MS method for the quantitation of metabolites of eight commonly-used synthetic cannabinoids in human urine – an Australian perspective. J Chromatogr B Analyt Technol Biomed Life Sci 897:22–31
27. Dowling G, Regan L (2011) A method for CP 47, 497 a synthetic non-traditional cannabinoid in human urine using liquid chromatography tandem mass spectrometry. J Chromatogr B Analyt Technol Biomed Life Sci 879:253–259
28. ElSohly MA, Gul W, Elsohly KM, Murphy TP, Madgula VL, Khan SI (2011) Liquid chromatography-tandem mass spectrometry analysis of urine specimens for K2 (JWH-018) metabolites. J Anal Toxicol 35:487–495
29. Hutter M, Broecker S, Kneisel S, Auwarter V (2012) Identification of the major urinary metabolites in man of seven synthetic cannabinoids of the aminoalkylindole type present as adulterants in 'herbal mixtures' using LC-MS/MS techniques. J Mass Spectrom 47:54–65
30. Sobolevsky T, Prasolov I, Rodchenkov G (2012) Detection of urinary metabolites of AM-2201 and UR-144, two novel synthetic cannabinoids. Drug Test Anal 4:745–753
31. Paine MF, Hart HL, Ludington SS, Haining RL, Rettie AE, Zeldin DC (2006) The human intestinal cytochrome P450 "pie". Drug Metab Dispos 34:880–886
32. Pavanello S, Fedeli U, Mastrangelo G, Rota F, Overvad K, Raaschou-Nielsen O, Tjonneland A, Vogel U (2012) Role of CYP1A2 polymorphisms on lung cancer risk in a prospective study. Cancer Genet 205:278–284
33. Dresen S, Ferreiros N, Putz M, Westphal F, Zimmermann R, Auwarter V (2010) Monitoring of herbal mixtures potentially containing synthetic cannabinoids as psychoactive compounds. J Mass Spectrom 45:1186–1194
34. Kacinko SL, Xu A, Homan JW, McMullin MM, Warrington DM, Logan BK (2011) Development and validation of a liquid chromatography-tandem mass spectrometry method for the identification and quantification of JWH-018, JWH-073, JWH-019, and JWH-250 in human whole blood. J Anal Toxicol 35:386–393
35. Chimalakonda KC, Moran CL, Kennedy PD, Endres GW, Uzieblo A, Dobrowolski PJ, Fifer EK, Lapoint J, Nelson LS, Hoffman RS, James LP, Radominska-Pandya A, Moran JH (2011) Solid-phase extraction and quantitative measurement of omega and omega-1 metabolites of JWH-018 and JWH-073 in human urine. Anal Chem 83:6381–6388
36. Seely KA, Prather PL, James LP, Moran JH (2011) Marijuana-based drugs: innovative therapeutics or designer drugs of abuse? Mol Interv 11:36–51
37. Huffman JW, Thompson AL, Wiley JL, Martin BR (2008) Synthesis and pharmacology of 1-deoxy analogs of CP-47,497 and CP-55,940. Bioorg Med Chem 16:322–335
38. Huffman JW, Padgett LW (2005) Recent developments in the medicinal chemistry of cannabimimetic indoles, pyrroles and indenes. Curr Med Chem 12:1395–1411
39. Frost JM, Dart MJ, Tietje KR, Garrison TR, Grayson GK, Daza AV, El-Kouhen OF, Yao BB, Hsieh GC, Pai M, Zhu CZ, Chandran P, Meyer MD (2010) Indol-3-ylcycloalkyl ketones: effects of N1 substituted indole side chain variations on CB(2) cannabinoid receptor activity. J Med Chem 53:295–315
40. Griffin G, Atkinson PJ, Showalter VM, Martin BR, Abood ME (1998) Evaluation of cannabinoid receptor agonists and antagonists using the guanosine-5′-O-(3-[35S]thio)-triphosphate binding assay in rat cerebellar membranes. J Pharmacol Exp Ther 285:553–560
41. Huffman JW, Szklennik PV, Almond A, Bushell K, Selley DE, He H, Cassidy MP, Wiley JL, Martin BR (2005) 1-Pentyl-3-phenylacetylindoles, a new class of cannabimimetic indoles. Bioorg Med Chem Lett 15:4110–4113
42. Uchiyama N, Kawamura M, Kikura-Hanajiri R, Goda Y (2013) URB-754: a new class of designer drug and 12 synthetic cannabinoids detected in illegal products. Forensic Sci Int 227:21–32
43. Jankovics P, Varadi A, Tolgyesi L, Lohner S, Nemeth-Palotas J, Balla J (2012) Detection and identification of the new potential synthetic cannabinoids 1-pentyl-3-(2-iodobenzoyl)indole and 1-pentyl-3-(1-adamantoyl)indole in seized bulk powders in Hungary. Forensic Sci Int 214:27–32

44. Atwood BK, Lee D, Straiker A, Widlanski TS, Mackie K (2011) CP47,497-C8 and JWH073, commonly found in 'Spice' herbal blends, are potent and efficacious CB(1) cannabinoid receptor agonists. Eur J Pharmacol 659:139–145
45. Howlett AC, Champion TM, Wilken GH, Mechoulam R (1990) Stereochemical effects of 11-OH-delta 8-tetrahydrocannabinol-dimethylheptyl to inhibit adenylate cyclase and bind to the cannabinoid receptor. Neuropharmacology 29:161–165
46. Lindigkeit R, Boehme A, Eiserloh I, Luebbecke M, Wiggermann M, Ernst L, Beuerle T (2009) Spice: a never ending story? Forensic Sci Int 191:58–63
47. Hoffman AF, Lycas MD, Kaczmarzyk JR, Spivak CE, Baumann MH, Lupica CR (2016) Disruption of hippocampal synaptic transmission and long-term potentiation by psychoactive synthetic cannabinoid 'Spice' compounds: comparison with Δ9-tetrahydrocannabinol. Addict Biol. doi:10.1111/adb.12334/full (in press)
48. Fan F, Compton DR, Ward S, Melvin L, Martin BR (1994) Development of cross-tolerance between delta-9-tetrahydrocannabinol, CP 55,940 and WIN 55,212. J Pharmacol Exp Ther 271:1383–1390
49. Hirvonen J, Goodwin RS, Li CT, Terry GE, Zoghbi SS, Morse C, Pike VW, Volkow ND, Huestis MA, Innis RB (2012) Reversible and regionally selective downregulation of brain cannabinoid CB1 receptors in chronic daily cannabis smokers. Mol Psychiatry 17:642–649
50. Sim-Selley LJ, Schechter NS, Rorrer WK, Dalton GD, Hernandez J, Martin BR, Selley DE (2006) Prolonged recovery rate of CB1 receptor adaptation after cessation of long-term cannabinoid administration. Mol Pharmacol 70:986–996
51. Villares J (2007) Chronic use of marijuana decreases cannabinoid receptor binding and mRNA expression in the human brain. Neuroscience 145:323–334
52. Zimmermann US, Winkelmann PR, Pilhatsch M, Nees JA, Spanagel R, Schulz K (2009) Withdrawal phenomena and dependence syndrome after the consumption of "spice gold". Dtsch Arztebl Int 106:464–467
53. Kochanowski M, Kala M (2005) Tetrahydrocannabinols in clinical and forensic toxicology. Przegl Lek 62:576–580
54. Brents LK, Zimmerman SM, Saffell AR, Prather PL, Fantegrossi WE (2013) Differential drug-drug interactions of the synthetic cannabinoids JWH-018 and JWH-073: implications for drug abuse liability and pain therapy. J Pharmacol Exp Ther 346:350–361
55. Fattore L, Fratta W (2011) Beyond THC: the new generation of cannabinoid designer drugs. Front Behav Neurosci 5:60
56. Lapoint J, James LP, Moran CL, Nelson LS, Hoffman RS, Moran JH (2011) Severe toxicity following synthetic cannabinoid ingestion. Clin Toxicol 49:760–764
57. Patton AL, Chimalakonda KC, Moran CL, McCain KR, Radominska-Pandya A, James LP, Kokes C, Moran JH (2013) K2 toxicity: fatal case of psychiatric complications following AM2201 exposure. J Forensic Sci 58:1676–1680
58. Seely KA, Lapoint J, Moran JH, Fattore L (2012) Spice drugs are more than harmless herbal blends: a review of the pharmacology and toxicology of synthetic cannabinoids. Prog Neuropsychopharmacol Biol Psychiatry 39:234–243
59. Seely KA, Patton AL, Moran CL, Womack ML, Prather PL, Fantegrossi WE, Radominska-Pandya A, Endres GW, Channell KB, Smith NH, McCain KR, James LP, Moran JH (2013) Forensic investigation of K2, Spice, and "bath salt" commercial preparations: a three-year study of new designer drug products containing synthetic cannabinoid, stimulant, and hallucinogenic compounds. Forensic Sci Int 233(1–3):416–422
60. Klein TW, Newton C, Larsen K, Lu L, Perkins I, Nong L, Friedman H (2003) The cannabinoid system and immune modulation. J Leukoc Biol 74:486–496
61. Patel KD, Davison JS, Pittman QJ, Sharkey KA (2010) Cannabinoid CB(2) receptors in health and disease. Curr Med Chem 17:1393–1410
62. Onaivi ES, Ishiguro H, Gong JP, Patel S, Meozzi PA, Myers L, Perchuk A, Mora Z, Tagliaferro PA, Gardner E, Brusco A, Akinshola BE, Hope B, Lujilde J, Inada T, Iwasaki S,

Macharia D, Teasenfitz L, Arinami T, Uhl GR (2008) Brain neuronal CB2 cannabinoid receptors in drug abuse and depression: from mice to human subjects. PLoS One 3:e1640
63. Gamaleddin I, Zvonok A, Makriyannis A, Goldberg SR, Le Foll B (2012) Effects of a selective cannabinoid CB2 agonist and antagonist on intravenous nicotine self administration and reinstatement of nicotine seeking. PLoS One 7:e29900
64. Xi ZX, Peng XQ, Li X, Song R, Zhang HY, Liu QR, Yang HJ, Bi GH, Li J, Gardner EL (2011) Brain cannabinoid CB(2) receptors modulate cocaine's actions in mice. Nat Neurosci 14:1160–1166
65. Haj-Dahmane S, Shen RY (2011) Modulation of the serotonin system by endocannabinoid signaling. Neuropharmacology 61:414–420
66. Franklin JM, Carrasco GA (2013) G-protein receptor kinase 5 regulates the cannabinoid receptor 2-induced up-regulation of serotonin 2A receptors. J Biol Chem 288:15712–15724
67. Franklin JM, Vasiljevik T, Prisinzano TE, Carrasco GA (2013) Cannabinoid 2 receptor- and beta arrestin 2-dependent upregulation of serotonin 2A receptors. Eur Neuropsychopharmacol 23:760–767
68. Ghisleni G, Kazlauckas V, Both FL, Pagnussat N, Mioranzza S, Rocha JB, Souza DO, Porciuncula LO (2008) Diphenyl diselenide exerts anxiolytic-like effect in Wistar rats: putative roles of GABAA and 5HT receptors. Prog Neuropsychopharmacol Biol Psychiatry 32:1508–1515
69. Morgan D, Kondabolu K, Kuipers A, Sakhuja R, Robertson KL, Rowland NE, Booth RG (2013) Molecular and behavioral pharmacology of two novel orally-active 5HT2 modulators: potential utility as antipsychotic medications. Neuropharmacology 72:274–281
70. Fantegrossi WE, Murnane KS, Reissig CJ (2008) The behavioral pharmacology of hallucinogens. Biochem Pharmacol 75:17–33
71. Gunderson EW, Haughey HM, Ait-Daoud N, Joshi AS, Hart CL (2012) "Spice" and "K2" herbal highs: a case series and systematic review of the clinical effects and biopsychosocial implications of synthetic cannabinoid use in humans. Am J Addict 21:320–326
72. Celofiga A, Koprivsek J, Klavz J (2014) Use of synthetic cannabinoids in patients with psychotic disorders: case series. J Dual Diagn 10(3):168–173
73. Seng KC, Seng CK (2008) The success of the genome-wide association approach: a brief story of a long struggle. Eur J Hum Genet 16(5):554–564
74. Shirley KL, Hon YY, Penzak SR, Lam YW, Spratlin V, Jann MW (2003) Correlation of cytochrome P450 (CYP) 1A2 activity using caffeine phenotyping and olanzapine disposition in healthy volunteers. Neuropsychopharmacology 28(5):961–966
75. Hiratsuka M (2012) In vitro assessment of the allelic variants of cytochrome P450. Drug Metab Pharmacokinet 27(1):68–84

Tripping with Synthetic Cannabinoids ("Spice"): Anecdotal and Experimental Observations in Animals and Man

Torbjörn U.C. Järbe and Jimit Girish Raghav

Abstract The phenomenon of consuming synthetic cannabinoids ("Spice") for recreational purposes is a fairly recent trend. However, consumption of cannabis dates back millennia, with numerous accounts written on the experience of its consumption, and thousands of scientific reports published on the effects of its constituents in laboratory animals and humans. Here, we focus on consolidating the scientific literature on the effects of "Spice" compounds in various behavioral assays, including assessing abuse liability, tolerance, dependence, withdrawal, and potential toxicity. In most cases, the behavioral effects of "Spice" compounds are compared with those of Δ^9-tetrahydrocannabinol. Methodological aspects, such as modes of administration and other logistical issues, are also discussed. As the original "Spice" molecules never were intended for human consumption, scientifically based information about potential toxicity and short- and long-term behavioral effects are very limited. Consequently, preclinical behavioral studies with "Spice" compounds are still in a nascent stage. Research is needed to address the addiction potential and other effects, including propensity for producing tissue/organ toxicity, of these synthetic cannabimimetic "Spice" compounds.

Keywords Cannabinoid • Cannabinoid receptor 1 • Marijuana • 'Spice' • Synthetic marijuana • THC

Contents

T.U.C. Järbe (✉) and J.G. Raghav
Department of Pharmaceutical Sciences, Center for Drug Discovery (CDD), Northeastern University, 116 Mugar Hall, 360 Huntington Ave, Boston, MA 02115, USA
e-mail: t.jarbe@neu.edu

Curr Topics Behav Neurosci (2017) 32: 263–282
DOI 10.1007/7854_2016_16
Published Online: 8 June 2016

Abbreviations

11-OH-THC	11-Hydroxy-Δ^9-tetrahydrocannabinol
2-AG	2-Arachidonoyl glycerol
AM	Alexandros Makriyannis
CB1R	Cannabinoid receptor type-1
CB2R	Cannabinoid receptor type-2
CBD	Cannabidiol
CP	Compound Pfizer
CPP	Conditioned place preference
ECS	Endocannabinoid system
ER	Emergency room
HU	Hebrew University
i.p.	Intraperitoneal
i.v.	Intravenous
ICSS	Intra-cranial self-stimulation
JWH	John W. Huffman
MFB	medial forebrain bundle
SA	Self-administration
THC	Δ^9-Tetrahydrocannabinol

1 The Cannabis Plant

The relationship between cannabis and man has a long and varied history spanning millennia. The plant has been used for hemp production, a food source, as a medicine, and as a recreational drug. Although its exact geographical origin is unknown, many taxonomists have suggested its origins to be central Asia. The original plant composition likely no longer exists, as humans have greatly influenced the genome through selective breeding. One can grossly divide the plant into two main sub-strains, one emphasizing its use for fiber (cordage) production (high cannabidiol, CBD, and content) and the other strain emphasizing its use as an intoxicant, the latter primarily due to the phytocannabinoid chemical Δ^9-tetrahydrocannabinol (THC). The amount and ratios of cannabinoids, terpenes, and other chemicals also show regional variations [1].

2 Personal Testimonies

Many personal accounts surrounding the mind altering effects of cannabis are dispersed throughout history, including those of members of the hashish eaters club in Paris around the early to mid-seventeenth century. One likely member of the club was the French psychiatrist J. J. Moreau de Tours (1804–1884) who, in his youth, had traveled abroad extensively including visiting the Middle East where he had encountered hashish (cannabis resin). Moreau was fascinated with the topic of the mind and psychosis/mental illness. He believed that valuable insights into the "psychotic state of mind" could be achieved by inducing a temporary "insanity" episode through the means of taking mind altering drugs. He and his students ingested various amounts of hashish resin and systematically recorded their observations. Such observations/recordings formed the foundation for the publication "Du Hachisch et de l'Alienation Mentale; Etudes Psychologiques" (Hashish and Mental Illness; Psychological studies). The 400-page book by Moreau was originally published in 1845 and an English translation was reissued in 1973 by Raven Press. Given Moreau's belief in the usefulness for psychiatry of the revelations of the workings of the brain by mind altering drugs (model psychosis), many scholars view him as the founding father of the discipline we today refer to as Psychopharmacology [2–4]. Description(s) of the mental and physiological effects of high-dose ingestion of hashish ("A young physician, terrified, pressed his head with both hands as if to keep it from bursting, crying: I am lost; I have lost my head; I am going crazy!"; p. 83, English version) are akin to descriptions related to the more recent phenomenon of using synthetic cannabinoids for recreational purposes, here collectively referred to as "Spice" ("My heart starts pounding so fast and hard and doesn't feel real. As of that moment, I no longer know who I am, where I am and what is real"; The Day AM-2201 Ruined My Life, Anonymous testimony, Erowid.org. Aug 11, 2015).

3 Synthetic Cannabinoids

AM-2201 is a synthetic chemical capable of activating brain cannabinoid receptors (CB1R and CB2R) with high receptor binding affinity and potency. While its effects mimic those induced by THC in rat drug discrimination studies, AM-2201

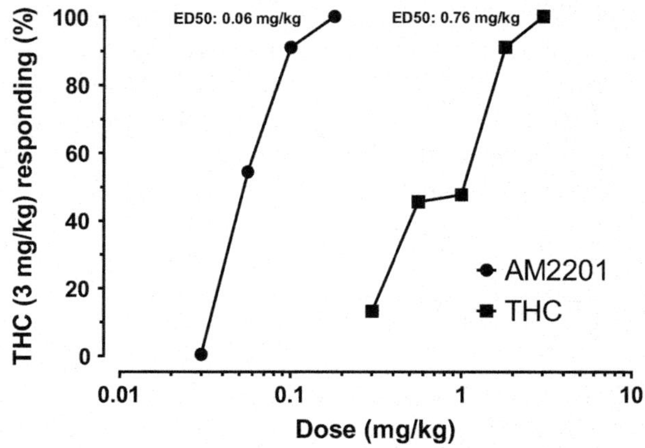

Fig. 1 Substitution test data for AM-2201 and Δ^9-THC in rats trained to discriminate between THC (3 mg/kg) and vehicle 20 min i.p. post-administration. Graph modified from Järbe et al. [5]

is 5–12 times (see Fig. 1) more potent than THC and its duration of action appears shorter [5, 6]. (Note: Potency estimates will depend on the particular endpoint being examined.)

Originally, AM-2201 and similar indole-based ligands (see Fig. 2 for examples) were developed to gain insight into the workings of the endocannabinoid system (ECS), a modulatory signaling system present in brain as well as peripheral body tissues. Endogenous ligands include anandamide and 2-arachidonoyl glycerol (2-AG), but other fatty acid related constituents also have been identified in brain [1]; for chemical structures of the phytocannabinoids THC and CBD, as well as the endogenous ligands anandamide and 2-AG, see Fig. 3. Endocannabinoids bind to and activate both CB1R and CB2R, but the cannabis/THC produced "high" is primarily mediated through activation of CB1R; the contribution by CB2R, if any, remains elusive. Thus, clandestine manufacturing has targeted compounds that activate CB1R. Offerings through the internet, "head-shops," convenience stores, and other venues aim to provide alternatives to cannabis that are not detected by analytical screening assays. As forensic chemists have developed assays to detect synthetic cannabinoids, the clandestine manufacturers have made tweaks in the chemical structures to continue drug trafficking and to evade legal restrictions after previous chemicals have been banned. In addition, the package wrapping usually contains a disclaimer which states that the herbal incense "is not for human consumption." Although the initial wave of recreational synthetic cannabinoids borrowed design and synthetic routes that were already available in the scientific literature, many more recent chemicals are novel and have not been described before; for references, see [7, 8]. Information on the biological effects of many clandestine synthetic cannabinoids is scant and mostly limited to receptor binding affinities for the two receptors. Accounts on the evolution of this relatively recent phenomenon of clandestine production of "Spice" synthetic cannabinoids have been provided by many scholars, e.g., [9]. In addition to synthetic cannabinoids, synthetic drugs mimicking the effects of psychomotor stimulants (cathinone,

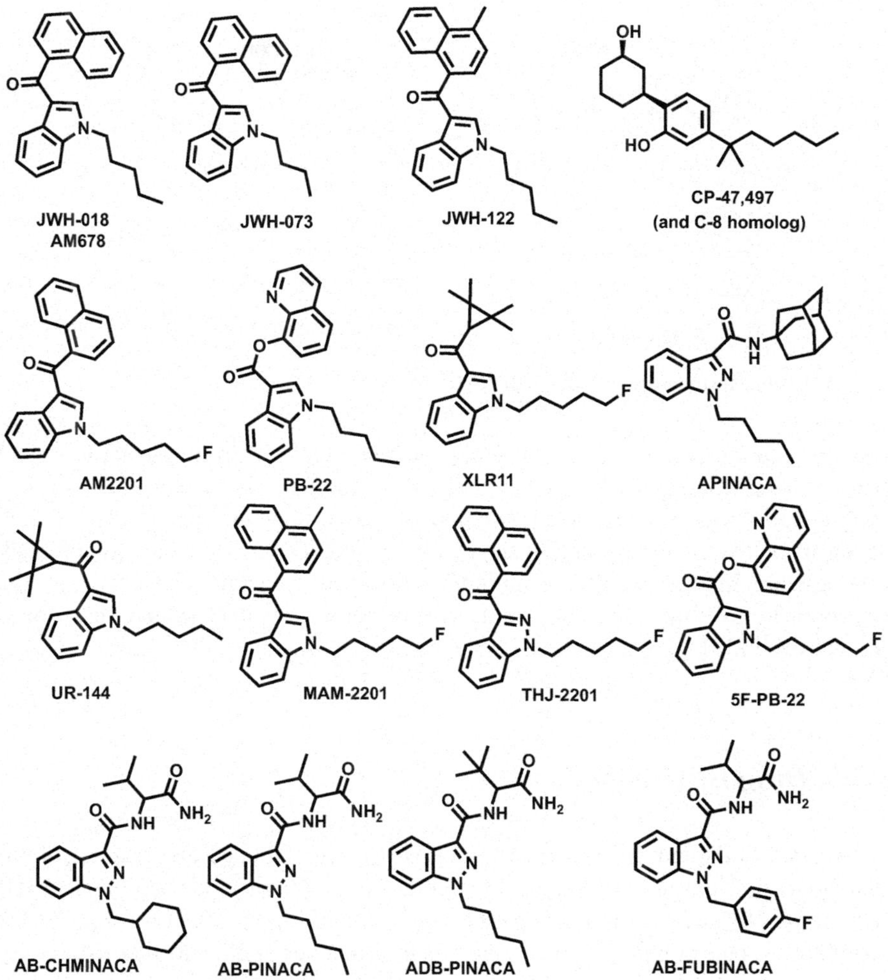

Fig. 2 Examples of chemical structures of cannabinergic indoles detected in "Spice" concoctions

"Bath salts"), hallucinogens/psychedelics (e.g., "N-bomb"), and opiates (e.g., "Krokodile") are also available through the internet [10]; see also [11].

4 Common Cannabinoid Screens: Tetrad and Drug Discrimination

4.1 Tetrad

Two common procedures for detecting CB1R activation in vivo are a tetrad battery of pharmacological tests in rodents and drug discrimination. The "tetrad" is a composite bioassay developed primarily for mice and rats [12], consisting of

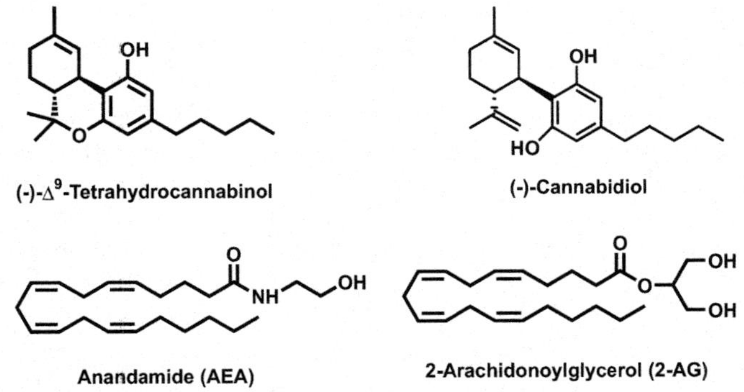

Fig. 3 Chemical structures of THC, CBD, AEA, and 2-AG

(a) nociception testing, (b) rectal body temperature, (c) locomotor activity, and (d) increased catalepsy (muscle rigidity), where the four components are assessed sequentially after a single administration of the compound. Psychoactive CB1R agonists increase antinociception and catalepsy and decrease locomotion and body temperature in these tests. To minimize the occurrence of false positives, the importance of including all four components in the screen has been emphasized [13].

4.2 *Drug Discrimination*

Compared to the tetrad, drug discrimination is more elaborate and pharmacologically specific, but also is a more time-consuming approach for assessing CB1R activation [14]. The approach is based on the idea that drug effects can serve as a stimulus or cue to guide choice behavior in a manner similar to that of sensory events such as sound or light. For example, a food restricted rat is injected with a training drug (e.g., THC) or with vehicle, and is placed into an operant box with the task of selecting one of two levers to gain access to food. If the rat was injected with the training drug, pressing one lever (e.g., left) will deliver food; conversely, if the rat was injected with vehicle, pressing the other (right) lever will deliver food. Repeated pairings of the two training conditions and associated differential responding required to access food are the basis for stimulus control of the choice behavior, i.e., the presence/absence of the training drug effects guides the subjects' choice performance. After asymptotic performance is reached, new chemicals can be tested for generalization/substitution. If the new chemical mimics the pharmacological effects of the training drug, the subject will press the drug-associated lever; if not, the vehicle-associated lever will be chosen. Co-administration of a presumed receptor blocker with the training drug will evaluate antagonism [15, 16].

4.3 Drug Discrimination and "Spice"

Using cannabinoids (primarily THC) as the training drug, discriminative control over choice behavior has been studied in various species (pigeons, gerbils, mice, rats, monkeys, and humans) [17]. Several synthetic "Spice" compounds have been found to mimic the effects of THC in drug discrimination in rodents, as evidenced by full substitution for THC [5, 6, 18–24]. Additionally, CB1R mediation has been verified by reversal of the THC-like effects of synthetic cannabinoids with rimonabant, a selective CB1R antagonist/inverse agonist. More recent studies have employed synthetic cannabinoids from the initial waves of clandestine offerings such as JWH-018 (also known as AM-678) as discriminative stimuli in monkeys and rodents [25, 26]. Cross-substitution with THC has been demonstrated in these studies. Given the rapid proliferation of "Spice" chemicals, more cannabimimetic drugs have been tried by humans than have been tested in preclinical studies in animals. So far, drug discrimination studies in animals have validated the general cannabimimetic nature of "Spice" compounds without exception. Of course, that does not mean that "Spice" compounds are more or less safe than THC/marijuana. It only means that activation of CB1R is a shared mechanism of action in producing the in vivo "high."

5 Logistical Issues

5.1 "Spice" Toxicity

As summarized previously, e.g. [27, 28], the use of "Spice" compounds has resulted in increased cannabinoid-related calls/visits to emergency room (ER) facilities and sometimes has been associated with life-threatening medical consequences. Given the unregulated "Spice" market, the exact amounts of biologically active components contained in the "herbal incense" products are unclear. Although independent chemical analysis of samples of two "Spice" drugs (JWH-018 and JWH-073) purchased from China were of reasonable purity [29], the degree to which these samples represent the bulk of material contained in "Spice" products is unknown. Consideration should be given to possible impurities remaining from the synthesis process, as well as residual organic solvents, as contributors to toxicity of "Spice" concoctions. In addition, the sequence of events required to mix bulk chemicals with plant material to prepare products for sale is unregulated and often not known. The limited pharmacological/toxicological data gathered thus far do not seem to implicate any particular feature of synthetic cannabimimetic drugs that distinguish them from THC/marijuana, except potency and efficacy. Generally, most "Spice" compounds are full agonists at CB1R (as well as CB2R) compared to THC, which is a partial agonist. Given the abundance of CB1R in brain and the putative role of the ECS in fine-tuning neurotransmission and homeostasis, it should be no surprise that global activation of CB1R will have ripple effects across the nervous system with consequences for regulatory control of the body [30].

5.2 Metabolism

Another potentially significant difference between THC and "Spice" drugs concerns their metabolism. Whereas metabolism of THC essentially yields only one major psychotropic metabolite (i.e., 11-OH-THC) before excretion [31, 32], the metabolism of, e.g., JWH-018/AM-678 yields more biologically active intermediaries before elimination [33]. Furthermore, depending on the specific synthetic cannabinoid in question, there may also be differences between THC and "Spice" (as well as among "Spice" molecules) regarding metabolic pathways as well as interactions with potassium, nicotinic, and serotonin receptors, in addition to CB1R [34–36]. For example, based on the metabolite profiles, it has been observed that the major human cytochrome P450 enzymes (liver and recombinant) involved in the oxidative metabolism of AB-CHMINACA differ when compared to those that are involved in the biotransformation and elimination of JWH-018/AM678 [34, 37]. This could be important as the degrading enzymes involved with AB-CHMINACA are also involved in the metabolism of several marketed medications as well as hormones [34].

5.3 Mode of Administration

Members of the hashish eaters club took hashish by eating it in the form of a paste, which is less common today. Rather, inhalation as smoke or vapor of cannabis products is currently the preferred way of consuming the intoxicant. Inhalation is also the most common route of administration for "Spice" compounds. Delivery through the lungs allows the user to titrate the dose compared to oral intake, as absorption of THC through the digestive system is quite variable [31]. Furthermore, inhalation [as well as intravenous (i.v.) infusion] will circumvent first-pass metabolism in the liver and delivers the drug directly to the brain. Few studies have compared the effects following inhalation of "Spice" with the more commonly used intraperitoneal (i.p.) injection procedure in rodents. One study involving JWH-018 suggested comparable results in the tetrad [38], whereas another study [39] noted absence of catalepsy after inhalation of the "Spice" compounds JWH-073 and JWH-018 as well as THC. The authors of the latter study speculated that the different outcomes might be due to a lack of active metabolites after inhalation of the cannabinergics compared to the i.p. route. Additionally, as a cursory note, combustion due to heating may change drug composition, as heat apparently converts some of the parent compound AM-2201 into the closely related JWH-018/AM-678 and JWH-022 "Spice" compounds [40]. The two latter

"Spice" molecules, especially JWH-018/AM-678, also exhibit higher potency compared to THC in drug discrimination using rodents [5, 6, 21, 26, 39, 41] and monkeys [25, 42]. Further research is needed to better delineate the details and generality of such findings.

5.4 Methodological/Procedural Considerations

For the most part, cannabinergics are not water soluble and therefore require organic solvents as carriers in order to prepare suspensions suitable for parenteral administration. A common vehicle for injecting ligands acting at CB1R in experimental animals consists of ethanol:cremophor:saline in the proportion of 1:1:18. However, there are many variations of this scheme and comparisons regarding efficiency in uptake and distribution of the injected materials are scant, making comparisons of results across laboratories more difficult [17]. Regarding inhalation (smoke, vapor) of cannabinergics, scientists have designed their own delivery devices. Hence, there are no uniform standards, although such devices have been shown to deliver sufficient concentrations of cannabinergics to the organism to produce relevant pharmacological outcomes, e.g., [43]. As animals have not voluntarily inhaled the combustion products, the consequences of the imposed restraint necessary for the exposure may by itself alter subsequent test performance; see, e.g., [39].

The injection-to-test interval is an important consideration, as some exogenous cannabinergics such as HU-210 have a slow onset as well as an extended duration of action in pigeons, rats, and monkeys [44, 45]; for other examples of cannabinergics with similar in vivo time-course profile, see [46–48]. The slow onset of certain cannabimimetic drugs may initially be miscued as low potency, and more material will be ingested to achieve the "high." Under such circumstances, when the effect of the drug finally kicks in, a much stronger than desired effect may occur. Such delayed onset of effect has also been described for edible marijuana [49]. As HU-210 was present in some early confiscated batches of "Spice," it is now a Schedule I drug in the USA.

When dealing with live experimental animals, environmental and procedural factors can have profound effects on responsiveness to drugs. It has been argued that the effects of exogenous cannabinergics and endogenous cannabinoid-like signaling molecules may be particularly susceptible to such influences given the purported involvement of the ECS in the regulatory modulation of emotion and cognition processes [50]; see also [51]. In essence, this boils down to the importance of the concepts of "set" (the state of the organism at drug exposure) and "setting" (the prevailing external conditions surrounding the drug experience).

6 Drug Reinforcement

Drugs considered to have high abuse potential in humans oftentimes also serve as reinforcers for animals, as examined mostly in laboratory rodents and monkeys. Thus, animals press a lever to get access to i.v. infusions of opioids (e.g., heroin), psychomotor stimulants (e.g., cocaine), and other drugs (self-administration, SA); they display preference (or aversion) for distinct environments associated with positive drug effects (conditioned place preference, CPP); and they exhibit facilitation (or not) of intra-cranial electrical stimulation (ICSS). These are the three most common techniques in experimental animals for assessing abuse liability of drugs in vivo.

6.1 SA and Cannabinergics

Early attempts to establish THC as a reinforcer in animals met with limited success using the SA paradigm [52]. However, more recent developments have suggested that i.v. infusions of WIN 55,212-2, an aminoalkyl indole that is chemically related to many "Spice" compounds, may sustain SA in rodents; for review, see [53]. In agreement with non-human primate data (see below), the endogenous ligand 2-AG seemed to maintain robust SA in rats [54] whereas similar infusion protocols applied to rodents responding for the "Spice" aminoalkyl indole JWH-018/AM-678 produced more variable SA performances [55], more akin to SA performances seen with WIN 55,212-2 [53]. However, other indole cannabinergics (JWH-073, JWH-081, and JWH-210) as well as THC did not serve as reinforcers using SA protocols in mice [56].

An intriguing finding from one laboratory was that when THC was substituted for the maintenance drug WIN 55,212-2, responding for i.v. infusions stopped, suggesting that infusions of THC are less rewarding than infusions of WIN 55,212-2 [57]. Using a relatively unconventional choice procedure, Braida et al. [58, 59] found biphasic functions for THC as well as the potent cannabinergic CP-55,940, i.e., there was an increased rate of responding after low doses and a decrease after higher doses, where the drugs were delivered by an intracerebroventricular route of administration.

Using squirrel monkeys, THC as well as anandamide and its longer acting analog methanandamide and 2-AG as well as the putative anandamide transporter ligand AM-404 have all been found to serve as reinforcers, exhibiting the typical inverted dose-response function, with middle doses producing the highest response rate; for reviews, see [52, 60, 61]. As a cautionary note, there are no published replications of these findings with monkeys from other laboratories except to note that rhesus monkeys did not self-administer THC when THC was presented alone to these animals, originally trained to self-administer heroin [62].

Based on above examples as well as other studies reviewed elsewhere [63], it seems clear that cannabinergics can serve as reinforcers for experimental animals. However, the many reported failures of finding SA with cannabinergics, especially THC, also suggest that the conditions under which SA occurs with cannabinergics are more limited compared to abused drugs like cocaine or heroin [64, 65]. Given the more positive findings with WIN 55,212-2, and the observation that THC suppressed responding in rats maintained on WIN 55,212-2 SA, might suggest that cannabinergics may differ in their reinforcing effects and hence in their capacity to subserve maintenance of SA. Fast onset and a relatively short duration of action for most of the cannabinergic "Spice" indoles may be contributing factors.

6.2 CPP and Cannabinergics

The CPP procedure may be used for assessing the valence of drug effects (preferred, neutral, or non-preferred) in rodents. The hedonic value is inferred by measuring the relative time spent in two contextually distinct compartments after explicit repeated pairings of the drug effects with one of the two contexts. Often the two environments are separated by a smaller box from which the two distinct compartments can be accessed (start-area) [66, 67]. It is imperative that a range of doses be examined as the window of detecting hedonia (reinforcing effects) can be quite limited as seems to be the case with cannabinergics. Such studies have been reviewed elsewhere [53] and the authors emphasized that dose, time of testing, rodent strain, and other variables are all critical determinants for the outcome of CPP studies. For example, 2–4 mg/kg doses of THC resulted in a place preference in rats at the 24 h post-injection test exposure whereas an aversion occurred when examined 48 h post-injection, i.e., animals spent more time in the alternate compartment compared to controls [68]. Familiarity with the drug effects may facilitate detecting hedonic or reinforcing effects of the drug in question. Thus, priming with an administration of THC the day before the first drug-context pairing resulted in context-preference with lower doses of THC as opposed to non-primed animals, which exhibited only context-aversion with the same doses; higher doses resulted in non-preference irrespective of priming or not [69]. Genetic variants may include "spontaneously" hypertensive rats, as they exhibited preference as adolescents in a CPP protocol with doses of WIN 55,212-2 which produced context/place aversion in controls [70]. This genetic pre-disposition has been proposed as a rodent model for human attention deficit disorder. As such, it may point to a possible pre-dispositional vulnerability in developing drug abuse disorders.

Thus, traditionally examined cannabinergics (THC, WIN 55,212-4, CP-55,940 as well as HU-210) may exhibit hedonic valence in rodents in the CPP model of reinforcement but appear to do so under more limited conditions compared to heroin and cocaine.

Studies employing "Spice" compounds do not indicate any major shift in our understanding of outcomes of cannabinergics and CPP. For example, a history of pre-exposure to THC facilitated detection of place preference with one early "Spice" compound (JWH-018/AM-678) at the lower dose range whereas non-primed rodents displayed aversion [71]. Additional studies indicated that JWH-175, a weaker but chemically related compound, resulted in place preference in mice without pre-exposure [72]. Additional related "Spice" chemicals (JWH-073, JWH-081, and JWH-210) demonstrated preference at lower doses and aversion at higher doses [56]. Hence, dose is a crucial factor in CPP studies examining cannabinergic compounds detected in "Spice" products.

6.3 ICSS and Cannabinergics

Like drug self-administration, ICSS is an operant procedure whereby pressing a lever, or by other means, animals can deliver electrical currents through electrodes implanted in brain areas associated with pleasure. The neural circuitry sustaining ICSS is referred to as medial forebrain bundle and involves brain areas connecting ventral tegmentum, lateral hypothalamus, nucleus accumbens, and surrounding areas with further projection(s) into the pre-frontal cortex [73].

Traditionally studied cannabinergics such as THC, CP-55,940, and WIN 55,212-2 have produced mixed outcomes [53]. It seems fair to conclude that the cannabinergics studied thus far have not reliably produced a facilitation of brain thresholds in comparison to the facilitation more robustly seen with heroin and cocaine [73]. Whether "Spice" drugs will pose a challenge to this general consensus is an open question.

7 Tolerance, Dependence, and Withdrawal

Upon repeated administration, tolerance to the effects of cannabinergics is likely to occur [74]. The extent and rate of tolerance development is dependent upon a variety of factors, including the endpoint being examined. For example, attenuation of hypothermia in rodents is marked and swift, occurring within days of repeated drug exposure [75]. Tolerance development to the effects of cannabinergics is primarily due to pharmacodynamic, rather than pharmacokinetic changes, e.g., [76]. Proposed neural mechanisms underlying the adaptation are receptor downregulation and/or receptor desensitization, i.e., reduced expression of receptors and/or a diminished responsiveness of the receptors to their ligands [77–79]. Published case reports suggest that profound tolerance can develop with repeated exposure to "Spice" cannabimimetics and that this tolerance may lead to an escalation of consumption of the drug in order to achieve desired effects. Adjustment to high "Spice" doses likely will be accompanied by withdrawal reactions upon termination of drug ingestion. Experimental animals also display withdrawal phenomena after continuous exposure to cannabinergics; for references, see [80]. Tolerance and cross-tolerance to and between "Spice" drugs (JWH-018/AM-678 and JWH-073) and THC suggested rapid tolerance development to hypothermia and cross-tolerance between "Spice" and THC in mice. As measured, drug-induced locomotor suppression did not indicate tolerance development for any of these three compounds. The extent by which the distinction between full and partial agonism is helpful in explaining these outcomes was discussed [81]; for further elaboration on this topic, see also [78].

8 "Spice": Neurological and Sensomotor Aspects

Published case reports indicate that "Spice" products are associated with increased incidences of severe psychiatric and neurological reactions, rarely seen after intake of cannabis preparations [82, 83]. Given the paucity of information related to safety issues about "Spice" chemicals, there are many unknowns. For example, are certain chemical/structural features more linked to severe reactions? An article about an acute delirium outbreak in users exposed to the chemical AB-CHIMNACA might suggest such a possibility [28]. Alternatively, can such an outbreak be understood basically in terms of excessive cannabinoid receptor stimulation, resulting in homeostatic collapse(s) leading to physiological dysregulation and organ failure (s)? [30]. Science has only recently begun to address such questions and thus far only limited data are available.

Examining a variety of more recently appearing "Spice" cannabimimetics, Banister and colleagues [7, 8] observed dose-related hypothermia and bradycardia in mice – though duration of action differed among compounds. Another study [83] examined effects of halogenated derivatives of JWH-018/AM-678 and found that the derivatives appeared to exhibit a more benign side-effect profile compared to the parent compound. Whereas administration of JWH-018/AM-678 was associated with a high incidence of seizures, myoclonia, and hyperreflexia, the bromo analog exhibited much reduced or no such effects at all; the chloro analog showed intermediate activity. All compounds, including THC, produced a typical constellation of cannabinergic behavioral effects (decreased locomotion, hypothermia, and catalepsy). All effects were attenuated by chemical CB1R antagonist blockade with AM-251. A related study [82] examined sensomotor functions (visual, auditory, and tactile) as well as neurological changes (convulsion, myoclonia, and hyperreflexia), comparing the effects of THC and JWH-018/AM-678. At lower doses, both drugs produced similar effects. At higher doses, JWH-018/AM-678 stood out as it resulted in pronounced neurological changes and marked decreases in sensomotor functions. All effects, including those of THC, were prevented by administration of the CB1R antagonist/inverse agonist AM-251. The related cannabimimetic indoles JWH-081 and JWH-210 dose-dependently affected locomotor activity and rotarod performance; unlike the indoles described above, convulsion was absent although no direct comparison with above indoles was made; tremor, however, was noted [84]. Histological analysis of hippocampal tissue suggested histopathology although no changes in overt performance in the Morris water maze task were reported [84]. Another study, examining JWH-081 reported, however, that the drug negatively affected performance in a Y-maze task, accompanied by molecular changes in hippocampal activity. Higher doses of this ligand impaired object recognition in an open-field test and interfered with spontaneous alternation in the Y-maze test [85]. These results were observed only in wild-type, but not in CB1R knock-out mice, hence implicating CB1R mediation and perhaps also absence of an involvement of "off-target" effects in generating these outcomes. Although above findings do not allow for a general conclusion about the potential toxicity of "Spice" products, they represent initial efforts by the scientific community to assist society in its challenge to deal with this relatively recent phenomenon of "high-jacking" science for unregulated commercial endeavors [86].

The previous quote from Moreau mentioned in the introduction contains a final sentence which reads: "Fortunately, his fears were promptly replaced by the wildest kind of joy." (p. 83, English version). However, based on the findings presented in this chapter, tripping with synthetic "Spice" cannabinoids is no laughing matter.

Acknowledgements Preparation of this manuscript was defrayed in part by the National Institute on Drug Abuse grant 5RO1DA 009064-19 as well as other NIH/NIDA funds awarded to the CDD. We thank Dr K. Vemuri for his help with chemistry related issues, including preparing Figs. 2 and 3 and Dr. J.L. Wiley for editorial suggestions. "Disclosure Statement": Authors declare that there is no actual or potential conflict of interest related to this manuscript. "Role of the funding source (s)": Authors declare that the study sponsor did not have any role in study design; in the collection, analysis, and interpretation of data; in the writing of the report; and in the decision to submit the paper for publication.

References

1. Hanus LO (2009) Pharmacological and therapeutic secrets of plant and brain (endo)cannabinoids. Med Res Rev 29:213–271
2. Badiani A (2010) History of psychopharmacology. In: Stolerman IP (ed) Encyclopedia of psychopharmacology. Springer, Berlin, Heidelberg, pp 593–603
3. Baruk H (1970) The Society Moreau de Tours and Psychopharmacology. Med Klin 65:1803–1805
4. Mora G (1989) Historical antecedents of modern psychopharmacology. Psychiatr J Univ Ott 14:279–281
5. Järbe TUC, Gifford RS, Zvonok A, Makriyannis A (2015) Δ9-Tetrahydrocannabinol discriminative stimulus effects of AM2201 and related aminoalkylindole analogs in rats. Behav Pharmacol
6. Gatch MB, Forster MJ (2014) Delta9-Tetrahydrocannabinol-like discriminative stimulus effects of compounds commonly found in K2/Spice. Behav Pharmacol 25:750–757
7. Banister SD, Moir M, Stuart J, Kevin RC, Wood KE, Longworth M, Wilkinson SM, Beinat C, Buchanan AS, Glass M, Connor M, McGregor IS, Kassiou M (2015) Pharmacology of indole and indazole synthetic cannabinoid designer drugs AB-FUBINACA, ADB-FUBINACA, AB-PINACA, ADB-PINACA, 5F-AB-PINACA, 5F-ADB-PINACA, ADBICA, and 5F-ADBICA. ACS Chem Neurosci 6:1546–1559

8. Banister SD, Stuart J, Kevin RC, Edington A, Longworth M, Wilkinson SM, Beinat C, Buchanan AS, Hibbs DE, Glass M, Connor M, McGregor IS, Kassiou M (2015) Effects of bioisosteric fluorine in synthetic cannabinoid designer drugs JWH-018, AM-2201, UR-144, XLR-11, PB-22, 5F-PB-22, APICA, and STS-135. ACS Chem Neurosci
9. Fattore L, Fratta W (2011) Beyond THC: the new generation of cannabinoid designer drugs. Front Behav Neurosci 5:60 (1-12)
10. Baumann MH, Solis E Jr, Watterson LR, Marusich JA, Fantegrossi WE, Wiley JL (2014) Baths salts, spice, and related designer drugs: the science behind the headlines. J Neurosci 34:15150–15158
11. Castaneto MS, Gorelick DA, Desrosiers NA, Hartman RL, Pirard S, Huestis MA (2014) Synthetic cannabinoids: epidemiology, pharmacodynamics, and clinical implications. Drug Alcohol Depend 144:12–41
12. Martin BR, Compton DR, Thomas BF, Prescott WR, Little PJ, Razdan RK, Johnson MR, Melvin LS, Mechoulam R, Ward SJ (1991) Behavioral, biochemical, and molecular modeling evaluations of cannabinoid analogs. Pharmacol Biochem Behav 40:471–478
13. Wiley JL, Razdan RK, Martin BR (2006) Evaluation of the role of the arachidonic acid cascade in anandamide's in vivo effects in mice. Life Sci 80:24–35
14. Järbe TUC, Gifford RS (2014) "Herbal incense": designer drug blends as cannabimimetics and their assessment by drug discrimination and other in vivo bioassays. Life Sci 97:64–71
15. Järbe TUC, Swedberg MDB (1982) A conceptualization of drug discrimination learning. In: Colpaert FC, Slangen JL (eds) Drug discrimination: application in CNS pharmacology. Elsevier/North-Holland Biomed Press, Amsterdam, pp 327–341
16. Solinas M, Panlilio LV, Justinova Z, Yasar S, Goldberg SR (2006) Using drug-discrimination techniques to study the abuse-related effects of psychoactive drugs in rats. Nat Protoc 1:1194–1206
17. Järbe TUC (2011) Perceptual drug discriminative aspects of the endocannabinoid signaling system in animals and man. In: Glennon RA, Young R (eds) Drug discrimination: applications to medicinal chemistry and drug studies. Wiley, Hoboken, pp 241–285
18. Brents LK, Zimmerman SM, Saffell AR, Prather PL, Fantegrossi WE (2013) Differential drug-drug interactions of the synthetic cannabinoids JWH-018 and JWH-073: implications for drug abuse liability and pain therapy. J Pharmacol Exp Ther 346:350–361
19. Gatch MB, Forster MJ (2015) Delta9-Tetrahydrocannabinol-like effects of novel synthetic cannabinoids found on the gray market. Behav Pharmacol 26:460–468
20. Gatch MB, Forster MJ (2016) Delta-Tetrahydrocannabinol-like effects of novel synthetic cannabinoids in mice and rats. Psychopharmacology 233(10):1901–10
21. Järbe TUC, Deng H, Vadivel SK, Makriyannis A (2011) Cannabinergic aminoalkylindoles, including AM678=JWH018 found in 'Spice', examined using drug (Δ^9-THC) discrimination for rats. Behav Pharmacol 22:498–507
22. Samano KL, Poklis A, Lichtman AH (2013) Preclinical investigation of the abused synthetic cannabinoid CP47,497. College on Problems of Drug Dependence 75th Annual Meeting. College on Problems of Drug Dependence, San Diego, CA, abstract # 555
23. Wiley JL, Marusich JA, Lefever TW, Antonazzo KR, Wallgren MT, Cortes RA, Patel PR, Grabenauer M, Moore KN, Thomas BF (2015) AB-CHMINACA, AB-PINACA, and FUBIMINA: affinity and potency of novel synthetic cannabinoids in producing Delta9-tetrahydrocannabinol-like effects in mice. J Pharmacol Exp Ther 354:328–339
24. Wiley JL, Marusich JA, Lefever TW, Grabenauer M, Moore KN, Thomas BF (2013) Cannabinoids in disguise: Delta9-tetrahydrocannabinol-like effects of tetramethylcyclopropyl ketone indoles. Neuropharmacology 75:145–154
25. Rodriguez JS, McMahon LR (2014) JWH-018 in rhesus monkeys: differential antagonism of discriminative stimulus, rate-decreasing, and hypothermic effects. Eur J Pharmacol 740:151–159

26. Wiley JL, Lefever TW, Cortes RA, Marusich JA (2014) Cross-substitution of Delta9-tetrahydrocannabinol and JWH-018 in drug discrimination in rats. Pharmacol Biochem Behav 124:123–128
27. Muller HH, Kornhuber J, Sperling W (2015) The behavioral profile of spice and synthetic cannabinoids in humans. Brain Res Bull S0361–9230(15):30052–6
28. Tyndall JA, Gerona R, De Portu G, Trecki J, Elie MC, Lucas J, Slish J, Rand K, Bazydlo L, Holder M, Ryan MF, Myers P, Iovine N, Plourde M, Weeks E, Hanley JR, Endres G, Germaine DS, Dobrowolski PJ, Schwartz M (2015) An outbreak of acute delirium from exposure to the synthetic cannabinoid AB-CHMINACA. Clin Toxicol 53(10):950–6
29. Ginsburg BC, McMahon LR, Sanchez JJ, Javors MA (2012) Purity of synthetic cannabinoids sold online for recreational use. J Anal Toxicol 36:66–68
30. Lu H-C, Mackie K (2016) An introduction to the endogenous cannabinoid system. Biol Psychiatry 79(7):516–25
31. Huestis MA (2007) Human cannabinoid pharmacokinetics. Chem Biodivers 4:1770–1804
32. Matsuda LA, Lolait SJ, Brownstein MJ, Young AC, Bonner TI (1990) Structure of a cannabinoid receptor and functional expression of the cloned cDNA. Nature 346:561–564
33. Su MK, Seely KA, Moran JH, Hoffman RS (2015) Metabolism of classical cannabinoids and the synthetic cannabinoid JWH-018. Clin Pharmacol Ther 97:562–564
34. Erratico C, Negreira N, Norouzizadeh H, Covaci A, Neels H, Maudens K, van Nuijs AL (2015) In vitro and in vivo human metabolism of the synthetic cannabinoid AB-CHMINACA. Drug Test Anal 7:866–876
35. Gurney SM, Scott KS, Kacinko SL, Presley BC, Logan BK (2014) Pharmacology, toxicology, and adverse effects of synthetic cannabinoid drugs. Forensic Sci Rev 26:53–78
36. Pertwee RG (2010) Receptors and channels targeted by synthetic cannabinoid receptor agonists and antagonists. Curr Med Chem 17:1360–1381
37. Chimalakonda KC, Seely KA, Bratton SM, Brents LK, Moran CL, Endres GW, James LP, Hollenberg PF, Prather PL, Radominska-Pandya A, Moran JH (2012) Cytochrome P450-mediated oxidative metabolism of abused synthetic cannabinoids found in K2/Spice: identification of novel cannabinoid receptor ligands. Drug Metab Dispos 40:2174–2184
38. Wiebelhaus JM, Poklis JL, Poklis A, Vann RE, Lichtman AH, Wise LE (2012) Inhalation exposure to smoke from synthetic "marijuana" produces potent cannabimimetic effects in mice. Drug Alcohol Depend 126:316–323
39. Marshell R, Kearney-Ramos T, Brents LK, Hyatt WS, Tai S, Prather PL, Fantegrossi WE (2014) In vivo effects of synthetic cannabinoids JWH-018 and JWH-073 and phytocannabinoid Delta9-THC in mice: inhalation versus intraperitoneal injection. Pharmacol Biochem Behav 124:40–47
40. Donohue K, Steiner R (2012) JWH-018 and JWH-022 as combustion products of AM2201. Microgram J 9:52–56
41. Järbe TUC, Li C, Vadivel SK, Makriyannis A (2010) Discriminative stimulus functions of methanandamide and Δ^9-THC in rats: tests with aminoalkylindoles (WIN55,212-2 and AM678) and ethanol. Psychopharmacology (Berl) 208:87–98
42. Ginsburg BC, Schulze DR, Hruba L, McMahon LR (2012) JWH-018 and JWH-073: Delta(9)-tetrahydrocannabinol-like discriminative stimulus effects in monkeys. J Pharmacol Exp Ther 340:37–45
43. Vann RE, Walentiny DM (2011) Preclinical evaluation of marijuana's appetitive, rewarding and psychoactive properties. The 21st Annual Symposium of the International Cannabinoid Research Society. International Cannabinoid Research Society, Research Triangle Park, NC, USA, Pheasant Run, St. Charles, IL, USA
44. Hruba L, McMahon LR (2014) The cannabinoid agonist HU-210: pseudo-irreversible discriminative stimulus effects in rhesus monkeys. Eur J Pharmacol 727:35–42
45. Järbe TUC, Hiltunen AJ, Mechoulam R (1989) Stereospecificity of the discriminative stimulus functions of the dimethylheptyl homologs of 11-hydroxy-Δ^8-tetrahydrocannabinol in rats and pigeons. J Pharmacol Exp Ther 250:1000–1005

46. Devane WA, Breuer A, Sheskin T, Järbe TUC, Eisen MS, Mechoulam R (1992) A novel probe for the cannabinoid receptor. J Med Chem 35:2065–2069
47. Järbe TUC, Swedberg MD, Mechoulam R (1981) A repeated test procedure to assess onset and duration of the cue properties of (-) Δ^9-THC, (-) Δ^8-THC-DMH and (+) Δ^8-THC. Psychopharmacology (Berl) 75:152–157
48. Järbe TUC, Tai S, Lemay BJ, Nikas SP, Shukla VG, Zvonok A, Makriyannis A (2012) AM2389, a high-affinity, in vivo potent CB(1)-receptor-selective cannabinergic ligand as evidenced by drug discrimination in rats and hypothermia testing in mice. Psychopharmacology (Berl) 220:417–426
49. Hancock-Allen JB, Barker L, VanDyke M, Holmes DB (2015) Notes from the field: death following ingestion of an Edible Marijuana Product--Colorado, March 2014. MMWR Morb Mortal Wkly Rep 64:771–772
50. Zanettini C, Panlilio LV, Alicki M, Goldberg SR, Haller J, Yasar S (2011) Effects of endocannabinoid system modulation on cognitive and emotional behavior. Front Behav Neurosci 5:57
51. Tan H, Ahmad T, Loureiro M, Zunder J, Laviolette SR (2014) The role of cannabinoid transmission in emotional memory formation: implications for addiction and schizophrenia. Front Psychiatry 5:73
52. Panlilio LV, Goldberg SR, Justinova Z (2015) Cannabinoid abuse and addiction: clinical and preclinical findings. Clin Pharmacol Ther 97:616–627
53. Panagis G, Mackey B, Vlachou S (2014) Cannabinoid regulation of brain reward processing with an emphasis on the role of CB1 receptors: a step back into the future. Front Psychiatry 5:92
54. De Luca MA, Valentini V, Bimpisidis Z, Cacciapaglia F, Caboni P, Di Chiara G (2014) Endocannabinoid 2-arachidonoylglycerol self-administration by Sprague-Dawley rats and stimulation of in vivo dopamine transmission in the nucleus accumbens shell. Front Psychiatry 5:140
55. De Luca MA, Bimpisidis Z, Melis M, Marti M, Caboni P, Valentini V, Margiani G, Pintori N, Polis I, Marsicano G, Parsons LH, Di Chiara G (2015) Stimulation of in vivo dopamine transmission and intravenous self-administration in rats and mice by JWH-018, a Spice cannabinoid. Neuropharmacology 99:705–714
56. Cha HJ, Lee KW, Song MJ, Hyeon YJ, Hwang JY, Jang CG, Ahn JI, Jeon SH, Kim HU, Kim YH, Seong WK, Kang H, Yoo HS, Jeong HS (2014) Dependence potential of the synthetic cannabinoids JWH-073, JWH-081, and JWH-210: in vivo and in vitro approaches. Biomol Ther (Seoul) 22:363–369
57. Lefever TW, Marusich JA, Antonazzo KR, Wiley JL (2014) Evaluation of WIN 55,212-2 self-administration in rats as a potential cannabinoid abuse liability model. Pharmacol Biochem Behav 118:30–35
58. Braida D, Iosue S, Pegorini S, Sala M (2004) Delta9-tetrahydrocannabinol-induced conditioned place preference and intracerebroventricular self-administration in rats. Eur J Pharmacol 506:63–69
59. Braida D, Pozzi M, Parolaro D, Sala M (2001) Intracerebral self-administration of the cannabinoid receptor agonist CP 55,940 in the rat: interaction with the opioid system. Eur J Pharmacol 413:227–234
60. Schindler CW, Scherma M, Redhi GH, Vadivel SK, Makriyannis A, Goldberg SR, Justinova Z (2016) Self-administration of the anandamide transport inhibitor AM404 by squirrel monkeys. Psychopharmacology 233(10):1867–77
61. Vlachou S, Panagis G (2014) Regulation of brain reward by the endocannabinoid system: a critical review of behavioral studies in animals. Curr Pharm Des 20:2072–2088
62. Li JX, Koek W, France CP (2012) Interactions between Delta(9)-tetrahydrocannabinol and heroin: self-administration in rhesus monkeys. Behav Pharmacol 23:754–761
63. Panlilio LV, Justinova Z, Goldberg SR (2010) Animal models of cannabinoid reward. Br J Pharmacol 160:499–510

64. Cooper ZD, Haney M (2008) Cannabis reinforcement and dependence: role of the cannabinoid CB1 receptor. Addict Biol 13:188–195
65. Justinova Z, Goldberg SR, Heishman SJ, Tanda G (2005) Self-administration of cannabinoids by experimental animals and human marijuana smokers. Pharmacol Biochem Behav 81:285–299
66. Bardo MT, Bevins RA (2000) Conditioned place preference: what does it add to our preclinical understanding of drug reward? Psychopharmacology (Berl) 153:31–43
67. Tzschentke TM (2007) Measuring reward with the conditioned place preference (CPP) paradigm: update of the last decade. Addict Biol 12:227–462
68. Lepore M, Vorel SR, Lowinson J, Gardner EL (1995) Conditioned place preference induced by delta 9-tetrahydrocannabinol: comparison with cocaine, morphine, and food reward. Life Sci 56:2073–2080
69. Valjent E, Maldonado R (2000) A behavioural model to reveal place preference to delta 9-tetrahydrocannabinol in mice. Psychopharmacology (Berl) 147:436–438
70. Pandolfo P, Vendruscolo LF, Sordi R, Takahashi RN (2009) Cannabinoid-induced conditioned place preference in the spontaneously hypertensive rat-an animal model of attention deficit hyperactivity disorder. Psychopharmacology (Berl) 205:319–326
71. Hyatt WS, Fantegrossi WE (2014) Delta9-THC exposure attenuates aversive effects and reveals appetitive effects of K2/'Spice' constituent JWH-018 in mice. Behav Pharmacol 25:253–257
72. Tampus R, Yoon SS, de la Pena JB, Botanas CJ, Kim HJ, Seo JW, Jeong EJ, Jang CG, Cheong JH (2015) Assessment of the abuse liability of synthetic cannabinoid agonists JWH-030, JWH-175, and JWH-176. Biomol Ther (Seoul) 23:590–596
73. Negus SS, Miller LL (2014) Intracranial self-stimulation to evaluate abuse potential of drugs. Pharmacol Rev 66:869–917
74. Desai RI, Thakur GA, Vemuri VK, Bajaj S, Makriyannis A, Bergman J (2013) Analysis of tolerance and behavioral/physical dependence during chronic CB1 agonist treatment: effects of CB1 agonists, antagonists, and noncannabinoid drugs. J Pharmacol Exp Ther 344:319–328
75. Singh H, Schulze DR, McMahon LR (2011) Tolerance and cross-tolerance to cannabinoids in mice: schedule-controlled responding and hypothermia. Psychopharmacology (Berl) 215:665–675
76. Fan F, Tao Q, Abood M, Martin BR (1996) Cannabinoid receptor down-regulation without alteration of the inhibitory effect of CP 55,940 on adenylyl cyclase in the cerebellum of CP 55,940-tolerant mice. Brain Res 706:13–20
77. D'Souza DC, Cortes-Briones JA, Ranganathan M, Thurnauer H, Creatura G, Surti T, Planeta B, Neumeister A, Pittman B, Normandin M, Kapinos M, Ropchan J, Huang Y, Carson RE, Skosnik PD (2016) Rapid changes in CB1 receptor availability in cannabis dependent males after abstinence from cannabis. Biol Psychiatry Cogn Neurosci Neuroimaging 1:60–67
78. Gifford RS, Raghav JG, Järbe TUC (2016) Does agonist efficacy alter the in vivo effects of cannabinoids: girl, could we get much "higher"? In: Preedy VR (ed) Neuropathology of drug addictions and substance misuse, vol 1. Elsevier Inc. doi:10.1016/B978-0-12-800213-1.00067-5
79. Tai S, Fantegrossi WE (2014) Synthetic cannabinoids: pharmacology, behavioral effects, and abuse potential. Curr Addict Rep 1:129–136
80. Tai S, Nikas SP, Shukla VG, Vemuri K, Makriyannis A, Jarbe TU (2015) Cannabinoid withdrawal in mice: inverse agonist vs neutral antagonist. Psychopharmacology (Berl) 232:2751–2761
81. Tai S, Hyatt WS, Gu C, Franks LN, Vasiljevik T, Brents LK, Prather PL, Fantegrossi WE (2015) Repeated administration of phytocannabinoid Delta-THC or synthetic cannabinoids JWH-018 and JWH-073 induces tolerance to hypothermia but not locomotor suppression in mice, and reduces CB1 receptor expression and function in a brain region-specific manner. Pharmacol Res 102:22–32

82. Ossato A, Vigolo A, Trapella C, Seri C, Rimondo C, Serpelloni G, Marti M (2015) JWH-018 impairs sensorimotor functions in mice. Neuroscience 300:174–188
83. Vigolo A, Ossato A, Trapella C, Vincenzi F, Rimondo C, Seri C, Varani K, Serpelloni G, Marti M (2015) Novel halogenated derivates of JWH-018: behavioral and binding studies in mice. Neuropharmacology 95:68–82
84. Cha HJ, Seong YH, Song MJ, Jeong HS, Shin J, Yun J, Han K, Kim YH, Kang H, Kim HS (2015) Neurotoxicity of synthetic cannabinoids JWH-081 and JWH-210. Biomol Ther (Seoul) 23:597–603
85. Basavarajappa BS, Subbanna S (2014) CB1 receptor-mediated signaling underlies the hippocampal synaptic, learning, and memory deficits following treatment with JWH-081, a new component of spice/K2 preparations. Hippocampus 24:178–188
86. Wiley JL, Marusich JA, Huffman JW, Balster RL, Thomas BF (2011) Hijacking of basic research: the case of synthetic cannabinoids. Methods Rep RTI Press. doi:10.3768/rtipress.2011.op.0007.1111

Pharmacology and Toxicology of *N*-Benzylphenethylamine ("NBOMe") Hallucinogens

Adam L. Halberstadt

Abstract Serotonergic hallucinogens induce profound changes in perception and cognition. The characteristic effects of hallucinogens are mediated by 5-HT_{2A} receptor activation. One class of hallucinogens are 2,5-dimethoxy-substituted phenethylamines, such as the so-called 2C-X compounds 2,5-dimethoxy-4-bromophenethylamine (2C-B) and 2,5-dimethoxy-4-iodophenethylamine (2C-I). Addition of an *N*-benzyl group to phenethylamine hallucinogens produces a marked increase in 5-HT_{2A}-binding affinity and hallucinogenic potency. *N*-benzylphenethylamines ("NBOMes") such as *N*-(2-methoxybenzyl)-2,5-dimethoxy-4-iodophenethylamine (25I-NBOMe) show subnanomolar affinity for the 5-HT_{2A} receptor and are reportedly highly potent in humans. Several NBOMEs have been available from online vendors since 2010, resulting in numerous cases of toxicity and multiple fatalities. This chapter reviews the structure–activity relationships, behavioral pharmacology, metabolism, and toxicity of members of the NBOMe hallucinogen class. Based on a review of 51 cases of NBOMe toxicity reported in the literature, it appears that rhabdomyolysis is a relatively common complication of severe NBOMe toxicity, an effect that may be linked to NBOMe-induced seizures, hyperthermia, and vasoconstriction.

Keywords Head twitch response • Locomotor activity • Psychedelic • Research chemical • Serotonin syndrome

Contents

A.L. Halberstadt (✉)
Department of Psychiatry, University of California San Diego, 9500 Gilman Drive, La Jolla, CA 92093-0804, USA
Research Service, VA San Diego Healthcare System, San Diego, CA, USA
e-mail: ahalbers@ucsd.edu

Curr Topics Behav Neurosci (2017) 32: 283–312
DOI 10.1007/7854_2016_64
Published Online: 1 January 2017

1 Introduction

Serotonergic hallucinogens, also known as classical hallucinogens or psychedelics, produce marked alterations of perception, mood, and cognition (reviewed by Halberstadt [1] and Nichols [2]). Indoleamines and phenylalkylamines are the two main structural classes of hallucinogens. The phenylalkylamines can be divided into two categories: *phenethylamines*, including mescaline and the 2C-X compounds 2,5-dimethoxy-4-bromophenethylamine (2C-B) and 2,5-dimethoxy-4-iodophenethylamine (2C-I), and *phenylisopropylamines* ("amphetamines") such as 2,5-dimethoxy-4-bromoamphetamine (DOB) and 2,5-dimethoxy-4-methylamphetamine (DOM). In contrast to indoleamine hallucinogens, which are relatively nonselective for serotonin (5-HT) receptors, phenylalkylamine hallucinogens are highly selective for 5-HT_2 sites.

The characteristic effects of serotonergic hallucinogens are believed to be mediated by activation of 5-HT_{2A} receptors. Pretreatment with the 5-HT_{2A} antagonist ketanserin blocks the hallucinogenic effects of psilocybin and the botanical hallucinogen *ayahuasca* in human volunteers [3–5]. Similarly, most of the behavioral effects of hallucinogens in laboratory animals are linked to 5-HT_{2A} activation [1]. There is also a significant correlation between 5-HT_{2A} affinity and hallucinogen potency [6, 7].

Although designer drugs are not a new phenomenon, the number and availability of new psychoactive substances (NPS) are unprecedented and have increased dramatically over the last 5 years. At least 299 different NPS were available across Europe in 2013, with an additional 101 new drugs appearing in 2014 [8]. The four main classes of NPS are cannabinoids, psychostimulants, opioids, and hallucinogens. Cannabinoids and psychostimulants are the most commonly abused NPS, but hallucinogenic NPS are also very popular and their proliferation is causing a significant public health problem.

Internet vendors have been marketing a class of hallucinogens known as *N*-benzylphenethylamines (NBOMes) as NPS since 2010 [9, 10]. *N*-(2-methoxybenzyl)-2,5-dimethoxy-4-iodophenethylamine (25I-NBOMe) was the first NBOMe to appear on the illicit market, followed by *N*-(2-methoxybenzyl)-2,5-dimethoxy-4-bromophenethylamine (25B-NBOMe) and *N*-(2-methoxybenzyl)-2,5-dimethoxy-4-chlorophenethylamine (25C-NBOMe) [11, 12]. Numerous other NBOMes have now been detected [13–15]. The most common method of NBOMe distribution is on blotter paper, but powdered material, solutions, and pills are also available. 25I-NBOMe is reportedly a highly potent hallucinogen in humans, with typical doses ranging between 0.5 and

1 mg. NBOMes are reportedly not active orally and are usually taken sublingually or buccally.

2 Structure–Activity Relationships of NBOMe Hallucinogens

For phenylalkylamine hallucinogens, *N*-alkyl substitution results in a marked reduction of 5-HT_{2A} affinity and behavioral potency (Fig. 1). For example, 2,5-dimethoxy-4-methylamphetamine (DOM; $K_i = 100$ nM) has higher affinity than *N*-methyl-DOM ($K_i = 414$ nM) for 5-HT_{2A} sites labeled by [^{3}H]ketanserin in rat frontal cortex homogenates [17]. When tested in rats trained to discriminate DOM (1.0 mg/kg, IP) from saline [18], *N*-methyl-DOM ($ED_{50} = 3.99$ mg/kg, IP) was found to be ninefold less potent than DOM ($ED_{50} = 0.44$ mg/kg, IP). It has also been reported that addition of an *N*-methyl group to DOM produces a tenfold reduction of hallucinogenic potency in humans [19]. The presence of a longer alkyl group is apparently even more detrimental; compared to 2,5-dimethoxy-4-bromoamphetamine (DOB; $K_i = 63$ nM; [17]), *N*-propyl-DOB has much lower affinity for 5-HT_{2A} sites ($K_i = 1{,}930$ nM [16]).

Surprisingly, however, the presence of an *N*-benzyl group can actually *increase* 5-HT_{2A} affinity. Glennon first reported in 1994 [20] that *N*-benzyl-2,5-dimethoxy-4-bromophenethylamine (25B-NB; $K_i = 16$ nM vs. [^{3}H]ketanserin) has higher binding affinity than the unsubstituted parent compound 2,5-dimethoxy-4-bromophenethylamine (2C-B; $K_i = 36$ nM) for 5-HT_{2A} receptors labeled with [^{3}H]ketanserin (see Fig. 2). Although Glennon et al. did not report functional data, further work, published in abstract format, confirmed that *N*-benzylphenthylamines act as potent 5-HT_{2A} agonists [21, 22].

Nichols and colleagues subsequently conducted a detailed investigation of the effects of *N*-benzyl substitution on the 5-HT_{2A} receptor-binding affinity and efficacy of phenylalkylamine hallucinogens [23]. These investigations revealed several important findings. First, although *N*-benzyl substitution consistently increases the 5-HT_{2A} affinity of phenethylamine hallucinogens, compounds having low-to-moderate affinity tend to be the most sensitive to the substitution. For example, for

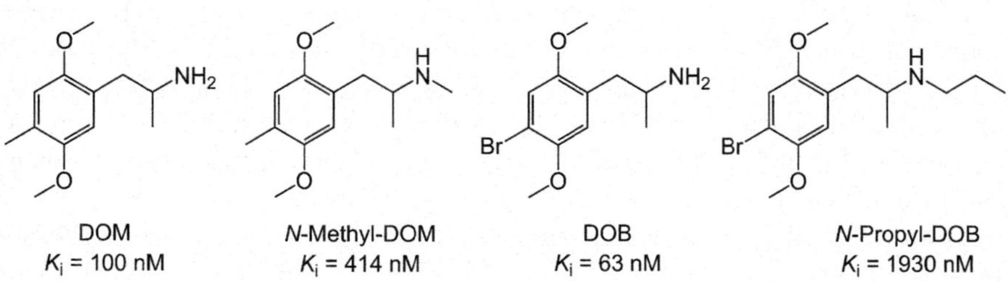

Fig. 1 Effect of *N*-alkyl substitution of the binding affinity of 2,5-dimethoxy-4-methylamphetamine (DOM) and 2,5-dimethoxy-4-bromoamphetamine (DOB) for 5-HT_{2A} receptors labeled with [^{3}H] ketanserin in rat brain homogenates [16, 17]

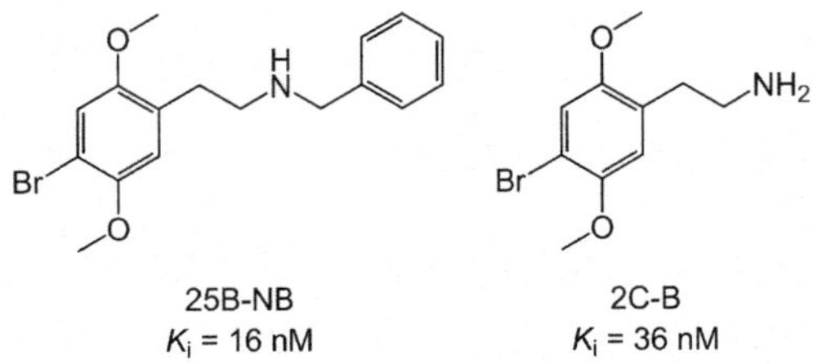

Fig. 2 Comparison of the affinities of *N*-benzyl-2,5-dimethoxy-4-bromophenethylamine (25B-NB) and 2,5-dimethoxy-4-bromophenethylamine (2C-B) for 5-HT_{2A} receptors labeled with [^{3}H]ketanserin in rat frontal cortex homogenates [20]

2,5-dimethoxyphenethylamine (2C-H), which binds to 5-HT_{2A} sites with moderate affinity, the addition of an *N*-benzyl group increased affinity five- to tenfold (see Table 1). By contrast, the effect of *N*-benzylation on the affinity of 2,5-dimethoxy-4-iodophenethylamine (2C-I), which has relatively high affinity for 5-HT_{2A} sites, is comparatively modest. Second, addition of an *N*-benzyl group with an oxygenated substituent at the ortho position results in an even greater increase in 5-HT_{2A} affinity. As shown in Table 1, 25I-NBOMe, 25I-NBOH, and 25I-NBMD have higher 5-HT_{2A} affinity than 25I-NB.

Homology modelling [23] indicates that the presence of an *N*-benzyl moiety increases 5-HT_{2A} receptor affinity because the benzyl ring is stabilized by aromatic stacking with Phe^{339} in transmembrane domain 6 (TM6). Indeed, mutagenesis of Phe^{339} does not normally alter the binding affinity of 5-HT_{2A} agonists [24] but does detrimentally affect the affinity and agonist activity of 25I-NBOMe and other *N*-benzyl-substituted phenethylamines [23]. Replacement of the *N*-benzyl ring in 25B-NBOMe with an electron-deficient heterocyclic ring (e.g., *N*-pyridinyl) reportedly produces a marked reduction in 5-HT_{2A} affinity (see Fig. 3), which is consistent with the evidence that electron-deficient ring systems produce relatively weak aromatic stacking interactions [27]. One explanation proposed to account for the effect of an oxygenated *N*-benzyl ring on affinity is that it may serve as a hydrogen-bond (H-bond) acceptor. Indeed, in silico homology models predict that the 2-position oxygen can form a H-bond with Tyr^{370} in TM7 [28, 29]. Some attempts to investigate the predicted interaction, however, have yielded conflicting findings. For example, according to an unpublished study, mutation of Tyr^{370} to Phe (which cannot form a H-bond) does not significantly alter the affinity of 25I-NBOMe [30]. Other studies indicate that Tyr^{370} plays a general role in 5-HT_{2A} signal transduction [24] and is unlikely to interact directly with the *N*-benzyl moiety. As an alternative to Tyr^{370}, the 2-position oxygen may interact with a different H-bond donor, e.g., with an amine moiety in the protein backbone.

In rodents, 5-HT_{2A} receptor activation induces the head twitch response (HTR), a rapid paroxysmal head rotation [31–33]. The HTR is widely used as a behavioral proxy in rodents for hallucinogen effects in humans and is one of the few behaviors that can reliably be used to distinguish hallucinogenic and non-hallucinogenic 5-HT_{2A} agonists [34]. Although the HTR is usually assessed by direct observation, we have

Table 1 Effect of *N*-benzyl substitution of the 5-HT_{2A} affinity (K_i, nM) of phenethylamine hallucinogens at 5-HT_{2A} receptors labeled with [^{125}I]DOI or [^{3}H]ketanserin

Ligand name	Structure	Rat 5-HT_{2A} [^{125}I]DOI	Human 5-HT_{2A} [^{125}I]DOI	Human 5-HT_{2A} [^{3}H]ketanserin
2C-H	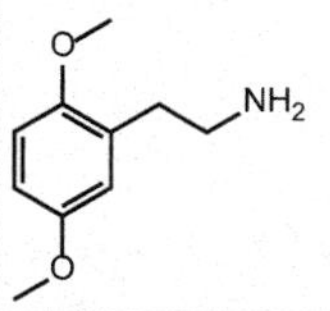	227	377	1,999
25H-NB	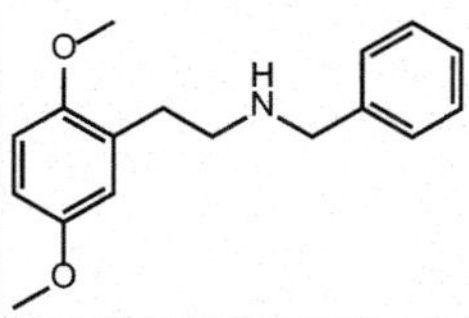	17.5	68.1	184
2C-I	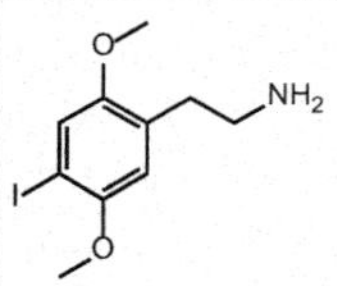	0.62	0.73	4.52
25I-NB	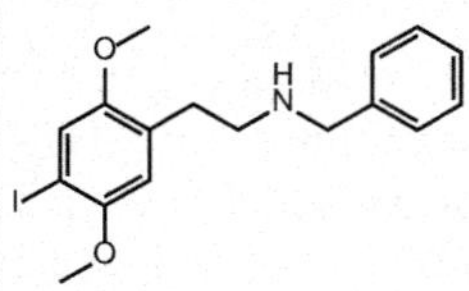	0.31	0.25	0.28
25I-NBOMe	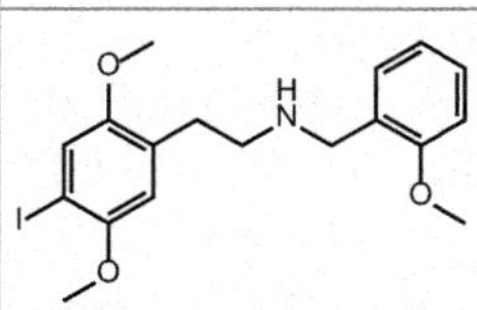	0.087	0.044	0.15
25I-NBOH	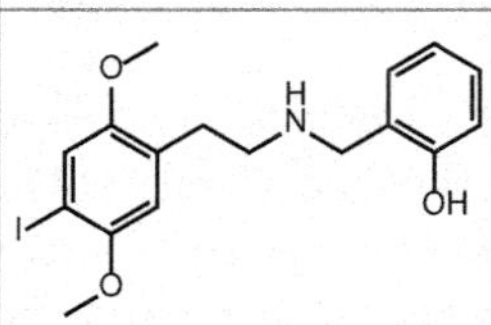	0.12	0.061	0.068
25I-NBMD	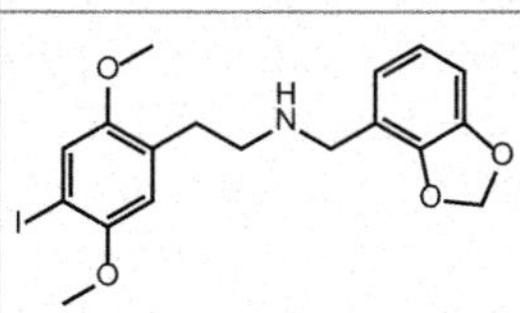	0.19	0.049	0.21

Data from Braden et al. [23]

developed a magnetometer coil-based system for recording head movements that can detect the behavior with high sensitivity and reliability [33]. The HTR detection

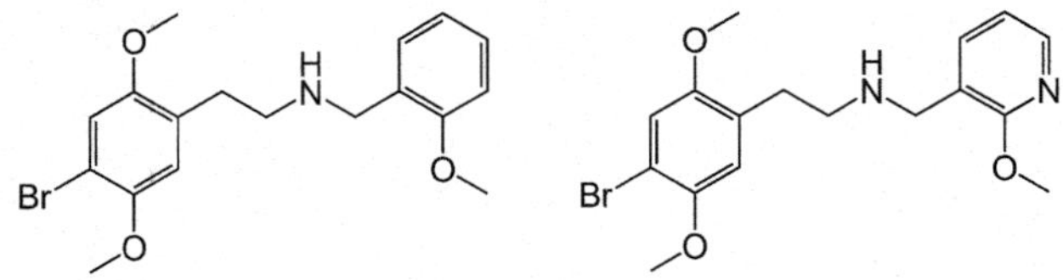

Fig. 3 Comparison of the 5-HT_{2A} affinities of *N*-(2-methoxybenzyl)-2,5-dimethoxy-4-bromophenethylamine (25B-NBOMe) [25] and its 3-pyridinyl analog [26]. Affinity was assessed at human 5-HT_{2A} receptors labeled with [^{3}H]ketanserin

system has been used to assess the HTR induced by a variety of hallucinogens and analogs [35–39].

We have used the HTR to assess whether *N*-benzyl substitution alters the behavioral pharmacology of phenethylamines. Our experiments demonstrated that 25I-NBOMe produces a robust HTR in C57BL/6J mice [37]. After subcutaneous (SC) administration, 25I-NBOMe induced the HTR with an ED_{50} of 78 μg/kg (0.17 μmol/kg), making it slightly less potent than LSD, which induces the HTR with an ED_{50} of 53 μg/kg (0.13 μmol/kg; [33]). Additionally, as shown in Fig. 4, 25I-NBOMe has tenfold higher potency than 2C-I ($ED_{50} = 830$ μg/kg; 2.42 μmol/kg), which is consistent with their relative binding affinities at 5-HT_{2A} receptor sites (Table 1). According to another study, 25B-NBOMe induces the HTR in mice with similar potency to that of 25I-NBOMe [40]. Pretreatment with M100907, a 5-HT_{2A} receptor antagonist that is highly selective versus 5-HT_{2C} sites, produced a dose-dependent blockade of the HTR induced by 25I-NBOMe [37]. Hence, the results of our studies are consistent with anecdotal evidence that 25I-NBOMe acts as a hallucinogen in humans with potency approaching that of LSD. 25I-NBMD also induces the HTR but its potency ($ED_{50} = 1.13$ mg/kg; 2.36 μmol/kg) is not as high as would be anticipated based on its 5-HT_{2A} affinity.

There appear to be strict steric and positional constraints on the *N*-benzyl moiety in NBOMes. As shown in Fig. 5, the effect of the *N*-benzyl ring H-bond acceptor on 5-HT_{2A} affinity depends on its position. Moving the *ortho*-methoxy group in 25I-NBOMe to the meta position (25I-NB3OMe) or the para position (25I-NB4OMe) progressively reduces 5-HT_{2A} affinity [39]. H-bond formation is dependent on the distance between acceptor and donor; hence, the diminished 5-HT_{2A} affinities of the meta and para isomers of 25I-NBOMe are consistent with the methoxy group on the *N*-benzyl ring being moved away from a H-bond donor in the binding pocket. However, steric factors may also contribute to the reduction of 5-HT_{2A} affinity that occurs with *para*-methoxy substitution. Substitution of bromine in the para position in the *N*-benzyl ring (25I-NB4Br) produces a >tenfold reduction in affinity compared with substitution in the ortho (25I-NB2Br) or meta (25I-NB3Br) positions (see Fig. 5) [39]. These findings indicate that the region of the binding pocket proximal to the para position of the *N*-benzyl ring may be sterically constrained. The existence of such steric constraints may

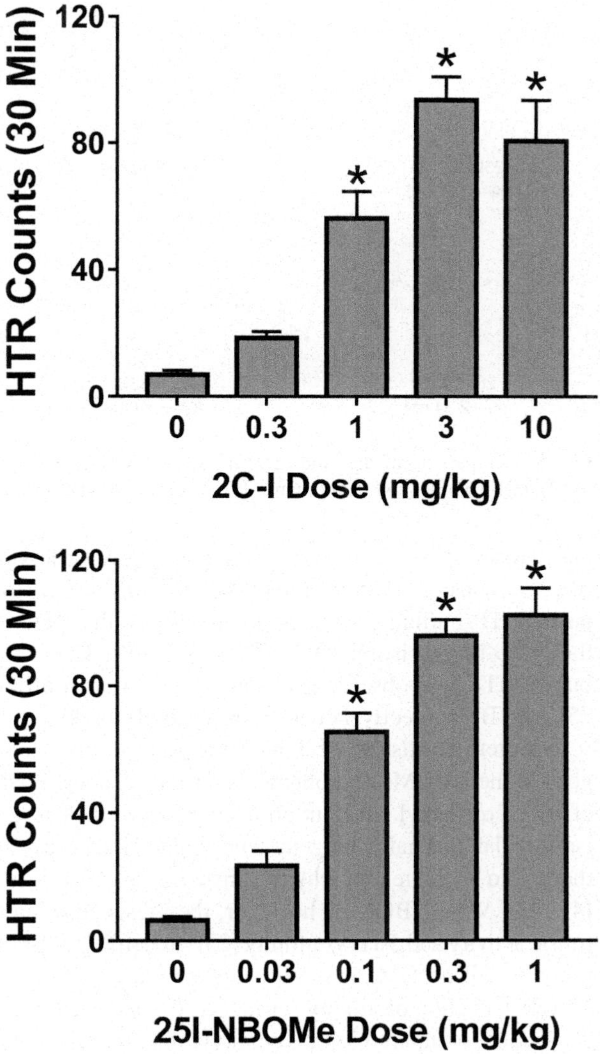

Fig. 4 Head twitch response (HTR) induced by 2C-I (*top*) and 25I-NBOMe (*bottom*) in male C57BL/6J mice. Data are mean ± SEM. *$p < 0.01$ vs. vehicle control group. Redrawn from Halberstadt and Geyer [37]

explain why replacement of the *N*-benzyl group in 25I-NB with a bulky *N*-naphthyl group reduces 5-HT_{2A} affinity by a factor of ten- to 20-fold [23].

We have compared the behavioral potencies of the compounds shown in Fig. 5 using the HTR assay [39]. For the methoxy-substituted regioisomers, moving the methoxy group from the ortho position (25I-NBOMe) to the meta position (25I-NB3OMe) produced a significant drop in potency ($ED_{50} = 4.34$ mg/kg; 9.36 μmol/kg), whereas the para isomer (25I-NB4OMe) was inactive at doses up to 30 mg/kg SC. Although it is not clear why the potency of the meta regioisomer 25I-NB3OMe is so low compared to 25I-NBOMe, 25I-NB3OMe was observed to have a relatively brief duration of action in mice (data not shown), meaning the clearance rate of 25I-NB3OMe in mice may limit

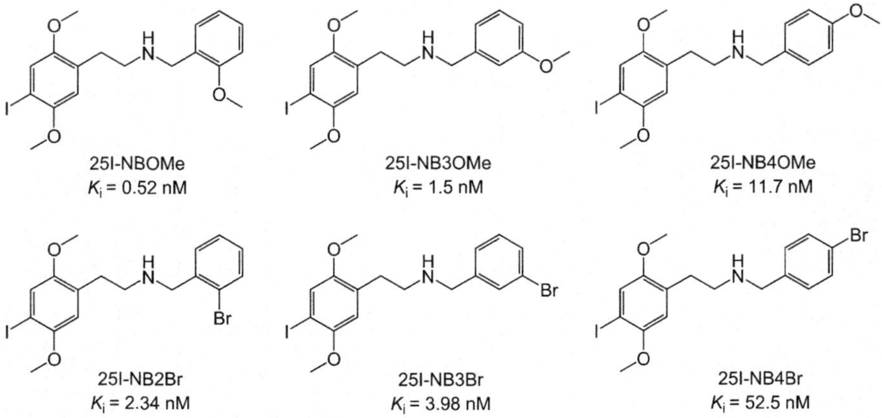

Fig. 5 Comparison of the binding affinities of the regioisomers of 25I-NBOMe (*top row*) and 25I-NB2Br (*bottom row*) at human 5-HT_{2A} receptors labeled with [^{3}H]ketanserin [39]

the magnitude of its behavioral response. Nevertheless, the relative potencies of the methoxy-substituted regioisomers (ortho > meta > para) are consistent with their relative 5-HT_{2A} affinities. According to studies with the bromine-substituted regioisomers, the *ortho*-bromo isomer 25I-NB2Br is active (ED_{50} = 2.31 mg/kg; 4.50 μmol/kg) but no HTR was observed with the *meta*- or *para*-bromo isomers (25I-NB3Br and 25I-NB4Br, respectively) at doses up to 30 mg/kg.

Differences exist between the structure–activity relationships (SAR) of hallucinogens in the NBOMe and phenylalkylamine classes. First, there is a difference in the effect of α-methyl substitution. Compared to their α-desmethyl congeners, phenylisopropylamine hallucinogens have higher intrinsic activities at 5-HT_{2A}, which is thought to be the reason why the phenylisopropylamines have higher potency in vivo [41, 42]. With NBOMes, however, the presence of an α-methyl group reduces intrinsic activity and 5-HT_{2A} affinity [23]. According to Braden et al., adding an α-methyl group to 25I-NBOMe reduced its efficacy (E_{max}) from 78% to 43% and produced a 12-fold reduction of affinity for rat 5-HT_{2A} receptors labeled with [^{125}I]DOI.

Second, compared to phenylalkylamine hallucinogens, NBOMes are less sensitive to the loss of an oxygenated five-position substituent. It is well established that 2,5-dimethoxy substitution is optimal for activity in phenylalkylamine hallucinogens. The 5-methoxy group is believed to interact with Ser^{239} in TM5 of the 5-HT_{2A} receptor [43]. Based on the results of mutagenesis studies performed by Braden and Nichols, it appears that Ser^{239} donates a H-bond to the 5-methoxy in DOM. Indeed, removal of the 5-methoxy group in DOM produces a 28-fold reduction in 5-HT_{2A} affinity and a 130-fold reduction in agonist potency [43]. Interestingly, however, removal of the 5-methoxy group from 25B-NBOMe produces only a tenfold reduction in 5-HT_{2A} affinity and does not appreciably alter agonist potency [44]. Hence, the 5-methoxy group in NBOMes may not play an essential role in 5-HT_{2A} binding and activation.

3 Discovery of NBOMes That Are Selective for 5-HT_{2A} Vs. 5-HT_{2C} Receptors

5-HT_{2A} and 5-HT_{2C} receptors display parallel structure–affinity relationships for ligand binding [20, 45, 46]. Hallucinogens display nonselective agonist activity at 5-HT_{2A} and 5-HT_{2C} receptors [46, 47]. The cloning of rat and human 5-HT_{2A} and 5-HT_{2C} receptors [48, 49] revealed that these 5-HT receptor subtypes exhibit significant sequence homology, especially in the α-helical regions where the ligand-binding sites are located. Depending upon the species examined, the seven transmembrane domains of 5-HT_{2A} and 5-HT_{2C} receptors display between 79 and 80% sequence conservation. Because of the high degree of structural homology shared by 5-HT_{2A} and 5-HT_{2C} receptors, it is not surprising that most ligands that bind to 5-HT_{2A} sites are also capable of interacting with 5-HT_{2C} sites.

Similar to other classes of hallucinogens, NBOMes act as 5-HT_{2C} receptor agonists and are relatively nonselective for 5-HT_{2A} vs. 5-HT_{2C} sites [39]. Given their high affinity and efficacy at 5-HT_{2A} receptors, NBOMes have been developed as 5-HT_{2A} agonist radioligands [50] and as PET tracers [51, 52]. Such work has also encouraged development of NBOMes exhibiting selectivity for 5-HT_{2A} receptors compared with 5-HT_{2C} sites. In contrast to other phenylisopropylamines, 2,5-dimethoxy-4-cyanoamphetamine (DOCN, Fig. 6) exhibits moderate (22-fold) selectivity for human 5-HT_{2A} ($K_i = 45.7$ nM) vs. 5-HT_{2C} ($K_i = 1,011$ nM) sites labeled with [^{125}I]DOI [46], indicating that 4-cyano substitution represents a potential strategy to augment 5-HT_{2A} selectivity. Applying this strategy to the NBOMe class led to the discovery of N-(2-hydroxybenzyl)-2,5-dimethoxy-4-cyanophenethylamine (25CN-NBOH), which is reportedly 100-fold selective for 5-HT_{2A} receptors ($K_i = 1.3$ nM vs. [^{3}H]ketanserin) compared with 5-HT_{2C} receptors ($K_i = 132$ nM vs. [^{3}H]mesulergine) [25]. However, according to a more recent investigation using the same antagonist radioligands, 25CN-NBOH is only 23-fold selective for 5-HT_{2A} receptors ($K_i = 2.2$ nM) relative to 5-HT_{2C} sites ($K_i = 49.8$ nM) [38]. Although the selectivity of 25CN-NBOH may be less than was initially believed, it still exhibits moderate 5-HT_{2A} selectivity. Importantly, we have confirmed that SC administration of 25CN-NBOH induces the HTR in mice with moderate potency ($ED_{50} = 0.36$ mg/kg; 1.03 μmol/kg [38]). Another group has confirmed that the HTR induced by 25CN-NBOH is blocked by pretreatment with 0.01 mg/kg M100907 [53]. The latter study also showed that pretreatment with 25CN-NBOH produces a dose-dependent blockade of the HTR induced by R-(−)-2,5-

Fig. 6 Structures of the 4-cyano-substituted phenylalkylamines DOCN and 25CN-NBOH

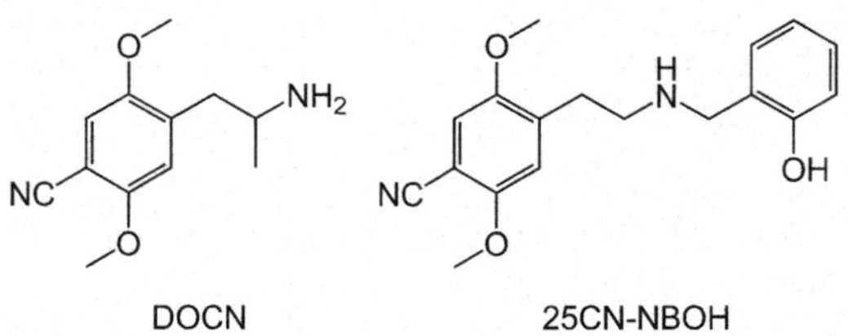

dimethoxy-4-iodoamphetamine (*R*-(−)-DOI), indicating that 25CN-NBOH acts as a partial agonist.

NBOMes exhibit a high degree of conformational flexibility and could potentially adopt a range of active binding poses. In order to identify the active conformation, Nichols and colleagues synthesized a series of rigid analogues of 25B-NBOMe [54]. Of the nine structurally constrained compounds tested, (±)-*trans*-DMBMPP (Fig. 7) was the most potent, binding to human 5-HT_{2A} receptors with a K_i of 5.3 nM. Interestingly, the affinity of (±)-*trans*-DMBMPP for human 5-HT_{2C} sites is significantly lower in comparison, making it 98-fold selective for 5-HT_{2A} receptors. The (*S*,*S*) enantiomer of DMBMPP, resolved by derivatization with a chiral auxiliary, has even higher 5-HT_{2A} affinity (K_i = 2.5 nM) and is reportedly 124-fold selective for 5-HT_{2A} vs. 5-HT_{2C} receptors. By contrast, (*R*,*R*)-DMBMPP has μM affinity for 5-HT_{2A} receptors (Fig. 7). It appears that the structural configuration of (*S*,*S*)-DMBMPP closely mirrors the active binding conformation of NBOMes.

4 Other Behavioral Studies with NBOMes

Our previous studies have shown that phenylalkylamine hallucinogens produce dose-dependent effects on locomotor activity in C57BL/6J mice, increasing activity at low-to-moderate doses and reducing activity at higher doses [55, 56]. For example, DOI and DOM increase locomotor activity at 1 mg/kg and reduce activity at 10 mg/kg [32, 56]. The hyperactivity produced by DOI and other phenylalkylamines is blocked by M100907 and is absent in 5-HT_{2A} knockout mice, indicating mediation by 5-HT_{2A} receptors [55, 56]. To determine whether 25I-NBOMe produces similar effects on locomotor activity, we conducted dose-response studies in C57BL/6J mice after IP and SC administration. Administration of 25I-NBOMe by the IP route had no effect on locomotor activity (Fig. 8a), although there was a trend toward a main effect of drug treatment ($F(5,54) = 2.08$, $p < 0.09$) and a significant interaction between drug treatment and time block ($F(25,270) = 2.22$, $p < 0.001$). By contrast, when

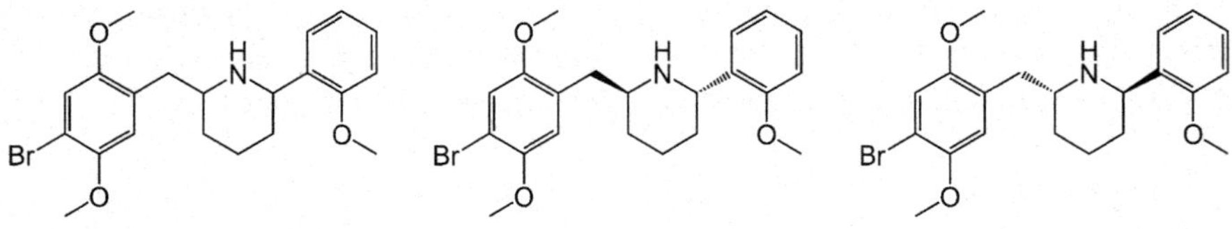

(±)-*trans*-DMBMPP
5-HT_{2A} K_i = 5.3 nM
5-HT_{2C} K_i = 520 nM

(*S*,*S*)-DMBMPP
5-HT_{2A} K_i = 2.5 nM
5-HT_{2C} K_i = 310 nM

(*R*,*R*)-DMBMPP
5-HT_{2A} K_i = 2,100 nM
5-HT_{2C} K_i = 28,600 nM

Fig. 7 Structures of racemic *trans*-2-(2,5-dimethoxy-4-bromobenzyl)-6-(2-methoxyphenyl)piperidine (±-*trans*-DMBMPP) and its *S*,*S* and *R*,*R* enantiomers. Binding affinities were assessed at human 5-HT_{2A} and 5-HT_{2C} receptors labeled with [^{3}H]ketanserin and [^{3}H]mesulergine, respectively [54]

administered SC as in all the HTR experiments, 25I-NBOMe produced effects on locomotor activity that mimicked those induced by phenylalkylamine hallucinogens (drug effect: $F(5,52) = 5.16, p < 0.001$; drug × time: $F(25,260) = 2.26, p < 0.001$). As shown in Fig. 8b, 0.1 mg/kg 25I-NBOMe increased locomotor activity during

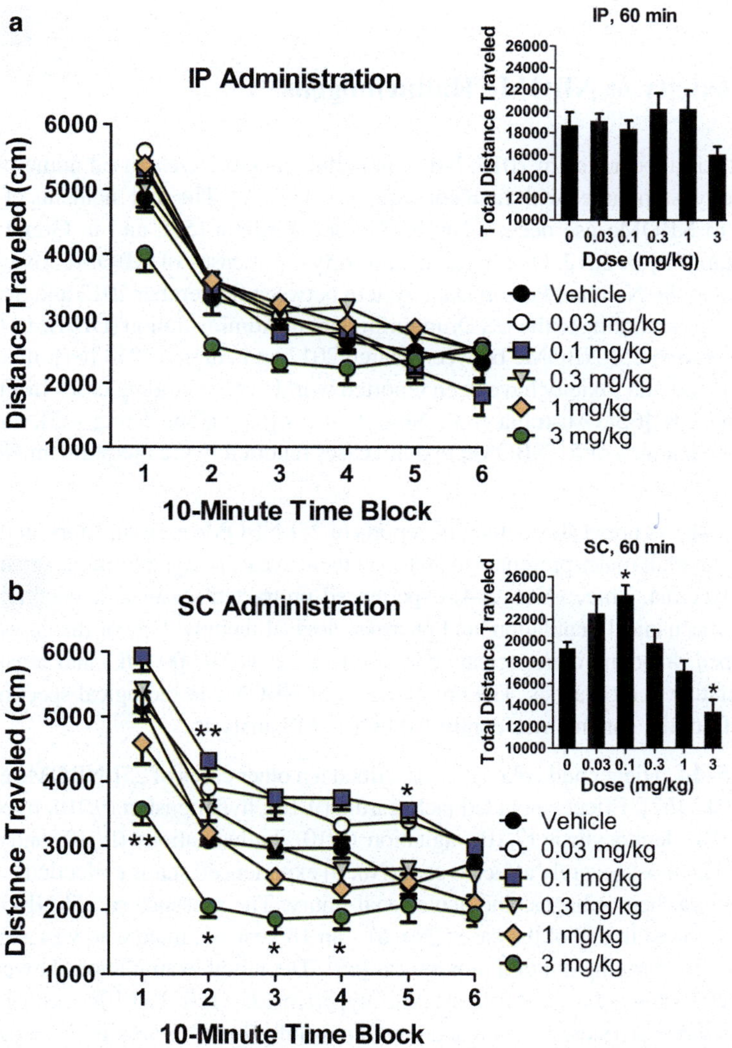

Fig. 8 Effect of 25I-NBOMe on locomotor activity after intraperitoneal (IP) (**a**) or subcutaneous (SC) (**b**) administration. Male C57BL/6J mice ($n = 9$–11, 60 total in experiment a; $n = 8$–10, 58 total in experiment b) were treated with vehicle or 25I-NBOMe (0.03, 0.1, 0.3, 1, and 3 mg/kg) and activity recorded in the mouse behavioral pattern monitor for 60 min. Data are presented as group means ± SEM for successive 10-min intervals, or group means ± SEM for the entire 60-min test session (inset histograms). *$p < 0.05$, **$p < 0.01$, significant difference from vehicle control group

time blocks 2 and 5 ($p < 0.01$, 0.05, Tukey's test), and 3 mg/kg 25I-NBOMe reduced locomotor activity during blocks 1, 2, 3, and 4 ($p < 0.01$, 0.05, Tukey's test). In summary, the effects of 25I-NBOMe on locomotor activity in mice mirror those produced by phenylalkylamine hallucinogens, but 25I-NBOMe appears to be unusually potent in comparison.

5 Toxicity of NBOMe Hallucinogens

Since their appearance in 2010, NBOMe hallucinogens have caused numerous cases of toxicity, sometimes with fatal consequences [11, 57]. The first incidents of toxicity linked to NBOMe use occurred in Richmond, Virginia [58], and in Grand Rapids, Michigan [59] in 2012. One hundred and forty-eight cases of NBOMe toxicity were reported to the National Poison Data System between September 2012 and September 2014 [60]. According to the US Drug Enforcement Administration (DEA), 19 fatalities were linked to 25I-NBOMe between March 2012 and August 2013 [61]. Incidents of NBOMe use and toxicity have been reported worldwide, including cases from Europe [57], the UK [62], Australia [63], New Zealand [64], Hong Kong [65], and Japan [66]. The features of 51 NBOMe toxicity cases reported in the literature are described below.

Cases 1–4 As noted above, the first reports of 25I-NBOMe exposure appeared in 2012 [59]. Four adult males presented to an ED with tachycardia, hypertension, agitation, and hyperglycemia. Three (**Nos. 2–4**) experienced protracted seizures, necessitating intubation, mechanical ventilation, and pharmacological therapy. One of the cases (**No. 4**) developed rhabdomyolysis (creatine kinase (CK) level 30,000 U/L) and renal failure, necessitating hemodialysis. The presence of 25I-NBOMe in biological specimens collected from the patients was confirmed by LC-MS analysis.

Cases 5–14 Hieger and colleagues described ten other cases of 25I-NBOMe exposure from 2012 [67]. Effects included tachycardia (9/10), hypertension (9/10), hyperglycemia (9/10), leukocytosis (7/10), agitation (7/10), hallucinations (5/10), and seizures (2/10). The most severely affected case (**No. 5**) experienced status epilepticus, multiple intracerebral hemorrhages, and acute renal injury. The presence of 25I-NBOMe was only confirmed in one of the cases (**No. 6**) – an 18-year-old male who was admitted to the ED after jumping out of a moving car [68]. The initial examination showed severe agitation, tachycardia (>150 bpm), and hypertension (150–170/100 mmHg). Treatment included physical restraint and continuous infusion of lorazepam. Symptoms improved over 48 h although lorazepam and dexmedetomidine had to be administered on the second day of hospitalization due to continuing episodes of agitation. 25I-NBOMe was detected (LC-MS/MS) in serum at a concentration of 0.76 μg/L.

Cases 15–21 Hill described seven cases of 25I-NBOMe toxicity that occurred in the UK in January 2013 [62]. The first case (**No. 15**) was a 29-year-old male who injected an unknown quantity of 25I-NBOMe intravenously. The initial examination showed agitation, aggression, self-harm, seizures, tachycardia (160 bpm), hypertension (187/171 mmHg), tachypnea (58 breaths/min), low oxygen saturation (94%), and hyperthermia (39.0°C). He also presented with leukocytosis (WBC 23.5 × 10^9/L), respiratory and metabolic acidosis (pH 7.20, $PaCO_2$ 66 mmHg), rhabdomyolysis (CK 15,424 U/L), and impaired renal function. Treatment included intubation, mechanical ventilation, sedation/neuromuscular blockade, and administration of fluids and antibiotics. He subsequently developed anuria and renal impairment, and there was evidence of pulmonary injury. Normalization of renal and pulmonary function required 43 days of hospitalization, including ICU admission for 38 days where he received hemodialysis. In the second case (**No. 16**), a 20-year-old-male collapsed and had convulsions after ingesting a powder containing 25I-NBOMe. In addition to agitation, he exhibited tachycardia (126 bpm), hypertension (170/90 mmHg), hyperthermia (38.8°C), tachypnea (24 breaths/min), low oxygen saturation (92%), and urinary retention. He was anesthetized, intubated, and artificially ventilated. Sustained clonus and ocular clonus were noted 5 h after ED admission, necessitating treatment with cyproheptadine. His CK level was initially elevated (peak value of 550 U/L) but renal function was restored by fluid replacement. He was extubated 3 days after admission and discharged on the fifth day. The clonus was likely due to serotonin syndrome; the patient had a history of depression and was being treated with fluoxetine. The third case (**No. 17**, a 19-year-old male) insufflated a powder containing 25I-NBOMe. He became agitated and violent, with tachycardia (110 bpm) and hypertension (138/100 mmHg). He also exhibited leukocytosis (WBC 18.9 × 10^9/L). Recovery occurred after treatment with diazepam. The fourth case (**No. 18**) was a 22-year-old male who insufflated 25I-NBOMe and then had a tonic-clonic seizure. When he arrived at the ED he was agitated and aggressive and had to be sedated with diazepam. He presented with mild tachycardia (104 bpm) and elevated CK levels, peaking at 633 U/L. Recovery occurred without further intervention and he was discharged from the hospital on the same day he was admitted. The other three cases (**Nos. 19–21**) were young adult males who experienced more moderate symptoms after insufflating or ingesting 25I-NBOMe. In addition to visual and auditory hallucinations, all three cases exhibited tachycardia and hypertension. One of those cases (**No. 19**) exhibited agitation and aggressive behavior severe enough to require sedation; he also showed hyperthermia (38.4°C) and elevated CK levels (598 U/L). Another case (**No. 20**) had inducible ankle clonus. LC-MS/MS analysis of plasma samples confirmed that all the individuals had taken 25I-NBOMe.

Case 22 In another reported case of 25I-NBOMe overdose, an 18-year-old female became confused and agitated and had a grand mal seizure after taking an unknown amount of the drug sublingually at a party [69]. The initial examination revealed tachycardia (145 bpm) and hypertension (145/100 mmHg). Hyperreflexia was also present. Her condition normalized after administration of intravenous fluids and lorazepam. LS-MS/MS analysis of a urine sample confirmed the presence of 25I-NBOMe and suspected *O*-desmethyl metabolites, as well as small amounts of 25H-NBOMe and 2C-I.

Case 23 Umemura described the case of a 17-year-old female who died after ingesting blotter paper containing 25I-NBOMe [70]. She was hospitalized with status epilepticus and subsequently developed hyperthermia, metabolic acidosis, rhabdomyolysis, and kidney injury. The patient survived for 7 days, but brain death occurred due to cerebral edema. The presence of 25I-NBOMe in whole blood collected antemortem was confirmed by LC-TOF-MS. Therapeutic levels of lithium were also present in the blood sample.

Cases 24–25 Another two deaths linked to 25I-NBOMe were reported by Walterscheid and colleagues [71]. In both cases, the decedents began to "flail about" before becoming unresponsive. The first victim (**No. 24**, a 21-year-old male) became violent while driving, damaging the interior of the vehicle. Cardiopulmonary recitation was attempted but was unsuccessful. The second victim (**No. 25**, a 15-year-old female) had ingested a liquid containing 25I-NBOMe; death occurred due to cardiac failure soon after arriving at the ED. Hyperthermia was noted in the second victim (39.9°C). 25I-NBOMe was detected in heart blood and urine collected from both decedents. Analysis also revealed evidence of marijuana use in both individuals.

Case 26 A behavioral fatality linked to 25I-NBOMe has been reported [72]. In that case, a 19-year-old male ingested blotter paper containing "acid." The man exhibited bizarre behavior prior to falling multiple stories from an apartment balcony. 25I-NBOMe was detected (LC-MS/MS) in samples of heart blood (410 ng/L), peripheral blood (405 ng/L), and brain tissue (2,780 pg/g) collected 7 h postmortem.

Case 27 A 15-year-old male had multiple seizures and lost consciousness after ingesting hallucinogenic mushrooms in combination with a liquid containing 25I-NBOMe [73]. He developed renal and liver failure. Death ultimately occurred following cardiopulmonary arrest. An antemortem blood sample contained 0.76 μg/L 25I-NBOMe.

Case 28 A 16-year-old male was found dead after consuming 25I-NBOMe on a piece of blotter paper [74]. There was evidence that his death was preceded by violent behavior (broken glass was found at the scene and the decedent had multiple contusions and abrasions). Analysis of heart blood collected postmortem revealed 19.8 μg/L 25I-NBOMe.

Case 29 Another death linked to 25I-NBOMe was described by Keuppers and Cooke [63]. In that case, a 23-year-old female insuffulated a powder purported to be "synthetic LSD." She soon became severely agitated and aggressive, and then had a seizure before collapsing. Cardiopulmonary resuscitation (CPR) was attempted but was unsuccessful. 25I-NBOMe (28 μg/L), 25H-NBOMe (1 μg/L), and 25C-NBOMe (0.7 μg/L) were detected in aortic blood postmortem. Methamphetamine (0.39 mg/L) and THC (3.4 μg/L) were also present.

Case 30 An unsuccessful suicide attempt following 25I-NBOMe ingestion has been reported [75]. After taking "two hits of acid" sublingually, the 18-year-old male had a panic attack, and stabbed himself in the throat and chest. A serum sample, collected ~11 h post-ingestion, contained 34 ng/L 25I-NBOMe (LC-MS/MS).

Case 31 Tang described one case [65] where a 17-year-old male was hospitalized in a confused and agitated state. He had a seizure after being admitted to the ED. Other effects included hypertension (215/94 mmHg), tachycardia (140 bpm), hyperthermia (38.4°C), and diaphoresis. Examination also showed sinus tachycardia. He was initially treated with intravenous fluids and diazepam. Later, he was intubated and admitted to the pediatric ICU, where he was given sedatives/neuromuscular blockade and treated with cyproheptadine. After the patient regained consciousness, he admitted ingesting a pill containing an "NBOMe." LC-MS/MS analysis of the patient's urine confirmed the presence of 25B-NBOMe.

Case 32 Another case reported by Tang involved a 31-year-old male who presented with confusion, agitation, hypertension (160/123 mmHg), sinus tachycardia (162 bpm), hyperthermia (39.6°C), and diaphoresis [65]. Troponin I (0.38 ng/L) and lactate (8.6 mmol/L) were also elevated. Treatment included administration of fluids and lorazepam, as well as cooling measures. He subsequently developed rhabdomyolysis (CK levels peaked at 11,066 U/L), impaired renal function, and altered liver function. Rhabdomyolysis was treated with fluids and bicarbonate. The patient left the hospital on day 3 against medical advice. Prior to release, he admitted sublingual use of a drug called "Holland film." His urine was positive for 25B-NBOMe and 25C-NBOMe.

Case 33 A 19-year-old male who had taken 25B-NBOMe subsequently had generalized grand mal seizures and became unresponsive [76]. Examination showed hyperthermia (40°C), tachycardia (152 bpm), hypertension (145/90 mmHg), agitation, diaphoresis, and respiratory distress. He was intubated and mechanically ventilated; sedatives and a neuromuscular blocking agent were administered to control agitation and seizures. Laboratory values indicated hyperglycemia (286 mg/L), leukocytosis (WBC 26.1×10^9/L), and respiratory and metabolic acidosis (pH 6.9, pCO_2 89 mmHg, HCO_3^- 19.3 mEq/L, base deficit 13 mmol/L). The patient subsequently developed signs of rhabdomyolysis, with CK levels peaking at 11,645 U/L. Recovery required 6 days of ICU treatment. Serum from the patient contained 180 ng/L 25B-NBOMe.

Case 34 Isbister described the case of a 15-year-old male who ingested blotter paper purportedly containing LSD [77]. He had three seizures before arriving at the ED and one following admission. He presented with acute respiratory acidosis (venous pH 6.93, $PvCO_2$ 120 mmHg, base excess −7 mEq/L). Seizures were treated with midazolam; he was intubated, ventilated, sedated and paralyzed, and transferred to the ICU. Examination in the ICU showed leukocytosis (WBC 16.3×10^9/L). On day 2, he began to show signs of rhabdomyolysis; his CK level peaked at 34,778 U/L on day 3. He recovered and was discharged on day 5. Analysis of blood collected 22 h post-dosing confirmed the presence of 25B-NBOMe at 0.089 μg/L.

Case 35 Yoshida recounted the case of a male, approximately 20 years old, who exhibited violent behavior and convulsions after ingesting blotter paper containing 25B-NBOMe [66]. He was hospitalized in a comatose state; other effects included tachycardia (156 bpm), tachypnea (48 breaths/min), hyperthermia (41.5°C), and systolic hypotension (90 mmHg). He also presented with thrombocytopenia, rhabdomyolysis, acidosis, and multi-organ failure. Myoclonus and tendon hyperreflexia were also observed. Despite supportive therapy, the patient died 3 days later. A plasma sample collected when the man was admitted to the hospital (2–3 h after drug intake) contained 3.15 μg/L 25B-NBOMe and 0.433 μg/L 25C-NBOMe, as well as benzodiazepines.

Case 36 An 18-year-old man died after consuming two squares of blotter paper containing 25B-NBOMe [74]. His death was preceded by destructive behavior and loss of consciousness. Autopsy revealed pulmonary edema and aspiration of gastric contents. 25B-NBOMe (1.59 μg/L) and cannabinoids were present in postmortem heart blood.

Case 37 Laskowski described a 15-year-old male who became agitated ~6 h after sublingual administration of blotter paper impregnated with 25B-NBOMe [78]. After ED admission, the patient had multiple tonic-clonic seizures, which were treated with IV lorazepam. He also exhibited hypertension (177/93 mmHg), tachycardia (111 bpm), diaphoresis, and mild hyperthermia (37.7°C). Laboratory tests showed hyperglycemia (224 mg/dL), leukocytosis (WBC 17×10^9/L), and acidosis (HCO_3^- 13 mEq/L, lactate 7.3 mmol/L). Additionally, CK levels were elevated (429 U/L). He recovered after being transferred to the pediatric ICU and was discharged <48 h after drug intake. Serum collected from the patient when he arrived at the ED contained 1.2 μg/L 25B-NBOMe (LC-MS/MS).

Cases 38–47 Gee reported ten cases where recreational use of 25B-NBOMe resulted in adverse effects [64]. Agitation, hallucinations, tachycardia, and hypertension were present in all of the cases. Diaphoresis occurred in eight cases and hyperthermia in four cases. The toxicity was not severe in nine of the cases; two of those patients recovered with limited medical intervention and the other seven required only physical restraint and/or sedation with benzodiazepines. The toxicity was more severe in the remaining case – a 24-year-old male (**No. 47**) who snorted an unknown amount of 25B-NBOMe. He was hospitalized due to agitation and self-injurious behavior. The patient continued to struggle despite physical restraint and administration of haloperidol and a large dose of midazolam. Heart rate, blood pressure, and temperature peaked at 175 bpm, 200/90 mmHg, and 38.5°C, respectively. He was eventually anesthetized in the ED, intubated, and transferred to the ICU. CK and troponin I levels peaked at 18,361 U/L and 399 ng/L, respectively. The patient gradually improved after being anesthetized; he was discharged from the hospital 60 h after admission. Three of the patients, including **No. 47**, reportedly exhibited inducible clonus; one of the other cases exhibited tremor and hyperreflexia. According to LC-MS/MS analysis, plasma levels of 25B-NBOMe in the ten patients ranged from 0.7 to 10.7 μg/L.

Case 48 In one case reported by Grautoff and Kähler [79], a 19-year-old male had a generalized seizure and lost consciousness 2 h after snorting 2 mg of 25C-NBOMe (identity confirmed by GC-MS). Other presenting features included tachycardia (120 bpm) and hypertension, as well as low oxygen saturation (50%), which necessitated intubation and mechanical ventilation. Despite supportive care, the patient developed rhabdomyolysis (CK levels peaked at 5,533 U/L), renal failure, pulmonary hypertension, and evidence of lung injury. He required hemodialysis and administration of multiple antihypertensive agents. The patient remained in the ICU for 13 days before making a full recovery.

Case 49 A 24-year-old woman became confused and agitated 30 min after consuming three squares of blotter paper impregnated with 25C-NBOMe [80]. Examination revealed tachycardia (140 bpm), tachypnea (32 breaths/min), and skin that was "moist and hot to the touch." After being physically restrained, she was given intravenous fluids and lorazepam. Full recovery occurred within 10 h. The identity of the drug was confirmed by LC-TOF/MS.

Case 50 A fatality due to 25C-NBOMe has also been described [81]. A 22-year-old man snorted an unknown amount of 25C-NBOMe. Over the next few hours he acted agitated and incoherent. He had lost consciousness by the time an ambulance arrived. The initial examination showed low oxygen saturation (80%) and generalized seizures, so he was intubated and treated with multiple sedatives and neuromuscular blockade. When he arrived at the ED 30 min later he exhibited hyperthermia (40°C), tachycardia (140 bpm), bleeding from mucous membranes, rhabdomyolysis, respiratory and metabolic acidosis (pH 6.69, $PaCO_2$ 78 mmHg, lactate 28 mmol/L), hyperkalemia, and low BP. His CK and troponin I levels eventually peaked at >42,670 U/L and 3,513 ng/L, respectively. Although the patient was placed in a medically induced coma and cooling measures were attempted, he died of multi-organ failure approximately 12 h after drug intake. A blood sample collected antemortem contained 0.81 μg/kg 25C-NBOMe (LS-MS/MS).

Case 51 A 16-year-old female was hospitalized due to tonic-clonic seizures and altered mental status after sublingual use of blotter paper purportedly containing 25I-NBOMe [78]. She presented with hypertension (130/73 mmHg) and tachycardia (146 bpm). There was also evidence of hyperglycemia (glucose 170 mg/dL), hypernatremia (serum Na^+ 149 mEq/L), leukocytosis (WBC 19×10^9/L), and acidosis (HCO_3^- 11 mEq/L, anion gap 21 mmol/L). Seizures were treated with intravenous lorazepam. Ankle clonus was observed when the patient was examined in the pediatric ICU. CK levels were elevated, peaking at 47,906 U/L on the third day in the hospital. Her mental status normalized 24 h after drug intake and she was discharged 7 days later. Analysis of serum confirmed the presence of 25C-NBOMe; no 25I-NBOMe was detected.

As noted elsewhere, cases of NBOMe toxicity can be divided into two general categories based on their severity [57, 82]. Hallucinations, agitation, confusion, diaphoresis, hypertension, and tachycardia are common features of NBOMe toxicity. These symptoms usually resolve spontaneously or with minimal medical intervention. Twenty-four of the cases described above fall into the less severe category. By contrast, the features associated with severe cases of NBOMe toxicity include seizures, rhabdomyolysis, metabolic acidosis, renal failure, multi-organ failure, and coma. Death may occur, especially in the absence of supportive care. Twenty-three of the cases were of the latter type (see Table 2).

Table 2 Clinical features observed in moderate-to-severe cases of NBOMe toxicity[a]

Case	Seizures	Hyperthermia	Rhabdomyolysis	Renal injury	Acidosis	Death	Pharmacological therapies
2	YES						"Aggressive pharmacologic therapy"
3	YES						"Aggressive pharmacologic therapy"
4	YES		YES	YES			"Aggressive pharmacologic therapy"
5	YES			YES			Benzodiazepines
7	YES						Benzodiazepines
15	YES	YES	YES	YES	YES		Propofol, midazolam, atracurium
16	YES	YES	CK elevated				Diazepam, propofol, midazolam, atracurium, lorazepam, cyproheptadine
18	YES		CK elevated				Diazepam
19		YES	CK elevated				Diazepam, lorazepam, haloperidol
22	YES						Lorazepam
23	YES	YES	YES	YES	YES	YES	UNKNOWN
25		YES				YES	UNKNOWN
27	YES			YES		YES	UNKNOWN
29	YES					YES	NONE
31	YES	YES					Diazepam, midazolam, rocuronium, cyproheptadine
32		YES	YES	YES	YES		Lorazepam
33	YES	YES	YES		YES		Lorazepam, propofol, pancuronium, midazolam, vecuronium
34	YES		YES				Rocuronium, morphine, midazolam
35	YES	YES	YES	YES	YES	YES	Diazepam, fosphenytoin, vecuronium
37	YES	YES	YES		YES		Lorazepam
47		YES	YES				Midazolam, haloperidol, propofol
48	YES		YES	YES			Sedatives, antihypertensives, vasodialators, diuretics
50	YES	YES	YES	YES	YES	YES	Diazepam, midazolam, suxamethonium, ketamine, fentanyl
51	YES		YES	YES	YES		Lorazepam

[a]Cases were included if there was evidence of rhabdomyolysis or if seizures, renal failure, or metabolic acidosis occurred

5.1 Mild Cases of NBOMe Toxicity

The features of the less severe cases are not unique to NBOMes and are also induced by other serotonergic hallucinogens. Serotonergic hallucinogens can cause confusion, agitation, and panic attacks, especially during "bad trips" or after administration of high doses [83, 84]. LSD, DOM, mescaline, DMT, and psilocybin produce hypertension, tachycardia, and diaphoresis in humans [85–91]. The hypertension induced by DOI is mediated in the periphery by activation of vascular 5-HT_{2A} receptors, which produces vasoconstriction [92–95].

5.2 Moderate-to-Severe Cases of NBOMe Toxicity

The progression from rhabdomyolysis to metabolic acidosis and renal failure is a common complication of severe NBOMe toxicity. Of the 23 moderate-to-severe cases listed in Table 2, 14 had at least one of these features, and three others showed subclinical features such as elevated creatinine kinase (CK) levels (indicating that rhabdomyolysis may have developed in the absence of medical treatment). Rhabdomyolysis is caused by skeletal muscle damage, for example by prolonged muscle activity or compression of muscles. Ischemic muscle tissue produces lactic acid, resulting in metabolic acidosis. The disintegration of striated myocytes releases their intracellular contents, including CK, myoglobin, and potassium, into plasma. Myoglobin release can produce acute renal failure due to tubule obstruction, whereas hyperkalemia increases the risk of cardiac failure.

Increased muscle activity due to seizures appears to be the primary cause of rhabdomyolysis in cases of NBOMe toxicity, although agitation, hyperthermia, and ischemia due to peripheral vasoconstriction may also contribute. A similar constellation of factors (muscle hyperactivity due to agitation or seizures, hyperthermia, and vasoconstriction) are thought to underlie cocaine-induced rhabdomyolysis [96]. Seizures can induce rhabdomyolysis [97–101]. Seizure activity was present in 12/14 (85.7%) of the NBOMe toxicity cases featuring rhabdomyolysis (CK levels $\geq$ 1,000 U/L), metabolic acidosis, or renal failure (Table 2). Seizures did not always result in rhabdomyolysis but in many of the non-progressing cases the patients received medications that would minimize muscle tissue damage (e.g., benzodiazepines, anesthetics, or paralytic agents).

Similar to NBOMes, phenylalkylamines such as 2C-I, 2C-T-7, 2C-T-21, DOB, DOC, and bromo-dragonfly (1-(8-bromobenzo[1,2-*b*;4,5-*b*′]difuran-4-yl)-2-aminopropane) can also induce seizures [102–108]. Srisuma et al. [60] compared the clinical features of 148 cases of NBOMe exposure and 193 cases of 2C-X hallucinogen exposure reported to the National Poison Data System, a database of poison exposures. The features of NBOMe and 2C-X toxicity were virtually identical, with the exception of single-episode seizures, which were significantly more likely to occur with NBOMes (8.8% of cases) than with phenethylamines (3.1% of cases). Although it is not clear why NBOMes are more likely to induce seizures compared with other phenethylamines, this propensity may increase the likelihood of severe NBOMe toxicity. In contrast to phenylalkylamines and NBOMes, it is uncommon for LSD to induce seizures, even following massive overdoses [109], but reports of LSD-induced seizures have appeared in the literature [110–112].

Hyperthermia may contribute to the development of severe NBOMe toxicity. The hyperthermia produced by serotonergic hallucinogens is thought to reflect sympathetically mediated cutaneous vasoconstriction, which reduces heat dissipation [113]. Activation of 5-HT_{2A} receptors in the CNS increases sympathetic outflow by exciting bulbospinal neurons [114]. More than half of the severe cases (12/23) – including the four cases that did not feature seizure activity – showed evidence of hyperthermia (Table 2). Elevated body temperature is known to exacerbate the muscle tissue damage underlying rhabdomyolysis [115, 116]. However, hyperthermia was also reported to occur in many of the less severe cases that resolved spontaneously. Therefore, in contrast to seizures, the presence of hyperthermia does not reliably predict that NBOMe toxicity will result in serious physiological sequelae or death.

Direct effects on muscle tissue may also play a role in NBOMe-induced damage to myocytes. 5-HT_{2A} activation releases Ca^{2+} from the endoplasmic reticulum in cells including myocytes, increasing the intracellular concentration of Ca^{2+}. Additionally, 5-HT_{2A} receptor activation causes vascular smooth muscle contraction and reduces peripheral blood flow [117–120]. Phenylalkylamine hallucinogens can cause significant peripheral vasoconstriction and vasospasm, occasionally resulting in limb ischemia [121–123].

In summary, NBOMes produce several direct and indirect effects (hyperthermia, ischemia due to vasoconstriction, hyperactivity) that would likely exacerbate muscle tissue injury. Such effects would increase the likelihood that seizure activity would produce severe muscle damage. Hence, the high incidence of seizures with NBOMes compared to other hallucinogens may translate into an elevated risk for rhabdomolysis.

The features of serotonin syndrome were not present in most cases of NBOMe toxicity. The most important diagnostic criteria for serotonin toxicity is the presence of clonus [124]. In the absence of clonus, the co-occurrence of hyperreflexia and tremor is also evidence of serotonin toxicity. Clonus, or hyperreflexia and tremor, was rarely noted in cases of NBOMe toxicity (9/51 total cases), even in the most severe cases (5/23 cases). In at least one case of NBOMe toxicity where clonus occurred (**No. 16**), the patient had also taken the selective serotonin reuptake inhibitor (SSRI) fluoxetine. SSRIs are a known risk factor for serotonin toxicity and it is possible that combined use of NBOMes and SSRIs increases the risk for serotonin excess. Co-abuse of multiple substances is extremely common and it is possible that use of other recreational substances with serotonergic effects was a contributing factor in some of the cases featuring clonus or hyperreflexia. Although it is possible that symptoms of serotonin toxicity were present but undetected in other cases, the generally low rate at which such features occurred indicates that NBOMe toxicity is not due to serotonin excess. Indeed, classical serotonergic hallucinogens rarely produce serotonin toxicity.

6 Biotransformation of NBOMe Hallucinogens

NBOMes are extensively metabolized. Caspar et al. tentatively identified 37 phase I and 31 phase II metabolites of 25I-NBOMe in rat and human urine [125]. The primary metabolites of 25I-NBOMe are 2-*O*-desmethyl-25I-NBOMe, 5-*O*-desmethyl-25I-NBOMe, 25I-NBOH, and their glucuronic acid conjugates [125–128]. Similar findings have been reported for 25B-NBOMe and 25C-NBOMe [128–130]. CYP3A4 is the major cytochrome P450 isoenzyme responsible for the biotransformation of 25I-NBOMe, with CYP2C9 and CYP2C19 potentially also contributing [125, 126].

As noted above, NBOMes are reportedly inactive after oral administration. Leth-Petersen et al. assessed the microsomal stability of NBOMes and found that they have much higher clearance rates than the corresponding 2C-X phenethylamine hallucinogens [131]. For example, 2C-I has an intrinsic clearance rate of 0.20 L/kg/h, whereas the clearance rate for 25I-NBOMe is 4.1 L/kg/h. Because 25I-NBOMe clearance exceeds the hepatic blood flow rate (1.2 L/h/kg [132]), Leth-Petersen concluded that it is subject to extensive first-pass metabolism, potentially explaining why NBOMes are not active orally. In our locomotor studies, 25I-NBOMe altered activity in mice when administered SC but not IP, which is consistent with first-pass metabolism. CYP3A4, the CYP isoenzyme primarily responsible for metabolizing 25I-NBOMe, is expressed heavily in the gut and liver.

It is also possible that oral administration increases the *N*-dealkylation of NBOMes to their parent phenethylamines, which are generally an order of magnitude less potent. *N*-dealkylation of 25I-NBOMe to 2C-I is normally a relatively minor route of biotransformation [69, 127, 128]. For example, a urine specimen collected in a case of sublingual 25I-NBOMe exposure (case **No. 22**) contained 7.5 ng/mL of 25I-NBOMe and 1.8 ng/mL 2C-I [69]. The *N*-dealklyation route, however, may be more prominent after oral administration. Grumann reported a clinical case where a man inadvertently ingested a "sip" of a liquid containing 2.8 mg/mL 25I-NBOMe [133]. Analysis of serum from the patient showed that the level of 2C-I (290 ng/mL) greatly exceeded that of 25I-NBOMe (2.6 ng/mL). Because 2C-I is significantly less potent than 25I-NBOMe, *N*-dealkylaion after oral administration would produce a marked reduction of hallucinogenic potency, potentially contributing to the perceived inactivity of NBOMes when administered by that route.

Another unresolved question is whether metabolites contribute to the toxicity of NBOMes. The major metabolites of NBOMes are their 2- and 5-*O*-desmethyl derivatives. The *O*-desmethyl derivatives of 2,5-dimethoxy-substituted phenylalkylamine hallucinogens are known to be pharmacologically active [134, 135]. However, the 2- and 5-*O*-desmethyl derivatives of 25I-NBOMe, 25B-NBOMe, and 25C-NBOMe are rapidly conjugated with glucuronic acid [130], limiting their systemic exposure. A plasma sample collected 30 min after IV administration of 2 mg 25B-NBOMe to a Danish landrace pig contained 87 nM 5-*O*-desmethyl-25B-NBOMe glucuronide but only 0.63 nM 5-*O*-desmethyl-25B-NBOMe [130]. There may be a region of bulk tolerance associated with the five-position of 2,5-dimethoxy-substituted phenylalkylamines. The affinity of the 5-benzyloxy analog of DOB for the rat 5-HT_{2A} receptor

(K_i = 140 nM [136]) is only slightly lower than the affinity of DOB (see Fig. 1). Although substitution of bulky isopropoxy or 2-methoxyethoxy groups in the 5-position of 25B-NBOMe significantly reduces 5-HT_{2A} affinity, those ligands still bind to the receptor with affinities in the 10^{-9} M range [44]. If the 5-*O*-desmethyl NBOMe glucuronide conjugates bind the 5-HT_{2A} receptor with nM affinity then they would likely contribute to the in vivo response.

7 Conclusions

The hallucinogenic effects induced by LSD, mescaline, and related substances are mediated by activation of 5-HT_{2A} receptors. The 5-HT_{2A} receptor affinities of phenethylamine hallucinogens from the 2C-X class are markedly increased by addition of an *N*-benzyl group. Likewise, the presence of an *N*-benzyl group increases the behavioral potency of phenethylamine hallucinogens in laboratory animals. Anecdotal reports from recreational users confirm that *N*-benzylphenethylamines such as 25I-NBOMe and 25B-NBOMe are potent hallucinogens, active at sub-milligram doses. Unfortunately, use of NBOMe hallucinogens has been linked to cases of severe toxicity. Potential complications of NBOMe use include hyperthermia, seizures, metabolic acidosis, rhabdomyolysis, organ failure, and death. Prompt treatment is required to manage cases of NBOMe toxicity.

Acknowledgements This work was supported by grants from NIDA (R01 DA002925), NIMH (K01 MH100644), and the Veteran's Affairs VISN 22 MIRECC. I would like to than Drs. Mark Geyer and Susan Powell, as well as Mr. Landon Klein, for their support and assistance with these studies.

References

1. Halberstadt AL (2015) Recent advances in the neuropsychopharmacology of serotonergic hallucinogens. Behav Brain Res 277:99–120
2. Nichols DE (2016) Psychedelics. Pharmacol Rev 68:264–355
3. Quednow BB, Kometer M, Geyer MA, Vollenweider FX (2012) Psilocybin-induced deficits in automatic and controlled inhibition are attenuated by ketanserin in healthy human volunteers. Neuropsychopharmacology 37:630–640
4. Valle M, Maqueda AE, Rabella M, Rodríguez-Pujadas A, Antonijoan RM, Romero S, Alonso JF, Mañanas MÀ, Barker S, Friedlander P, Feilding A, Riba J (2016) Inhibition of alpha oscillations through serotonin-2A receptor activation underlies the visual effects of ayahuasca in humans. Eur Neuropsychopharmacol 26:1161–1175
5. Vollenweider FX, Vollenweider-Scherpenhuyzen MF, Babler A, Vogel H, Hell D (1998) Psilocybin induces schizophrenia-like psychosis in humans via a serotonin-2 agonist action. Neuroreport 9:3897–3902
6. Glennon RA, Titeler M, McKenney JD (1984) Evidence for 5-HT2 involvement in the mechanism of action of hallucinogenic agents. Life Sci 35:2505–2511

7. Titeler M, Lyon RA, Glennon RA (1988) Radioligand binding evidence implicates the brain 5-HT2 receptor as a site of action for LSD and phenylisopropylamine hallucinogens. Psychopharmacology (Berl) 94:213–216
8. EMCDDA (2015) New psychoactive substances in Europe. An update from the EU Early Warning System (March 2015). Publications Office of the European Union, Luxembourg
9. Poklis JL, Raso SA, Alford KN, Poklis A, Peace MR (2015) Analysis of 25I-NBOMe, 25B-NBOMe, 25C-NBOMe and other dimethoxyphenyl-N-[(2-methoxyphenyl) methyl]ethanamine derivatives on blotter paper. J Anal Toxicol 39:617–623
10. Zuba D, Sekula K, Buczek A (2013) 25C-NBOMe – new potent hallucinogenic substance identified on the drug market. Forensic Sci Int 227:7–14
11. Forrester MB (2014) NBOMe designer drug exposures reported to Texas poison centers. J Addict Dis 33:196–201
12. Lawn W, Barratt M, Williams M, Horne A, Winstock A (2014) The NBOMe hallucinogenic drug series: patterns of use, characteristics of users and self-reported effects in a large international sample. J Psychopharmacol 28:780–788
13. Johnson RD, Botch-Jones SR, Flowers T, Lewis CA (2014) An evaluation of 25B-, 25C-, 25D-, 25H-, 25I- and 25T2-NBOMe via LC-MS-MS: method validation and analyte stability. J Anal Toxicol 38:479–484
14. Poklis JL, Clay DJ, Poklis A (2014) High-performance liquid chromatography with tandem mass spectrometry for the determination of nine hallucinogenic 25-NBOMe designer drugs in urine specimens. J Anal Toxicol 38:113–121
15. Zuba D, Sekula K (2013) Analytical characterization of three hallucinogenic N-(2-methoxy) benzyl derivatives of the 2C-series of phenethylamine drugs. Drug Test Anal 5:634–645
16. Glennon RA, McKenney JD, Lyon RA, Titeler M (1986) 5-HT1 and 5-HT2 binding characteristics of 1-(2,5-dimethoxy-4-bromophenyl)-2-aminopropane analogues. J Med Chem 29:194–199
17. Shannon M, Battaglia G, Glennon RA, Titeler M (1984) 5-HT1 and 5-HT2 binding properties of derivatives of the hallucinogen 1-(2,5-dimethoxyphenyl)-2-aminopropane (2,5-DMA). Eur J Pharmacol 102:23–29
18. Glennon RA, Young R, Jacyno JM (1983) Indolealkylamine and phenalkylamine hallucinogens. Effect of alpha-methyl and N-methyl substituents on behavioral activity. Biochem Pharmacol 32:1267–1273
19. Shulgin A, Shulgin A (1994) PiHKAL: a chemical love story. Transform Press, Berkeley, CA
20. Glennon RA, Dukat M, el-Bermawy M, Law H, De los Angeles J, Teitler M, King A, Herrick-Davis K (1994) Influence of amine substituents on 5-HT2A versus 5-HT2C binding of phenylalkyl- and indolylalkylamines. J Med Chem 37:1929–1935
21. Pertz HH, Heim R, Elz S (2000) N-Benzylated phenylethanamines are highly potent partial agonists at 5-HT2A receptors. Arch Pharm Pharm Med Chem 333(Suppl 2):30
22. Pertz HH, Rheineck A, Elz S (1999) N-Benzylated derivatives of the hallucinogenic drugs mescaline and escaline as partial agonists at rat vascular 5-HT2A receptors. Naunyn Schmiedebergs Arch Pharmacol 359(Suppl 3):R29
23. Braden MR, Parrish JC, Naylor JC, Nichols DE (2006) Molecular interaction of serotonin 5-HT2A receptor residues Phe339(6.51) and Phe340(6.52) with superpotent N-benzyl phenethylamine agonists. Mol Pharmacol 70:1956–1964
24. Roth BL, Shoham M, Choudhary MS, Khan N (1997) Identification of conserved aromatic residues essential for agonist binding and second messenger production at 5-hydroxytryptamine2A receptors. Mol Pharmacol 52:259–266
25. Hansen M, Phonekeo K, Paine JS, Leth-Petersen S, Begtrup M, Brauner-Osborne H, Kristensen JL (2014) Synthesis and structure-activity relationships of N-benzyl phenethylamines as 5-HT2A/2C agonists. ACS Chem Nerosci 5:243–249
26. Hansen M, Jacobsen SE, Plunkett S, Liebscher GE, McCorvy JD, Brauner-Osborne H, Kristensen JL (2015) Synthesis and pharmacological evaluation of N-benzyl substituted 4-bromo-2,5-dimethoxyphenethylamines as 5-HT2A/2C partial agonists. Bioorg Med Chem 23:3933–3937

27. Geronimo I, Lee EC, Singh NJ, Kim KS (2010) How different are electron-rich and electron-deficient π interactions? J Chem Theory Comput 6:1931–1934
28. Isberg V, Balle T, Sander T, Jorgensen FS, Gloriam DE (2011) G protein- and agonist-bound serotonin 5-HT2A receptor model activated by steered molecular dynamics simulations. J Chem Inf Model 51:315–325
29. Silva ME, Heim R, Strasser A, Elz S, Dove S (2011) Theoretical studies on the interaction of partial agonists with the 5-HT2A receptor. J Comput Aided Mol Des 25:51–66
30. McCorvy JD (2012) Mapping the binding site of the 5-HT2A receptor using mutagenesis and ligand libraries: insights into the molecular actions of psychedelics. Purdue University, Lafayette
31. Canal CE, Morgan D (2012) Head-twitch response in rodents induced by the hallucinogen 2,5-dimethoxy-4-iodoamphetamine: a comprehensive history, a re-evaluation of mechanisms, and its utility as a model. Drug Test Anal 4:556–576
32. Halberstadt AL, Geyer MA (2011) Multiple receptors contribute to the behavioral effects of indoleamine hallucinogens. Neuropharmacology 61:364–381
33. Halberstadt AL, Geyer MA (2013) Characterization of the head-twitch response induced by hallucinogens in mice: detection of the behavior based on the dynamics of head movement. Psychopharmacology (Berl) 227:727–739
34. Gonzalez-Maeso J, Weisstaub NV, Zhou M, Chan P, Ivic L, Ang R, Lira A, Bradley-Moore M, Ge Y, Zhou Q, Sealfon SC, Gingrich JA (2007) Hallucinogens recruit specific cortical 5-HT (2A) receptor-mediated signaling pathways to affect behavior. Neuron 53:439–452
35. Brandt SD, Kavanagh PV, Westphal F, Elliott SP, Wallach J, Colestock T, Burrow TE, Chapman SJ, Stratford A, Nichols DE, Halberstadt AL (2016) Return of the lysergamides. Part II: Analytical and behavioural characterization of N6-allyl-6-norlysergic acid diethylamide (AL-LAD) and (2′S,4′S)-lysergic acid 2,4-dimethylazetidide (LSZ). Drug Test Anal. doi: 10.1002/dta.1985 (in press)
36. Brandt SD, Kavanagh PV, Westphal F, Stratford A, Elliott SP, Hoang K, Wallach J, Halberstadt AL (2016) Return of the lysergamides. Part I: analytical and behavioural characterization of 1-propionyl-d-lysergic acid diethylamide (1P-LSD). Drug Test Anal 8:891–902
37. Halberstadt AL, Geyer MA (2014) Effects of the hallucinogen 2,5-dimethoxy-4-iodophenethylamine (2C-I) and superpotent N-benzyl derivatives on the head twitch response. Neuropharmacology 77:200–207
38. Halberstadt AL, Sindhunata IS, Scheffers K, Flynn AD, Sharp RF, Geyer MA, Young JW (2016) Effect of 5-HT_{2A} and 5-HT_{2C} receptors on temporal discrimination by mice. Neuropharmacology 107:364–375
39. Nichols DE, Sassano MF, Halberstadt AL, Klein LM, Brandt SD, Elliott SP, Fiedler WJ (2015) N-Benzyl-5-methoxytryptamines as potent serotonin 5-HT2 receptor family agonists and comparison with a series of phenethylamine analogues. ACS Chem Nerosci 6:1165–1175
40. Ettrup A, Holm S, Hansen M, Wasim M, Santini MA, Palner M, Madsen J, Svarer C, Kristensen JL, Knudsen GM (2013) Preclinical safety assessment of the 5-HT2A receptor agonist PET radioligand [11C]Cimbi-36. Mol Imaging Biol 15:376–383
41. Nichols DE, Frescas S, Marona-Lewicka D, Huang X, Roth BL, Gudelsky GA, Nash JF (1994) 1-(2,5-Dimethoxy-4-(trifluoromethyl)phenyl)-2-aminopropane: a potent serotonin 5-HT2A/2C agonist. J Med Chem 37:4346–4351
42. Parrish JC, Braden MR, Gundy E, Nichols DE (2005) Differential phospholipase C activation by phenylalkylamine serotonin 5-HT 2A receptor agonists. J Neurochem 95:1575–1584
43. Braden MR, Nichols DE (2007) Assessment of the roles of serines 5.43(239) and 5.46(242) for binding and potency of agonist ligands at the human serotonin 5-HT2A receptor. Mol Pharmacol 72:1200–1209
44. Leth-Petersen S, Petersen IN, Jensen AA, Bundgaard C, Bæk M, Kehler J, Kristensen JL (2016) 5-HT2A/5-HT2C receptor pharmacology and intrinsic clearance of N-benzylphenethylamines modified at the primary site of metabolism. ACS Chem Neurosci 7:1614–1619

45. Glennon RA, Teitler M, Sanders-Bush E (1992) Hallucinogens and serotonergic mechanisms. NIDA Res Monogr 119:131–135
46. Nelson DL, Lucaites VL, Wainscott DB, Glennon RA (1999) Comparisons of hallucinogenic phenylisopropylamine binding affinities at cloned human 5-HT2A, -HT(2B) and 5-HT2C receptors. Naunyn Schmiedebergs Arch Pharmacol 359:1–6
47. Porter RH, Benwell KR, Lamb H, Malcolm CS, Allen NH, Revell DF, Adams DR, Sheardown MJ (1999) Functional characterization of agonists at recombinant human 5-HT2A, 5-HT2B and 5-HT2C receptors in CHO-K1 cells. Br J Pharmacol 128:13–20
48. Pritchett DB, Bach AW, Wozny M, Taleb O, Dal Toso R, Shih JC, Seeburg PH (1988) Structure and functional expression of cloned rat serotonin 5HT-2 receptor. EMBO J 7:4135–4140
49. Saltzman AG, Morse B, Whitman MM, Ivanshchenko Y, Jaye M, Felder S (1991) Cloning of the human serotonin 5-HT2 and 5-HT1C receptor subtypes. Biochem Biophys Res Commun 181:1469–1478
50. Nichols DE, Frescas SP, Chemel BR, Rehder KS, Zhong D, Lewin AH (2008) High specific activity tritium-labeled N-(2-methoxybenzyl)-2,5-dimethoxy-4-iodophenethylamine (INBMeO): a high-affinity 5-HT2A receptor-selective agonist radioligand. Bioorg Med Chem 16:6116–6123
51. Ettrup A, Hansen M, Santini MA, Paine J, Gillings N, Palner M, Lehel S, Herth MM, Madsen J, Kristensen J, Begtrup M, Knudsen GM (2011) Radiosynthesis and in vivo evaluation of a series of substituted 11C-phenethylamines as 5-HT (2A) agonist PET tracers. Eur J Nucl Med Mol Imaging 38:681–693
52. Ettrup A, Palner M, Gillings N, Santini MA, Hansen M, Kornum BR, Rasmussen LK, Nagren K, Madsen J, Begtrup M, Knudsen GM (2010) Radiosynthesis and evaluation of 11C-CIMBI-5 as a 5-HT2A receptor agonist radioligand for PET. J Nucl Med 51:1763–1770
53. Fantegrossi WE, Gray BW, Bailey JM, Smith DA, Hansen M, Kristensen JL (2015) Hallucinogen-like effects of 2-([2-(4-cyano-2,5-dimethoxyphenyl) ethylamino]methyl)phenol (25CN-NBOH), a novel N-benzylphenethylamine with 100-fold selectivity for 5-HT(2)A receptors, in mice. Psychopharmacology (Berl) 232:1039–1047
54. Juncosa JI Jr, Hansen M, Bonner LA, Cueva JP, Maglathlin R, McCorvy JD, Marona-Lewicka D, Lill MA, Nichols DE (2013) Extensive rigid analogue design maps the binding conformation of potent N-benzylphenethylamine 5-HT2A serotonin receptor agonist ligands. ACS Chem Nerosci 4:96–109
55. Halberstadt AL, Powell SB, Geyer MA (2013) Role of the 5-HT(2)A receptor in the locomotor hyperactivity produced by phenylalkylamine hallucinogens in mice. Neuropharmacology 70:218–227
56. Halberstadt AL, van der Heijden I, Ruderman MA, Risbrough VB, Gingrich JA, Geyer MA, Powell SB (2009) 5-HT(2A) and 5-HT(2C) receptors exert opposing effects on locomotor activity in mice. Neuropsychopharmacology 34:1958–1967
57. Wood DM, Sedefov R, Cunningham A, Dargan PI (2015) Prevalence of use and acute toxicity associated with the use of NBOMe drugs. Clin Toxicol (Phila) 53:85–92
58. Rose RS, Cumpston KL, Stromberg PE, Wills BK (2012) Severe poisoning following self-reported use of 25-I, a novel substituted amphetamine. Clin Toxicol 50:707–708
59. Kelly A, Eisenga B, Riley B, Judge B (2012) Case series of 25I-NBOMe exposures with laboratory confirmation. Clin Toxicol 50:702
60. Srisuma S, Bronstein AC, Hoyte CO (2015) NBOMe and 2C substitute phenylethylamine exposures reported to the National Poison Data System. Clin Toxicol (Phila) 53:624–628
61. DEA (2013) Three more synthetic drugs become illegal for at least two years. https://www.dea.gov/divisions/hq/2013/hq111513.shtml
62. Hill SL, Doris T, Gurung S, Katebe S, Lomas A, Dunn M, Blain P, Thomas SH (2013) Severe clinical toxicity associated with analytically confirmed recreational use of 25I-NBOMe: case series. Clin Toxicol (Phila) 51:487–492
63. Kueppers VB, Cooke CT (2015) 25I-NBOMe related death in Australia: a case report. Forensic Sci Int 249:e15–e18

64. Gee P, Schep LJ, Jensen BP, Moore G, Barrington S (2016) Case series: toxicity from 25B-NBOMe – a cluster of N-bomb cases. Clin Toxicol (Phila) 54:141–146
65. Tang MH, Ching CK, Tsui MS, Chu FK, Mak TW (2014) Two cases of severe intoxication associated with analytically confirmed use of the novel psychoactive substances 25B-NBOMe and 25C-NBOMe. Clin Toxicol (Phila) 52:561–565
66. Yoshida K, Saka K, Shintani-Ishida K, Maeda H, Nakajima M, Hara S, Ueno M, Sasaki K, Iwase H, Sakamoto T (2015) A case of fatal intoxication due to the new designer drug 25B-NBOMe. Forensic Toxicol 33:396–401
67. Hieger MA, Rose SR, Cumpston KL, Stromberg PE, Miller S, Wills BK (2015) Severe poisoning after self-reported use of 2-(4-iodo-2,5-dimethoxyphenyl)-N-[(2-methoxyphenyl)methyl]ethanamine, a novel substituted amphetamine: a case series. Am J Emerg Med 33 (1843):e1–e3
68. Rose SR, Poklis JL, Poklis A (2013) A case of 25I-NBOMe (25-I) intoxication: a new potent 5-HT2A agonist designer drug. Clin Toxicol (Phila) 51:174–177
69. Stellpflug SJ, Kealey SE, Hegarty CB, Janis GC (2014) 2-(4-Iodo-2,5-dimethoxyphenyl)-N-[(2-methoxyphenyl)methyl]ethanamine (25I-NBOMe): clinical case with unique confirmatory testing. J Med Toxicol 10:45–50
70. Umemura Y, Andrew T, Jacobs V, Giustini A, Lewis L, Hanowell J, Filiano J (2014) Fatal outcome of status epilepticus, hyperthermia, rhabdomyolysis, multi-organ failure, and cerebral edema after 25I-NBOMe ingestion (P1. 342). Neurology 82(10 Supplement):P1-342
71. Walterscheid JP, Phillips GT, Lopez AE, Gonsoulin ML, Chen HH, Sanchez LA (2014) Pathological findings in 2 cases of fatal 25I-NBOMe toxicity. Am J Forensic Med Pathol 35:20–25
72. Poklis JL, Devers KG, Arbefeville EF, Pearson JM, Houston E, Poklis A (2014) Postmortem detection of 25I-NBOMe [2-(4-iodo-2,5-dimethoxyphenyl)-N-[(2-methoxyphenyl)methyl]ethanamine] in fluids and tissues determined by high performance liquid chromatography with tandem mass spectrometry from a traumatic death. Forensic Sci Int 234:e14–e20
73. Lowe LM, Peterson BL, Couper FJ (2015) A case review of the first analytically confirmed 25I-NBOMe-related death in Washington State. J Anal Toxicol 39:668–671
74. Shanks KG, Sozio T, Behonick GS (2015) Fatal intoxications with 25B-NBOMe and 25I-NBOMe in Indiana during 2014. J Anal Toxicol 39:602–606
75. Suzuki J, Poklis JL, Poklis A (2014) "My friend said it was good LSD": a suicide attempt following analytically confirmed 25I-NBOMe ingestion. J Psychoactive Drugs 46:379–382
76. Poklis JL, Nanco CR, Troendle MM, Wolf CE, Poklis A (2014) Determination of 4-bromo-2,5-dimethoxy-N-[(2-methoxyphenyl)methyl]-benzeneethanamine (25B-NBOMe) in serum and urine by high performance liquid chromatography with tandem mass spectrometry in a case of severe intoxication. Drug Test Anal 6:764–769
77. Isbister GK, Poklis A, Poklis JL, Grice J (2015) Beware of blotting paper hallucinogens: severe toxicity with NBOMes. Med J Aust 203:266–267 e.1
78. Laskowski LK, Elbakoush F, Calvo J, Exantus-Bernard G, Fong J, Poklis JL, Poklis A, Nelson LS (2015) Evolution of the NBOMes: 25C- and 25B- sold as 25I-NBOMe. J Med Toxicol 11:237–241
79. Grautoff S, Kahler J (2014) Near fatal intoxication with the novel psychoactive substance 25C-NBOMe. Med Klin Intensivmed Notfmed 109:271–275
80. Armenian P, Gerona RR (2014) The electric Kool-Aid NBOMe test: LC-TOF/MS confirmed 2C-C-NBOMe (25C) intoxication at Burning Man. Am J Emerg Med 32(1444):e3–e5
81. Andreasen MF, Telving R, Rosendal I, Eg MB, Hasselstrom JB, Andersen LV (2015) A fatal poisoning involving 25C-NBOMe. Forensic Sci Int 251:e1–e8
82. Suzuki J, Dekker MA, Valenti ES, Arbelo Cruz FA, Correa AM, Poklis JL, Poklis A (2015) Toxicities associated with NBOMe ingestion-a novel class of potent hallucinogens: a review of the literature. Psychosomatics 56:129–139
83. Miller PL, Gay GR, Ferris KC, Anderson S (1992) Treatment of acute, adverse psychedelic reactions: "I've tripped and I can't get down". J Psychoactive Drugs 24:277–279

84. Strassman RJ (1984) Adverse reactions to psychedelic drugs. A review of the literature. J Nerv Ment Dis 172:577–595
85. Gouzoulis-Mayfrank E, Thelen B, Habermeyer E, Kunert HJ, Kovar KA, Lindenblatt H, Hermle L, Spitzer M, Sass H (1999) Psychopathological, neuroendocrine and autonomic effects of 3,4-methylenedioxyethylamphetamine (MDE), psilocybin and d-methamphetamine in healthy volunteers. Results of an experimental double-blind placebo-controlled study. Psychopharmacology (Berl) 142:41–50
86. Grob CS, Danforth AL, Chopra GS, Hagerty M, McKay CR, Halberstadt AL, Greer GR (2011) Pilot study of psilocybin treatment for anxiety in patients with advanced-stage cancer. Arch Gen Psychiatry 68:71–78
87. Hollister LE, Macnicol MF, Gillespie HK (1969) An hallucinogenic amphetamine analog (DOM) in man. Psychopharmacologia 14:62–73
88. Schmid Y, Enzler F, Gasser P, Grouzmann E, Preller KH, Vollenweider FX, Brenneisen R, Muller F, Borgwardt S, Liechti ME (2015) Acute effects of lysergic acid diethylamide in healthy subjects. Biol Psychiatry 78:544–553
89. Snyder SH, Faillace L, Hollister L (1967) 2,5-dimethoxy-4-methyl-amphetamine (STP): a new hallucinogenic drug. Science 158:669–670
90. Strassman RJ, Qualls CR (1994) Dose-response study of N,N-dimethyltryptamine in humans. I. Neuroendocrine, autonomic, and cardiovascular effects. Arch Gen Psychiatry 51:85–97
91. Wolbach AB Jr, Isbell H, Miner EJ (1962) Cross tolerance between mescaline and LSD-25, with a comparison of the mescaline and LSD reactions. Psychopharmacologia 3:1–14
92. Alper RH (1990) Hemodynamic and renin responses to (+—)-DOI, a selective 5-HT2 receptor agonist, in conscious rats. Eur J Pharmacol 175:323–332
93. Dabire H, Chaouche-Teyara K, Cherqui C, Fournier B, Laubie M, Schmitt H (1989) Characterization of DOI, a putative 5-HT2 receptor agonist in the rat. Eur J Pharmacol 168:369–374
94. Dabire H, Chaouche-Teyara K, Cherqui C, Fournier B, Schmitt H (1989) DOI is a mixed agonist-antagonist at postjunctional 5-HT2 receptors in the pithed rat. Eur J Pharmacol 170:109–111
95. Rittenhouse PA, Bakkum EA, Van de Kar LD (1991) Evidence that the serotonin agonist, DOI, increases renin secretion and blood pressure through both central and peripheral 5-HT2 receptors. J Pharmacol Exp Ther 259:58–65
96. Nolte KB (1991) Rhabdomyolysis associated with cocaine abuse. Hum Pathol 22:1141–1145
97. Gupta P, Singh VP, Chatterjee S, Agarwal AK (2010) Acute renal failure resulting from rhabdomyolysis following a seizure. Singapore Med J 51:e79–e80
98. Liguori R, Giannoccaro MP, Pasini E, Riguzzi P, Valentino ML, Comi GP, Carelli V, Bresolin N, Michelucci R (2013) Acute rhabdomyolysis induced by tonic-clonic epileptic seizures in a patient with glucose-6-phosphate dehydrogenase deficiency. J Neurol 260:2669–2671
99. Os I, Lyngdal PT (1989) General convulsions and rhabdomyolysis. Case reports. Acta Neurol Scand 79:246–248
100. Park HJ, Chang Y, Lee JE, Koo H, Oh J, Chol YC, Park KD (2016) Recurrent episodes of rhabdomyolysis after seizures in a patient with glycogen storage disease type V. J Clin Neurol 12:373–375
101. Winocour PH, Waise A, Young G, Moriarty KJ (1989) Severe, self-limiting lactic acidosis and rhabdomyolysis accompanying convulsions. Postgrad Med J 65:321–322
102. Balikova M (2005) Nonfatal and fatal DOB (2,5-dimethoxy-4-bromoamphetamine) overdose. Forensic Sci Int 153:85–91
103. Bosak A, LoVecchio F, Levine M (2013) Recurrent seizures and serotonin syndrome following "2C-I" ingestion. J Med Toxicol 9:196–198
104. Buhrich N, Morris G, Cook G (1983) Bromo-DMA: the Australasian hallucinogen? Aust N Z J Psychiatry 17:275–279
105. Burish MJ, Thoren KL, Madou M, Toossi S, Shah M (2015) Hallucinogens causing seizures? A case report of the synthetic amphetamine 2,5-dimethoxy-4-chloroamphetamine. Neurohospitalist 5:32–34

106. Dean BV, Stellpflug SJ, Burnett AM, Engebretsen KM (2013) 2C or not 2C: phenethylamine designer drug review. J Med Toxicol 9:172–178
107. Winek CL, Collom WD, Bricker JD (1981) A death due to 4-bromo-2,5-dimethoxyamphetamine. Clin Toxicol 18:267–271
108. Wood DM, Looker JJ, Shaikh L, Button J, Puchnarewicz M, Davies S, Lidder S, Ramsey J, Holt DW, Dargan PI (2009) Delayed onset of seizures and toxicity associated with recreational use of Bromo-dragonFLY. J Med Toxicol 5:226–229
109. Klock JC, Boerner U, Becker CE (1974) Coma, hyperthermia and bleeding associated with massive LSD overdose. A report of eight cases. West J Med 120:183–188
110. Fisher DD, Ungerleider JT (1967) Grand mal seizures following ingestion of LSD. Calif Med 106:210–211
111. Picker W, Lerman A, Hajal F (1992) Potential interaction of LSD and fluoxetine. Am J Psychiatry 149:843–844
112. Rosenberg J, Pentel P, Pond S, Benowitz N, Olson K (1986) Hyperthermia associated with drug intoxication. Crit Care Med 14:964–969
113. Blessing WW, Seaman B (2003) 5-hydroxytryptamine(2A) receptors regulate sympathetic nerves constricting the cutaneous vascular bed in rabbits and rats. Neuroscience 117:939–948
114. Clement ME, McCall RB (1990) Studies on the site and mechanism of the sympathoexcitatory action of 5-HT2 agonists. Brain Res 515:299–302
115. Sprague JE, Brutcher RE, Mills EM, Caden D, Rusyniak DE (2004) Attenuation of 3,4-methylenedioxymethamphetamine (MDMA, Ecstasy)-induced rhabdomyolysis with alpha1-plus beta3-adrenoreceptor antagonists. Br J Pharmacol 142:667–670
116. Sprague JE, Moze P, Caden D, Rusyniak DE, Holmes C, Goldstein DS, Mills EM (2005) Carvedilol reverses hyperthermia and attenuates rhabdomyolysis induced by 3,4-methylenedioxymethamphetamine (MDMA, Ecstasy) in an animal model. Crit Care Med 33:1311–1316
117. Cohen ML, Fuller RW, Wiley KS (1981) Evidence for 5-HT2 receptors mediating contraction in vascular smooth muscle. J Pharmacol Exp Ther 218:421–425
118. Florian JA, Watts SW (1998) Integration of mitogen-activated protein kinase kinase activation in vascular 5-hydroxytryptamine2A receptor signal transduction. J Pharmacol Exp Ther 284:346–355
119. McKune CM, Watts SW (2001) Characterization of the serotonin receptor mediating contraction in the mouse thoracic aorta and signal pathway coupling. J Pharmacol Exp Ther 297:88–95
120. Zhang L, Dyer DC, Hembrough FB, Isla M (1991) Effect of R(-)2,5-dimethoxy-4-methylamphetamine on uterine and umbilical blood flow in conscious pregnant sheep. Eur J Pharmacol 199:179–184
121. Bowen JS, Davis GB, Kearney TE, Bardin J (1983) Diffuse vascular spasm associated with 4-bromo-2,5-dimethoxyamphetamine ingestion. JAMA 249:1477–1479
122. Personne M, Hultén P (2008) Bromo-dragonfly, a life threatening designer drug. Clin Toxicol 46:379–380
123. Thorlacius K, Borna C, Personne M (2008) Bromo-dragon fly – life-threatening drug. Can cause tissue necrosis as demonstrated by the first described case. Lakartidningen 105:1199–1200
124. Dunkley EJ, Isbister GK, Sibbritt D, Dawson AH, Whyte IM (2003) The Hunter Serotonin Toxicity Criteria: simple and accurate diagnostic decision rules for serotonin toxicity. QJM 96:635–642
125. Caspar AT, Helfer AG, Michely JA, Auwarter V, Brandt SD, Meyer MR, Maurer HH (2015) Studies on the metabolism and toxicological detection of the new psychoactive designer drug 2-(4-iodo-2,5-dimethoxyphenyl)-N-[(2-methoxyphenyl)methyl]ethanamine (25I-NBOMe) in human and rat urine using GC-MS, LC-MS(n), and LC-HR-MS/MS. Anal Bioanal Chem 407:6697–6719
126. Nielsen LM, Holm NB, Leth-Petersen S, Kristensen JL, Olsen L, Linnet K (2016) Characterization of the hepatic cytochrome P450 enzymes involved in the metabolism of 25I-NBOMe and 25I-NBOH. Drug Test Anal. doi: 10.1002/dta.2031 (in press)

127. Poklis JL, Dempsey SK, Liu K, Ritter JK, Wolf C, Zhang S, Poklis A (2015) Identification of metabolite biomarkers of the designer hallucinogen 25I-NBOMe in mouse hepatic microsomal preparations and human urine samples associated with clinical intoxication. J Anal Toxicol 39:607–616
128. Wohlfarth A, Roman M, Andersson M, Kugelberg FC, Diao X, Carlier J, Eriksson C, Wu X, Konradsson P, Josefsson M, Huestis MA, Kronstrand R (2016) 25C-NBOMe and 25I-NBOMe metabolite studies in human hepatocytes, in vivo mouse and human urine with high-resolution mass spectrometry. Drug Test Anal. doi: 10.1002/dta.2044 (in press)
129. Boumrah Y, Humbert L, Phanithavong M, Khimeche K, Dahmani A, Allorge D (2016) In vitro characterization of potential CYP- and UGT-derived metabolites of the psychoactive drug 25B-NBOMe using LC-high resolution MS. Drug Test Anal 8:248–256
130. Leth-Petersen S, Gabel-Jensen C, Gillings N, Lehel S, Hansen HD, Knudsen GM, Kristensen JL (2016) Metabolic fate of hallucinogenic NBOMes. Chem Res Toxicol 29:96–100
131. Leth-Petersen S, Bundgaard C, Hansen M, Carnerup MA, Kehler J, Kristensen JL (2014) Correlating the metabolic stability of psychedelic 5-HT(2)A agonists with anecdotal reports of human oral bioavailability. Neurochem Res 39:2018–2023
132. Davies B, Morris T (1993) Physiological parameters in laboratory animals and humans. Pharm Res 10:1093–1095
133. Grumann C, Franz F, Hermanns-Clausen M, Kithinji J, Auwaerter V (2015) 25I-NBOMe – case report and metabolism. In: 53th annual meeting of The International Association of Forensic Toxicologists, Florence, Italy
134. Eckler JR, Chang-Fong J, Rabin RA, Smith C, Teitler M, Glennon RA, Winter JC (2003) Behavioral characterization of 2-O-desmethyl and 5-O-desmethyl metabolites of the phenylethylamine hallucinogen DOM. Pharmacol Biochem Behav 75:845–852
135. Glennon RA, Young R, Rosecrans JA (1982) Discriminative stimulus properties of DOM and several molecular modifications. Pharmacol Biochem Behav 16:553–556
136. Lyon RA, Titeler M, Seggel MR, Glennon RA (1988) Indolealkylamine analogs share 5-HT2 binding characteristics with phenylalkylamine hallucinogens. Eur J Pharmacol 145:291–297

Clinical Pharmacology of the Synthetic Cathinone Mephedrone

Esther Papaseit, José Moltó, Robert Muga, Marta Torrens, Rafael de la Torre, and Magí Farré

Abstract 4-Methyl-*N*-methylcathinone (mephedrone) is a popular new psychoactive substance (NPS) that is structurally related to the parent compound cathinone, the β-keto analogue of amphetamine. Mephedrone appeared on the street drug market as a substitute for 3,4-methylenedioxy-*N*-methylamphetamine (MDMA, ecstasy) and was subsequently banned due to the potential health risks associated with its use. Nevertheless, mephedrone continues to be widely consumed among specific populations, with unique patterns of misuse. To date, most information about the biological effects of mephedrone comes from user experiences, epidemiological data, clinical cases, toxicological findings, and animal studies, whilst there are very few data regarding its human pharmacodynamics and pharmacokinetics. This chapter reviews the available published data on patterns of mephedrone use, its acute and

E. Papaseit, R. Muga, and M. Farré (✉)
Hospital Universitari Germans Trias i Pujol (IGTP), Badalona, Spain

Universitat Autònoma de Barcelona (UAB), Cerdanyola del Vallés, Spain
e-mail: mfarre.germanstrias@gencat.cat

J. Moltó
Hospital Universitari Germans Trias i Pujol (IGTP), Badalona, Spain

Fundació Lluita contra la Sida, Badalona, Spain

M. Torrens
IMIM-Hospital del Mar Medical Research Institute, Barcelona, Spain

Institut de Neuropsiquiatria i Adiccions, Barcelona, Spain

Universitat Autònoma de Barcelona (UAB), Cerdanyola del Vallés, Spain

R. de la Torre
IMIM-Hospital del Mar Medical Research Institute, Barcelona, Spain

Universitat Pompeu Fabra (UPF), Barcelona, Spain

Curr Topics Behav Neurosci (2017) 32: 313–332
DOI 10.1007/7854_2016_61
Published Online: 19 December 2016

chronic effects, and its pharmacokinetic properties. More human research is needed to elucidate the safety, toxicity, and addiction potential of mephedrone and related NPS.

Keywords 3,4-Methylenedioxy-*N*-methylamphetamine (MDMA ecstasy) • 4-Methyl-*N*-methylcathinone (4-MMC mephedrone) • New psychoactive substance (NSP) • Synthetic cathinones

Contents

1 Synthetic Cathinones and Mephedrone

Synthetic cathinones are man-made derivatives of the parent compound cathinone, a naturally occurring psychostimulant found in the khat plant, *Catha edulis*. Cathinones are phenethylamine derivatives, structurally and pharmacologically similar to amphetamine, 3,4-methylenedioxy-*N*-methylamphetamine (MDMA, ecstasy), and other related substances [1]. All cathinones possess a β-keto substituent in their chemical structure, which defines the properties of this class of compounds [2]. The most relevant compounds of the group are *N*-methylcathinone (ephedrone), 4-methyl-*N*-methylcathinone (mephedrone, 4-MMC) – considered to be the prototypical ring-substituted synthetic cathinone derivative, 3,4-methylenedioxypyrovalerone (MDPV, bath salts), and α-pyrrolidinovalerophenone (α-PVP, flakka) [3]. Synthetic cathinones represent the second largest group of new psychoactive substances (NPS) in the street drug market [4–6].

Mephedrone, initially described in 1929 by Saem de Burnaga, remained in obscurity for decades. In the early 2000s, however, clandestine chemists began altering the chemical structure of the cathinone template to synthesize unscheduled compounds [7], and mephedrone was rediscovered and first reported online in 2003. In subsequent years, the popularity of mephedrone increased due to a number of factors including competitive pricing, widespread online availability, and perceived low potential for harm, together with the shortage of MDMA [8, 9]. As a consequence, mephedrone rapidly became a replacement for MDMA and was marketed as a 'legal high' [10]. In this scenario, mephedrone was sold openly on the grey market (online vendors from surface and deep websites) and in head shops as a number of authentic commercial products. It was typically marketed and sold as 'bath salts', 'incense', and 'plant food/fertilizer', and advertised as 'not for human consumption' to circumvent potential legislative control [8, 11–13].

In 2008, mephedrone was first identified on internet web sites and in products confiscated by law enforcement [14]; soon after, the misuse of mephedrone became a particularly widespread and problematic concern in Australia, the United States (USA), and several European countries. For example, during 2009–2010 in the United Kingdom (UK), the distribution and consumption of mephedrone experienced a rapid increase which exceeded even that of ecstasy [3, 15, 16]. The higher prevalence of use was associated with reports of numerous hospital admissions and related overdose deaths. As a consequence of media attention and an official risk assessment, mephedrone was made illegal in the UK and classified as a Class B substance under the Misuse of Drugs Act on April 16, 2010. Subsequently, other European countries and the USA also adopted control measures [5, 17], and it was banned in 2010 and 2011, respectively.

2 Patterns of Use

In 2009–2010, a non-representative internet survey investigating mephedrone use among readers of *MixMag*, a UK clubbers' magazine, estimated a 42% life-time use and a 34% past-month use (2,295 respondents) [18]. One year later, life-time and past-month use of mephedrone increased to 61% and 51%, respectively (2,560 respondents) [16]. Immediately after mephedrone was banned, estimated overall rates of use remained low but stable, ranging from 0 to 6.3% life-time prevalence depending on age and country (Table 1). Nevertheless, mephedrone consumption among certain subgroups including drug users, club scene participants, readers of music and clubbing magazines, and the gay community continued to be considerable and much greater in comparison to the general population. In 2011–2012, past-year use of mephedrone was 19.5% among the UK *MixMag* survey respondents, and 30% in a subset of regular clubbers [19]. Mephedrone was the 11th most prevalent drug amongst clubbers, in terms of life-time use, when it was legal (42%), moving up to 4th place (61%) after being banned. In spite of the legislative ban on mephedrone, 42% of the respondents to an online survey of mephedrone users reported still trying to obtain it, and 53% said that the ban had not affected availability in their area [20]. Immediately after its prohibition in the UK, mephedrone had become the most popular drug in London among men who have sex with men, with over half the clubbers having tried it [12].

To date, the latest official epidemiological data indicate that mephedrone continues to be one of the most relevant NPS used in recreational nightlife settings according to indirect estimations. It is remarkable that mephedrone is currently involved in 50% of all hospital emergency presentations related to NPS misuse in Europe [5, 21], particularly in England where the number of individuals requiring treatment for mephedrone has more than doubled from 953 in 2010–2011 to 2,024 in 2014–2015 [22]. With respect to fatalities, in 2015 the number of cases in which mephedrone was implicated ($n = 34$, 1.5%) was comparable to that of MDMA ($n = 28$, 1.2%). According to these recent data, all mephedrone-associated deaths in 2015 were in men, 68% of whom were men who had sex with other men [23].

Table 1 National estimates of mephedrone use in the general population in European countries from 2010 to 2015

Year	Country	Prevalence (%) Age of population	Life-time Total	Previous year Total	Previous month Total	Source
2010	Slovakia	15–64	0.0	0.10	0.0	Annual report questionnaire
		15–19	1.7	n/a	n/a	2011 National Report to EMCDDA
2010	Spain	14–18	0.4	0.3	0.2	ESTUDES 2010
2010/2011	Ireland	15–64 15–34	2 4.3	1.1 2.2	0.1 0.1	Drug Prevalence Survey 2010/2011: Regional Drug Task Force and Health and Social Care Trust
2010/2011	United Kingdom	16–59	n/a	1.3	n/a	2010/2011 Crime Survey for England and Wales
		16–24	n/a	4.4	n/a	
	Malta	15–16	3.5	5.0	2.0	Annual report questionnaire
2011	Spain	15–64	0.1	0.0	0.0	EDADES 2011
		15–24	0.3	0.2	0.0	
	Hungary	16	6	n/a	n/a	Annual report questionnaire
2011/2012	United Kingdom	16–59	n/a	1.0	n/a	2011/2012 Crime Survey for England and Wales
		16–24	n/a	3.3	n/a	
2012	Spain	n/a	0.5	0.3	0.2	ESTUDES 2012
2012	Croatia	n/a	0.3	1.5	n/a	2012 National Report to EMCDDA
2012/2013	United Kingdom	16–59	1.9	0.5	n/a	2012/2013 Crime Survey for England and Wales
		16–24	4.5	1.6	n/a	
2013	Australia	≥14	n/a	0.4	n/a	National Drug Strategy Household Survey 2013
2013	Spain	15–64	0.1	0.6	0.0	EDADES 2013
		15–24	0.1	0.0	0.0	
2013/2014	United Kingdom	16–59	2.3	0.6	n/a	2013/2014 Crime Survey for England and Wales
		16–24	6.3	1.9	n/a	
2014	Spain	14–18	0.5	n/a	n/a	ESTUDES 2014
2014/2015	United Kingdom	16–59 16–24	2.2 5.3	0.5 1.9	0.2 0.5	2014/2015 Crime Survey for England and Wales

n/a not available

Furthermore, from the total amount of mephedrone seized in the European Union, almost all of it was in the UK in accordance with historically high estimates of its use in that country [5, 24].

Over the past few years, forensic evidence from confiscated drug products shows that mephedrone is typically available as a fine white, off-white or yellowish powder, or as tablets and capsules with a very high purity (around 99%). The powder can be directly taken by the intra-nasal route, or it can be easily dissolved in water for oral/rectal use and intravenous/intramuscular injection. Although mephedrone can be administered by any of these routes, it is predominantly consumed intra-nasally, orally, and by intravenous injection [12, 25]. Nasal insufflation ('snorting') is a common route of administration, but unwanted effects including nasal burning, clogging of nasal passages, nasal dripping, and nasal bleeding have led some users to switch to oral ingestion [26]. By the oral route, mephedrone is typically swallowed by 'dabbing' with a moistened finger or 'bombing' wrapped in thin cigarette paper, or ingested directly as tablets and capsules. The combination of both routes, nasal and oral, has been frequently reported as an alternative to avoid undesirable local reactions whilst maintaining sustained psychoactive effects [27, 28].

Mephedrone injection has become a common pattern of use among high-risk drug users, mainly experienced intravenous drug consumers, and more recently, among some sub-groups of men who have sex with men [4, 5, 29]. A new trend, known as 'slamming', consists of the intravenous injection of a combination of mephedrone with methamphetamine, gamma-hydroxybutyric acid (GHB), cocaine or sildenafil during the so-called chemsex parties which can last for many hours. The drug cocktail is injected as a means to sustain and enhance sexual experiences [30–32]. Consequently, intravenous mephedrone use in conjunction with unprotected sexual activity is associated with a highly elevated risk of blood-borne and sexually transmitted diseases, in addition to adverse effects of the drug itself [4, 33]. Anecdotally, other routes of administration such as intrapulmonary (smoking), rectal ('booty bumping' or 'plugging'), subcutaneous, and intramuscular have also been described [14, 34–36].

Based on recreational user reports, mephedrone is initially consumed in single doses although repeated doses (re-dosing or 'bingeing'), in a similar manner to MDMA, is common practice in order to maintain pleasurable effects [37]. A typical drug use session lasts for approximately 8–10 h during which 5–6 mephedrone doses are taken, equivalent to a total median dose of 1–1.9 g [18, 36, 37]. Mephedrone oral dosage ranges from 15 to 300 mg, whereas nasal insufflation dosage is somewhat lower and ranges from 5 to 200 mg. Intravenous/intramuscular injection has been reported at approximately half or one-third of oral dosage, whilst 100 mg is described as a usual rectal dose [26, 38]. The initial impact is felt by recreational users approximately 30 min after oral ingestion, with effects lasting for 2–5 h [39]; in contrast, intravenous and rectal administration produce earlier onset of action and shorter duration [40]. Like other classical drugs of abuse, intravenous injection produces the most intense acute pleasurable effects, which are characterized by an initial rush of approximately 5 min followed by euphoric effects which can last for 60 min [9].

3 Pharmacodynamics of Mephedrone in Human Subjects

Mephedrone, like other amphetamines and MDMA, has a chiral centre in its structure, so exists as two enantiomers, *R*-mephedrone and *S*-mephedrone. Importantly, racemic mephedrone is the most common form of the substance available in the street drug market [41]. The molecular mechanism of action for mephedrone is most comparable to MDMA. Like MDMA, mephedrone is a non-selective releaser and reuptake inhibitor at monoamine transporters in the brain and periphery [42–45]. Another chapter of this volume is focused on the mechanism of mephedrone and other synthetic cathinones at human monoamine transporters (see [46]).

Preclinical studies performed in rodents and invertebrate models have evaluated acute mephedrone effects [39, 47]. In rats, mephedrone induces locomotor hyperactivity, increases in blood pressure and heart rate, and changes in temperature [48–52]. There is a paucity of data on the pharmacodynamics of mephedrone after administration to humans in laboratory conditions, with the exception of a recently published study from our group [53]. Most information regarding psychological and subjective effects related to mephedrone is based on user reports posted on internet web sites and blogs, surveys and questionnaires, whilst data concerning acute clinical toxicity has been provided from emergency department cases and toxicological consultations [37, 54, 55].

Our group had the opportunity to evaluate the pharmacodynamics effects of mephedrone for the first time in humans in a randomized, double-blind, crossover, and placebo-controlled trial [53], where a single oral dose of 200 mg mephedrone was compared to 100 mg MDMA and placebo. The mephedrone dose was selected after a series of pilot studies, which tested doses ranging from 50 to 200 mg [56]. The administration of mephedrone by the oral route at a dose of 200 mg produced significant increases in arterial blood pressure and heart rate, a modest increase in pupil diameter, and slight changes in oral temperature. These effects were similar in intensity to those induced by 100 mg MDMA, except for mydriasis that was lower with mephedrone. Mephedrone induced subjective feelings of euphoria, wellbeing, pleasure, stimulant-like effects and mild changes in perception, but not hallucinations or psychotic symptoms, these effects were similar in magnitude to those induced by MDMA. Both substances were well tolerated and no serious adverse events were presented [53]. Under laboratory conditions, physiological and subjective effects produced by 200 mg of oral mephedrone started at 30–45 min and lasted 2–3 h, in the case of 100 mg of MDMA the effects started at 45 min–1 h and lasted 3–5 h. As a summary, the pharmacological effects of mephedrone were similar to MDMA but with a more rapid onset and a shorter duration of effects, probably related to its brief half-life (see Sect. 4). Mephedrone impaired short-term memory in a similar manner to MDMA, while it improved critical tracking tasks but did not change reaction times or divided attention tasks in laboratory tests assessing driving-related skills [57].

According to consumers' experiences and preferences, mephedrone effects are characterized by sympathomimetic, psychostimulant, and empathogen–entactogen reactions that resemble MDMA, amphetamines, and cocaine [14, 20, 36, 58, 59]. In spite of the differences in doses and routes of administration employed, recreational mephedrone consumers (according to surveys, questionnaires, interviews, forums, etc.) consistently describe pleasurable effects such as euphoria, increased energy, mood enhancement, talkativeness, increase music sensitivity, empathy, sociability, sensory enhancement, moderate sexual arousal, and perceptual distortions ([8, 11, 12, 14, 60, 61]). In contrast, the undesirable effects include jaw clenching, bruxism, body sweats, palpitations, anxiety, tremor in extremities, blurred vision, shortness of breath, headache, cold or numb extremities, nausea and vomiting, agitation, anxiety, aggressiveness, paranoia, and panic [37, 59, 60], most of which are slight or moderate and do not require medical assistance. Post-drug recovery effects include craving, decreased appetite, lack of motivation, paranoia, insomnia, and irritability, all of which ameliorate after a few days ('feeling normal' after 4 $\pm$ 2 days) [37, 60]. Such symptoms concur with those self-reported in counselling services and emergency department admissions.

Acute toxic effects induced by mephedrone include the enhancement of cardiovascular response and neuropsychiatric effects and, in some cases, life-threatening reactions requiring medical assistance/hospitalization. In this respect, multiple cases have been reported of hypertension, cardiac arrhythmia, chest pain, paranoia, psychosis, hallucinations, agitation, aggressive behaviour, and suicidal ideation attributed to mephedrone use ([62–70]; Sivagnanam et al. 2013), although only a few cases have had the presence of mephedrone confirmed by forensic analysis (Table 2).

Upon administration of mephedrone in controlled clinical settings or self-administration in recreational settings, mephedrone effects are characterized by a rapid onset and a short-lasting duration. The pharmacodynamic response to mephedrone reported after experimental administration in humans could partially justify the pattern of use among mephedrone consumers in real-life conditions [60].

Regarding the possible long-term toxicity of mephedrone, the fact that the drug possesses structural and pharmacological similarities to MDMA, amphetamines, and cathinone suggests the likelihood that repeated and/or prolonged use produces similar consequences on neurochemical and neuropsychological function. From the limited results to date, it should be pointed out that repeated mephedrone administration in experimental animals has not shown evidence of neurotoxicity to monoaminergic systems in the brain [42, 88–91]. It should be kept in mind that further research with mephedrone in humans is required to first establish safety aspects, and the potential for long-term toxicity remains an important research question.

Table 2 Mephedrone concentrations in blood from human subjects based on findings from human performance, toxicological assessments, and postmortem cases

Subjects	Concentration of mephedrone	Reference
N = 1	0.5 mg/L, 198 mg/L (urine)	[71]
N = 1[a]	0.15 mg/L (serum)	[68]
N = 1 (Case 2)	3.3 mg/L, 4.2 ng/mg, and 4.7 ng/mg (hair)	[72]
N = 1 (Case 1)	22 mg/L	
N = 1	13.2 µg/g (antemortem), 8.4 µg/g (postmortem)	[73]
N = 10	26.8 ng/mg (0.2–313.2) (hair)	[74]
N = 1	193 mg/L	[75]
N = 1 (Case 1)	0.98 mg/L	[64]
N = 1 (Case 2)	2.24 mg/L	
N = 1 (Case 3)	0.13 mg/L	
N = 1 (Case 4)	0.23 mg/L	
N = 1 (Case 1)	51 µg/kg, 560 µg/kg (urine)	[76]
N = 1 (Case 2)	28 µg/kg	
N = 1 (Case 3)	29 µg/kg	
N = 1 (Case 1)	1 µg/kg	
N = 1	0.5 mg/L, 14.8 mg/L (urine), 38 mg/L (gastric), 1.9 mg/L (bile)	[77]
N = 36[b]	1.586 mg/L (<0.01–22 mg/L)	[78]
N = 1	21.11 pg/mg	[79]
N = 32	0.21 mg/L (0.01–0.74)	[80]
N = 9	0.27 mg/L (0.08–0.66)	
Case 2	2.1 mg/L	
Case 6	1.94 mg/L	
N = 1	5,500 and 7,100 ng/mL (vitreous humor)	[81]
N = 6	161 ng/mL (39–370 ng/mL)	[82]
N = 1	1.33 mg/L, 144 mg/L (urine), 4.52 mg/L (gastric), 1.29 mg/L (bile), 0.89 mg/L (brain), 0.25 ng/mg (hair)	[83]
N = 1 (case 2)	412 ng/mL	[84]
N = 2	50–59 pg/mg (hair)	[85]
N = 5[c]	1,774 ng/mL (13–5,500)	[86]

(continued)

Table 2 (continued)

Subjects	Concentration of mephedrone	Reference
$N = 1$	2,600 ng/mL	
$N = 1$	692 ng/mL	
$N = 1$	52 ng/mL	
$N = 1$	13 ng/mL	
$N = 4$	1.7 mg/L (0.19–3.30)	[87]
$N = 8$	1.34 mg/L (0.07–2.24)	

[a]4 g (200 mg by oral and 3.8 g by intramuscular injection)
[b]2009–2011
[c]2012–2014

4 Pharmacokinetics of Mephedrone in Humans

Most of the information concerning the pharmacokinetics of mephedrone in humans has arisen from toxicological cases, been extrapolated from preclinical pharmacokinetic and toxicokinetic data, or come from in vitro and in vivo metabolism findings [92]. There are very few data available related to human mephedrone pharmacokinetics under controlled conditions. It is noteworthy that there is no widely available commercial test to rapidly determine the presence of mephedrone or other cathinones in urine and blood. However, there are sophisticated analytical methods, such as gas/liquid chromatography coupled to mass spectrometry, for the determination of mephedrone concentrations in human tissue specimens for toxicological services, drugged driving cases in police departments, and clinical case reports [93]. Among the suspected cases of mephedrone exposure (case reports, driving under the influence of drugs, and fatalities) the dose, purity, route of administration, pattern of use, and concentrations leading to intoxications or fatal clinical consequences are usually unknown (Table 2).

There are some exceptions, for example, in reports of fatalities directly attributed to mephedrone toxicity, and impaired driving cases with high mephedrone concentrations [80, 87]. These mephedrone-associated events occurred mainly in recreational drug users who had used mephedrone in combination with alcohol or other recreational drugs (cocaine, amphetamine, or MDMA) and among intravenous drug users combining mephedrone with heroin. Lately, confirmed mephedrone fatalities have been detected in combination with other drugs among men who have sex with men due to the widespread chemsex phenomenon. The collective evidence indicates that mephedrone is only directly implicated as a principal cause in a limited number of deaths, though it undoubtedly plays a role in fatalities secondary to poly-drug exposures.

Since saliva, sweat, and hair samples can be more easily obtained as alternatives to blood or urine, reliable analytical methods have been developed in these matrices to detect a number of drugs such as MDMA, amphetamines, and certain NPS. Mephedrone concentrations from recreational users have been determined in hair, whilst published analysis methods for its detection in saliva have not yet been applied to human specimens [94–97]. Mephedrone is a weak base with a relatively low molecular weight, which results in easy diffusion across cell membranes before incorporation into hair. In fact, after repeated consumption in a similar manner to amphetamines, mephedrone concentrations in hair have been detected in the range of nanogram per milligram [79, 85], in agreement with the amounts of the drug detected in other biological matrixes from intoxications and fatalities [72, 74, 83] (Table 2).

Regarding experimental studies in humans, we recently carried out the first controlled clinical trial to evaluate mephedrone pharmacokinetic parameters in humans [53]. The administration of an oral dose of 200 mg in 12 healthy male recreational drug users produced a mean maximal plasma concentration (C_{max}) value of 134.6 ng/mL (range 51.7–218.3 ng/mL). Concentrations decreased to one-half at 2 h and were undetectable at 24 h post-administration (Fig. 1). In spite of high inter-individual variability observed after oral administration, peak plasma concentration (T_{max}) was attained at 1.25 h, earlier than most psychostimulant drugs. Mean elimination half-life ($T_{1/2}$) was 2.15 h, significantly faster than MDMA, amphetamine, and methamphetamine ($T_{1/2}$ of 8, 10–12, and 15 h, respectively), and more similar to cathinone ($T_{1/2}$ of 4 h). The clinical pharmacokinetic data obtained after controlled mephedrone administration are congruent with the rapid onset and short duration of effects described after oral mephedrone use in recreational settings.

In same clinical trial mentioned above, urine samples were collected over a 4-h period after mephedrone administration to characterize the human metabolism of mephedrone [98]. These new in vivo data provide an important comparison to previous analyses performed in post-mortem samples [76] and in vitro models [99]. The findings confirmed the involvement of three main hepatic pathways of mephedrone metabolism: (1) *N*-demethylation to form 4-methylcathinone (nor-mephedrone), (2) phenyl ring hydroxylation to form 4-hydroxytolylmephedrone (4-OH-mephedrone), and (3) β-keto

Fig. 1 Time course of plasma mephedrone concentrations after controlled administration of an oral dose of 200 mg in one study participant [53]

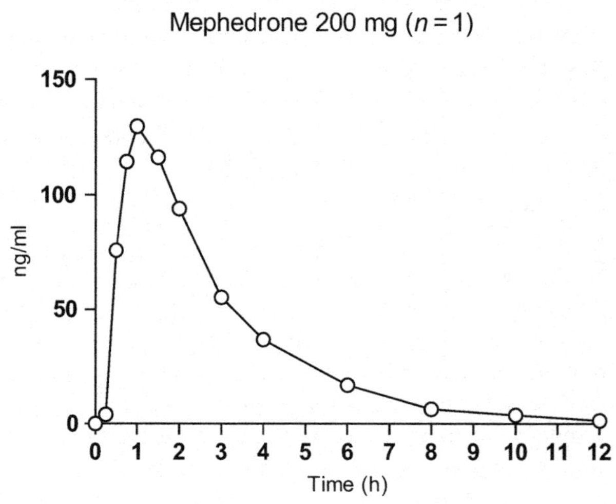

reduction to form dihydro-mephedrone. All of the identified phase I metabolites are reported to act as non-selective substrates at plasma membrane transporters for dopamine, norepinephrine, and serotonin, similar to mephedrone. Importantly, dihydro-mephedrone is a much weaker transporter substrate in comparison to nor-mephedrone and 4-OH-mephedrone [100]. In vitro studies suggest that mephedrone metabolic disposition (phase I) is regulated mainly by cytochrome P450 isoenzyme 2D6 (CYP2D6) [76]. For instance, human polymorphisms in CYP2D6, as well as co-administration of mephedrone with specific drugs also metabolized through CYP2D6 (or inhibitors), could increase potential risk of toxicity. At present, no studies have examined the interaction of mephedrone with other susceptible drugs commonly used among recreational drug users [e.g. MDMA, antiretroviral drugs, selective serotonin reuptake inhibitors (SSRIs)]. In contrast to the various metabolites reported in animals [101, 102], a number of unique phase I and phase II metabolites are found in humans including *N*-desmethyl-mephedrone-3 carboxylic acid, hydroxylmephedrone-3-*O*-glucoronide, *N*-succinyl nor-mephedrone, and hydroxyl nor-mephedrone-3-*O*-glucoronide (Fig. 2).

In forensic cases, nor-mephedrone and dihydro-mephedrone have been detected in blood and urine, whilst 4-OH-mephedrone, the most abundant mephedrone metabolite, has only been reported in urine. Taken together with in vitro data, it has been suggested that nor-mephedrone possesses lipophilic properties that may allow it to penetrate through the blood–brain barrier and act on the central nervous system, similar to the parent compound. In fact, in vivo studies in rats have confirmed that nor-mephedrone is the only bioactive mephedrone metabolite to exert effects on extracellular dopamine and locomotion [100]. Currently, the potential role of nor-mephedrone and other metabolites on human mephedrone effects is unknown and, consequently, more research on this topic is warranted.

Overall, the initial results regarding the human pharmacology of mephedrone under controlled administration confirm previous extrapolation from non-experimental human data and preclinical studies. However, it is still unclear whether the pharmacokinetics and/or metabolism of mephedrone, including its active metabolites, might have implications in the particular susceptibility of some subjects to develop acute mephedrone intoxication.

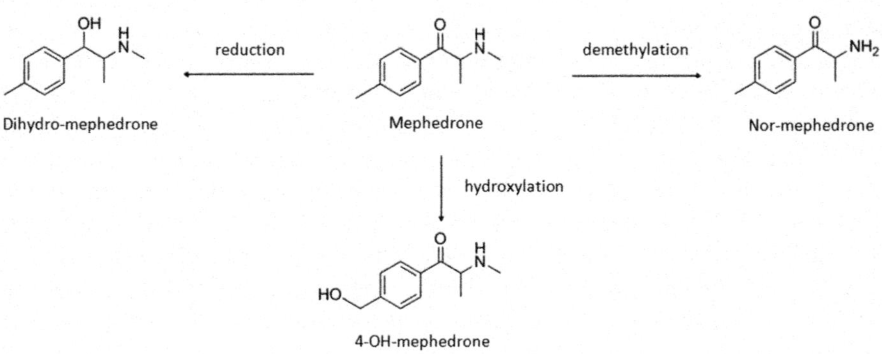

Fig. 2 Proposed pathways of mephedrone metabolism in humans

5 Abuse Liability

Predictions about the human abuse liability of mephedrone are based on user experiences, clinical reports, and extrapolation of preclinical studies, and have been recently confirmed by the first clinical trial described above [53]. To date, the potential abuse liability of mephedrone has been assessed in experimental models through a number of approaches and direct comparison with other well-known drugs of abuse (MDMA, amphetamine, cocaine) and/or NPS such as 3-methyl-*N*-methylcathinone (3-MMC), MDPV, methylone, naphyrone, flephedrone, and butylone [51, 103–109]. Systemic administration of mephedrone in laboratory animals produces elevations in the extracellular concentrations of dopamine, serotonin, and norepinephrine leading to self-administration, place preference, and locomotor hyperactivity [39, 43, 52, 110, 111]. Furthermore, mephedrone shares with other stimulant drugs of abuse the ability to activate the mesolimbic dopamine reward circuit as a main mechanism involved in the positive-reinforcing effects produced (euphoria, pleasure, well-being, happiness). At the same time, feelings of euphoria and stimulation can lead to high abuse liability similar in magnitude to that observed after controlled administration of MDMA, one of the drugs preferred by users in recreational settings.

In humans, data about the pharmacology of mephedrone are scant and findings are variable depending on the route of administration used. Evidence from questionnaires, online-surveys and even some case reports related to long-term mephedrone consumption have indicated addiction potential and dependency symptoms, cravings, and withdrawal syndromes [34, 37, 60, 112–115]. Some epidemiological data highlight the prevalence of self-reported tolerance, impaired control, continued mephedrone consumption despite physical and psychological problems, strong urges, and bingeing [36, 37, 60, 112], whilst tiredness, insomnia, nasal congestion, and impaired concentration are described as the most common withdrawal-related effects. Intranasal users considered mephedrone to be more addictive than cocaine [60]. The transition from nasal to injection route, and intravenous injection use, is associated with a compulsive consumption pattern. Hence, mephedrone addicts commonly report re-injecting of the drug with excessive binge use over long periods of time [9, 114, 116]. The potential of mephedrone for cravings, compulsive re-dosing, and uncontrollable binge use have been attributed to its self-reported high and short duration effects [8, 117]. For intranasal and/or intravenous administration, no human clinical trials have been conducted. As previously explained in other sections, relative mephedrone abuse liability has been assessed in a double blind, double dummy, placebo, and positive comparator controlled and crossover clinical trial [53]. Mephedrone produced similar effects to MDMA on standardized questionnaires developed to measure abuse potential in humans ([118]; Visual Analog Scale, 49-item Addiction Research Center Inventory-ARCI, and Evaluation of Subjective Effects of Substances with Abuse Potential questionnaire-VESSPA-SEE), but with a faster onset of the desired high and shorter duration of action [53]. Even though preclinical and human clinical findings are limited, overall data suggest that the shorter half-life and duration of effects induced by mephedrone may lead to more compulsive drug-taking behaviour in order to maintain euphoria.

6 Summary

To conclude, mephedrone is perhaps the most representative synthetic cathinone in the recent NPS phenomenon. The misuse of mephedrone in recreational settings has been related to serious medical problems which have negative impacts on global public health, especially after binge use, intravenous administration or in the context of chemsex scenarios. The known effects of mephedrone on dopamine and serotonin systems in the brain confer psychostimulant-like effects that resemble those of other amphetamine-compounds and MDMA. Further human clinical pharmacology research focused on pharmacodynamics and pharmacokinetics is essential to better understand mephedrone effects, as well as the potentially life-threatening medical complications associated with its consumption, in order to provide more effective interventions.

Acknowledgments This work was supported in part by grants from Instituto de Salud Carlos III (ISCIII, FIS-FEDER PI11/01961; Red de Trastornos Adictivos-RTA-FEDER RD16/0017/0003 and RD16/0017/0010; Juan Rodes fellowship JR16/00020) and The European Commission (Drugs Policy Initiatives, Justice Programme 2014–2020, Contract no. HOME/2014/JDRU/AG/DRUG/7082, PREDICT Project).

References

1. Valente MJ, Guedes de Pinho P, de Lourdes Bastos M, et al. (2014) Khat and synthetic cathinones: a review. Arch Toxicol 88:15–45
2. Hill SL, Thomas SH (2011) Clinical toxicology of newer recreational drugs. Clin Toxicol (Phila) 49:705–719
3. Karila L, Megarbane B, Cottencin O, et al. (2015) Synthetic cathinones: a new public health problem. Curr Neuropharmacol 13:12–20
4. European Monitoring Centre for Drugs and Drug Addiction (EMCDDA) (2014) European drug report 2014: trends and developments. http://www.emcdda.europa.eu/system/files/publications/963/TDAT14001ENN.pdf
5. European Monitoring Centre for Drugs and Drug Addiction (EMCDDA) (2016) 2016 EU drug markets report: strategic overview. http://www.emcdda.europa.eu/publications/eu-drug-markets/2016/strategic-overview
6. Papaseit E, Farré M, Schifano F, et al. (2014) Emerging drugs in Europe. Curr Opin Psychiatry 27:243–250
7. United Nations Office on Drugs and Crime (UNODC) (2016) Laboratory and scientific section portals. Synthetic cathinones. https://www.unodc.org/LSS/SubstanceGroup/Details/67b1ba69-1253-4ae9-bd93-fed1ae8e6802
8. Brunt TM, Poortman A, Niesink RJ, et al. (2011) Instability of the ecstasy market and a new kid on the block: mephedrone. J Psychopharmacol 25:1543–1547

9. Van Hout MC, Bingham T (2012) "A costly turn on": patterns of use and perceived consequences of mephedrone based head shop products amongst Irish injectors. Int J Drug Policy 23:188–197
10. Prosser JM, Nelson LS (2012) The toxicology of bath salts: a review of synthetic cathinones. J Med Toxicol 8:33–42
11. Deluca P, Schifano F, Davey Z, et al. (2009) Mephedrone report. Institute of Psychiatry, King's College London, London http://www.psychonautproject.eu
12. Measham F, Moore K, Newcombe Z, et al. (2010) Tweaking, bombing, dabbing and stockpiling: the emergence of mephedrone and the perversity of prohibition. Drugs Alcohol Today 10:14–21
13. Winstock AR, Mitcheson L, Marsden J (2010) Mephedrone: still available and twice the price. Lancet 376(9752):1537
14. Schifano F, Albanese A, Fergus S, Psychonaut Web Mapping, ReDNet Research Groups, et al. (2011) Mephedrone (4-methylmethcathinone; 'meow meow'): chemical, pharmacological and clinical issues. Psychopharmacology (Berl) 214:593–602
15. European Monitoring Centre for Drugs and Drug Addiction (EMCDDA) (2015) European drug report 2015. http://www.emcdda.europa.eu/attachements.cfm/att_239505_EN_TDAT15001ENN.pdf
16. Winstock AR (2011a) Drugs survey. MixMag 238:49–59
17. Advisory Council on the Misuse of Drugs (ACMD) (2015) ACMD report on definitions for the Psychoactive Substances Bill, 2015. https://www.gov.uk/government/uploads/system/uploads/attachment_data/file/454039/Definitions_report_final_14_august.pgl
18. Dick D, Torrance C (2010) Drugs survey. MixMag 225:44–53
19. Winstock A (2012) Global drug survey. MixMag 251:68–73
20. Carhart-Harris RL, King LA, Nutt DJ (2011) A web-based survey on mephedrone. Drug Alcohol Depend 118:19–22
21. Wood DM, Heyerdahl F, Yates CB, et al. (2014) The European Drug Emergencies Network (Euro-DEN). Clin Toxicol (Phila) 52:239–241
22. Public Health England (2015) Adult substance misuse statistics from the National Drug Treatment Monitoring System (NDTMS) 1 April 2014 to 31 March 2015. Public Health England, London https://www.ndtms.net/Publications/downloads/Adult%20Substance%20Misuse/adult-statistics-from-the-national-drug-treatment-monitoring-system-2014-2015.pdf
23. Hockenhull J, Murphy KG, Paterson S (2016) Mephedrone use is increasing in London. Lancet 387:1719–1720
24. United Nations Office on Drugs and Crime (UNODC) (2015) World drug report 2015. https://www.unodc.org/documents/wdr2015/World_Drug_Report_2015.pdf
25. European Monitoring Centre for Drugs and Drug Addiction (EMCDDA) (2011) Report on the risk assessment of mephedrone in the framework of the council decision on new psychoactive substances. http://www.emcdda.europa.eu/system/files/publications/571/TDAK11001ENC_WEB-OPTIMISED_FILE_280269.pdf
26. Van Hout MC (2014) An Internet study of user's experiences of the synthetic cathinone 4-methylethcathinone (4-MEC). J Psychoactive Drugs 46:273–286
27. Psychonaut Web Mapping Research Group (2009) Mephedrone report. Institute of Psychiatry, King's College London, London http://194.83.136.209/documents/reports/
28. The Vaults of Erowid (2016) https://www.erowid.org/chemicals/4_methylmethcathinone/4_methylmethcathinone.shtml
29. Péterfi A, Tarján A, Horváth GC, et al. (2014) Changes in patterns of injecting drug use in Hungary: a shift to synthetic cathinones. Drug Test Anal 6:825–831
30. Bourne A, Reid D, Hickson F, Torres-Rueda S (2015) "Chemsex" and harm reduction need among gay men in South London. Int J Drug Policy 26:1171–1176
31. Daskalopoulou M, Rodger A, Phillips AN, et al. (2014) Recreational drug use, polydrug use, and sexual behaviour in HIV-diagnosed men who have sex with men in the UK: results from the cross-sectional ASTRA study. Lancet HIV 1:e22–e31

32. McCall H, Adams N, Mason D, et al. (2015) What is chemsex and why does it matter? BMJ 351:h5790
33. Hope VD, Cullen KJ, Smith J, et al. (2016) Is the recent emergence of mephedrone injecting in the United Kingdom associated with elevated risk behaviours and blood borne virus infection? Euro Surveill 21(19). doi:10.2807/1560-7917.ES.2016.21.19.30225
34. Dargan PI, Sedefov R, Gallegos A, Wood DM (2011) The pharmacology and toxicology of the synthetic cathinone mephedrone (4-methylmethcathinone). Drug Test Anal 3:454–463
35. Russo R, Marks N, Morris K, et al. (2012) Life-threatening necrotizing fasciitis due to 'bath salts' injection. Orthopedics 35:e124–e127
36. Winstock AR, Mitcheson LR, Deluca P, et al. (2011b) Mephedrone, new kid for the chop? Addiction 106:154–161
37. Jones L, Reed P, Parrott A (2016) Mephedrone and 3,4-methylenedioxy-methamphetamine: comparative psychobiological effects as reported by recreational polydrug users. J Psychopharmacol (Epub a head)
38. World Health Organization (WHO) (2014) Mephedrone. Critical Review Report. 36th ECDD (2014) Agenda ítem 4.12. http://www.who.int/medicines/areas/quality_safety/4_12_review.pdf
39. Green AR, King MV, Shortall SE, et al. (2014) The preclinical pharmacology of mephedrone; not just MDMA by another name. Br J Pharmacol 171:2251–2268
40. Karch SB (2015) Cathinone neurotoxicity ("The "3Ms"). Curr Neuropharmacol 13:21–25
41. Gibbons S, Zloh M (2010) An analysis of the 'legal high' mephedrone. Bioorg Med Chem Lett 20:4135–4139
42. Baumann MH, Ayestas Jr MA, Partilla JS, et al. (2012) The designer methcathinone analogs, mephedrone and methylone, are substrates for monoamine transporters in brain tissue. Neuropsychopharmacology 37:1192–1203
43. Baumann MH, Partilla JS, Lehner KR (2013) Psychoactive "bath salts": not so soothing. Eur J Pharmacol 698:1–5
44. Saha K, Partilla JS, Lehner KR, et al. (2015) 'Second-generation' mephedrone analogs, 4-MEC and 4-MePPP, differentially affect monoamine transporter function. Neuropsychopharmacology 40:1321–1331
45. Simmler LD, Buser TA, Donzelli M, et al. (2013) Pharmacological characterization of designer cathinones in vitro. Br J Pharmacol 168:458–470
46. Simmler LD, Liechti ME (2016) Interactions of cathinone NPS with human transporters and receptors in transfected cells. Curr Top Behav Neurosci (Epub ahead of print)
47. Zawilska JB (2014) Mephedrone and other cathinones. Curr Opin Psychiatry 27:256–262
48. Meng H, Cao J, Kang J, et al. (2012) Mephedrone, a new designer drug of abuse, produces acute hemodynamic effects in the rat. Toxicol Lett 208:62–68
49. Shortall SE, Macerola AE, Swaby RT, et al. (2013) Behavioural and neurochemical comparison of chronic intermittent cathinone, mephedrone and MDMA administration to the rat. Eur Neuropsychopharmacol 23:1085–1095
50. Shortall SE, Green AR, Swift KM, et al. (2013) Differential effects of cathinone compounds and MDMA on body temperature in the rat, and pharmacological characterization of mephedrone-induced hypothermia. Br J Pharmacol 168:966–977
51. Varner KJ, Daigle K, Weed PF, et al. (2013) Comparison of the behavioral and cardiovascular effects of mephedrone with other drugs of abuse in rats. Psychopharmacology (Berl) 225:675–685
52. Wright Jr MJ, Angrish D, Aarde SM, et al. (2012) Effect of ambient temperature on the thermoregulatory and locomotor stimulant effects of 4-methylmethcathinone in Wistar and Sprague-Dawley rats. PLoS One 7:e44652
53. Papaseit E, Pérez-Mañá C, Mateus JA, et al. (2016) Human pharmacology of mephedrone in comparison to MDMA. Neuropsychopharmacology 41:2704–2713

54. Karila L, Billieux J, Benyamina A, et al. (2016) The effects and risks associated to mephedrone and methylone in humans: a review of the preliminary evidences. Brain Res Bull 126:61–67
55. Wood DM, Hill SL, Thomas SH, Dargan PI (2014) Using poisons information service data to assess the acute harms associated with novel psychoactive substances. Drug Test Anal 6:850–860
56. Farré M, Papaseit E, Pérez-Mañá C, Yubero-Lahoz S, et al. (2014) Pharmacokinetics and pharmacodynamics of mephedrone in humans. Res Adv Psychiatry 1(Suppl 1):19
57. de Sousa Fernandes Perna EB, Papaseit E, Pérez-Maná C, et al. (2016) Neurocognitive performance following acute mephedrone administration, with and without alcohol. J Psychopharmacol (Epub ahead of print)
58. Kapitány-Fövény M, Kertész M, Winstock A, et al. (2013) Substitutional potential of mephedrone: an analysis of the subjective effects. Hum Psychopharmacol 28:308–316
59. Wood DM, Davies S, Greene SL, et al. (2010) Case series of individuals with analytically confirmed acute mephedrone toxicity. Clin Toxicol (Phila) 48:924–927
60. Winstock A, Mitcheson L, Ramsey J, et al. (2011c) Mephedrone: use, subjective effects and health risks. Addiction 106:1991–1996
61. Deluca P, Davey Z, Corazza O, et al. (2012) Identifying emerging trends in recreational drug use; outcomes from the Psychonaut Web Mapping project. Prog Neuropsychopharmacol Biol Psychiatry 39:221–226
62. Durham M (2011) Ivory wave: the next mephedrone? Emerg Med J 28:1059–1060
63. James D, Adams RD, Spears R, National Poisons Information Service, et al. (2011) Clinical characteristics of mephedrone toxicity reported to the U.K. National Poisons Information Service. Emerg Med J 28:686–689
64. Maskell PD, De Paoli G, Seneviratne C, et al. (2011) Mephedrone (4-methylmethcathinone)-related deaths. J Anal Toxicol 35:188 Poster from The International Association of Forensic Toxicology
65. Nicholson PJ, Quinn MJ, Dodd JD (2010) Headshop heartache: acute mephedrone 'meow' myocarditis. Heart 96:2051–2052
66. Regan L, Mitchelson M, Macdonald C (2011) Mephedrone toxicity in a Scottish emergency department. Emerg Med J 28:1055–1058
67. Schifano F, Corkery J, Ghodse AH (2012) Suspected and confirmed fatalities associated with mephedrone (4-methylmethcathinone, "meow meow") in the United Kingdom. J Clin Psychopharmacol 32:710–714
68. Wood DM, Davies S, Puchnarewicz M, et al. (2010) Recreational use of mephedrone (4-methylmethcathinone, 4-MMC) with associated sympathomimetic toxicity. J Med Toxicol 6:327–330
69. Barrio P, Gaskell M, Goti J, et al. (2016) Persistent psychotic symptoms after long-term heavy use of mephedrone: a two-case series. Adicciones 28:154–157
70. Sivagnanam K, Chaudari D, Lopez P, et al. (2013) "Bath salts" induced severe reversible cardiomyopathy. Am J Case Rep 14:288–291
71. Dickson AJ, Vorce SP, Levine B, et al. (2010) Multiple-drug toxicity caused by the coadministration of 4-methylmethcathinone (mephedrone) and heroin. J Anal Toxicol 34:162–168
72. Torrance H, Cooper G (2010) The detection of mephedrone (4-methylmethcathinone) in 4 fatalities in Scotland. Forensic Sci Int 202:e62–e63
73. Wikström M, Thelander G, Nyström I, et al. (2010) Two fatal intoxications with the new designer drug methedrone (4-methoxymethcathinone). J Anal Toxicol 34:594–598
74. Martin M, Muller JF, Turner K, et al. (2012) Evidence of mephedrone chronic abuse through hair analysis using GC/MS. Forensic Sci Int 218:44–48
75. Mazoyer C, Boucher A, Gaillard Y et al (2011) Case report of driving under the influence of drugs (DUID) with mephedrone in France. Poster from Society of Forensic Toxicologists (SOFT) and The International Association of Forensic Toxicologists (TIAFT)

76. Pedersen AJ, Reitzel LA, Johansen SS, et al. (2013) In vitro metabolism studies on mephedrone and analysis of forensic cases. Drug Test Anal 5:430–438
77. Aromatario M, Bottoni E, Santoni M, et al. (2012) New "lethal highs": a case of a deadly cocktail of GHB and Mephedrone. Forensic Sci Int 223:e38–e41
78. Corkery JM, Schifano F, Ghodse H (2012) Mephedrone-related fatalities in the United Kingdom: contextual, clinical and practical issues. In: Gallelli L (ed) Pharmacology. InTech, ISBN: 978-953-51-0222-9. http://www.intechopen.com/books/pharmacology/mephedrone-related-fatalities-in-the-united-kingdom-contextual-clinical-and-practical-issues
79. Shah SA, Deshmukh NI, Barker J, et al. (2012) Quantitative analysis of mephedrone using liquid chromatography tandem mass spectroscopy: application to human hair. J Pharm Biomed Anal 61:64–69
80. Cosbey SH, Peters KL, Quinn A, et al. (2013) Mephedrone (methylmethcathinone) in toxicology casework: a Northern Ireland perspective. J Anal Toxicol 37:74–82
81. Adamowicz P, Tokarczyk B, Stanaszek R, et al. (2013) Fatal mephedrone intoxication – a case report. J Anal Toxicol 37:37–42
82. Burch HJ, Clarke EJ, Hubbard AM, et al. (2013) Concentrations of drugs determined in blood samples collected from suspected drugged drivers in England and Wales. J Forensic Leg Med 20:278–289
83. Gerace E, Petrarulo M, Bison F, et al. (2014) Toxicological findings in a fatal multidrug intoxication involving mephedrone. Forensic Sci Int 243:68–73
84. Maas A, Wippich C, Madea B, et al. (2015) Driving under the influence of synthetic phenethylamines: a case series. Int J Leg Med 129:997–1003
85. Salomone A, Gazzilli G, Di Corcia D, et al. (2016) Determination of cathinones and other stimulant, psychedelic, and dissociative designer drugs in real hair samples. Anal Bioanal Chem 408:2035–2042
86. Adamowicz P, Gieroń J, Gil D, et al. (2016) The prevalence of new psychoactive substances in biological material – a three-year review of casework in Poland. Drug Test Anal 8:64–71
87. Loi B, Corkery JM, Claridge H, Goodair C, et al. (2015) Deaths of individuals aged 16-24 years in the UK after using mephedrone. Hum Psychopharmacol 30:225–232
88. Angoa-Pérez M, Kane MJ, Francescutti DM, et al. (2012) Mephedrone, an abused psychoactive component of 'bath salts' and methamphetamine congener, does not cause neurotoxicity to dopamine nerve endings of the striatum. J Neurochem 120:1097–1107
89. Angoa-Pérez M, Kane MJ, Briggs DI, et al. (2013) Mephedrone does not damage dopamine nerve endings of the striatum, but enhances the neurotoxicity of methamphetamine, amphetamine, and MDMA. J Neurochem 125:102–110
90. den Hollander B, Rozov S, Linden AM, et al. (2013) Long-term cognitive and neurochemical effects of "bath salt" designer drugs methylone and mephedrone. Pharmacol Biochem Behav 103:501–509
91. Shortall SE, Green AR, Fone KC, et al. (2016) Caffeine alters the behavioural and body temperature responses to mephedrone without causing long-term neurotoxicity in rats. J Psychopharmacol 30:698–706
92. Ellefsen KN, Concheiro M, Huestis MA (2016) Synthetic cathinone pharmacokinetics, analytical methods, and toxicological findings from human performance and postmortem cases. Drug Metab Rev 1:1–29
93. Busardò FP, Kyriakou C, Napoletano S, et al. (2015) Mephedrone related fatalities: a review. Eur Rev Med Pharmacol Sci 19:3777–3790
94. Baciu T, Borrull F, Calull M, et al. (2016) Enantioselective determination of cathinone derivatives in human hair by capillary electrophoresis combined in-line with solid-phase extraction. Electrophoresis 37:2352–2362
95. de Castro A, Lendoiro E, Fernández-Vega H, et al. (2014) Liquid chromatography tandem mass spectrometry determination of selected synthetic cathinones and two piperazines in oral fluid. Cross reactivity study with an on-site immunoassay device. J Chromatogr A 1374:93–101

96. Mercolini L, Protti M, Catapano MC, et al. (2016) LC-MS/MS and volumetric absorptive microsampling for quantitative bioanalysis of cathinone analogues in dried urine, plasma and oral fluid samples. J Pharm Biomed Anal 123:186–194
97. Strano-Rossi S, Anzillotti L, Castrignanò E, et al. (2012) Ultra high performance liquid chromatography-electrospray ionization-tandem mass spectrometry screening method for direct analysis of designer drugs, "spice" and stimulants in oral fluid. J Chromatogr A 1258:37–42
98. Pozo ÓJ, Ibáñez M, Sancho JV, et al. (2015) Mass spectrometric evaluation of mephedrone in vivo human metabolism: identification of phase I and phase II metabolites, including a novel succinyl conjugate. Drug Metab Dispos 43:248–257
99. Khreit OI, Grant MH, Zhang T, et al. (2013) Elucidation of the Phase I and Phase II metabolic pathways of (±)-4′-methylmethcathinone (4-MMC) and (±)-4′-(trifluoromethyl) methcathinone (4-TFMMC) in rat liver hepatocytes using LC-MS and LC-MS^2. J Pharm Biomed Anal 72:177–185
100. Mayer FP, Wimmer L, Dillon-Carter O, et al. (2016) Phase I metabolites of mephedrone display biological activity as substrates at monoamine transporters. Br J Pharmacol 173:2657–2668
101. Martínez-Clemente J, López-Arnau R, Carbó M, et al. (2013) Mephedrone pharmacokinetics after intravenous and oral administration in rats: relation to pharmacodynamics. Psychopharmacology (Berl) 229:295–306
102. Meyer MR, Wilhelm J, Peters FT, et al. (2010) Beta-keto amphetamines: studies on the metabolism of the designer drug mephedrone and toxicological detection of mephedrone, butylone, and methylone in urine using gas chromatography-mass spectrometry. Anal Bioanal Chem 397:1225–1233
103. Aarde SM, Angrish D, Barlow DJ, et al. (2013) Mephedrone (4-methylmethcathinone) supports intravenous self-administration in Sprague-Dawley and Wistar rats. Addict Biol 18:786–799
104. Creehan KM, Vandewater SA, Taffe MA (2015) Intravenous self-administration of mephedrone, methylone and MDMA in female rats. Neuropharmacology 92:90–97
105. Gatch MB, Taylor CM, Forster MJ (2013) Locomotor stimulant and discriminative stimulus effects of 'bath salt' cathinones. Behav Pharmacol 24:437–447
106. Hadlock GC, Webb KM, McFadden LM, et al. (2011) 4-Methylmethcathinone (mephedrone): neuropharmacological effects of a designer stimulant of abuse. J Pharmacol Exp Ther 339:530–536
107. Karlsson L, Andersson M, Kronstrand R, et al. (2014) Mephedrone, methylone and 3,4-methylenedioxypyrovalerone (MDPV) induce conditioned place preference in mice. Basic Clin Pharmacol Toxicol 115:411–416
108. Lisek R, Xu W, Yuvasheva E, Chiu YT, et al. (2012) Mephedrone ('bath salt') elicits conditioned place preference and dopamine-sensitive motor activation. Drug Alcohol Depend 126:257–262
109. Vandewater SA, Creehan KM, Taffe MA (2015) Intravenous self-administration of entactogen-class stimulants in male rats. Neuropharmacology 99:538–545
110. Motbey CP, Clemens KJ, Apetz N, et al. (2013) High levels of intravenous mephedrone (4-methylmethcathinone) self-administration in rats: neural consequences and comparison with methamphetamine. J Psychopharmacol 27:823–836
111. Shortall SE, Spicer CH, Ebling FJ, et al. (2015) Contribution of serotonin and dopamine to changes in core body temperature and locomotor activity in rats following repeated administration of mephedrone. Addict Biol 21:1127–1139
112. Dargan PI, Albert S, Wood DM (2010) Mephedrone use and associated adverse effects in school and college/university students before the UK legislation change. QJM 103:875–879
113. Bajaj N, Mullen D, Wylie S (2010) Dependence and psychosis with 4-methylmethcathinone (mephedrone) use. BMJ Case Rep 2010. doi:10.1136/bcr.02.2010.2780

114. Dolengevich-Segal H, Rodríguez-Salgado B, Gómez-Arnau J, et al. (2016) Severe psychosis, drug dependence, and Hepatitis C related to slamming mephedrone. Case Rep Psychiatry 2016:8379562
115. Uosukainen H, Tacke U, Winstock AR (2015) Self-reported prevalence of dependence of MDMA compared to cocaine, mephedrone and ketamine among a sample of recreational poly-drug users. Int J Drug Policy 26:78–83
116. Kapitány-Fövény M, Mervó B, et al. (2015) Is there any difference in patterns of use and psychiatric symptom status between injectors and non-injectors of mephedrone? Hum Psychopharmacol 30:233–243
117. European Monitoring Centre for Drugs and Drug Addiction (EMCDDA) (2010) 2010 Annual report on the state of the drugs problem in Europe. http://www.emcdda.europa.eu/system/files/publications/926/EMCDDA_AR2010_EN_242720.pdf
118. Food and Drug Administration (FDA) (2010) Draft guidance "assessment of abuse potential of drugs". http://www.fda.gov/downloads/Drugs/GuidanceComplianceRegulatoryInformation/Guidances/UCM198650.pdf

Application of a Combined Approach to Identify New Psychoactive Street Drugs and Decipher Their Mechanisms at Monoamine Transporters

Felix P. Mayer, Anton Luf, Constanze Nagy, Marion Holy, Rainer Schmid, Michael Freissmuth, and Harald H. Sitte

Abstract Psychoactive compounds can cause acute and long-term health problems and lead to addiction. In addition to well-studied and legally controlled compounds like cocaine, new psychoactive substances (NPS) are appearing in street drug markets as replacement strategies and legal alternatives. NPS are effectively marketed as "designer drugs" or "research chemicals" without any knowledge of their underlying pharmacological mode of action and their potential toxicological effects and obviously devoid of any registration process. As of 2016, the knowledge of structure–activity relationships for most NPS is scarce, and predicting detailed pharmacological activity of newly emerging drugs is a challenging task. Therefore, it is important to combine different approaches and employ biological test systems that are superior to mere chemical analysis in recognizing novel and potentially harmful street drugs. In this chapter, we provide a detailed description of techniques to decipher the molecular mechanism of action of NPS that target the high-affinity transporters for dopamine, norepinephrine, and serotonin. In addition, this chapter provides insights into a

F.P. Mayer, M. Holy, and M. Freissmuth
Center of Physiology and Pharmacology, Institute of Pharmacology, Medical University Vienna, Waehringerstrasse 13a, A-1090 Vienna, Austria

A. Luf and R. Schmid
Clinical Department of Laboratory Medicine, Medical University of Vienna, Waehringer Guertel 10-20, 1090 Vienna, Austria

C. Nagy
checkit! – Suchthilfe Wien GmbH, Gumpendorfer Gürtel 8, 1060 Vienna, Austria

H.H. Sitte (✉)
Center of Physiology and Pharmacology, Institute of Pharmacology, Medical University Vienna, Waehringerstrasse 13a, A-1090 Vienna, Austria
Center for Addiction Research and Science – Medical University Vienna, Waehringerstrasse 13A, 1090 Vienna, Austria
e-mail: harald.sitte@meduniwien.ac.at

Curr Topics Behav Neurosci (2017) 32: 333–350
DOI 10.1007/7854_2016_63
Published Online: 26 December 2016

combined approach to identify and characterize new psychoactive street drugs of unknown content in a collaboration with the Austrian prevention project "checkit!."

Keywords Amphetamine • Analytical identification • Bath salts • Cocaine • Dopamine • Monoamine transporters • New psychoactive substances • Norepinephrine • Psychostimulants • Research chemicals • Serotonin

Contents

1 Introduction

An emerging problem in recent years is the increasing abuse of new psychoactive substances (NPS) – the so-called legal highs, bath salts, or research chemicals available on street drug markets. NPS are mostly distributed over the Internet and comprise failed pharmaceuticals, like the bath salt 3,4-methylenedioxypyrovalerone (MDPV, "cloud nine") [1], or synthetic substances of novel structure. The latter are based on known chemical structures that target receptors or transporters for neurotransmitters in the nervous system [2]. However, due to chemical modification of the parent drugs, NPS evade current drug control legislation [3]. A major challenge associated with NPS is the flexibility of vendors to adapt to changes within legislative boundaries. For instance, JWH-018, a former unregulated cannabinoid receptor agonist, was found to be an active ingredient of the "legal high" product "Spice." Only 4 weeks after the legislative ban on JWH-018, the unregulated analogue JWH-073 began appearing in "Spice" preparations [1, 4]. The replacement of JWH-018 with JWH-073 represents a perfect example of the often-cited "*cat-and-mouse game*" [1, 3] whereby the ban of a given substance inevitably results in the appearance of novel and uncontrolled substances as a replacement strategy. Strikingly, the number of NPS (in total: 251 in 2012) has already overtaken the number of controlled substances (in total: 234) (World drug report 2013; https://www.unodc.org/unodc/secured/wdr/wdr2013/World_Drug_Report_2013.pdf). The alarming increase in NPS available on recreational drug markets prompted the United Nations Office on Drug and Crime (UNODC) to launch an early warning advisory system on NPS in 2013 (UNODC 2013; https://www.unodc.org/unodc/en/press/releases/2013/September/unodc-early-warning-system-records-rapid-increase-in-legal-highs-in-2013.html) to provide a variety of up-to-date information on an international scale.

Unfortunately, knowledge about the pharmacology of most NPS is limited, if not missing altogether [5]. Moreover, high-dose or chronic exposure to NPS may result in severe medical conditions, including psychosis, tachycardia, and even death [5–8]. NPS satisfy a broad spectrum of individual demands for recreational drugs [9], a spectrum which ranges from legal alternatives for cannabinoids and hallucinogens to stimulants.

Stimulant-like NPS exert amphetamine- or cocaine-like pharmacological effects by interfering with monoaminergic signaling pathways [5, 6]. Simplified, psychostimulants come in two flavors as (1) cocaine-like uptake inhibitors or (2) amphetamine-like releasers. The first mentioned class, including cocaine and methylphenidate, exerts its effects by inhibiting the high-affinity monoamine transporters (MATs) for dopamine (dopamine transporter, DAT), norepinephrine (norepinephrine transporter, NET), and serotonin (serotonin transporter, SERT). As a consequence, the monoamine concentration increases in the synaptic cleft and activates pre- and postsynaptic receptors [10]. The second mentioned class, including amphetamine and its analogues like 3,4-methylenedioxymethamphetamine (MDMA, "ecstasy"), triggers a transporter-mediated reverse transport of monoamines from the cytoplasm into the extracellular space [11]. It is currently believed that releasers are transported as substrates by DAT, NET, and SERT and reverse the normal direction of transport flux [12]. Consequently, releasers elevate the extracellular concentration of monoamines independently from vesicular release events [13].

Nevertheless, MATs at the plasma membrane operate in concert with vesicular monoamine transporters (VMATs) to refill the vesicular storage pools [14]. In addition to their actions at MATs, amphetamine-like drugs (releasers) have been shown to interact with VMATs [15] and to release monoamines from synaptic vesicles into the cytosol [13]. The concomitant availability of monoamines for reverse transport, and the inverted direction of flux for plasmalemmal MATs, has been hypothesized to be crucial for the actions of amphetamine-like drugs [12, 16]. Most importantly, a link between amphetamine-like drugs and neurotoxicity has been established [5, 17, 18]. Hence, to assess the potential risk of psychostimulant NPS on neuronal function and overall health status, it is imperative to decipher their molecular mode of drug action at MATs.

Structure–activity relationship (SAR) studies have evolved from providing useful representations of docked molecules to – in many cases – promising tools to shape and form our understanding of the molecular mechanism of action of drugs, as these studies shed light onto the molecular determinants of these actions. Certainly, this evolution has been triggered by the development of more reliable (and biochemically verified) homology models [19] and the availability of more and more structures of bacterial transporter homologs [20, 21] and even structures derived from drosophila [22] and human species [23]. However, as of 2016, only a few SAR studies on psychostimulant NPS exist [24–27]. Future studies should be based and designed on the current state-of-the-art techniques and approaches to enhance our understanding of drug actions at MATs. This could certainly also help to bypass the tedious and time-consuming intermediate step of biological evaluation of each of the compounds found on the illicit drug market.

As a result of the current dynamics of the drug markets, society is flooded with a variety of modified substances. Previous studies reveal that chemical modifications of psychostimulants might switch their activity from amphetamine-like to cocaine-like or MDMA-like drugs [26]. Consequently, the structural modifications of known substances may result in ineffective or even toxic substances. A major threat for the physical and mental health of drug users is that NPS are rarely sold under their real name or in their pure form on the street.

To identify potential harmful substances, i.e., adulterants or drug combinations, the City of Vienna in Austria launched the prevention project known as "checkit!." Without the risk of criminalization, drug users can have their drugs anonymously tested for active ingredients and adulterants. In addition to a permanent location based in Vienna, "checkit!" offers "on-site" analysis [28]. To reach out to people at rave parties and music festivals, a mobile laboratory has been installed in a bus. This unique project not only reduces the occurrence of severe intoxications and offers harm reduction information, but also provides valuable insights into the current street drug market situation and allows for documenting market entries of NPS. The drug user provides a few milligrams of his/her purchased product, which is then analyzed by high-performance liquid chromatography coupled with mass spectrometry (HPLC-MS). Immediately after analysis by "checkit!," the user receives information on the content of the drug sample. If the analysis yields inconclusive results, the content is classified as "unknown" [29]. A combined approach of high-resolution mass spectrometry (HRMS) and biological activity assays performed in heterologous expression systems described herein can be used to identify the content of "unknown" test drugs and the underlying pharmacological activity profile [30]. On a monthly basis, "checkit!" publishes detailed warning lists, which are also forwarded to the European Drug Monitoring Centre for Drugs and Drug Addiction (EMCDDA; http://www.emcdda.europa.eu/edr2015) in Lisbon, Portugal. These lists contain information on high-dose drug preparations or adulterated and combined drug mixtures that should be handled with special care.

Recently, collaboration between the Sitte research group and the "checkit!" program has shed light on the reason for the ubiquitous use of levamisole as adulterant in cocaine. In addition to its bitter taste, a metabolite of levamisole, aminorex, exerts amphetamine-like effects at MATs [31]. In this chapter, we discuss techniques that have been established and successfully applied to identify substances that target MATs and experimental approaches to discriminate amphetamine-like drugs from non-transported inhibitors. Furthermore, we discuss the application of the described techniques to identify street drugs of unknown content.

2 Methods for Assessing Drug Actions at Transporters

2.1 HEK293 Cell Culture

Human embryonic kidney cells (HEK293) are cultured in 10-cm cell culture dishes (Sarstedt, Germany) in Dulbecco's modified Eagle's medium (DMEM; high glucose 4,500 mg/L), sodium bicarbonate, and L-glutamine (Sigma Aldrich, St. Louis, MO, USA), supplemented with 10% fetal bovine serum (FBS). To maintain the cells in a sub-confluent state, the cells are washed with phosphate-buffered saline (PBS) every 4 days and exposed to 1 mL of trypsin for 2–3 min at room temperature. After establishing a monocellular suspension by trituration, 0.7–1.0 $\times$ 10^6 cells are transferred into a new 10-cm dish and DMEM supplemented with FBS is added to a final volume of 10 mL. In the case of HEK293 cells stably expressing the desired MATs, the selection process is maintained by adding the appropriate antibiotics according to the protocol of the vector-supplying company. If needed, the cell culture medium may be further supplemented with penicillin (100 IU/100 mL) and streptomycin (100 μg/100 mL).

Transfection of HEK293 cells with the MAT of interest can be achieved by application of the $CaPO_4$ method [32], which is reliable and inexpensive. Prior to transfection, the cells should reach a confluency of 35–45%. This can be achieved by seeding 1.8 $\times$ 10^6 cells into a 10-cm dish 24 h prior to transfection. For each 10-cm dish, mix 20 μL of DNA (1 μg/μL) with 480 μL of 0.26 M $CaCl_2$ in H_2O. Subsequently, transfer the DNA-containing mixture into 500 μL of Hepes-buffered saline (HEBS) and let the mixture sit for 6 min at room temperature. After 6 min, a fine-grained DNA-Ca^{2+} precipitate should be visible. Add the solution with the DNA-precipitate dropwise onto the cells. Let cells sit for 3.5–6 h at 37°C, with exposure to 5% CO_2. Afterwards, aspirate the transfection medium and add 1 mL of glycerol (10 vol%) to the cells and remove it immediately. Wash the cells with 7.5 mL of PBS and add 10 mL of DMEM, supplemented with FBS. If MAT-cDNAs are used that carry a fluorescent protein tag such as green fluorescent protein (GFP) or any other fluorescent protein, expression can be monitored the next day by use of a fluorescence microscope. If the transfected MATs do not carry a tag allowing visual assessment of expression, perform a single-point uptake of tritiated substrate in absence and presence of specific inhibitors to verify correct MAT expression at the plasma membrane.

To generate cell lines stably expressing the MAT of choice, add the antibiotic listed in the datasheet of the applied expression vector 48 h after transfection. Maintain a high-selection pressure for up to 5 days until only viable single cells are present. To establish monoclonal cell lines, pick one to ten individual cells with a sterile 200 μL pipette tip. Expand the clones in individual cell culture dishes and repeat this step 2–3 times. Finally, test each individual clone for its transport characteristics, i.e., K_M and

V_{max}, for reference substrates (endogenous MAT substrates or MPP^+) [33]. For the uptake inhibition experiments, monocellular suspensions of HEK293 cells expressing the desired MAT are seeded at 40,000 cells per well onto poly-D-lysine-coated 96-well plates (Sarstedt) in a final volume of 200 μL/well 24 h prior to the experiment. For release experiments, poly-D-lysine-coated glass coverslips (5 mm in diameter) are placed into 96-well plates. Subsequently, 40,000 cells per well are seeded onto the glass coverslips in a final volume of 200 μL/well 24 h prior to the experiment.

2.2 Uptake Inhibition Assays

Sodium bicarbonate buffer containing DMEM is removed from the cells and replaced with Krebs-HEPES buffer (KHB, 25 mM HEPES, 120 mM NaCl, 5 mM KCl, 1.2 mM $CaCl_2$, and 1.2 mM $MgSO_4$, 5 mM D-glucose, pH adjusted to 7.3 with NaOH) at a final volume of 200 μL/well. Subsequently, cells are exposed to KHB containing various concentrations of test drug for 5 min to achieve equilibration. Afterwards, tritiated substrate ($[^3H]$-MPP^+ at a final concentration of 20 nM for DAT and NET or $[^3H]$-5-HT at a final concentration of 100 nM for SERT) is added. The uptake incubation is terminated after 180 (DAT and NET) or 60 s (SERT) by removing the tritiated substrate and washing the cells with 200 μL of ice-cold KHB. Finally, the cells are lysed with 1% SDS (100 μL/well) and transferred into counting vials filled with 2 mL of scintillation cocktail. Uptake of tritiated substrate is determined with a beta-scintillation counter. Nonspecific uptake via DAT, NET, or SERT is assessed in the presence of 10 μM mazindol, desipramine, or paroxetine, respectively, and subtracted from all data to yield specific uptake. Uptake in the absence of test drugs is defined as 100% and uptake in the presence of drugs is expressed as a percentage of control uptake. The half-maximal inhibitory concentration is determined by nonlinear regression fits according to the equation [Y = Bottom + (Top − Bottom)/(1 + 10^((LogIC50-X)*HillSlope))].

2.3 Release Assays

2.3.1 Dynamic Superfusion Assay

MAT-mediated reverse transport is assessed by use of a dynamic superfusion apparatus as described in detail elsewhere [34]. In brief, MAT-expressing cells grown on 5-mm glass coverslips are exposed to 0.1 μM $[^3H]$-MPP^+ (DAT and NET) or 0.4 μM $[^3H]$-5-HT (SERT) at 37°C for 20 min. Subsequently, the cells are transferred into 12 individual small cylindrical chambers (8 mm in diameter; volume 200 μL) and superfused with KHB at a flow rate of 0.7 mL/min for 40 min to establish a stable basal release of tritiated substrate. To ensure that the experiment is conducted at a constant temperature, the tubes delivering KHB to the individual chambers are placed into a water bath set to 25°C. After a 40-min wash-out phase, three 2-min fractions are collected which represent the untreated (i.e., basal) release. Afterwards, the cells are exposed to monensin (10 μM) or vehicle for four fractions, prior to the addition of test drugs in the presence or absence of monensin for five fractions (the reason for the addition of monensin will be discussed in detail below). Finally, the cells are superfused with 1% SDS for another three fractions and total radioactivity present in each fraction is determined by a beta-scintillation counter (Perkin Elmer, Waltham, MA, USA). For analysis, the release of preloaded substrate is expressed as fractional rate, i.e., the amount of released radioactivity per fraction is expressed as percentage of total radioactivity present in the cells at the beginning of that fraction [35].

2.3.2 Static Batch Release Assay

Originally described by Rudnick and coworkers for MAT-expressing cells in 1995, the static batch release assay serves as a technique to identify substrate-induced efflux [36]. HEK293 cells expressing the MAT of interest are preloaded with tritiated substrate (0.05 μM in KHB, 100 μL/well) for 20 min (37°C). Subsequently, the cells are washed three times with KHB (200 μL/well) at room temperature to rinse away any tritiated substrate free in solution, which improves the signal-to-noise ratio. Finally, the test drugs (in KHB, 100 μL/well) are added at a concentration that inhibits uptake via the respective MAT by 50%, and test drugs are always compared to an established MAT substrate, e.g., (+)amphetamine. To determine the specificity of drug-induced reverse transport, controls are performed in the presence of selective MAT inhibitors, e.g., 10 μM of mazindol for DAT and NET and 10 μM of paroxetine for SERT. After 10 min, the supernatant is transferred into liquid scintillation counting vials. The cells are lysed in 100 μL of 1% SDS and transferred into independent counting vials. Total radioactivity present in the supernatant and the cell lysate is set as 100%, and the amount of $[^3H]$-substrate present in the supernatant is expressed as percentage of the total.

3 Interpreting Data from Transporter Assays

3.1 Uptake Inhibition Assays

Drugs that target MATs inhibit the uptake of their cognate neurotransmitter substrates in a dose-dependent manner (Fig. 1). The apparent IC_{50} values may vary with expression levels and the cell system used (see below, Sect. 5). Hence, cocaine may be applied as an internal reference drug for comparison to the potencies of the test compounds under scrutiny. Figure 1 depicts the effects of cocaine and MDPV on DAT- and SERT-mediated uptake in HEK293 cells. As described previously [37, 38], MDPV inhibits

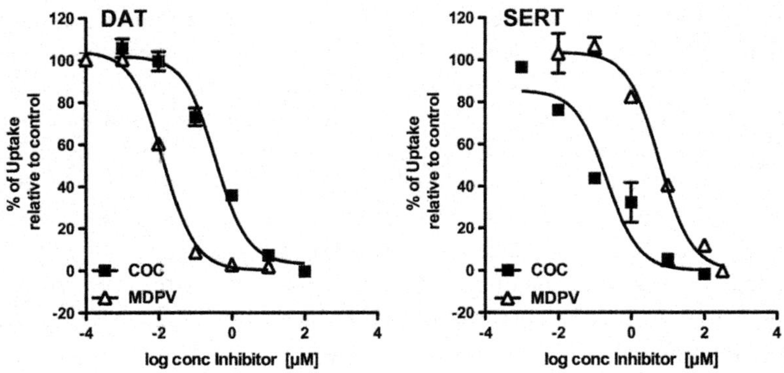

Fig. 1 Effects of MDPV and cocaine on DAT- and SERT-mediated uptake in HEK293 cells. The human isoforms of DAT (*left*) and SERT (*right*) were stably expressed in HEK293 cells, and incubated with increasing concentrations of cocaine or MDPV. Uptake of [^{3}H]-MPP$^+$ by DAT or [^{3}H]-5-HT by SERT is expressed as a percentage of uptake in the absence of inhibitors. Nonspecific uptake is determined in the presence of 10 μM mazindol or paroxetine for DAT or SERT, respectively

DAT-mediated uptake with much higher potency than cocaine (IC_{50} values of 0.015 vs. 0.36 μM for MDPV and cocaine, respectively). On the contrary, MDPV is strikingly less potent than cocaine as an inhibitor at SERT (IC_{50} ~6 μM for MDPV as compared to 0.22 μM for cocaine). As a negative control, one might examine the effects of test drugs at the GABA transporter [30]; this transporter should be essentially unaffected by any given MAT substrate or inhibitor.

Uptake inhibition assays allow for the identification of drugs that counteract MAT-mediated uptake. However, this assay lacks the ability to differentiate amphetamine-like substrates from non-transported inhibitors [37, 39]. Aside from triggering reverse transport, amphetamine-like drugs, i.e., drugs that act as "releasers" similar to (+)amphetamine, bind to the orthosteric site on transporters and are subsequently transported as substrates. However, it is worth mentioning that some drugs that share structural features with (+)amphetamine do not trigger transporter-mediated efflux and act as non-transported inhibitors, with methylphenidate being the most prominent example [12]. Hence, drugs acting as transporter substrates compete for the orthosteric binding sites with the natural neurotransmitter substrates, and engender dose-dependent reductions in uptake of tritiated substrate. As shown in Fig. 2, the former legal high and MAT substrate mephedrone [37, 39] inhibits DAT-mediated uptake with increasing concentrations.

The non-transported inhibitor MDPV and the MAT substrate mephedrone share a similar profile of activity in uptake inhibition assays. This illustrates that assays dedicated to reveal drug-induced reverse transport are needed to identify the precise mode of action for drugs which target transporters. Nevertheless, uptake inhibition assays provide a fast and reliable tool to identify the compounds that interact with MATs.

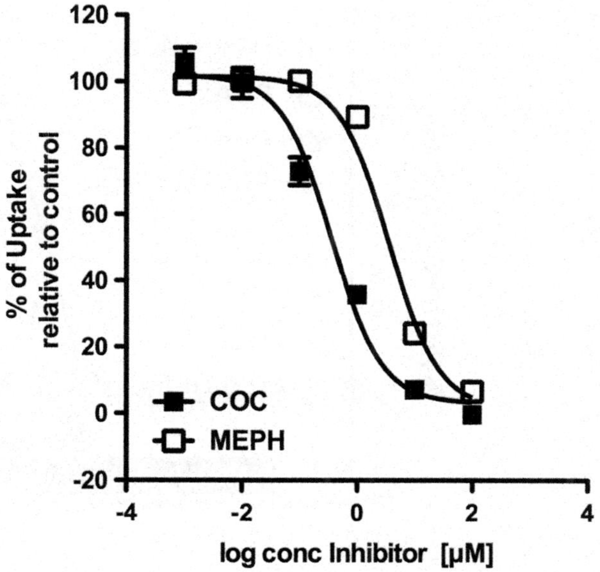

Fig. 2 Effects of mephedrone and cocaine on DAT-mediated uptake in HEK293 cells. DAT-expressing HEK293 cells were incubated with the indicated concentrations of cocaine or mephedrone (MEPH). Uptake of [^{3}H]-MPP^+ is expressed as percent of uptake without inhibitors present, and nonspecific uptake is defined in the presence of mazindol (10 μM)

3.2 Release Assays

The major advantage of the dynamic superfusion experiments is the elimination of back-and-forth movements of substrates by diffusion [40]. The constant flow rate ensures that released substrates are cleared from the cellular vicinity. This prevents transporter-mediated reuptake or diffusion events that might counteract reverse transport induced by test drugs. Additionally, and contrary to static release assays, the temporal resolution of superfusion enables deciphering the time course of effects for each drug. Monoamine transporters of the SLC6 family utilize the preexisting sodium gradient across cell membranes as a driving force [41]. Application of the selective Na^+/H^+ ionophore monensin [42] dissipates the sodium gradient by allowing sodium entry into the cytosol. As a result, the increase in intracellular sodium fosters MAT-mediated reverse transport and selectively augments substrate-induced reverse transport. As a consequence, only efflux triggered by "true" substrates is sensitive to enhancement by monensin, while the effects of non-transported inhibitors remain unaffected [27]. A representative experiment showing the effect of monensin on transporter-mediated efflux is given in Fig. 3. The addition of the MAT substrate *para*-chloroamphetamine (PCA, 3 μM) robustly elevates the basal release of tritiated substrate via SERT. The presence of monensin (10 μM) further augments PCA-triggered reverse transport. In striking contrast, application of the SERT inhibitor paroxetine (3 μM) does not result in tritium outflow, and no difference between paroxetine plus monensin or vehicle can be observed.

As exemplified in Fig. 3, only the effects of transporter substrates are sensitive to the presence of monensin. This property precludes the misinterpretation of "pseudo"-efflux events. Previous studies have shown that MAT inhibitors can unmask a basal

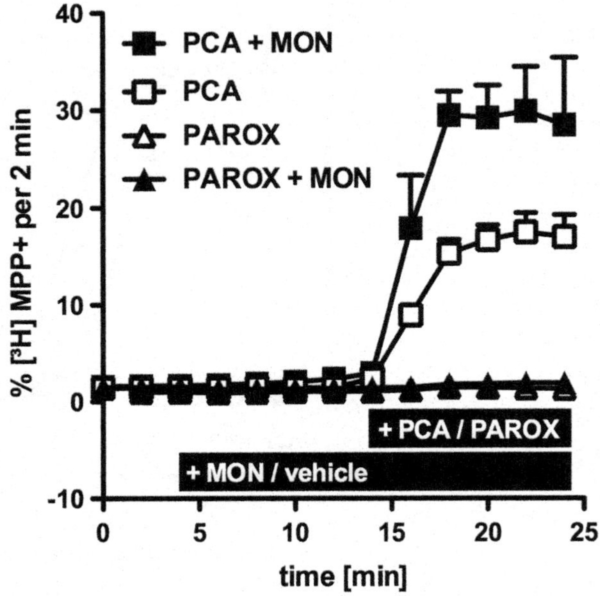

Fig. 3 The impact of monensin on SERT-mediated reverse transport in HEK293 cells. HEK293 cells stably expressing human SERT were preloaded with [^{3}H]-MPP$^+$ and superfused. Pretreatment with monensin (MON, 10 μM) augments the reverse transport triggered by para-chloroamphetamine (PCA, 3 μM) vs. the vehicle-pretreated condition. Application of the SERT inhibitor paroxetine (PAROX, 3 μM) does not result in elevated [^{3}H]-MPP$^+$ outflow and is insensitive to pretreatment with monensin. The presence of monensin, vehicle, PCA, and PAROX is indicated by *black boxes*

loss of tritiated substrate from the preloaded cells [35]. Despite the advantages of superfusion methods, limited yet measurable amounts of tritiated substrate can leak from cells via simple diffusion. This leak process is normally counteracted by the activity of MATs in transfected cells. However, in the presence of MAT inhibitors, the reuptake of extracellular substrates is precluded. Thus, the presence of MAT inhibitors can elevate the amount of tritiated substrate in the superfusate and might be interpreted as drug-induced efflux. As shown in Fig. 4a, by magnification of the *Y*-axis, the addition of MDPV elevates the basal release of [^{3}H]-MPP$^+$ from HEK293 cells stably expressing human DAT. Considered in isolation, it might be interpreted that MDPV acts as a weak amphetamine-like substrate, although the effect is rather modest and does not reach statistical significance. By contrast, when compared to the effects of PCA in Fig. 3, the presence of monensin reveals that MDPV does not induce reverse transport via the sodium-dependent DAT (Fig. 4b), which is in line with previous studies that conclusively substantiated that MDPV acts as non-transported inhibitor at DAT [37, 38].

This example highlights that observing the effects of test drugs in the absence *and* the presence of monensin is a direct and effective strategy to distinguish between drugs which act as MAT substrates and those which act as non-transported inhibitors.

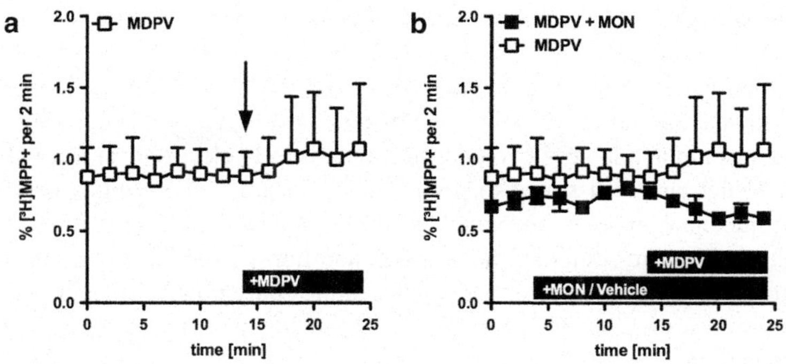

Fig. 4 Effects of MDPV in the absence and presence of monensin on DAT-mediated reverse transport in HEK293 cells. HEK293 cells stably expressing human DAT were preloaded with [^{3}H]-MPP$^+$ and superfused. (**a**) Addition of MDPV slightly elevates the basal release of tritiated substrate. (**b**) The presence of monensin (MON) does not augment the modest effect of MDPV. The addition of substances is indicated by *black bars and arrows*

However, it is important to stress potential pitfalls associated with the use of monensin. As a Na$^+$/H$^+$ ionophore, monensin dissipates sodium gradients but also alters the intracellular pH. Monoamine neurotransmitters (i.e., dopamine, norepinephrine, and 5-HT) are weak bases with pKa values greater than 8. Consequently, alterations in the intracellular concentration of H$^+$ ions can affect the ratio between protonated and unprotonated forms of these amines. In 2000, Scholze and coworkers demonstrated that the addition of monensin alone elevated the amount of tritium in superfusates when [^{3}H]-5-HT was used to preload HEK293 cells in the absence of any plasmalemmal transporter expression [40]. The increase in preloading time up to 1 h was enough to efficiently load the cells with tritiated 5-HT by a non-transporter-related mechanism. A likely explanation for this observation is that elevation of intracellular pH by monensin increases the unprotonated and more lipophilic form of 5-HT, and facilitates diffusion of 5-HT across cellular membranes [43]. Regarding the "pseudo-efflux" described by Scholze et al. [40] and Sitte et al. [35], the elevation of tritiated substrates in the superfusate by MAT blockers could be misinterpreted as the effect of an amphetamine-type releasing agent, as outlined above. If the results obtained with [^{3}H]-5-HT are unclear, the use of [^{3}H]-MPP$^+$ is an effective countermeasure. The permanent charge of [^{3}H]-MPP+ prevents passive diffusion across cellular membranes to a large extent. Therefore, MPP$^+$ reveals only transporter-mediated translocation events.

It is also noteworthy that transporter-mediated release induced by amphetamine-like substrates results in a bell-shaped dose–response curve. As extensively studied by Seidel and colleagues [44], high concentrations of test drugs may counteract transporter-mediated reverse transport, if the applied concentration of the drug of interest exceeds the determined IC_{50} by severalfold. Therefore, uptake inhibition assays not only serve as a tool to identify potential candidate drugs for further analysis but also provide the basis to identify the correct concentration range for superfusion studies, i.e., the half-maximal inhibitory concentration for MAT-mediated uptake.

A major disadvantage of dynamic superfusion experiments is that large amounts of buffer and substance are required. To bypass this obstacle – if only small amounts of test drugs are available – two possibilities exist: (1) electrophysiological investigations, which are well suited to discriminate substrates from inhibitors [27]; the electrogenic uptake process mediated by MATs clearly identifies substrates based on drug-induced inward sodium currents while non-transported inhibitors fail to induce such currents; (2) an alternative assay that requires only minute amounts of the drug under scrutiny – the static batch release assay. Analogous to the dynamic superfusion experiments, the experimental strategy in the static batch assay is based on the fact that MAT-mediated transport can run in reverse. Hence, MAT-expressing cells are preloaded with tritiated substrate and incubated with buffer containing the drug of interest. Subsequently, the amount of radioactivity present in the incubation buffer is determined and expressed as percentage of total radioactivity present (i.e., in cells and in buffer), as determined by disintegrating the cells with 1% SDS. The effect of test drugs needs to be determined in the presence and absence of specific MAT inhibitors. Specific MAT inhibitors markedly reduce the amount of tritium in the supernatant if the cells are incubated with substrate-type drugs (Fig. 5). On the contrary, the effects of drugs that act as non-transported inhibitors are insensitive to the presence of other MAT inhibitors. Even though the dynamic superfusion assay yields results of higher quality, the static batch assay provides a reliable readout to identify substrate-type drugs targeting MATs.

Another limitation of dynamic superfusion systems is that determining the half-maximal stimulatory concentration for release is cumbersome and time consuming. Initial characterization of the underlying pharmacology, i.e., inhibitor versus substrate, followed by dose-dependent release assays performed in rat brain synaptosomes as described in various publications by Baumann and coworkers [37–39], has proven to be a reliable and time-efficient strategy. Moreover, assays conducted in native tissues serve as an important physiologically relevant comparator system, since the preparations closely reflect the natural MAT environment; apart from synaptosomes, this also includes slice preparations from brain tissue [45, 46].

4 Biological Assays to Identify Street Drugs of Unknown Content

To identify the mechanism of action for purchased stimulants that were classified as "unknown" by "checkit!," pharmacological "fingerprints" for well-established reference compounds were first generated. These reference fingerprints include concentration-response curves for AMPH, MDMA, D-fenfluramine, methamphetamine, mephedrone, MDPV, and cocaine, obtained from uptake inhibition experiments performed with HEK293 cells expressing the human isoforms of MATs.

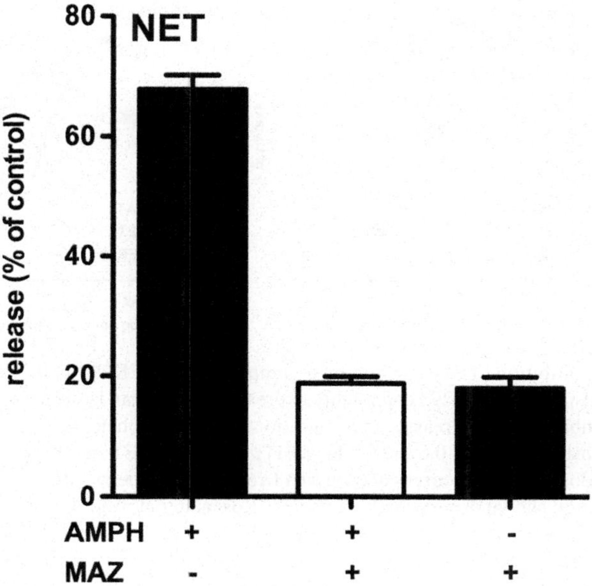

Fig. 5 Effect of AMPH in the absence and presence of mazindol on NET-mediated reverse transport in HEK293 cells in the static batch release assay. HEK293 cells stably expressing human NET were exposed to 50 nM of [^{3}H]-MPP$^+$ for 20 min. After washing the cells with buffer, the cells were exposed to buffer containing AMPH (2 μM) in the presence or absence of mazindol (MAZ, 10 μM) for 10 min. Subsequently, the cells were lysed in 1% SDS. For analysis, radioactivity present in the supernatant and the cell lysate was determined by liquid scintillation counting. "Release" is expressed as percentage of total radioactivity present within one well, i.e., the sum of radioactivity present in the supernatant and the cell lysate. Data are shown as modified version of Rosenauer et al. [30]

Each substance reveals a unique profile of selectivity for uptake inhibition at DAT, NET, and SERT (Fig. 6). For example, the data in Fig. 6 demonstrate that methamphetamine displays potent inhibition of uptake at DAT and NET but not SERT, whereas methylone is much less selective in this regard. For the initial evaluation of an unknown compound, the test drug is examined for its ability to inhibit uptake in a dilution series covering six orders of magnitude. In addition to human DAT, NET, and SERT, rat GABA transporter 1 (rGAT1) is included in the experimental series. rGAT1 serves as negative control since amphetamines do not function as substrates at this member of the neurotransmitter: sodium symporter family [44]. Data on uptake inhibition reveal the selectivity of test drugs for the individual MATs. Comparing the profile of activity for an unknown drug with the various reference "fingerprint" drugs allows for identifying the drug under investigation, or at least narrowing down the choice of potential candidate drugs. If amphetamine-like transporter substrates are suspected, static batch release assays are performed to verify substrate-like activity at MATs. Similar to the uptake inhibition experiments, each drug is characterized for its ability to serve as a substrate at DAT, NET, and SERT. The combination

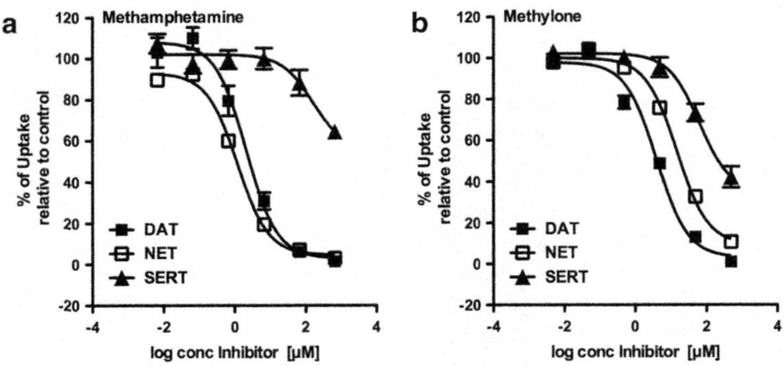

Fig. 6 Uptake-inhibition profile of reference compounds in HEK293 cells expressing human monoamine transporters. HEK293 cells stably expressing the human isoforms of DAT, NET, and SERT were incubated with increasing concentrations of methamphetamine (**a**) or methylone (**b**) and tritiated substrate (SERT: 0.03 μM [^{3}H]-5-HT; DAT and NET: 0.05 μM [^{3}H]-MPP$^+$). The synopsis of the dose–response curves at each MAT reveals a unique profile of selectivity for each substance. Data are shown as modified versions of Rosenauer et al. [30]

of uptake inhibition and release experiments provides the basis to identify drugs based on their pharmacological fingerprints.

5 Choosing the Appropriate Expression System

Heterologous expression systems provide a powerful strategy to investigate MATs in vitro. These systems bypass the possible impact of vesicular storage mechanisms and presynaptic autoreceptors on the effects of test drugs. Furthermore, native tissue preparations normally contain more than one MAT. Hence, the use of specific inhibitors is a prerequisite to eliminate "off-target" effects of test drugs when using tissue preparations. Consequently, expression of MATs in heterologous systems ensures that the chosen MAT mediates the observed effects per se. However, it is noteworthy that a reduction in Na^+/K^+-ATPase levels in nonneuronal cells and different membrane compositions can bias transporter function. Another important issue is choosing between stable and transient expression. Stable expression systems display constant expression levels. However, expression levels correlate with the relative potencies of test drugs [47] and high expression levels have been shown to result in steep dose–response curves [48]. To exclude data misinterpretation, it is critical to assess the kinetic parameters (K_M and V_{max}) of the chosen cell lines. Additionally, the inclusion of internal standard drugs with known pharmacology (e.g., cocaine) is a necessary step when estimating the relative potencies of new drugs.

6 Discussion

Various methodologies have been employed to assess the mechanism of action of drugs at MATs. These include electrophysiological recordings [27]; efflux of preloaded [^{3}H] substrates in static batch release assays [36] and native tissue preparations [39, 49]; and dynamic superfusion experiments [35]. The ultimate goal of applying these various methods is to identify the precise molecular mechanism of action for drugs interacting at MATs: uptake inhibitors versus substrates. Drug-induced carrier-mediated release events have been observed and studied for over five decades [50]. However, the exact mechanism and subsequent intracellular cascades involved in transporter-mediated release are far from being fully understood. Amphetamine-like transporter substrates can cause internalization events [51], alter the activity of kinases [52], and even affect second messengers [53] and the transcriptome [54]. Hence, it is of tremendous importance to unravel the mode of action of NPS at plasmalemmal transporters as they serve as a gateway to the intracellular compartment. Once located in the cytosol, NPS potentially trigger a variety of additional effects. For instance, a major target of amphetamine-like drugs appears to be VMAT2. Neurotoxicity of drugs can be linked to their substrate-like activity at the plasmalemmal and vesicular MATs [18]. After deciphering the mode of drug action at DAT, NET, and SERT, substrates can be tested for their activities at VMAT2. The current problem with NPS is that the drug markets are flooded with substances of unknown pharmacology. The rapid emergence and disappearance of NPS complicate long-term studies about their toxicity in vivo. Studies that examine the mode of action of psychostimulant NPS in vitro provide the basis to estimate the potential threat of individual drugs to public health. The techniques outlined in this chapter enable fast and reliable identification of MAT substrates that require further analysis in more depth. Most importantly, the combination of biological assays and chemical identification by HPLC-MS by "checkit!" is of tremendous clinical relevance. The current volatility of the street drug markets makes it cumbersome and challenging for medical professionals to treat adverse effects when the underlying pharmacology of ingested drugs is unknown. Maintaining an up-to-date database on the pharmacology and toxicology of newly emerging drugs is essential for formulating effective responses to the problem of NPS. As outlined above, street drugs are rarely sold in their pure form. The current "checkit!" warning system collects information on adulterants and potentially harmful drug combinations. In addition, elucidating the chemical nature and pharmacological fingerprints for new drugs provides the fundamental knowledge required to regulate and ban problematic drugs of abuse before they become well-established members of the drug market.

References

1. Baumeister D, Tojo LM, Tracy DK (2015) Legal highs: staying on top of the flood of novel psychoactive substances. Ther Adv Psychopharmacol 5:97–132
2. Lewin AH, Seltzman HH, Carroll FI, Mascarella SW, Reddy PA (2014) Emergence and properties of spice and bath salts: a medicinal chemistry perspective. Life Sci 97:9–19
3. Brandt SD, King LA, Evans-Brown M (2014) The new drug phenomenon. Drug Test Anal 6:587–597
4. Lindigkeit R, Boehme A, Eiserloh I, Luebbecke M, Wiggermann M, Ernst L et al (2009) Spice: a never ending story? Forensic Sci Int 191:58–63
5. Baumann MH, Solis E Jr, Watterson LR, Marusich JA, Fantegrossi WE, Wiley JL (2014) Baths salts, spice, and related designer drugs: the science behind the headlines. J Neurosci 34:15150–15158
6. Miliano C, Serpelloni G, Rimondo C, Mereu M, Marti M, De Luca MA (2016) Neuropharmacology of new psychoactive substances (NPS): focus on the rewarding and reinforcing properties of cannabimimetics and amphetamine-like stimulants. Front Neurosci 10:153
7. Prosser JM, Nelson LS (2012) The toxicology of bath salts: a review of synthetic cathinones. J Med Toxicol 8:33–42
8. Ross EA, Watson M, Goldberger B (2011) "Bath salts" intoxication. N Engl J Med 365:967–968
9. Soussan C, Kjellgren A (2016) The users of novel psychoactive substances: online survey about their characteristics, attitudes and motivations. Int J Drug Policy 32:77–84
10. Torres GE, Gainetdinov RR, Caron MG (2003) Plasma membrane monoamine transporters: structure, regulation and function. Nat Rev Neurosci 4:13–25
11. Sitte HH, Freissmuth M (2010) The reverse operation of Na(+)/Cl(−)-coupled neurotransmitter transporters – why amphetamines take two to tango. J Neurochem 112:340–355
12. Sitte HH, Freissmuth M (2015) Amphetamines, new psychoactive drugs and the monoamine transporter cycle. Trends Pharmacol Sci 36:41–50
13. Sulzer D, Sonders MS, Poulsen NW, Galli A (2005) Mechanisms of neurotransmitter release by amphetamines: a review. Prog Neurobiol 75:406–433
14. Egana LA, Cuevas RA, Baust TB, Parra LA, Leak RK, Hochendoner S et al (2009) Physical and functional interaction between the dopamine transporter and the synaptic vesicle protein synaptogyrin-3. J Neurosci 29:4592–4604
15. Schuldiner S, Steiner-Mordoch S, Yelin R, Wall SC, Rudnick G (1993) Amphetamine derivatives interact with both plasma membrane and secretory vesicle biogenic amine transporters. Mol Pharmacol 44:1227–1231
16. Freyberg Z, Sonders MS, Aguilar JI, Hiranita T, Karam CS, Flores J et al (2016) Mechanisms of amphetamine action illuminated through optical monitoring of dopamine synaptic vesicles in Drosophila brain. Nat Commun 7:10652
17. Baumann MH, Bulling S, Benaderet TS, Saha K, Ayestas MA, Partilla JS et al (2014) Evidence for a role of transporter-mediated currents in the depletion of brain serotonin induced by serotonin transporter substrates. Neuropsychopharmacology 39:1355–1365
18. Pifl C, Reither H, Hornykiewicz O (2015) The profile of mephedrone on human monoamine transporters differs from 3,4-methylenedioxymethamphetamine primarily by lower potency at the vesicular monoamine transporter. Eur J Pharmacol 755:119–126
19. Stockner T, Montgomery TR, Kudlacek O, Weissensteiner R, Ecker GF, Freissmuth M et al (2013) Mutational analysis of the high-affinity zinc binding site validates a refined human dopamine transporter homology model. PLoS Comput Biol 9(2):e1002909
20. Penmatsa A, Gouaux E (2014) How LeuT shapes our understanding of the mechanisms of sodium-coupled neurotransmitter transporters. J Physiol 592:863–869
21. Yamashita A, Singh SK, Kawate T, Jin Y, Gouaux E (2005) Crystal structure of a bacterial homologue of Na+/Cl−-dependent neurotransmitter transporters. Nature 437:215–223
22. Penmatsa A, Wang KH, Gouaux E (2013) X-ray structure of dopamine transporter elucidates antidepressant mechanism. Nature 503:85–90

23. Coleman JA, Green EM, Gouaux E (2016) X-ray structures and mechanism of the human serotonin transporter. Nature 532:334–339
24. Bonano JS, Banks ML, Kolanos R, Sakloth F, Barnier ML, Glennon RA et al (2015) Quantitative structure-activity relationship analysis of the pharmacology of para-substituted methcathinone analogues. Br J Pharmacol 172:2433–2444
25. Kolanos R, Sakloth F, Jain AD, Partilla JS, Baumann MH, Glennon RA (2015) Structural modification of the designer stimulant alpha-pyrrolidinovalerophenone (alpha-PVP) influences potency at dopamine transporters. ACS Chem Nerosci 6:1726–1731
26. Saha K, Partilla JS, Lehner KR, Seddik A, Stockner T, Holy M et al (2015) 'Second-generation' mephedrone analogs, 4-MEC and 4-MePPP, differentially affect monoamine transporter function. Neuropsychopharmacology 40:1321–1331
27. Sandtner W, Stockner T, Hasenhuetl PS, Partilla JS, Seddik A, Zhang YW et al (2016) Binding mode selection determines the action of ecstasy homologs at monoamine transporters. Mol Pharmacol 89:165–175
28. Brunt TM, Nagy C, Bücheli A, Martins D, Ugarte M, Beduwe C, Ventura Vilamala M (2016) Drug testing in Europe: monitoring results of the Trans European Drug Information (TEDI) project. Drug Test Anal. doi:10.1002/dta.1954 [Epub ahead of print]
29. Ostermann KM, Luf A, Lutsch NM, Dieplinger R, Mechtler TP, Metz TF et al (2014) MALDI Orbitrap mass spectrometry for fast and simplified analysis of novel street and designer drugs. Clin Chim Acta 433:254–258
30. Rosenauer R, Luf A, Holy M, Freissmuth M, Schmid R, Sitte HH (2013) A combined approach using transporter-flux assays and mass spectrometry to examine psychostimulant street drugs of unknown content. ACS Chem Nerosci 4:182–190
31. Hofmaier T, Luf A, Seddik A, Stockner T, Holy M, Freissmuth M et al (2014) Aminorex, a metabolite of the cocaine adulterant levamisole, exerts amphetamine like actions at monoamine transporters. Neurochem Int 73:32–41
32. Chen C, Okayama H (1987) High-efficiency transformation of mammalian cells by plasmid DNA. Mol Cell Biol 7:2745–2752
33. Hilber B, Scholze P, Dorostkar MM, Sandtner W, Holy M, Boehm S et al (2005) Serotonin-transporter mediated efflux: a pharmacological analysis of amphetamines and non-amphetamines. Neuropharmacology 49:811–819
34. Pifl C, Drobny H, Reither H, Hornykiewicz O, Singer EA (1995) Mechanism of the dopamine-releasing actions of amphetamine and cocaine: plasmalemmal dopamine transporter versus vesicular monoamine transporter. Mol Pharmacol 47:368–373
35. Sitte HH, Scholze P, Schloss P, Pifl C, Singer EA (2000) Characterization of carrier-mediated efflux in human embryonic kidney 293 cells stably expressing the rat serotonin transporter: a superfusion study. J Neurochem 74:1317–1324
36. Wall SC, Gu H, Rudnick G (1995) Biogenic amine flux mediated by cloned transporters stably expressed in cultured cell lines: amphetamine specificity for inhibition and efflux. Mol Pharmacol 47:544–550
37. Baumann MH, Partilla JS, Lehner KR (2013) Psychoactive "bath salts": not so soothing. Eur J Pharmacol 698:1–5
38. Baumann MH, Partilla JS, Lehner KR, Thorndike EB, Hoffman AF, Holy M et al (2013) Powerful cocaine-like actions of 3,4-methylenedioxypyrovalerone (MDPV), a principal constituent of psychoactive 'bath salts' products. Neuropsychopharmacology 38:552–562
39. Baumann MH, Ayestas MA Jr, Partilla JS, Sink JR, Shulgin AT, Daley PF et al (2012) The designer methcathinone analogs, mephedrone and methylone, are substrates for monoamine transporters in brain tissue. Neuropsychopharmacology 37:1192–1203
40. Scholze P, Zwach J, Kattinger A, Pifl C, Singer EA, Sitte HH (2000) Transporter-mediated release: a superfusion study on human embryonic kidney cells stably expressing the human serotonin transporter. J Pharmacol Exp Ther 293:870–878
41. Rudnick G, Clark J (1993) From synapse to vesicle: the reuptake and storage of biogenic amine neurotransmitters. Biochim Biophys Acta 1144:249–263

42. Mollenhauer HH, Morre DJ, Rowe LD (1990) Alteration of intracellular traffic by monensin; mechanism, specificity and relationship to toxicity. Biochim Biophys Acta 1031:225–246
43. Rudnick G, Kirk KL, Fishkes H, Schuldiner S (1989) Zwitterionic and anionic forms of a serotonin analog as transport substrates. J Biol Chem 264:14865–14868
44. Seidel S, Singer EA, Just H, Farhan H, Scholze P, Kudlacek O et al (2005) Amphetamines take two to tango: an oligomer-based counter-transport model of neurotransmitter transport explores the amphetamine action. Mol Pharmacol 67:140–151
45. Scholze P, Norregaard L, Singer EA, Freissmuth M, Gether U, Sitte HH (2002) The role of zinc ions in reverse transport mediated by monoamine transporters. J Biol Chem 277:21505–21513
46. Steinkellner T, Yang JW, Montgomery TR, Chen WQ, Winkler MT, Sucic S et al (2012) Ca(2+)/calmodulin-dependent protein kinase IIalpha (alphaCaMKII) controls the activity of the dopamine transporter: implications for Angelman syndrome. J Biol Chem 287:29627–29635
47. Ukairo OT, Ramanujapuram S, Surratt CK (2007) Fluctuation of the dopamine uptake inhibition potency of cocaine, but not amphetamine, at mammalian cells expressing the dopamine transporter. Brain Res 1131:68–76
48. Shoichet BK (2006) Interpreting steep dose-response curves in early inhibitor discovery. J Med Chem 49:7274–7277
49. Crespi D, Mennini T, Gobbi M (1997) Carrier-dependent and Ca(2+)-dependent 5-HT and dopamine release induced by (+)-amphetamine, 3,4-methylendioxymethamphetamine, p-chloroamphetamine and (+)-fenfluramine. Br J Pharmacol 121:1735–1743
50. Glowinski J, Axelrod J (1965) Effect of drugs on the uptake, release, and metabolism of H3-norepinephrine in the rat brain. J Pharmacol Exp Ther 149:43–49
51. Wheeler DS, Underhill SM, Stolz DB, Murdoch GH, Thiels E, Romero G et al (2015) Amphetamine activates Rho GTPase signaling to mediate dopamine transporter internalization and acute behavioral effects of amphetamine. Proc Natl Acad Sci U S A 112:E7138–E7147
52. Giambalvo CT (1992) Protein kinase C and dopamine transport – 1. Effects of amphetamine in vivo. Neuropharmacology 31:1201–1210
53. Gnegy ME, Khoshbouei H, Berg KA, Javitch JA, Clarke WP, Zhang M et al (2004) Intracellular Ca2+ regulates amphetamine-induced dopamine efflux and currents mediated by the human dopamine transporter. Mol Pharmacol 66:137–143
54. Allen JK, Wilkinson M, Soo EC, Hui JP, Chase TD, Carrey N (2010) Chronic low dose Adderall XR down-regulates cfos expression in infantile and prepubertal rat striatum and cortex. Neuroscience 169:1901–1912

NPS: Medical Consequences Associated with Their Intake

Fabrizio Schifano, Laura Orsolini, Duccio Papanti, and John Corkery

Abstract Over the last decade, the 'traditional' drug scene has been supplemented – but not replaced – by the emergence of a range of novel psychoactive substances (NPS), which are either newly created or existing drugs, including medications, now being used in novel ways. By the end of 2014, in excess of 500 NPS had been reported by a large number of countries in the world. Most recent data show, however, that synthetic cathinones, synthetic cannabinoids, and psychedelics/ phenethylamines account for the largest number of NPS.

The present chapter aims at providing an overview of the clinical and pharmacological issues relating to these most popular NPS categories. Given the vast range of medical and psychopathological issues associated with the molecules here described, it is crucial for health professionals to be aware of the effects and toxicity of NPS. A general overview of the acute management of NPS adverse events is provided as well, although further studies are required to identify a range of evidence-based, index molecule-focused, treatment strategies. The rapid pace of change in the NPS online market constitutes a major challenge to the provision of current and reliable scientific knowledge on these substances.

Keywords Drug misuse • Hallucinogenic drugs • Hallucinogens • Novel psychoactive substances • Phenethylamines • Psychiatric disturbances • Synthetic cannabimimetics • Synthetic cathinones

F. Schifano (✉), L. Orsolini, D. Papanti, and J. Corkery
Psychopharmacology, Drug Misuse and Novel Psychoactive Substances Research Unit, School of Life and Medical Sciences, University of Hertfordshire, College Lane Campus, Hatfield, Herts AL10 9AB, UK
e-mail: F.Schifano@herts.ac.uk

Curr Topics Behav Neurosci (2017) 32: 351–380
DOI 10.1007/7854_2016_15
Published Online: 12 July 2016

Contents

1 Introduction

Over the last decade, the 'traditional' drug scene has been supplemented – but not replaced – by the emergence of a range of psychoactive substances, which are either newly created or existing drugs, including medications, now being used in novel ways. These molecules are referred to as 'new/novel psychoactive substances' (NPS). Commonly, the term 'legal high' has been used to describe such substances. However, this is misleading since often these molecules are already subject to regulation. Moreover, because they are 'legal' they are incorrectly assumed to be safe.

The speed with which NPS emerge onto the drug market(s) has been accelerating over this period. The United Nations Office for Drugs and Crime (UNODC) suggests that between 2009 and 2014 the number of NPS reported increased from 126 to 450. By the end of 2014, 541 substances had been reported by 95 countries through the global Synthetics Monitoring: Analysis, Reporting and Trends (SMART) programme [1]. Data from the European Centre for Drugs and Drug Addiction (EMCDDA) provides a more complete assessment of these trends. The agency has been systematically monitoring NPS since 2005. Their latest report [2] shows that of the 101 NPS identified during 2014, 31 were cathinones and 30 were synthetic cannabinoids; these, together with phenethylamines, account for the largest number of NPS. In line with UNODC data, by 1st February 2016 the number of NPS notified to the agency was 561. At present, we see no reason why this situation of increasing availability of NPS should change in the immediate future. This is a reason of concern, since there is a general lack of information on newly created molecules and the possibility of serious health risks (physiological and mental) is typically based only on analogy with remaining chemically related molecules/drugs.

The present chapter aims to provide an overview of the clinical and pharmacological issues relating to the most popular NPS categories, with a particular focus on the medical and psychopathological issues associated with their ingestion.

2 Synthetic Cannabimimetics

Synthetic cannabimimetics (SC) were first detected in Europe toward the end of 2008 and, at the time of writing, constitute the largest group of substances (about 30%) monitored by the EMCDDA [3]. This situation probably reflects the overall demand for cannabis within the region, as well as the rapidity with which manufacturers can produce and supply new cannabimimetics to avoid ever-changing drug controls [2]. 'Spice' preparations are composed by both a dried plant base, to mimic the 'grass effect' of female cannabis dried inflorescences, and a mixture of SC, which is sprayed onto the plant material. SC dispersed in the grass preparation look like hashish, with capsules and e-liquid formulations being available as well. SC have also been found in tablets and sprayed on cannabis joints [3].

2.1 Market and Use

SC can be found on the web (e.g. both the surface-web and the deepweb/darknet), either as wholesale or retail products, but also from 'head-shops', gas stations, and from an ever expanding range of other outlets [4]. SC can be acquired as well as street drugs [5]. Apart from curiosity and the wish to get high – motivations frequently mentioned in the context of illicit drug use – legal availability is considered an important motivation for consumption [6]. Many users, on the other hand, do not seem to be aware of the serious adverse effects related to SC misuse, since these compounds are being perceived to be somehow equivalent to marijuana and hence 'safe' and 'natural' [7]. Many of the hundreds of SC have been synthesized for research purposes, but most had never been tested in animals/humans prior to being identified in products confiscated from human users [8].

The intake of SC usually occurs by inhalation from a joint/bong/pipe or utilizing a vaporizer. Other ways of intake include insufflation, oral ingestion, rectal administration, and injection [9]. Users are often young males, choosing Spice instead of cannabis for its low cost, easy availability, and undetectability in routine urine screening tests [10–12]. Generally, highly refined analytical techniques are needed to detect the presence of even small amounts of SC in Spice products, or to identify the presence of their metabolites in the body fluids [13]. The fact that SC are not detected by standard toxicology screens makes these substances an attractive alternative to marijuana for sub-populations undergoing regular drug tests (e.g. patients of: forensic wards, withdrawal clinics, residential treatments for substance use disorders, and acute psychiatric wards; prisoners/clients on probation), and for those subjected to workplace testing such as military personnel, mine workers, athletes, and driving license re-granting candidates [14].

2.2 *Neuropharmacology*

SC include compounds with diverse chemical structures, presenting with distinct physical properties, potency, pharmacokinetics, and pharmacodynamics when compared to the cannabis phytocannabinoid Δ^9-tetrahydrocannabinol (THC). THC is a partial agonist at cannabinoid receptor subtypes, CB-1 and CB-2, but the 'cannabis high' is associated with the binding to CB-1 site. With respect to THC itself, SC contained in Spice products are typically full-efficacy agonists with very high affinity for the CB-1 and CB-2 receptors, hence they elicit maximal activity at those receptors. Furthermore, the effects of THC in cannabis are modulated/dampened by the presence of other natural compounds such as cannabidiol and cannabivarin [15], but no such 'modulating' compounds are generally detected in Spice products [16]. Several SC have been found to interact with a range of non-cannabinoid receptors, including: 5-HT, nicotinic acetylcholine, glycine, and/or ionotropic glutamate (NMDA), and it is possible that these non-cannabinoid receptors contribute to the complex effects the synthetic substances [16–18].

Spice products are almost always laced with multiple SC in a single preparation [9]. As such, there is a potential for drug–drug interaction between multiple SC in a single product, and this may contribute to the abuse-related and dangerous synergistic effects of these compounds. Some SC metabolites display high affinity and efficacy for CB-1 receptors, thus prolonging and intensifying receptor activation, and contributing to the toxicity of the products [19]. A number of SC compounds incorporate indole-derived moieties, as components of the structure or as substituents [20]. Indoles are structurally similar to 5-HT, hence they can activate 5-HT receptors, which are typically associated with the effects of indoleamine hallucinogens such as dimethyltryptamine [21]. From this point of view, it could be argued that ingestion of indole SC compounds may be associated with significant levels of activation of 5-HT receptors [22]. It has been suggested that, at high doses, SC compounds may also possess some monoamine oxidase inhibitory properties [23]. These complex pharmacodynamic actions may further increase the risk of serotonin syndrome in SC misusers [14, 15]. The recent trend of SC fluorination, commonly applied in medicinal chemistry, may increase the compounds' lipophilicity, hence promoting the absorption through biological membranes/ blood brain barrier [24, 25], possibly enhancing the SC overall toxicity [19].

Other factors potentially contributing to negative side effects associated with use of SC include: the pharmacological activity of further molecules, other than SC, sprayed on the plant mixtures [15], the presence of contaminants, side-products, and solvents [14], the total lack of product quality control leading to significant differences in concentration ('hot-spots') of SC present in herbal incenses or e-liquids [22, 26], and the increased vulnerability to adverse effects due to pre-existing conditions of drug users, or concomitant intake of other psychoactives [19, 27].

2.3 *Desired Effects*

Although the 'cannabimimetic high' induced by SC presents with some similarities to that of cannabis, the effects of Spice products have been anecdotally described by users as intense and euphoric, with hallucinatory experiences at higher levels of intake [15]. In comparison with cannabis, use of SC may be associated with quicker 'kick off' effects, significantly shorter duration of action, larger levels of hangover effects, and more frequent/intense paranoid feelings [28]. Other Spice effects include: feelings of well-being, calmness, relaxation, increased creativity, mild perceptual alterations, and mild memory/attentional impairments [29].

2.4 *Adverse Effects*

At lower dosages, SC intake seems to be associated with anxiolytic effects, whilst at higher doses the effects are anxiogenic and associated with a propensity for continued ingestion [30]. Adverse side effects associated with SC use are more severe than observed after marijuana, and the related acute toxicity/intoxication seems to be similar to the one seen with the use of stimulant/sympathomimetic recreational drugs [11, 31, 32]. It has been suggested that the effects of SC are more significant in individuals with lower levels of previous exposure to cannabis, and especially those who are drug naïve [33].

The acute SC intoxication is characterized by a short-standing clinical picture with reported signs/symptoms of elevated heart rate/blood pressure levels, visual/ auditory hallucinations, mydriasis, agitation/anxiety, hyperglycaemia, dyspnoea/ tachypnea, and nausea/vomiting [9, 33, 34].

Other psychiatric and neurological effects include: suicidal ideation/self-injurious behaviour, aggressive behaviour, panic attacks, thought disorganization, psychosis, agitated/excited delirium, nystagmus, seizures, hyperemesis, encephalopathy, coma, and stroke [27, 35–39].

Severe cardiotoxic effects have been described as well, including: dysrhythmias, cardiac arrest, chest pain, and myocardial infarction [9, 11, 40]. Other potentially serious effects such as hypokalaemia, toxic hepatitis/liver failure, acute kidney injury, rhabdomyolysis, hyperthermia, and serotonin syndrome have been reported [9, 37, 40–46]. A chronic cough has been described in habitual SC users [47, 48], with pneumothorax, pneumomediastinum, and diffuse pulmonary infiltrates having been reported as well [47, 49].

Tolerance and dependence related to SC use have been described [29, 48]. Long-term SC misuse may also be associated with a severe prolonged withdrawal syndrome, characterized by drug craving, tachycardia, tremor, profuse sweating, diarrhoea, nightmares/insomnia, headache, anxiety/irritability, mood swings, feelings of emptiness/depressive symptoms, and somatic complaints [9, 50]. Natural cannabis has been reported to be associated with an increased risk of developing psychosis in users, depending on both the THC concentration and frequency of use [51, 52]. Similarly, the intake of SC has been associated with a range of psychotomimetic disturbances (e.g. paranoia, delusions, and hallucinations), the occurrence of florid/acute transient psychosis, relapse/worsening of a pre-existing psychosis, and persistent psychotic disorder/'Spiceophrenia' [27]. In association with Spice use, a range of further psychopathological disturbances have been described, including: behavioural dyscontrol and agitation [53], dysphoria, mood swings, suicide attempts/suicidal ideation [54], manic-like symptoms [55], and relapse of a pre-existing bipolar disorder [56, 57]. Four completed suicides following SC intake have been described [58–61]. Finally, a number of deaths have been related to SC ingestion, either on their own or in combination with other substances, in analytically confirmed reports [9, 40, 62].

3 Clinical and Adverse Effects of Synthetic Cathinones

3.1 Pharmacology/Neuropharmacology

Synthetic cathinones, such as ethylone and methylone, are β-keto-phenethylamines structurally similar to amphetamines (including methamphetamine and 3,4-methylenedioxymethamphetamine, or MDMA) and catecholamines, with subtle variations that alter their chemical properties, potency, pharmacokinetics, and pharmacodynamics [9]. Some cathinones used recreationally are analogues of pyrovalerone, e.g. 3,4-methylenedioxypyrovalerone (MDPV); naphyrone; 3,4-methylenedioxy-α-pyrrolidinobutiophenone (MDPBP); and α-pyrrolidinovalerophenone (α-PVP) [63]. Mephedrone and methylone are monoamine releasers

Table 1 Classification of synthetic cathinones [68]

Category	Examples
Substrates for DAT, SERT, and NET with MDMA-like profiles	Benzedrone, butylone, ethylone, 4-methylethcathinone (4-MEC)
Monoamine transporter substrates with DAT-selective profiles similar to amphetamine and methamphetamine	Cathinone, methcathinone, flephedrone. Naphyrone and 1-naphyrone have very high potencies and some degree of selectivity for DAT
Non-substrate transporter inhibitors	MDPV

DAT dopamine transporter, *SERT* serotonin transporter, *NET* norepinephrine transporter

and possess similar abilities to release dopamine (DA) and 5-HT, hence presenting with patterns of drug abuse akin to those of MDMA [64]. Conversely, methcathinone (ephedrone) selectively generates a release of DA greater than 5-HT. MDPV is a DA selective uptake inhibitor, but selectively blocks the uptake of DA greater than 5-HT, and presents with high abuse potential [65]. Similar to cocaine, MDPV inhibits monoamine uptake at the DA transporters (DAT) and norepinephrine (NE) transporters (NET) [66, 67]. Simmler et al. [68] have proposed classifying cathinones into three categories (Table 1).

Whilst mephedrone and MDPV have behavioural effects, respectively, similar to methamphetamine and cocaine, methylone's effects are closer to those of MDMA [69]. Individual cathinones have variable effects and potency levels on the DA, NE, and 5-HT pathways, but all typically possess sympathomimetic and/or amphetamine-like effects [70, 71]. The potency of different molecules to inhibit at DAT and NET levels corresponds to the human recreational dosage whilst their potency at SERT level does not [68]. However, the amphetamine-type subjective effects produced by these agents in humans appear to be more correlated with their potency in releasing NE rather than DA, therefore suggesting that NE may contribute to the profile of stimulants effects [72]. Mephedrone, methcathinone, MDPV, and naphyrone are potent NET inhibitors. Whilst increased brain NE levels are not associated with the intoxicant effect of such molecules, they could contribute to peripheral sympathomimetic effects leading to undesired side effects and even fatal cardiovascular consequences [73].

Mephedrone is metabolized in a similar way to ring-substituted amphetamines. Its half-life is as short as 1 h, hence the re-dosing risk [74]. MDPV is thought to have a half-life of 3–5 h.

3.2 International Control and Therapeutic Uses

In addition to cathinone and cathine, the only cathinones under international control (United Nations Convention on Psychotropic Substances 1971) are amfepramone, mephedrone, methcathinone, and pyrovalerone. Methcathinone, first synthesized prior to 1928 [75], was patented as an analeptic [76]. Mephedrone

(4-methylmethcathinone) appears to have been first synthesized in 1929 [77]. First synthesized in 1964, pyrovalerone was marketed as an appetite suppressant and for treating chronic fatigue; it was subsequently withdrawn due to abuse and dependency issues [78]. MDPV was first synthesized in 1969 [79]. Amfepramone (diethylcathinone) is also used as an appetite suppressant. Methylone has been patented as an antidepressant and anti-Parkinsonian agent. Bupropion has been licensed for use as an antidepressant and for treating nicotine dependence. Furthermore, its analogues were also tested as potential pharmacotherapies for cocaine in preclinical studies in rodents [80]. Bupropion inhibits the reuptake of DA, 5-HT, and NE, but both bupropion and its metabolites also substitute in amphetamine-trained animals [81]. It elicits a cocaine-like cue [82].

The large number of seizures of synthetic cathinones by EU law enforcement agencies reflects the demand for stimulants in the region, with many of them not only being employed as replacements for amphetamine, cocaine, and ecstasy (MDMA), but also used in conjunction with such substances [2]. At the time of writing, synthetic cathinones are the second largest group of substances (18%) monitored by the EMCDDA, after having first appeared in 2008 [83]. Their rapidly growing popularity was driven by the lack of availability, or the poor purity of 'traditional' drugs combined with little, if any, legal restrictions [9].

Of the nearly 100 cathinones notified so far to the EMCDDA, perhaps the principal ones of concern, based on adverse health effects, hospital admissions, and deaths are: 4-MEC, alpha-PVP (α-pyrrolidinopentiophenone), flephedrone, MDPBP, MDPV, mephedrone, methedrone, methylone, naphyrone, pentedrone, and pyrovalerone.

3.3 Administration

Typically, synthetic cathinones are sold as pills, capsules, and powders. They are commonly insufflated (snorted/sniffed), ingested orally by 'bombing' (swallowing the powder wrapped in a cigarette paper), mixed in a drink, or injected intravenously (see also below). MDPV has also been administered sub-lingually, intramuscularly, rectally, by smoking, and through vaporization (inhalation) [84].

3.4 What Else Is Being Used with Individual Cathinones and Why?

Internet user fora often discuss the use of other substances to enhance the effects of specific drug molecules or to reduce their harmful effects. For example, Coppola and Mondola [84] describe, with respect to MDPV, that users report ingesting: alcohol, propranolol, and other beta-blockers to counteract tachycardia, GHB/GBL as an aphrodisiac, zopiclone to produce visual hallucinations, kratom/Mitragyna speciosa, hallucinogens, and amphetamines to enhance stimulant and entactogenic effects, pregabalin, omeprazole, and domperidone to treat stomach pain, and benzodiazepines and cannabis to reduce anxiety.

Several different cathinones are often used together. This may cause synergistic effects, e.g. as with mephedrone and MDPV [73]. Based on mortality data from the UK, Hungary and Italy collated as part of the EU-MADNESS project (www.eumadness.eu), it is clear that synthetic cathinones are often present in post-mortem toxicology and/or implicated in deaths involving NPS. Other stimulants such as amphetamine, methamphetamine, cocaine, and MDMA are often used, as are piperazines (especially BZP). Central nervous system depressants such as alcohol, benzodiazepines, and opioids are not uncommon. In some sub-populations GHB/GBL and ketamine are used as well.

3.5 *Desired and Adverse Effects*

Consumers of synthetic cathinones use them for a variety of reasons including euphoria, stimulation, increased energy, empathy, openness, mood enhancement, mental clarity, hallucinogenic experiences, and increased libido.

Whereas cardiac, psychiatric, and neurological signs are some of the adverse effects reported by synthetic cathinone users, the most common symptom identified from medical observations is agitation, ranging from mild agitation to severe psychosis [85]. Patients under the apparent influence of mephedrone have also shown that synthetic cathinones present similar sympathomimetic effects (including tachycardia and hypertension as well as psychoactive effects) to amphetamine derivatives. More than half of those in a student survey who had taken mephedrone reported adverse effects associated with the central nervous, nasal/respiratory, and cardiovascular systems [86]. Abdominal pain, flushing, sweating, chills, restlessness, and anxiety can be observed as well [9, 70, 71, 87, 88]. Additional reported serious effects include hyperthermia, rhabdomyolysis, renal failure, and seizures.

Cathinone-related psychoactive effects include increased alertness, euphoria, excited delirium, hallucinations, agitation, and aggression, associated with tachycardia, hypertension, and dilated pupils. Mood disturbances and paranoid ideation have been observed in chronic users of cathinones, both natural and synthetic [87–90].

A large proportion of users of synthetic cathinones reports tolerance, dependence, and withdrawal symptoms [91]. The potential risk for long-term psychiatric problems is suggested by some abstinent methcathinone users presenting with decreased striatal DA transporter density on positron emission tomography scans [92].

3.6 Morbidity/Clinical Issues: Cathinones' Injecting Issues

An increasing area of concern in Europe is the injecting of NPS, especially synthetic cathinones, at a time when injection of heroin is falling in some countries. The self-injection of synthetic cathinones has emerged among specific segments of the population in Austria, Belgium, the Czech Republic, France, Germany, Ireland, Poland, Romania, Spain, and the United Kingdom [93]. In some parts of Europe (e.g. Graz in Austria and Bucharest in Romania), users injecting cathinones account for more than half of all drug injectors, with problem drug users now switching from heroin. The intake of cathinones may be associated with a high frequency of injection (up to 10–20 times per day). Mephedrone, MDPV, and 4-MEC are all reported to be injected. The injection of MDPV by problem drug users has been reported in a number of countries, including Hungary, Finland, and Romania, including those transitioning from amphetamine [1, 93]. A more recent development is the injecting of α-PVP in Ireland [94]. A death resulting from injecting α-PVP has been reported from Australia [95].

Behavioural data show that half (48%) of those currently injecting NPS reported sharing syringes [96], hence increasing public health risks [93]. Cathinones may act in concert with HIV, leading to glial and neuronal toxicity more significant than the neurotoxicity observed with either the cathinones or HIV individually [97].

Within the Chemsex context (e.g. performing sexual activities while under the influence of drugs, often involving group sex or a high number of partners in one session), synthetic cathinones/mephedrone, either on their own or together with methamphetamine and/or GHB/GBL, were frequently identified as possessing a significant influence on the men who have sex with men (MSM) risk-taking behaviour. Some men appeared to describe drugs as having 'myopic' properties, in that they altered their ability to perceive the wider consequences of their actions [98]. Consequently, this form of use is associated with a highly elevated risk of the spread of blood-borne and sexually transmitted diseases [96]. An increased number of sexual partners may also increase the risk of acquiring other sexually transmitted infections. Data from service users suggest an average of five sexual partners per session, with unprotected sex being the norm [99, 100]. Mephedrone injecting within the male gay community is also increasing in cities like London [99].

3.7 Deaths

Deaths have been associated with a range of synthetic cathinones, including: mephedrone [87, 88, 90, 101, 102], methylone and butylone [103], ethylone [104], bupropion [82], α-PVP [95, 105, 106], MDPV [93, 107–109], and methedrone [110].

In England and Wales, 83 deaths involving cathinones had been registered by the end of 2014 (ONS 2015); 15 such deaths were registered between 1 January 2013 and 30 June 2015 in Northern Ireland, together with 14 in Scotland (unpublished data, EU-MADNESS project). The majority of these involved mephedrone. Up to the end of 2014, 63 mephedrone-related deaths were registered in England and Wales (ONS 2015), 12 in Scotland (unpublished data, National Records of Scotland), and 13 mephedrone/cathinone deaths in Northern Ireland [111].

Up to 41% of hangings or other mechanical suicides examined by one forensic agency between 2010 and 2012 involved cathinones [112]. This confirms other UK reports of suicides involving hangings by mephedrone users [87, 88].

Unpublished data from Hungary collated for the EU-MADNESS project covering the period 1 January 2014 to 30 June 2015 indicates 16 deaths where α-PVP was found in the post-mortem toxicology; of these, only one was exclusively considered to have involved the drug on its own. However, many others were deemed to be drug intoxication/overdose. In many instances, benzodiazepines such as alprazolam and clonazepam were also present. In some cases, other α-PVP analogues, such as α-PVT (α-pyrrolidinopentiothiophenone) and α-PHP (α-pyrrolidinohexiophenone), were identified. Some of these fatalities had occurred after injecting. There were also 11 deaths involving pentedrone, including injectors. The circumstances of α-PVP and pentedrone deaths resembled those relating to mephedrone described in the UK, e.g. falls from heights, drowning, hangings, as well as overdoses [87].

4 Clinical and Adverse Effects of Synthetic Hallucinogens

The class of drugs known as hallucinogens are able to alter consciousness by both distorting the perception of time, motion, colour, sound and self, and by inducing sensory and perceptual disturbances. Hallucinogens may induce hallucinations (i.e. perceptions in the absence of external stimuli), illusions (i.e. perceptual distortion of normal environmental stimuli), and 'pseudo-hallucinations' (hallucinations recognized by the patient not to be the result of external stimuli) [113–115], together with intense emotional responses and thoughts that may influence the human psyche [116].

Hallucinogens are also called '*psychedelics*' (a term also describing the 'classical hallucinogen' such as LSD (*N,N*-diethyl-D-lysergamide) and psilocybin, '*psychotomimetics*' (a term emphasizing their effects that mimic psychotic symptoms), and '*entheogens*' (due to the mystical-type experiences these drugs may induce; [117]).

In general, the term classical '*hallucinogen*' is used to connote all drugs acting as agonists at the 5-HT_{2A} receptor. Beyond these, we include the '*synthetic hallucinogens*', mostly belonging to the NPS category (Table 2).

Along with the classical hallucinogens, other drug molecules (Table 3) may produce some hallucinogenic effects, even though they are not classified as 'serotonergic hallucinogens'.

Table 2 Classification of synthetic serotonergic hallucinogens

Category	Examples
Lysergamides	LSD, LSA, 1P-LSD, ALD-52, ETH-LAD, Pro-LAD, AL-LAD, LSZ and LSD-like structures, etc.
Tryptamines	Psilocybin, Psilocin, DMT, αMT, 5-MeO-DALT, DiPT, 5-MeO-DiPT, Ibogaine, etc.
Phenethylamines	Mescaline, 2C-series and their derivatives, DOx series and their derivatives, tetrahydro-diphenyl compounds like 2C-B-Fly, Bromo-DragonFly, etc.

Table 3 Classification of remaining hallucinogens

Category	Examples
Synthetic cannabinoids	
MDMA and MDMA-related drugs	
Dissociative anaesthetics	Ketamine, PCP (phencyclidine; aka 'angel dust', 'amp', 'embalming fluid', 'boat', 'zoom', 'belladonna', 'amoeba')
Psychoactive mushrooms	Amanita muscaria, Amanita pantherina
Herbal highs	Salvia divinorum

4.1 Pharmacology/Neuropharmacology

From a pharmacological point of view, classical hallucinogens share the ability to function as full agonists or partial agonists at 5-HT_2 receptors (particularly 5-HT_{2A} and/or other 5-HT_2 receptors; [21, 113, 118]). LSD display high affinity for various 5-HT receptor subtypes including 5-HT_{1B}, 5-HT_{1D}, 5-HT_7, 5-HT_6, and 5-HT_{2A} [119, 120]. However, the pharmacological potency of LSD and other hallucinogens mainly depends on their affinity for the 5-HT_{2A} receptors [21, 113], with mescaline characterized by the lowest potency and LSD being the most potent hallucinogen. Importantly, the NBOMe and 'Fly' series of drugs, which recently emerged in the online recreational market, are considered to possess a higher potency compared to remaining hallucinogens [2].

A number of studies have implicated non-5-HT receptors in the actions of hallucinogens, including sigma-1, NMDA, μ-opioid, muscarinic, and DA subtypes [120–123].

4.2 Prevalence of Use

The overall prevalence of use of hallucinogenic mushrooms and LSD in Europe has been generally low and stable for a number of years. Users of hallucinogens are typically young adults (15–34 years) who use a wide repertoire of other 'clubdrugs' [124]. Moreover, from the anecdotal discussions on online fora/blogs, 'psychonauts' are more likely to use hallucinogens, especially in combination with novel stimulants [124]. National surveys report last-year prevalence estimates of less than 1% for hallucinogens [2]. The use of hallucinogens appears to be of particular concern among youngsters. According to the 'Monitoring the Future Study' [125], the lifetime prevalence in hallucinogen use is about 2%, 5%, and 6.30%, respectively, among the 8th, 10th, and 12th graders. In particular, lifetime prevalence of LSD use ranged from 1.10% to 3.70%. Moreover, according to the National Survey on Drug Use and Health [126], the prevalence trend for hallucinogens use was around 2.50% (aged 12–17), 16.60% (aged 18–25), and 16.20% (aged >26).

4.3 Administration

Hallucinogens are typically ingested orally, sometimes through small blotter paper saturated with drug (i.e. 'tabs') held in the mouth to allow absorption through the oral mucosa. Other routes of administrations include insufflation, smoking, rectal, and injection (intravenous and intramuscular). The route of administration may influence the effects, their onset, and duration.

4.4 Desired and Adverse Effects

In some cultures, and particularly where shamanic practices are popular, hallucinogens are important tools used to enhance spiritual experiences [127, 128]. The experience of an altered state of consciousness facilitates a belief that a person is able to see beyond the boundaries of reality.

Hallucinogens are usually taken in combination with stimulant drugs, cannabis, cocaine, amphetamines, alcohol, prescribed drugs, and other NPS (www.erowid.org, accessed 16 Nov 2015). In general, reported effects include euphoria, mild stimulation, enhanced appreciation of music/light, visually appealing distortions, intensification of sensual/sexual feelings, and altered sense of time and space. However, each hallucinogen has distinct characteristics, and a large variability in multiple sensory and emotional dimensions has been described [123]. Furthermore, non-pharmacological variables such as expectations, personality, environment, and emotional state appear to have a much greater influence on the effects of hallucinogens than with other drugs [129]. The effects of hallucinogens are usually dose-dependent, highly context-dependent, and user-specific [119, 130]. Some of these molecules are ingested in order to reach a particular 'religious'/'spiritual'/introspective/meditation state [128].

Short-term hallucinogenic effects are associated with an increase in blood pressure, heart rate, body temperature, dizziness, sleeplessness, loss of appetite, dry mouth, sweating, impulsiveness, rapid emotional shifts from fear to euphoria, numbness, weakness, and tremors. Long-term effects may include the onset of a persistent psychosis (i.e. visual disturbances, disorganized thinking, paranoia, and mood disturbances) and/or of a hallucinogen-persisting perception disorder (HPPD). HPPD is a syndrome characterized by prolonged or recurring perceptual symptoms, reminiscent of acute hallucinogenic effects. The symptomatology mainly includes visual disorders (i.e. geometric pseudo-hallucinations, halos, flashes of colours/lights, motion-perception deficits, after-images, micropsy, and more acute awareness of floaters), at times associated with depressive symptoms and thought disorders.

4.5 Lysergamides

Lysergamides are polycyclic amides which possess both the phenethylamine and tryptamine groups embedded within their structure. Amongst these, LSD (aka '*acid*', '*A-tab*', '*Blotter*', '*Geltabs*', '*Windowpane*', and '*Microdots*') is the most popular [131]. Its effects appear approximately 1 h after oral ingestion. LSD is currently a Schedule I drug under the Controlled Substances Act of 1970 [132].

A range of LSD derivatives have recently become Class A drugs in the UK. These molecules include: LSZ (lysergic acid 2,4-dimethylazetidide); 1-P-LSD (1-propionyl-D-lysergic acid diethylamide hemitartrate) [133]; LSA (D-lysergic acid amide, sold as '*Morning Glory seeds*'/'*Hawaiian baby wood rose seeds*'; [134]); ALD-52 (1-acetyl-*N*,*N*-diethyllysergamide, previously known as '*Orange Sunshine Acid*'); ETH-LAD (6-ethyl-6-nor-lysergic acid diethylamide); PRO-LAD (6-propyl-6-nor-lysergic acid diethylamide); and AL-LAD (6-allyl-6-nor-lysergic acid diethylamide) [2]. They produce effects similar to those of LSD and present with a similar pharmacological profile by acting on 5-HT_{2A}-receptors, even though they may possess different potencies, onset, and duration of effects.

4.6 Tryptamines

These include a number of different substances that are derivatives of the controlled tryptamines and are designed to have predominantly hallucinogenic effects [113, 119]. Some tryptamines are compounds naturally produced by humans; these include 5-HT and melatonin (*N*-acetyl-5-methoxytryptamine), as well as a range of psychoactive methylated tryptamines whose biological functions remain unclear. These include: bufotenin (*N*,*N*-dimethylserotonin, or 5-OH-DMT, or 5-hydroxy-*N*,*N*-dimethyltryptamine), DMT (*N*,*N*-dimethyltryptamine), and 5-MeO-DMT [135–137]. Many of the tryptamines are psychoactive hallucinogens naturally found in plants, fungi, or animals: DMT and 5-MeO-DMT have been identified in some *Delosperma* plant species, psilocin (4-OH-DMT, 4-hydroxy-*N*,

N-dimethyltryptamine) and psilocybin (O-phosphoryl-4-hydroxy-*N,N*-dimethyl-tryptamine) are found in certain fungi (aka '*magic shrooms*' or '*mushies*'), whilst bufotenin and 5-hydroxy-indolethylamines are common constituents of the venoms from members of the genera *Hyla*, *Leptodactylus*, *Rana*, and *Bufo* [138–142]. Finally, other tryptamines have been synthesized for pharmaceutical/medical purposes (e.g. sumatriptan and zolmitriptan for migraine) [143]. Details on the synthesis and effects of 55 tryptamine-related compounds have been made available [123].

The use of psilocybin became widespread in the late 1950s in the USA [144], but synthetic tryptamines appeared in the illicit drug markets only during the 1990s. Tryptamine derivatives dominated the online drug market until 2007, when they were listed as 'narcotics' or 'designated substances' and were quickly replaced by cathinones, phenethylamines, and piperazines [2, 145].

Nevertheless, according to recent reports [146–148], a range of novel tryptamines continue to appear on the online drug market as 'legal highs'; these include: 5-MeO-DALT (*N,N*-diallyl-5-methoxytryptamine), AMT (alpha-methyltryptamine), 5-MeO-AMT (5-methoxy-α-methyltryptamine), 4-HO-DALT (*N,N*-Diallyl-4-hydroxytryptamine), and 5-IT (5-(2-aminopropyl)indole) [2, 149]. In 2013, around 2% of seizures of NPS reported to the EU Early Warning System were classified as tryptamine compounds [2].

DMT, psilocin, bufotenin, and DET (*N,N*-diethyltryptamine) were originally listed as Class A drugs in Part 1(a) of Schedule 2 of [150]. As esters and ethers of Class A drugs are also controlled, 5-MeO-DMT (5-methoxy-*N,N*-dimethyltryptamine) and psilocybin (the phosphate ester of psilocin) were also put under control (Drugs Act [150] c.17) together with AMT, 5-MeO-DALT, and etryptamine (alpha-ethyl tryptamine; AET).

Tryptamines are likely to be metabolized by monoamine oxidase (MAO) [151, 152]. The predominant clinical effects produced by tryptamine exposure consist of visual hallucinations, mediated by agonism at $5HT_{1A}$, $5HT_{2A}$, and $5HT_{2C}$ receptors [113, 120, 153], although they exhibit smaller levels of selectivity and affinity for $5HT_{2A}$ receptors if compared to phenethylamines [119]. Other transporters/receptors implicated in the effects of tryptamines include the vesicular monoamine transporter 2 (VMAT2), σ-1 receptors, serotonin transporter (SERT), and trace amine-associated receptors (TAAR) [151, 153–156].

Visual hallucinations are common for all tryptamines, whilst for DiPT (*N,N*-diisopropyltryptamine) auditory hallucinations are predominantly reported ([123]; www.erowid.org). Other clinical effects vary depending on the index compound and may include alterations in sensory perception, intensification of colours, distortion of body image, depersonalization, marked mood lability, euphoria, relaxation, entactogenic properties, and anxiety, ranging from mild apprehension to panic disorder ([123]; www.erowid.org; [157]). Untoward effects include agitation, tachyarrhythmia, hyperpyrexia, serotonergic neurotoxicity, and death [151]. There are small numbers of confirmed post-mortem toxicology reports on tryptamines rising from 1 in 2009 to 4 in 2013. AMT has the highest number of fatalities recorded in the UK to date, with 4 reported in 2012 and 3 in 2013 (NPSAD 2014).

Natural tryptamines are commonly available in preparations of dried or brewed mushrooms, while tryptamine derivatives are usually sold in capsules, tablets, powders, or in liquid formulations. Tryptamines are generally swallowed, sniffed, smoked, or injected. Street names for some tryptamines include 'Foxy-Methoxy' (5-MeO-DIPT, 5-methoxy-*N*,*N*-diisopropyltryptamine), 'alpha-O', 'alpha' and 'O-DMS' (5-MeO-AMT), '5-MEO' (5-MeO-DMT), 'spice' or 'changa' (DMT), 'T-9' (DET) (www.erowid.org; www.bluelight.com, accessed 22 Nov 2015).

DMT (*N*,*N*-dimethyltryptamine, aka '*Dimitri*') was first synthesized in 1931, but occurs naturally as well in many species of plants which are used in several South American shamanic practices. Ayahuasca and Yagé are decoctions that include DMT-containing plants together with *B. caapi*, containing a monoamine oxidase inhibitor which in turn allows DMT to be orally bioavailable. DMT has strong psychedelic properties, producing effects similar to those of LSD. Since DMT is inactive after oral administration, it is usually injected, snorted, or smoked. It can produce powerful entheogenic experiences, intense visual hallucinations, and euphoria. If DMT is smoked, peak effects last for a short period of time (5–30 min). The onset after inhalation is very fast (less than 45 s) and peak effects are reached within a minute.

Bufotenin (also known as cinobufotenine, mappin, *N*,*N*-dimethylserotonin, DM5-HT, 5-OH-DMT), a positional isomer of psilocin, is found in the skin of various species of the toad *Bufo* genus, in mushrooms such as Amanita, in plants such as *Anadenanthera peregrina* and *Piptoderma peregrina* [158]. Its psychoactivity is mainly due to its enzymatic conversion to 5-MeO-DMT [159]. It acts on 5-HT_{2A} receptors, as suggested by in vitro studies [160]. Its use has been reported to occur mainly by smoking crystals obtained by drying the liquid extracted from the frogs. It is also reported an intravenous use of bufotenin (www.bluelight.com).

Recently identified synthetic tryptamines include MET (*N*-methyl-*N*-ethyltryptamine, structurally related to DMT), and associated ring-substituted substances such as 4-AcO-MET (4-acetoxy-*N*-methyl-*N*-ethyltryptamine), 4-OH-MET (4-hydroxy-*N*-methyl-*N*-ethyl tryptamine), and 5-MeO-MET (*N*-ethyl-5-methoxy-*N*-methyltryptamine); 5-MeO-DALT; 5-MeO-MALT (*N*-[2-(5-methoxy-1H-indol-3-yl) ethyl]-*N*-methyl-prop-2-en-1-amine); 2-MeO-DMT (*N*,*N*-dimethyl-2-(2-methyl-1H-indol-3-yl)ethanamine); 5-MeO-EIPT (*N*-ethyl-*N*-isopropyl-5-methoxytryptamine); 5-IT (5-(2-aminopropyl)indole); 4-AcO-DPT (4-acetoxy-*N*,*N*-dipropyltryptamine); AMT; 5-MeO-NiPT (*N*-[2-(5-methoxy-1H-indol-3-yl)ethyl]-propan-2-amine); 5-MeO-AMT; and many others [2].

4.7 Hallucinogenic Phenethylamines

In the current drug market, the most recent and popular phenethylamines with hallucinogenic properties include both the so-called 2C series (i.e. 2C-B/'Nexus'; 2C-I; 2C-E) [161–163] and the NBOMe series drugs [2]. In 2013, around 8% of seizures of NPS reported to the EU Early Warning System were phenethylamines [2]. Overall, a range of serotonergic and sympathomimetic toxic effects can be observed after intake of these drugs, including tachycardia, hypertension, metabolic acidosis, convulsions, coma, rhabdomyolysis, mydriasis, vomiting/diarrhoea, and thrombocytopenia, while acute renal failure and hyperthermia are a reason for particular concern [9, 91, 164].

The 2C-hallucinogens are phenethylamines with methoxy substitutions at the 2- and 5-positions, structurally related to mescaline, producing psychological and somatic effects common to serotonergic hallucinogens. Most 2C-series compounds are usually ingested as MDMA substitutes, and show affinity for 5-HT_{2A} receptors [161–164], whilst some of them inhibit the reuptake of DA/NE/5-HT as well (for a thorough review, see [164]).

2C-B (aka '*Nexus*'/'*Bees*'/'*Venus*'/'*Bromo Mescaline*'/'*BDMPEA*') is a ring-substituted phenethylamine which was first synthesized by Shulgin in 1974 and then marketed as an MDMA replacement after its schedule in 1985. Its structural features are associated with stimulant and hallucinogenic activities. It is considered to be somewhat 'smoother' than LSD, being less frequently associated with panic attacks/'bad trips' at recreational dosages. Typically reported effects include intense body sensations (e.g. pleasurable energy, 'sense of being in the body', and unpleasant 'buzzing'). Untoward effects include gastrointestinal distress, anxiety, frightening thoughts, and visual perceptual disturbances. Other popular compounds of this class include 2C-D (2,5-dimethoxy-4-methylphenethylamine), 2C-E (2,5-dimethoxy-4-ethylphenethylamine), 2C-N (2,5-dimethoxy-4-nitrophenethylamine), 2C-H (2,5-dimethoxyphenethylamine), and *N*-ethyl-2C-B (*N*-ethyl-2C-B) [2].

All NPS belonging to the so-called fly series (the term is used to connote their molecular structure resembling an insect), and particularly 2C-B-Fly (8-bromo-2,3,6,7-benzo-dihydro-difuran-ethylamine), and 'Bromo-DragonFly'/'B-fly' (1-(4-Bromofuro[2,3-*f*] [1]benzofuran-8-yl)propan-2-amine), have been described as powerful and long lasting drugs, with effects lasting for up to 3 days and including: hallucinations, mood elevation, paranoid ideation, confusion, anxiety, and flashbacks [165].'B-fly' has been associated with a number of acute intoxications and fatalities in the EU [165].

NBOMe compounds, originally synthesized in 2003 to aid in the mapping of 5-HT receptors in the brain, started to be used recreationally in 2010 [9, 71]. The NBOMe market has recently increased in parallel with the declining availability of LSD. These molecules produce similar effects to LSD, but possessing a higher potency [2, 148]. NBOMes have been detected in Europe, North America, and Japan [148, 166, 167]. A growing number of related fatalities and hospitalizations, both in the UK and internationally [2, 168–171], have been described.

The 25X-NBOMe series include the *N*-methoxybenzyl substituted 2C-class of hallucinogens, initially synthetized for research purposes [172]. They act at a range of receptors, although they show a significantly higher affinity at the 5-HT_{2A} receptor level [173]. A few of them are amphetamine analogues; they are typically sold in 'blotters', powder or in liquid form. They are labelled as 'Bomb', 'Smiles' and are distributed, like LSD, in a colourful and painted bottle [174]. Their main routes of administration include sub-lingual/buccal and insufflation. The effects start within 15 min after oral intake, with the onset being even faster after insufflation. Desired effects include mental and physical stimulation, increase in associative and creative thinking, increased awareness, and appreciation of music, spiritual experiences, and euphoria (www.erowid.org).

There are several reports of related acute toxicity events, most being related to 25I-NBOMe (2-(4-iodo-2,5-dimethoxyphenyl)-*N*-[(2-methoxyphenyl)methyl] ethanamine; [175–177]). 25I-NBOMe shares some structural similarity with 2C-I (2,5-dimethoxy-4-iodophenethylamine). It is a partial agonist at 5-HT_{2A} receptors [9, 70]. Its effects are powerful and unpredictable, including mood lift, euphoria, colour shifts, brightening, erotic/sexual thoughts and sensations, feelings of love/ empathy, mydriasis, confusion, nausea, insomnia, paranoia, vasoconstriction, peripheral numbness, and swelling (www.erowid.org). A number of 25I-NBOMe-related fatalities and hospitalizations have been reported (Shanks et al. 2014; [2, 175–177]). Overall, the most commonly reported symptoms of acute toxicity include tachycardia (85%), hypertension (65%), agitation/aggression (85%), and seizures (40%). The most common abnormalities in laboratory tests include elevated level of creatinine kinase (45%), leucocytosis (25%), and hyperglycaemia (20%; [178]).

Other molecules of this class include 25B-NBOMe (2-(4-bromo-2,5-dimethoxyphenyl)-*N*-[(2-methoxyphenyl)methyl]ethanamine) and 25C-NBOMe ((2-(4-chloro-2,5-dimethoxyphenyl)-*N*-[(2-methoxyphenyl)methyl]ethanamine)) [174, 179]. A case of fatal intoxication following the use of 25B-NBOMe has been reported as well [180]. The effects of 25C-NBOMe (aka 'N-Bomb' or 'Pandora') include aggression, unpredictable violent episodes, dissociation, and anxiety.

5 Treatment Options: Physical and Psychiatric

5.1 Acute Management of Adverse Events

Consumers of NPS may present to the Accident and Emergency Departments without providing information about the substances(s) ingested and it is likely that standard drug tests will show negative results. Indeed, it is problematic to draft a good-for-all NPS treatment/management plan to cope with the behavioural and psychopathological disturbances related to the intake of the virtually few hundreds of substances currently available [70, 91]. Those individuals presenting with less severe symptoms should be assessed and managed as for any other users of psychoactive substances, and may simply need reassurance, support, and observation. When a medication may be needed, given the complex/unknown pharmacology of the substances arguably ingested, benzodiazepines may be the agents of choice [9, 70]. They may, however, need frequent re-dosing to achieve adequate sedative effect, and this may be a problem whilst in presence of alcohol. Benzodiazepines may be particularly useful for the treatment of the stimulant/synthetic cathinone-related agitation [181–183]. This approach may be useful as well to stop seizures [184, 185]. Where patients cannot be controlled with benzodiazepines alone, propofol and/or antipsychotics may be considered, although drugs such as haloperidol, olanzapine, or ziprasidone can lower seizure thresholds, limit the levels of heat dissipation [185], and contribute to dysrhythmias [186]. People with underlying cardiac, neurological, and psychiatric conditions, especially those on medication, are likely to be at greatest risk of serious adverse events [187]. For coronary ischaemia following the use of stimulants, consideration should be given to conventional treatment with nitroglycerin, morphine, and antiplatelet drugs [186]. Conversely, beta-blockers should be avoided as they could exacerbate symptoms, including worsening coronary vasoconstriction and hypertension [186]. Hyperthermia needs to be evaluated and treated aggressively, and this typically involves cooling measures and i.v. fluid administration for rhabdomyolysis concern. Appropriate sedation paralysis and assisted ventilation may at times be needed [185]. The intake of serotonergic drugs (e.g. phenethylamines, hallucinogens, NBOMe compounds, etc.) may be associated with the occurrence of the serotonin syndrome, to be managed using both benzodiazepines and cyproheptadine [9].

5.2 Longer Term Therapeutic Psychological and Harm Reduction Approaches

Little is known about the potential neurotoxicity or long-term consequences of mephedrone misuse, so only common sense advice about the use of any psychoactive stimulant can been provided [187, 188]. This may include taking small dosages per session, avoiding regular use to delay developing tolerance, not using the drug in combination with other stimulants or large amounts of alcohol and other depressants, not insufflating/injecting the drug, remaining well hydrated when using the drug, and avoiding becoming overheated. Both a brief motivational intervention and appropriately adapted psychosocial intervention have been suggested to treat mephedrone addiction [187].

6 Concluding Remarks

The rapidly evolving NPS phenomenon represents a challenge for medicine, and especially so for emergency physicians and mental health professionals. Indeed, NPS intake is typically associated with the imbalance of a range of neurotransmitter pathways/receptors, and consequently with a significant risk of psychopathological disturbances [9]. Vulnerable subjects, including both children/adolescents and psychiatric patients, may be exposed to a plethora of pro drug web pages, from which unpublished/anecdotal levels of knowledge related to the NPS are typically provided by the 'e-psychonauts' (e.g. drug fora/blog communities' members; [124]).

Although current general population surveys suggest relatively low levels of NPS use, at least if compared with classical scheduled substances such as THC, cocaine, and heroin, this may change. Indeed, future studies should provide better levels of NPS-clinical pharmacological-related knowledge, so that better tailored management/treatment strategies and guidelines can be made available.

Because of the large range of medical and psychopathological issues associated with the NPS intake here described, it is crucial for health professionals to be aware of the effects and toxicity of NPS, and especially the most popular ones here discussed, e.g. SC, synthetic cathinones, and the most recent hallucinogenic drugs. Finally, future approaches should consider the role of web-based preventative strategies in targeting youngsters/vulnerable individuals at risk of approaching the NPS market.

Acknowledgements/Funding Statements This paper was supported in part by grants of the European Commission (Drug Prevention and Information Programme 2014-16, contract no. JUST/2013/DPIP/AG/4823, *EU-MADNESS project*). Further financial support was provided by the EU Commission-targeted call on cross border law enforcement cooperation in the field of drug trafficking – DG Justice/DG Migrations and Home Affairs (JUST/2013/ISEC/DRUGS/AG/6429) *Project EPS/NPS* (Enhancing Police Skills concerning Novel Psychoactive Substances; NPS).

References

1. UNODC (United Nations Office on Drugs and Crime) (2015) World Drug Report. https://www.unodc.org/documents/wdr2015/World_Drug_Report_2015.pdf. Accessed 30 Nov 2015
2. EMCDDA (European Monitoring Centre for Drugs and Drug Addiction) (2015) New psychoactive substances in Europe - an update from the EU Early Warning System. Publications Office of the European Union, Luxembourg. http://www.emcdda.europa.eu/publications/2015/new-psychoactive-substances. Accessed 15 Nov 2015
3. EMCDDA (European Monitoring Centre for Drugs and Drug Addiction) (2015) Synthetic cannabinoids in Europe. http://www.emcdda.europa.eu/topics/pods/synthetic-cannabinoids. Accessed 29 Nov 2015
4. Daly M (2013) Streets legal. Druglink 28:17

5. Shlosberg D, Zalsman G, Shoval G (2014) Emerging issues in the relationship between adolescent substance use and suicidal behavior. Isr J Psychiatry Relat Sci 51:262–267
6. Auwarter V, Dargan PI, Wood DM (2013) Synthetic cannabinoid receptor agonists. In: Dargan PI, Wood DM (eds) Novel psychoactive substances classification, pharmacology and toxicology. Elsevier, London, pp 317–343
7. Schifano F, Corazza O, Davey Z, Di Furia L, Farre M, Flesland L, Mannonen M, Pagani S, Peltoniemi T, Pezzolesi C, Scherbaum N, Siemann H, Skutle A, Torrens M, Van Der Kreeft P (2009) Psychoactive drug or mystical incense? Overview of the online available information on Spice products. Int J Cult Ment Health 2:137–144
8. Thomas BF, Wiley JL, Pollard GT, Grabenauer M (2014) Cannabinoid designer drugs: effects and forensics. In: Pertwee RG (ed) Handbook of cannabis. Oxford University Press, Oxford, pp 710–729
9. Schifano F, Orsolini L, Papanti GD, Corkery J (2015) Novel psychoactive substances of interest for psychiatry. World Psychiatry 14:15–26
10. Martinotti G, Lupi M, Acciavatti A, Cinosi E, Santacroce R, Signorelli MS, Bandini L, Lisi G, Quattrone D, Ciambrone P, Aguglia A, Pinna F, Calò S, Janiri L, di Giannantonio M (2014) Novel psychoactive substances in young adults with and without psychiatric comorbidities. Biomed Res Int. doi:10.1155/2014/815424
11. Naviglio S, Papanti D, Moressa V, Ventura A (2015) An adolescent with an altered state of mind. BMJ 350:h299
12. Vento AE, Martinotti G, Cinosi E (2014) Substance use in the club scene of Rome: a pilot study. Biomed Res Int 2014:617546
13. Appendino G, Minassi A, Taglialatela-Scafati O (2014) Recreational drug discovery: natural products as lead structures for the synthesis of smart drugs. Nat Prod Rep 31:880–904
14. Abdulrahim D, Bowden-Jones O, on behalf of the NEPTUNE Expert Group (2015) Overview of the effects and harms of club drugs. In: Guidance on the Management of Acute and Chronic Harms of Club Drugs and Novel Psychoactive Substances. Novel Psychoactive Treatment UK Network (NEPTUNE), London.
15. Papanti D, Orsolini L, Francesconi G, Schifano F (2014) 'Noids' in a nutshell: everything you (don't) want to know about synthetic cannabimimetics. Adv Dual Diagn 7:137–148
16. Brents LK, Prather PL (2014) The K2/Spice Phenomenon: emergence, identification, legislation and metabolic characterization of synthetic cannabinoids in herbal incense products. Drug Metab Rev 46:72–85
17. Pertwee RG, Howlett AC, Abood ME, Alexander SP, Di Marzo V, Elphick MR, Greasley PJ, Handsen HS, Kunos G, Mackie K, Mechoulam R, Ross RA (2010) International Union of Basic and Clinical Pharmacology. LXXIX. Cannabinoid receptors and their ligands: beyond CB1 and CB2. Pharmacol Rev 62:588–631
18. Pertwee RG (2009) Receptors and channels targeted by synthetic cannabinoid receptor agonists and antagonists. Curr Med Chem 17:1360–1381
19. Fantegrossi WE, Moran JH, Radominska-Pandya A, Prather PL (2013) Distinct pharmacology and metabolism of K2 synthetic cannabinoids compared to Δ9-THC: mechanism underlying greater toxicity? Life Sci 97:45–54
20. Lewin AH, Seltzman HH, Carroll FI, Mascarella SW, Reddy PA (2013) Emergence and properties of spice and bath salts: a medicinal chemistry perspective. Life Sci 97:9–19
21. Halberstadt AL, Geyer MA (2011) Multiple receptors contribute to the behavioural effects of indoleamine hallucinogens. Neuropharmacology 61:364–381
22. Wells DL, Ott CA (2011) The new marijuana. Ann Pharmacother 45:414–417
23. Fisar Z (2010) Inhibition of monoamine oxidase activity by cannabinoids. Naunyn Schmiedebergs Arch Pharmacol 381:563–572
24. Ismail F (2002) Important fluorinated drugs in experimental and clinical use. J Fluor Chem 118:27–33
25. Wilkinson SM, Banister SD, Kassiou M et al (2015) Bioisosteric fluorine in the clandestine design of synthetic cannabinoids. Aust J Chem 68:4–8

26. Baggaley K (2015) Drugs by design: corrupt chemists tweak compounds faster than law enforcement can call them illegal. Sci News 187:22–25
27. Papanti D, Schifano F, Botteon G, Bertossi F, Mannix J, Vidoni D, Impagnatiello M, Pascolo-Fabrici E, Bonavigo T (2013) "Spiceophrenia': a systematic overview of 'Spice'-related psychopathological issues and a case report. Hum Psychopharmacol 28:379–389
28. Winstock AR, Barratt MJ (2013) Synthetic cannabis: a comparison of patterns of use and effect profile with natural cannabis in a large global sample. Drug Alcohol Depend 1:106–111
29. Spaderna M, Addy PH, D'Souza DC (2013) Spicing things up: synthetic cannabinoids. Psychopharmacology (Berl) 228:525–540
30. Wessinger WD, Moran JH, Seely KA (2015) Synthetic cannabinoid effects on behavior and motivation. In: Campolongo P, Fattore L (eds) Cannabinoid modulation of emotion, memory, and motivation. Springer, New York, pp 205–224
31. Wood DM, Dargan PI (2012) Mephedrone (4-methylmethcathinone): what is new in our understanding of its use and toxicity. Prog Neuropsychopharmacol Biol Psychiatry 39:227–233. doi:10.1016/j.pnpbp.2012.04.020
32. Wood DM, Dargan PI (2012) Novel psychoactive substances: how to understand the acute toxicity associated with the use of these substances. Ther Drug Monit 34:363–367
33. Hermanns-Clausen M, Kneisel S, Szabo B, Auwärter V (2013) Acute toxicity due to the confirmed consumption of synthetic cannabinoids: clinical and laboratory findings. Addiction 108:534–544
34. Winstock AR, Barratt MJ (2013) The 12-month prevalence and nature of adverse experiences resulting in emergency medical presentations associated with the use of synthetic cannabinoid products. Hum Psychopharmacol 28:390–393
35. Freeman MJ, Rose DZ, Myers MA (2013) Ischemic stroke after use of the synthetic marijuana 'Spice'. Neurology 81:2090–2093
36. Hopkins CY, Gilchrist BL (2013) A case of cannabinoid hyperemesis syndrome caused by synthetic cannabinoids. J Emerg Med 45:544–546
37. Louh IK, Freeman WD (2014) A 'spicy' encephalopathy: synthetic cannabinoids as cause of encephalopathy and seizure. Crit Care 18:553
38. Mir A, Obafemi A, Young A et al (2011) Myocardial infarction associated with use of the synthetic cannabinoid K2. Pediatrics 128:e1622–e1627
39. Rose DZ, Guerrero WR, Mokin MV (2015) Haemorrhagic stroke following use of the synthetic marijuana "spice". Neurology 85:1177–1179
40. Tait RJ, Caldicott D, Mountain D, Hill SL, Lenton S (2015) A systematic review of adverse events arising from the use of synthetic cannabinoids and their associated treatment. Clin Toxicol 54:1–13. doi:10.3109/15563650.2015.1110590
41. CDC (2012) Acute kidney injury associated with synthetic cannabinoid use – multiple states. Centers for Disease Control and Prevention (CDC). MMWR Morb Mortal Wkly Rep 62:93–98
42. Kasper AM, Ridpath AD, Arnold JK, Chatham-stephens K, Morrison M, Olayinka O, Parker C, Galli R, Cox R, Pteacely N, Anderson J, Kyle PB, Gerona R, Martin C, Schier J, Wolkin A, Dobbs T (2015) Severe illness associated with reported use of synthetic cannabinoids - Mississippi, April 2015. MMWR Morb Mortal Wkly Rep 64:1121–1122
43. Sheikh A, Lukšič M, Ferstenberg R, Culpepper-Morgan JA (2014) SPICE/K2 synthetic Marijuana-induced toxic hepatitis treated with N-acetylcysteine. Am J Case Rep 15:584–588
44. Spengler E, LaBrecque D (2014) A case of acute liver failure associated with synthetic cannabis use. American College of Gastroenterology Annual Scientific Meeting Abstracts, Philadelphia
45. Sweeney B, Talebi S, Toro D, Gonzalez K, Menoscal JP, Shaw R, Hassen GW (2015) Hyperthermia and severe rhabdomyolysis from synthetic cannabinoids. Am J Emerg Med 34:121.e1-2; doi:10.1016/j.ajem.2015.05.052

46. Tyndall JA, Gerona R, De Portu G et al (2015) An outbreak of acute delirium from exposure to the synthetic cannabinoid AB-CHMINACA. Clin Toxicol 3:950–956. doi:10.3109/15563650.2015.1100306
47. Alhadi S, Tiwari A, Vohra R (2013) High times, low SATS: diffuse pulmonary infiltrates associated with chronic synthetic cannabinoid use. J Med Toxicol 9:199–206
48. Gunderson EW, Haughey HM, Ait-Daoud N, Joshi AS, Hart CL (2012) 'Spice' and 'K2' herbal highs: a case series and systematic review of the clinical effects and biopsychosocial implications of synthetic cannabinoid use in humans. Am J Addict 21:320–326
49. Froberg BA, Bauer BD (2012) Pneumorachis, pneumomediastinum, and subcutaneous emphysema after synthetic cannabinoid use. Clin Toxicol 50:705
50. Macfarlane V, Christie G (2015) Synthetic cannabinoid withdrawal: a new demand on detoxification services. Drug Alcohol Rev 34:147–153
51. Di Forti M, Marconi A, Carra E, Fraietta S, Trotta A, Bonomo M, Bianconi F, Gardner-Sood P, O'Connor J, Russo M, Stilo SA, Margues TR, Mondelli V, Dazzan P, Pariante C, David AS, Gaughran F, Atakan Z, Iyegbe C, Powell J, Morgan C, Lynskey M, Murray RM (2015) Proportion of patients in south London with first-episode psychosis attributable to use of high potency cannabis: a case-control study. Lancet Psychiatry 2:233–238
52. Di Forti M, Morgan C, Dazzan P, Pariante C, Mondelli V, Marques TR, Handley R, Luzi S, Russo M, Paparelli A, Butt A, Stilo SA, Wiffen B, Powell J, Murray RM (2009) High-potency cannabis and the risk of psychosis. Br J Psychiatry 195:488–491
53. Brakoulias V (2012) Products containing synthetic cannabinoids and psychosis. Aust N Z J Psychiatry 46:281–282
54. Glue P, Al-Shaqsi S, Hancock D, Gale C, Strong B, Schep L (2013) Hospitalisation associated with use of the synthetic cannabinoid K2. N Z Med J 126:18–23
55. Celofiga A, Koprivsek J, Klavz J (2014) Use of synthetic cannabinoids in patients with psychotic disorders: case series. J Dual Diagn 10:168–173
56. Oluwabusi OO, Lobach L, Akhtar U, Youngman B, Ambrosini PJ (2012) Synthetic cannabinoid-induced psychosis: two adolescent cases. J Child Adolesc Psychopharmacol 22:393–395
57. Ustundag MF, Ozhan Ibis E, Yucel A, Ozcan H (2015) Synthetic cannabis-induced mania. Case Rep Psychiatry 2015:310930. doi:10.1155/2015/310930
58. Lászik A, Törő K, Vannai M, Sára-Klausz g, Kócs t, Farkas R, Keller É, Róna K (2015) Self inflicted fatal injuries in association with synthetic cannabinoid abuse. 24th International Meeting on Forensic Medicine Alpe-Adria-Pannonia, AAP 2015 Programme and Abstract Book
59. Patton AL, Chimalakonda KC, Moran CL, McCain KR, Radominska-Pandya A, James LP, Kokes C, Moran JH (2013) K2 toxicity: fatal case of psychiatric complications following AM2201 exposure. J Forensic Sci 58:1676–1680. doi:10.1111/1556-4029.12216
60. Rosenbaum CD, Carreiro SP, Babu KM (2012) Here today, gone tomorrow…and back again? A review of herbal marijuana alternatives (K2, Spice), synthetic cathinones (bath salts), kratom, Salvia divinorum, methoxetamine, and piperazines. J Med Toxicol 8:5–32. doi:10.1007/s13181
61. Shanks KG, Dahn T, Terrell AR (2012) Detection of JWH-018 and JWH-073 by UPLC-MS-MS in postmortem whole blood casework. J Anal Toxicol 36:145–152. doi:10.1093/jat/bks013
62. Trecki J, Gerona RR, Schwartz MD (2015) Synthetic cannabinoid-related illnesses and deaths. N Engl J Med 373:103–107
63. Rickli A, Hoener MC, Liechti ME (2015) Monoamine transporter and receptor interaction profiles of novel psychoactive substances: para-halogenated amphetamines and pyrovalerone cathinones. Eur Neuropsychopharmacol 25:365–376. doi:10.1016/j.euroneuro.2014.12.012
64. McElrath K, O'Neill C (2011) Experiences with mephedrone pre-and post-legislative controls: perceptions of safety and sources of supply. Int J Drug Policy 22:120–127. doi:10.1016/j.drugpo.2010.11.001

65. De Felice LJ, Glennon RA, Negus SS (2014) Synthetic cathinones: chemical phylogeny, physiology, and neuropharmacology. Life Sci 97:20–26. doi:10.1016/j.lfs.2013.10.029
66. Baumann MH, Ayestas MA Jr, Partilla JS, Sink JR, Shulgin AT, Daley PF, Brandt SD, Rothman RB, Ruoho AE, Cozzi NV (2012) The designer methcathinone analogs, mephedrone and methylone, are substrates for monoamine transporters in brain tissue. Neuropsychopharmacology 37:1192–1203. doi:10.1038/npp.2011.304
67. Baumann MH, Partilla JS, Lehner KR, Thorndike EB, Hoffman AF, Holy M et al (2013) Powerful cocaine-like actions of 3,4-methylenedioxypyrovalerone (MDPV), a principal constituent of psychoactive 'bath salts' products. Neuropsychopharmacology 38:552–562. doi:10.1038/npp.2012.204
68. Simmler LD, Buser TA, Donzelli M, Schramm Y, Dieu LH, Huwyler J, Chaboz S, Hoener MC, Liechti ME (2013) Pharmacological characterization of designer cathinones in vitro. Br J Pharmacol 168:458–470. doi:10.1111/j.1476-5381.2012.02145.x
69. Gregg RA, Rawls SM (2014) Behavioral pharmacology of designer cathinones: a review of the preclinical literature. Life Sci 97:27–30. doi:10.1016/j.lfs.2013.10.033
70. Schifano F (2015) Novel psychoactive substances (NPS): clinical and pharmacological issues. Drugs Alcohol Today 15:21–27. doi:10.1108/DAT-10-2014-0035
71. Schifano F (2014) Novel psychoactive substances also known as 'legal highs'. In: Davies SC (ed) Annual report of the Chief Medical Officer 2013. Public mental health priorities: investing in the evidence. Department of Health, London. https://www.gov.uk/government/uploads/system/uploads/attachment_data/file/413196/CMO_web_doc.pdf (2014). Accessed 22 Nov 2015
72. Rothman RB, Baumann MH, Dersch CM, Romero DV, Rice KC, Carroll FI, Partilla JS (2001) Amphetamine-type central nervous system stimulants release norepinephrine more potently than they release dopamine and serotonin. Synapse 39:32–41
73. Iversen L, White M, Treble R (2014) Designer psychostimulants: pharmacology and differences. Neuropharmacology 87:59–65. doi:10.1016/j.neuropharm.2014.01.015
74. Farrè M, Papaseit E, Pérez-Mañá C, Pujadas M, Fonseca F, Torrens M, De La Torre R, Farre M (2014) Human pharmacology of mephedrone: a dose-finding pilot study. College on Problems of Drug Dependence Meeting, San Juan
75. Hyde JF, Browning E, Adams R (1928) Synthetic homologs of d,I-Ephedrine. J Am Chem Soc 50:2287–2292. doi:10.1021/ja01395a032
76. L'Italien YJ, Park H, Rebstock LC (1957) Methylaminopropiophenone compounds and methods for producing the same. US Patent 2:802–865
77. de Burnaga Sanchez SJ (1929) Sur un homologue de l'ephedrine. Bull Soc Chim Fr 45:284–286
78. Meltzer P, Butler D, Deschamps JR, Madras BK (2006) 1-(4-methylphenyl)-2-pyrrolidin-1-yl-pentan-1-one (Pyrovalerone) analogues: a promising class of monoamine uptake inhibitors. J Med Chem 49:1420–1432
79. Boehringer. Boehringer Ingelheim Patent for MDPV (1969) http://catbull.com/alamut/Bibliothek/Boehringer_MDPV_Patent.htm. Accessed 20 Nov 2015
80. Carroll FI, Blough BE, Abraham P, Mills AC, Holleman JA, Wolckenhauer SA, Decker AM, Landavazo A, McElroy KT, Navarro HA, Gatch MB, Forster MJ (2009) Synthesis and biological evaluation of bupropion analogues as potential pharmacotherapies for cocaine addiction. J Med Chem 52:6768–6781. doi:10.1021/jm901189z
81. Bondarev ML, Bondareva TS, Young R, Glennon RA (2003) Behavioral and biochemical investigations of bupropion metabolites. Eur J Pharmacol 474:85–93
82. Vento AE, Schifano F, Gentili F, Pompei F, Corkery JM, Kotzalidis GD et al (2013) Bupropion perceived as a stimulant by two patients with a previous history of cocaine misuse. Ann Ist Super Sanita 49:402–405. doi:10.4415/ANN_13_04_14
83. Deluca P, Davey Z, Corazza O, Di Furia L, Farre M, Flesland LH, Mannonen M, Majava A, Peltoniemi T, Pasinetti M, Pezzolesi C, Scherbaum N, Siemann H, Skutle A, Torrens M, van der Kreeft P, Iversen E, Schifano F (2012) Identifying emerging trends in recreational drug

use; outcomes from the Psychonaut Web Mapping Project. Prog Neuropsychopharmacol Biol Psychiatry 39:221–226. doi:10.1016/j.pnpbp.2012.07.011

84. Coppola M, Mondola R (2012) 3,4-Methylenedioxypyrovalerone (MDPV): chemistry, pharmacology and toxicology of a new designer drug of abuse marketed online. Toxicol Lett 208:12–15. doi:10.1016/j.toxlet.2011.10.002
85. Prosser JM, Nelson LS (2012) The toxicology of bath salts: a review of synthetic cathinones. J Med Toxicol 8:33–42. doi:10.1007/s13181-011-0193-z
86. Dargan PI, Albert S, Wood DM (2010) Mephedrone use and associated adverse effects in school and college/university students before the UK legislation change. QJM 103:875–879. doi:10.1093/qjmed/hcq134
87. Corkery JM, Schifano F, Ghodse AH (2012) Mephedrone-related fatalities in the United Kingdom: contextual, clinical and practical issues. In: Gallelli L (ed) Pharmacology. InTech, Rijeka, Croatia, pp 355–380. doi:10.5772/32935. http://www.intechopen.com/books/pharmacology/mephedrone-related-fatalities-in-the-united-kingdom-contextual-clinical-and-practical-issues. Accessed 15 Nov 2015
88. Schifano F, Corkery J, Ghodse AH (2012) Suspected and confirmed fatalities associated with mephedrone (4-methylmethcathinone; 'meow meow') in the UK. J Clin Psychopharmacol 32:710–714. doi:10.1097/JCP.0b013e318266c70c
89. Corkery JM, Schifano F, Oyefeso A, Ghodse AH, Tonia T, Naidoo V, Button J (2011) 'Bundle of fun' or 'bunch of problems'? Case series of khat-related deaths in the UK. Drugs Educ Prev Policy 18:408–425. doi:10.3109/09687637.2010.504200
90. Loi B, Corkery JM, Claridge H, Goodair C, Chiappini S, Gimeno Clemente C, Schifano F (2015) Deaths of individuals aged 16–24 years in the UK after using mephedrone. Hum Psychopharmacol 30:225–232. doi:10.1002/hup.2423
91. Schifano F, Albanese A, Fergus S, Stair JL, Deluca P, Corazza O, Davey Z, Corkery J, Siemann H, Scherbaum N, Farre' M, Torrens M, Demetrovics Z, Ghodse AH; Psychonaut Web Mapping; ReDNet Research Groups (2011) Mephedrone (4-methylmethcathinone; 'meow meow'): chemical, pharmacological and clinical issues. Psychopharmacology (Berl) 214:593–602. doi:10.1007/s00213-010-2070-x
92. McCann UD, Wong DF, Yokoi F, Villemagne V, Dannals RF, Ricaurte GA (1998) Reduced striatal dopamine transporter density in abstinent methamphetamine and methcathinone users: evidence from positron emission tomography studies with [11C]WIN-35,428. J Neurosci 18:8417–8422
93. EMCDDA (European Monitoring Centre for Drugs and Drug Addiction) (2014) Perspectives on drugs: injection of synthetic cathinones. Perspectives on Drugs Series, Lisbon, 27 May. http://www.emcdda.europa.eu/topics/pods/synthetic-cathinones-injection. Accessed 15 Nov 2015
94. Giese C, Igoe D, Gibbons Z, Hurley C, Stokes S, McNamara S, Ennis O, O'Donnell K, Keenan E, De Gascun C, Lyons F, Ward M, Danis K, Glynn R, Waters A, Fitzgerald M, outbreak control team (2015) Injection of new psychoactive substance snow blow associated with recently acquired HIV infections among homeless people who inject drugs in Dublin, Ireland, 2015. Eurosurveill 20; issue 40; art 1. doi:10.2807/1560-7917.ES.2015.20.40.30036
95. Sellors K, Jones A, Chan B (2014) Death due to intravenous use of α-pyrrolidinopentiophenone. Med J Aust 201:601–603
96. EMCDDA (European Monitoring Centre for Drugs and Drug Addiction) (2015) Injection of synthetic cathinones. http://www.emcdda.europa.eu/topics/pods/synthetic-cathinones-injection. Accessed 29 Nov 2015
97. Gannon BM, Reichard EE, Fantegrossi WE (2014) Psychostimulant abuse and HIV infection: cocaine, methamphetamine, and "bath salts" cathinone analogues. Curr Addict Rep 1:237–242
98. Bourne A, Reid D, Hickson F, Rueda ST, Weatherburn P (2014) The Chemsex study: drug use in sexual settings among gay & bisexual men in Lambeth, Southwark & Lewisham. Sigma Research, London School of Hygiene & Tropical Medicine, London

99. Kirby T, Thornber-Dunwell M (2013) High-risk drug practices tighten grip on London gay scene. Lancet 381:101–102
100. McCall H, Adams N, Mason D, Willis J (2015) What is chemsex and why does it matter? It needs to become a public health priority. BMJ 351:h5790
101. Maskell PD, De Paoli G, Seneviratne C, Pounder DJ (2011) Mephedrone (4-methylmethcathinone)-related deaths. J Anal Toxicol 35:188–191
102. Torrance H, Cooper G (2010) The detection of mephedrone (4-methylmethcathinone) in 4 fatalities in Scotland. Forensic Sci Int 202:e62–e63. doi:10.1016/j.forsciint.2010.07.014
103. Warrick BJ, Wilson J, Hedge M, Freeman S, Leonard K, Aaron C (2012) Lethal serotonin syndrome after methylone and butylone ingestion. J Med Toxicol 8:65–68. doi:10.1007/s13181-011-0199-6
104. Lee D, Chronister CW, Hoyer J, Goldberger BA (2015) Ethylone-related deaths: toxicological findings. J Anal Toxicol 39:567–571. doi:10.1093/jat/bkv053
105. Dy I, Lankford S (2015) Bath salt thrombosis. J Clin Toxicol 5:230. doi:10.4172/2161-0495.1000230
106. EMCDDA (European Monitoring Centre for Drugs and Drug Addiction) (2015) EMCDDA–Europol Joint Report on a new psychoactive substance: 1-phenyl-2-(1-pyrrolidinyl)-1-pentanone (α-PVP). Publications Office of the European Union, Luxembourg. http://www.emcdda.europa.eu/attachements.cfm/att_242501_EN_TDAS15001ENN.pdf. Accessed 29 Nov 2015
107. Murray BL, Murphy CM, Beuhler MC (2012) Death following recreational use of designer drug "bath salts" containing 3,4-methylenedioxypyrovalerone (MDPV). J Med Toxicol 8:69–75. doi:10.1007/s13181-011-0196-9
108. Ojanperä IA, Heikman PK, Rasanen IJ (2011) Urine analysis of 3,4-methylenedioxypyrovalerone in opioid-dependent patients by gas chromatography-mass spectrometry. Ther Drug Monit 33(2):257–263. doi:10.1097/FTD.0b013e318208b693
109. Young AC, Schwarz ES, Velez LI, Gardner M (2013) Two cases of disseminated intravascular coagulation due to "bath salts" resulting in fatalities, with laboratory confirmation. Am J Emerg Med 31:445.e3-5. doi:10.1016/j.ajem.2012.05.032
110. Wikström M, Thelander G, Nyström I, Kronstrand R (2010) Two fatal intoxications with the new designer drug methedrone (4-methoxymethcathinone). J Anal Toxicol 34:594–598
111. NISRA (2015) Drug related and drug misuse deaths 2004–2014. Northern Ireland Statistics & Research Agency, Belfast. http://www.nisra.gov.uk/archive/demography/publications/drug_deaths/Drug_Tables_14.xls. Accessed 30 Nov 2015
112. Elliott S, Evans J (2014) A 3-year review of new psychoactive substances in casework. Forensic Sci Int 243:55–60. doi:10.1016/j.forsciint.2014.04.017
113. Fantegrossi WE, Murnane KS, Reissig CJ (2008) The behavioral pharmacology of hallucinogens. Biochem Pharmacol 75:17–33
114. Halpern JH, Suzuki J, Huertas PE, Passie T (2011) Hallucinogens. In: Addiction medicine. Springer, pp 1083–1098
115. Sinke C, Halpern JH, Zedler M, Neufeld J, Emrich HM, Passie T (2012) Genuine and drug-induced synesthesia: a comparison. Conscious Cogn 21:1419–1434
116. Chan RHA, Mendelson JE (2014) Hallucinogens. In: Madras B, Kuhar M (eds) The effects of drug abuse on the human nervous system. Elsevier
117. Jacob P, Shulgin AT (1994) Structure-activity relationships of classic hallucinogens and their analogs. NIDA Res Monogr 146:74–91
118. Glennon RA (1996) Classical hallucinogens. In: Schuster CR, Kuhar MJ (eds) Handbook of experimental pharmacology: pharmacological aspects of drug dependence. Springer, Basel, Switzerland, pp 343–371
119. Nichols DE (2004) Hallucinogens. Pharmacol Ther 101:131–181
120. Ray TS (2010) Psychedelics and the human receptorome. PLoS One 5, e9019
121. Bulling S, Schicker K, Zhang YW, Steinkellner T, Stockner T, Gruber CW, Boehm S, Freissmuth M, Rudnick G, Sitte HH, Sandtner W (2012) The mechanistic basis for

noncompetitive ibogaine inhibition of serotonin and dopamine transporters. J Biol Chem 287:18524–18534. doi:10.1074/jbc.M112.343681
122. Ede F, Szabo A, Winkelman MJ, Luna LE, McKenna DJ (2013) A possibly sigma-1 receptor mediated role of dimethyltryptamine in tissue protection, regeneration, and immunity. J Neural Transm 120:1295–1303
123. Shulgin A, Shulgin A (1997) TIHKAL: the continuation. Transform Press. http://www.erowid.org/library/books_online/tihkal. Accessed 15 Sept 2014
124. Orsolini L, Papanti GD, Francesconi G, Schifano F (2015) Mind navigators of chemicals' experimenters? A web-based description of e-psychonauts. Cyberpsychol Behav Soc Netw 18:296–300. doi:10.1089/cyber.2014.0486
125. Monitoring the Future Study (2014) Trends in prevalence of various drugs for 8th graders, 10th graders, and 12th graders. http://www.drugabuse.gov/drugs-abuse/hallucinogens. Accessed 20 Nov 2015
126. National Survey on Drug Use and Health: Trends in prevalence of various drugs for ages 12 or older, ages 12 to 17, ages 18 to 25, and ages 26 or older. http://www.drugabuse.gov/drugs-abuse/hallucinogens (2014). Accessed 20 Nov 2015
127. Heinrich C (2002) Magic mushrooms in religion and alchemy. Park Street Press, Rochester
128. Winkelman MJ, Roberts TB (2007) Psychedelic medicine: new evidence for hallucinogenic substances as treatments. Praeger Publishers, Westport
129. Studerus EA (2012) Prediction of psilocybin response in healthy volunteers. PLoS One 7, e30800
130. Gable RS (2007) Risk assessment of ritual use of oral dimethyltryptamine (DMT) and harmala alkaloids. Addiction 102:24–34
131. Hoffman A (1980) LSD: my problem child. McGraw-Hill, London
132. US Department of Justice, Drug Enforcement Administration. LSD in the United States. http://www.usdoj.gov/dea/pubs/lsd/toc.htm. Accessed 13 Nov 2015
133. Brandt SD, Kavanagh PV, Westphal F, Stratford A, Elliott SP, Hoang K, Wallach J, Halbertstadt AL (2015) Return of the lysergamides. Part I: analytical and behavioural characterization of 1-propionyl-d-lysergic acid diethylamide (1P-LSD). Drug Test Anal. doi:10.1002/dta.1884
134. Flodrops H, Houdon L, Harvet G, Enaud L, Renouil M (2006) Voluntary intoxication by D-lysergic acid amide (LSA) in a 15-year-old teenager via Internet: new fact of society. Arch Pediatr 13:1158–1159
135. Barker SA, Monti JA, Christian ST (1981) N,N-Dimethyltryptamine: an endogenous hallucinogen. Int Rev Neurobiol 22:83–110
136. Guichhait RB (1976) Biogenesis of 5-methoxy-N,N-dimethyltryptamine in human pineal gland. J Neurochem 26:187–190
137. Kärkkäinen J, Räisäinen M, Naukarinen H, Spoov J, Rimon R (1988) Urinary excretion of free bufotenin by psychiatric patients. Biol Psychiatry 24:441–446
138. Cimino G, De Stefano S (1978) Chemistry of Mediterranean Gorgonians. Simple indole derivatives from Paramuricea chamaeleon. Comptes Rendus Biochem Physiol Ser C 61:361–362
139. Collins M (2011) Some new psychoactive substances: precursor chemicals and synthesis-driven end-products. Drug Test Anal 3:404–416
140. Koike Y, Wada K, Kusano G, Nozoe S, Yokoyama K (1981) Isolation of psilocybin from Psilocybe argentipes and its determination in specimens of some mushrooms. Lloydia 44:362–365
141. Mckenna DJ, Towers GHN (1984) Biochemistry and pharmacology of tryptamines and beta-carbolines: a minireview. J Psychoactive Drugs 16:347–358
142. Schultes RE (1979) Hallucinogenic plants: their earliest botanical descriptions. J Psychedelic Drugs 11:13–24

143. Brandes JL, Kudrow D, Stark SR, O'Carroll CP, Adelman JU, O'Donnell FJ, Alexander WJ, Spruill SE, Barrett PS, Lener SE (2007) Sumatriptan-naproxen for acute treatment of migraine: a randomized trial. JAMA 297:1443–1454
144. Hoffmann A, Heim R, Brack A, Kobel H, Frey A, Ott H, Petrzilka T, Troxler F (1959) Psilocybin und Psilocin, zwei psychotrope Wirkstoffe aus mexikanischen Rauschpilzen. Helv Chim Acta 42:1557–1572
145. Sanders B, Lankenau SE, Bloom JJ, Hathazi D (2008) 'Research chemicals': tryptamine and phenylamine use among high-risk youth. SUM 43:389–402
146. UNODC (United Nations Office on Drugs and Crime) (2013) The challenge of new psychoactive substances. United Nations Office on Drugs and Crime, Vienna
147. UNODC (United Nations Office on Drugs and Crime) (2013) Early Warning Advisory on New Psychoactive Substances (NPS). October 2013 update
148. UNODC (United Nations Office on Drugs and Crime) (2014) Global synthetic drugs assessment. https://www.unodc.org/documents/scientific/2014_Global_Synthetic_Drugs_Assessment_web.pdf. Accessed 8 Dec 2015
149. Arunotayanun W, Dalley JW, Huang XP, Setola V, Treble R, Iversen L, Roth BL, Gibbons S (2013) An analysis of the synthetic tryptamines AMT and 5-MeO-DALT: emerging 'Novel Psychoactive Drugs'. Bioorg Med Chem Lett 23:3411–3415
150. The Misuse of Drugs Act 1971 (Modification) Order 2005. Office of Public Sector Information. Retrieved 23 September 2014
151. Dargan PI, Wood DM (2013) Novel psychoactive substances: classification, pharmacology and toxicology. Elsevier
152. Lessin AW, Long RF, Parkes MW (1965) Central stimulant actions of α-alkyl substituted tryptamine in mice. Br J Pharmacol 24:49–67
153. Pierce PA, Peroutka SJ (1989) Hallucinogenic drug interactions with neurotransmitter receptor binding sites in human cortex. Psychopharmacology 97:118–122
154. Cozzi NV, Gopalakrishnan A, Anderson LL, Feih JT, Shuldin AT, Daley PF, Ruoho AE (2009) Dimethyltryptamine and other hallucinogenic tryptamines exhibit substrate behavior at the serotonin uptake transporter and the vesicle monoamine transporter. J Neural Transm 116:1591–1599
155. Fontanilla D, Johannessen M, Hajipour AR, Cozzi NV, Jackson MB, Ruoho AE (2009) The hallucinogen N,N-dimethyltryptamine (DMT) is an endogenous sigma-1 receptor regulator. Science 323:934–937
156. Su TP, Hayashi T, Vaupel DB (2009) When the endogenous hallucinogenic trace amine N, N-dimethyltryptamine meets the sigma-1 receptor. Sci Signal 2:12
157. Sogawa C, Sogawa N, Tagawa J, Fujino A, Ohyama K, Asanuma M, Funada M, Kitayama S (2007) 5-Methoxy-N,N-diisopropyltryptamine (Foxy), a selective and high affinity inhibitor of serotonin transporter. Toxicol Lett 170:75–82
158. Lyttle T, Goldstein D, Gartz J (1996) Bufo. Toads and bufotenine: fact and fiction surrounding an alleged psychedelic. J Psychoactive Drugs 28:267–290
159. Shen HW, Jiang XL, Winter JC, Yu AM (2010) Psychedelic 5-methoxy-N,N-dimethyltryptamine: metabolism, pharmacokinetics, drug interactions, and pharmacological actions. Curr Drug Metab 11:659–666
160. Ujvary I (2014) Psychoactive natural products: overview of recent developments. Ann Ist Super Sanità 50:12–27
161. Ambrose JB, Bennett HD, Lee HS, Josephson SA (2010) Cerebral vasculopathy after 4-bromo-2,5-dimethoxyphenethylamine ingestion. Neurologist 16:199–202
162. Bosak A, LoVecchio F, Levine M (2013) Recurrent seizures and serotonin syndrome following "2C-I" ingestion. J Med Toxicol 9:196–198
163. Topeff JM, Ellsworth H, Willhite LA, Cole JB, Edwards EM (2011) A case series of symptomatic patients, including one fatality, following 2C-E exposure. Clin Toxicol 49:526
164. Nelson ME, Bryant SM, Aks SE (2014) Emerging drugs of abuse. Emerg Med Clin N Am 1–28. doi:10.1016/j.emc.2013.09.001

165. Corazza O, Schifano F, Farre M, Deluca P, Davey Z, Torrens M, Demetrovics Z, Di Furia L, Flesland L, Siemann H, Skutle A, Van Der Kreeft P, Scherbaum N (2011) Designer drugs on the internet: a phenomenon out-of-control? The emergence of hallucinogenic drug Bromo-Dragonfly. Curr Clin Pharmacol 6:125–129
166. DEA US (Drug Enforcement Administration) (2013) Midyear Report 2013. Office of Diversion Control: National Forensic Laboratory Information System. US Drug Enforcement Administration, Springfield, VA. Accessed 1 Oct 2015
167. Uchiyama N, Shimokawa Y, Matsuda S, Kawamura M, Kikura-Hanajiri R, Goda Y (2014) Two new synthetic cannabinoids, AM-2201 benzimidazole analog (FUBIMINA) and (4-methylpiperazin-1-yl)(1-pentyl-1H-indol-3-yl)methanone (MEPIRAPIM), and three phenethylamine derivatives, 25H-NBOMe 2,4,5-trimethoxybenzyl analog, 25B-NBOMe, and 2C-N-NBOMe, identified in illegal products. Forensic Toxicol 32:105–115
168. Caldicott DG, Bright SJ, Barratt MJ (2013) NBOMe – a very different kettle of fish. Med J Aust 199:322–323
169. Poklis JI, Devers KG, Arbefeville EF, Pearson JM, Houston E, Poklis A (2014) Postmortem detection of 25I-NBOMe [2-(4-iodo-2,5-dimethoxyphenyl)-N-[(2-methoxyphenyl)methyl] ethanamine] in fluids and tissues determined by high performance liquid chromatography with tandem mass spectrometry from a traumatic death. Forensic Sci Int 234:e14–e20
170. Ralston N, Davies I (2015) Teen jumps to his death after $1.50 drug hit 2013. http://www.smh.com.au/nsw/teen-jumps-to-his-death-after-150-drug-hit-20130606-2nrpe.html#ixzz35BP7Ymmx. Accessed 1 Oct 2015
171. Walterscheid JP, Phillips GT, Lopez AE, Gonsoulin MI, Chen HH, Sanchez LA (2014) Pathological findings in 2 cases of fatal 25I-NBOMe toxicity. Am J Forensic Med Pathol 35:20–25
172. Braden MR, Parrish JC, Naylor JC, Nichols DE (2006) Molecular interaction of serotonin 5-HT2A receptor residues Phe339(6.51) and Phe340(6.52) with superpotent N-benzyl phenethylamine agonists. Mol Pharmacol 70:1956–1964. doi:10.1124/mol.106.028720
173. Halberstadt AL, Geyer MA (2014) Effects of the hallucinogen 2,5-dimethoxy-4-iodophenethylamine (2C-I) and superpotent N-benzyl derivatives on the head twitch response. Neuropharmacology 77(February):200–207. doi:10.1016/j.neuropharm.2013.08.025
174. Wood DM, Sedefov R, Cunningham A, Dargan PI (2015) Prevalence of use and acute toxicity associated with the use of NBOMe drugs. Clin Toxicol (Phila) 53:85–92. doi:10.3109/15563650.2015.1004179
175. Kueppers VB, Cooke CT (2015) 25I-NBOMe related death in Australia: a case report. Forensic Sci Int 249:e15–e18. doi:10.1016/j.forsciint.2015.02.010
176. Lowe LM, Peterson BL, Couper FJ (2015) A case review of the first analytically confirmed 25I-NBOMe-related death in Washington State. J Anal Toxicol 39:668–671. doi:10.1093/jat/bkv092
177. Suzuki J, Poklis JL, Poklis A (2014) "My friend said it was good LSD": a suicide attempt following analytically confirmed 25I-NBOMe ingestion. J Psychoactive Drugs 46:379–382. doi:10.1080/02791072.2014.960111
178. Suzuki J, Dekker MA, Valenti ES, Arbelo Cruz FA, Correa AM, Poklis JL, Poklis A (2015) Toxicities associated with NBOMe ingestion-a novel class of potent hallucinogens: a review of the literature. Psychosomatics 56:129–139. doi:10.1016/j.psym.2014.11.002
179. Lawn W, Barratt M, Williams M, Horne A, Winstock A (2014) The NBOMe hallucinogenic drug series: patterns of use, characteristics of users and self-reported effects in a large international sample. J Psychopharmacol 28:780–788. doi:10.1177/0269881114523866
180. Shanks KG, Sozio T, Behonick GS (2015) Fatal intoxications with 25B-NBOMe and 25I-NBOMe in Indiana during 2014. J Anal Toxicol 39:602–606. doi:10.1093/jat/bkv058
181. Wood DM, Davies S, Puchnarewicz M, Button J, Archer R, Ovaska H, Ramsey J, Lee T, Holt DW, Dargan PI (2010) Recreational use of mephedrone (4-methylmethcathinone, 4-MMC)

with associated sympathomimetic toxicity. J Med Toxicol 6:327–330. doi:10.1007/s13181-010-0018-5
182. Wood DM, Davies S, Puchnarewicz M, Button J, Archer R, Ramsey J, Lee T, Holt DW, Dargan PI (2009) Recreational Use of 4-methylmethcathinone (4-MMC) presenting with sympathomimetic toxicity and confirmed by toxicological screening. Clin Toxicol 47:733
183. Wood DM, Greene SL, Dargan PI (2010) Clinical pattern of toxicity associated with the novel synthetic cathinone mephedrone. Emerg Med J 28:280–282
184. Fass JA, Fass AD, Garcia AS (2012) Synthetic cathinones (bath salts): legal status and patterns of abuse. Ann Pharmacother 46:436–441. doi:10.1345/aph.1Q628
185. Wilson B, Tavakoli H, DeCecchis D, Mahadev V (2013) Synthetic cannabinoids, synthetic cathinones, and other emerging drugs of abuse. Psychiatr Ann 43:558–564
186. Mas-Morey P, Visser MH, Winkelmolen L, Touw DJ (2013) Clinical toxicology and management of intoxications with synthetic cathinones ("bath salts"). J Pharm Pract 26:353–357. doi:10.1177/0897190012465949
187. Winstock AR, Marsden J, Mitcheson L (2010) What should be done about mephedrone? BMJ 340:c1605. doi:10.1136/bmj.c1605
188. Newcombe R (2009) Mephedrone: the use of mephedrone (M-cat,meow) in Middlesbrough. Lifeline Publications, Manchester. http://www.lifeline.org.uk/articles/the-use-of-mephedrone-m-cat-meow-in-middlesbrough/. Accessed 29 Nov 2015

The manufacturer's authorised representative in the EU is Springer
Nature Customer Service Centre GmbH, Europaplatz 3, 69115 Heidelberg,
Germany. If you have any concerns regarding our products, please
contact ProductSafety@springernature.com

Printed and bound by CPI Group (UK) Ltd, Croydon, CR0 4YY
01/10/2026
02225870-0014

The manufacturer's authorised representative in the EU is Springer Nature Customer Service Centre GmbH, Europaplatz 3, 69115 Heidelberg, Germany. If you have any concerns regarding our products, please contact ProductSafety@springernature.com

Printed and bound by CPI Group (UK) Ltd, Croydon, CR0 4YY
05/10/2026
02228127-0003